Basic College Mathematics

MEDIA UPDATE EDITION

Julie Miller
Daytona State College

Molly O'Neill
Daytona State College

Nancy Hyde
Formerly of Broward College

McGraw Hill
*Connect
Learn
Succeed*™

BASIC COLLEGE MATHEMATICS: MEDIA UPDATE, SECOND EDITION

Published by McGraw-Hill, a business unit of The McGraw-Hill Companies, Inc., 1221 Avenue of the Americas, New York, NY 10020. Copyright © 2013 by The McGraw-Hill Companies, Inc. All rights reserved. Printed in the United States of America. Previous editions © 2009 and 2007. No part of this publication may be reproduced or distributed in any form or by any means, or stored in a database or retrieval system, without the prior written consent of The McGraw-Hill Companies, Inc., including, but not limited to, in any network or other electronic storage or transmission, or broadcast for distance learning.

Some ancillaries, including electronic and print components, may not be available to customers outside the United States.

This book is printed on acid-free paper.

2 3 4 5 6 7 8 9 0 DOW/DOW 1 0 9 8 7 6 5 4 3

ISBN 978–0–07–340632–9
MHID 0–07–340632–5

ISBN 978–0–07–754355–6 (Annotated Instructor's Edition)
MHID 0–07–754355–6

Vice President, Editor-in-Chief: *Marty Lange*
Vice President, EDP: *Kimberly Meriwether David*
Senior Director of Development: *Kristine Tibbetts*
Editorial Director: *Stewart K. Mattson*
Executive Editor: *Dawn R. Bercier*
Sponsoring Editor: *Mary Ellen Rahn*
Developmental Editor: *Emily Williams*
Marketing Manager: *Peter A. Vanaria*
Lead Project Manager: *Peggy J. Selle*
Senior Buyer: *Sherry L. Kane*
Senior Media Project Manager: *Sandra M. Schnee*
Senior Designer: *Laurie B. Janssen*
Cover Illustration: *Imagineering Media Services Inc.*
Lead Photo Research Coordinator: *Carrie K. Burger*
Compositor: *Aptara®, Inc.*
Typeface: *10/12 Times Ten Roman*
Printer: *R. R. Donnelley*

All credits appearing on page or at the end of the book are considered to be an extension of the copyright page.

Library of Congress Cataloging-in-Publication Data

Miller, Julie, 1962–
 Basic college mathematics : media update / Julie Miller, Molly O'Neill, Nancy Hyde. — 2nd ed.
  p. cm.
 Includes index.
 ISBN 978–0–07–340632–9 —ISBN 0–07–340632–5 (hard copy : alk. paper)
 1. Mathematics—Textbooks. I. O'Neill, Molly, 1953– II. Hyde, Nancy. III. Title.
 QA37.3.M55 2013
 510—dc23
                2011030991

Hosted by **ALEKS Corp.**

Get Better Results with high-quality digital content and an easy-to-use platform!

The Miller/O'Neill/Hyde series now offers a complete digital solution for your course needs! Introducing Connect Math Hosted by ALEKS Corporation, McGraw-Hill's premier eLearning system.

With Connect Math Hosted by ALEKS Corporation instructors and students will experience:

▶ Intuitive and easy-to-use platform

▶ High-quality digital exercises

▶ Comprehensive Guided Solutions and Solve It examples that are consistent with the authors' approach, terminology, and methodology

▶ Integrated, Media-Rich eBook

▶ Integrated ALEKS® Assessment to identify strengths and weaknesses

▶ Flexible gradebook and assignment creation

▶ Flexible gradebook including reports on time spent working in Connect Math Hosted by ALEKS Corporation.

▶ Administer course consistency with Master Templates

Julie Miller's Lecture Videos and Dynamic Math Animations

Julie Miller's NEW lecture videos and dynamic math animations are also available within Connect Math Hosted by ALEKS Corporation. Students can learn and review the material right alongside Julie Miller as she narrates and teaches the learning objectives and brings the math concepts to life!

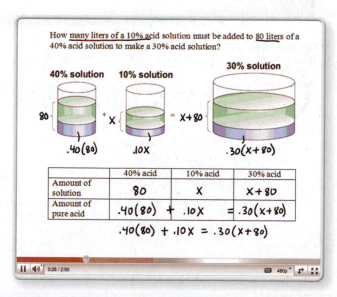

For more information, please contact your McGraw-Hill representative at **www.mhhe.com.**

ALEKS is a registered trademark of ALEKS Corporation.

McGraw Hill connect®
MATH
Hosted by ALEKS Corp.

Connect Math Hosted by ALEKS Corporation is an exciting, new ehomework platform combining the strengths of McGraw-Hill Higher Education and ALEKS Corporation. Connect Math Hosted by ALEKS Corporation is the first platform on the market to combine an artificially-intelligent, diagnostic assessment with an intuitive ehomework platform designed to meet your needs.

Connect Math Hosted by ALEKS Corporation is the culmination of a one-of-a-kind market development process involving full-time and adjunct Math faculty at every step of the process. This process enables us to provide you with a solution that best meets your needs.

Connect Math Hosted by ALEKS Corporation is built by Math educators for Math educators!

1 *Your students want a well-organized homepage where key information is easily viewable.*

Modern Student Homepage

▶ This homepage provides a dashboard for students to immediately view their assignments, grades, and announcements for their course. (Assignments include HW, quizzes, and tests.)

▶ Students can access their assignments through the course Calendar to stay up-to-date and organized for their class.

Modern, intuitive, and simple interface.

2 *You want a way to identify the strengths and weaknesses of your class at the beginning of the term rather than after the first exam.*

Integrated ALEKS® Assessment

▶ This artificially-intelligent (AI), diagnostic assessment identifies precisely what a student knows and is ready to learn next.

▶ Detailed assessment reports provide instructors with specific information about where students are struggling most.

▶ This AI-driven assessment is the only one of its kind in an online homework platform.

Recommended to be used as the first assignment in any course.

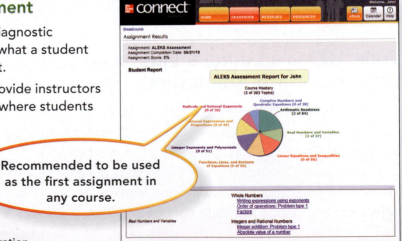

ALEKS is a registered trademark of ALEKS Corporation.

Resources for Online Homework

3 Your students want an assignment page that is easy to use and includes lots of extra help resources.

Efficient Assignment Navigation

- Students have access to immediate feedback and help while working through assignments.
- Students have direct access to a media-rich eBook for easy referencing.
- Students can view detailed, step-by-step solutions written by instructors who teach the course, providing a unique solution to each and every exercise.

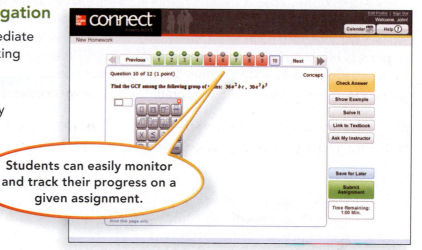

Students can easily monitor and track their progress on a given assignment.

4 You want a more intuitive and efficient assignment creation process because of your busy schedule.

Assignment Creation Process

- Instructors can select textbook-specific questions organized by chapter, section, and objective.
- Drag-and-drop functionality makes creating an assignment quick and easy.
- Instructors can preview their assignments for efficient editing.

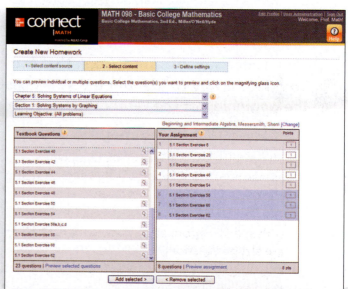

5 *Your students want an interactive eBook with rich functionality integrated into the product.*

connect plus+

|MATH

Hosted by **ALEKS Corp.**

Integrated Media-Rich eBook

▶ A Web-optimized eBook is seamlessly integrated within ConnectPlus Math Hosted by ALEKS Corp. for ease of use.

▶ Students can access videos, images, and other media in context within each chapter or subject area to enhance their learning experience.

▶ Students can highlight, take notes, or even access shared instructor highlights/notes to learn the course material.

▶ The integrated eBook provides students with a cost-saving alternative to traditional textbooks.

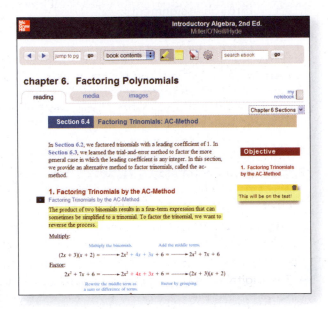

6 *You want a flexible gradebook that is easy to use.*

Flexible Instructor Gradebook

▶ Based on instructor feedback, Connect Math Hosted by ALEKS Corp.'s straightforward design creates an intuitive, visually pleasing grade management environment.

▶ Assignment types are color-coded for easy viewing.

▶ The gradebook allows instructors the flexibility to import and export additional grades.

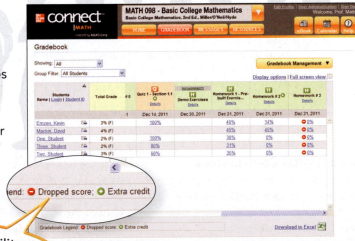

Instructors have the ability to drop grades as well as assign extra credit.

Built by Math Educators for Math Educators

 7 You want algorithmic content that was developed by math faculty to ensure the content is pedagogically sound and accurate.

Digital Content Development Story

As the usage of online homework progresses and evolves, McGraw-Hill understands the need to involve instructors while developing the digital content to ensure that what students see in the online homework system is consistent with what they see in their textbooks. For the Miller, O'Neill, and Hyde developmental math series, we partnered with instructors that have taught from the series to help ensure a seamless transition from print to digital offerings.

The development of McGraw-Hill's Connect Math Hosted by ALEKS Corporation content involved collaboration between McGraw-Hill, our authors, experienced instructors, and ALEKS Corporation, a company known for its high-quality digital content. The result of this process, outlined below, is accurate content created with your students in mind. It is available in a simple-to-use interface with all the functionality tools needed to manage your course.

1. McGraw-Hill selected experienced instructors to work as Digital Contributors.

2. The Digital Contributors selected the textbook exercises to be included in the algorithmic content to ensure appropriate coverage of the textbook content.

3. The Digital Contributors created detailed, stepped-out solutions for use in the Guided Solution and Solve It features, matching the voice of the authors.

4. The Digital Contributors provided detailed instructions for authoring the algorithm specific to each exercise to maintain the original intent and integrity of each unique exercise.

5. Each algorithm was reviewed by the Contributor, went through a detailed quality control process by ALEKS Corporation, and was copyedited prior to being posted live.

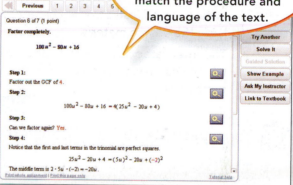

Solutions in Connect Math Hosted by ALEKS Corp. match the procedure and language of the text.

Result = Truly Vetted, Consistent Digital Content That Is Built By Math Educators And Supported By ALEKS Corporation.

Lead Digital Contributors
Donna Gerken, *Miami Dade College*
Nicole Lloyd, *Lansing Community College*
Stephen Toner, *Victor Valley College*

Digital Contributors
Jody Harris, *Broward College*
Lizette Hernandez Foley, *Broward College*
Linda Schott, *Ozarks Technical Community College*
Michael Larkin, *Pacific University*
Alina Coronel, *Miami Dade College*

www.connectmath.com

ALEKS 360: A Total Course Solution

A cost-effective total course solution: fully integrated, interactive eBook combined with ALEKS individualized assessment and learning.

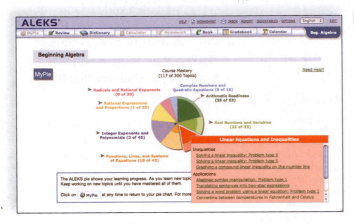

Individualized Learning

- The ALEKS Pie summarizes a student's current knowledge and provides individualized learning on the exact topics the student is **ready to learn**

- Artificial intelligence successfully targets gaps by assessing precisely a student's knowledge and periodically reassessing for long-term retention

- Adaptive, open-response environment avoids multiple-choice and includes problems, explanations, and realistic answer input tools

Interactive eBook

- eBook access provides worked examples, videos, and additional support

- Robust virtual features include highlighting, bookmarking, and note-taking capabilities

- Students can easily access the eBook, multimedia resources, and their notes from within their ALEKS Student Accounts

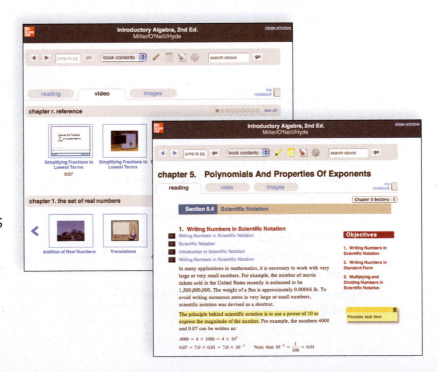

Learn More: www.aleks.com/highered/math/aleks360

ALEKS Course Management Tools

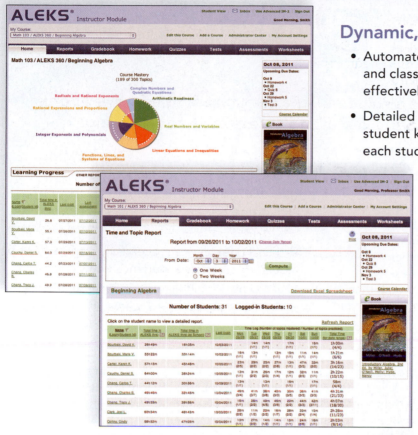

Dynamic, Automated Reporting

- Automated reports dynamically track student and class learning progress so instructors can effectively direct classroom instruction

- Detailed reports identify precisely what each student knows, and more importantly, what each student is ready to learn next

- Time and Topic Report offers up-to-the-minute daily progress, including time logged, topics attempted, and topics mastered

Course Control and Customization

- Align ALEKS topics with a textbook or course syllabus
- Create and customize course objectives and modules
- Set due dates for course objectives to pace student progress
- Assign automatically-graded homework, quizzes, and tests
- Seamlessly track and adjust student scores with the customizable gradebook

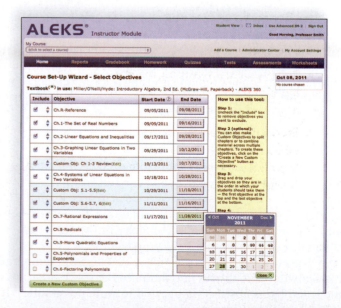

About the Authors

Julie Miller has been on the faculty of the Mathematics Department at Daytona State College for 19 years, where she has taught developmental and upper level courses. Prior to her work at Daytona State College, she worked as a software engineer for General Electric in the area of flight and radar simulation. Julie earned a bachelor of science in applied mathematics from Union College in Schenectady, New York, and a master of science in mathematics from the University of Florida. In addition to this textbook, she has authored several course supplements for college algebra, trigonometry, and precalculus, as well as several short works of fiction and nonfiction for young readers.

"My father is a medical researcher, and I got hooked on math and science when I was young and would visit his laboratory. I can remember using graph paper to plot data points for his experiments and doing simple calculations. He would then tell me what the peaks and features in the graph meant in the context of his experiment. I think that applications and hands-on experience made math come alive for me and I'd like to see math come alive for my students."

—*Julie Miller*

Molly O'Neill is also from Daytona State College, where she has taught for 21 years in the Mathematics Department. She has taught a variety of courses from developmental mathematics to calculus. Before she came to Florida, Molly taught as an adjunct instructor at the University of Michigan–Dearborn, Eastern Michigan University, Wayne State University, and Oakland Community College. Molly earned a bachelor of science in mathematics and a master of arts and teaching from Western Michigan University in Kalamazoo, Michigan. Besides this textbook, she has authored several course supplements for college algebra, trigonometry, and precalculus and has reviewed texts for developmental mathematics.

"I differ from many of my colleagues in that math was not always easy for me. But in seventh grade I had a teacher who taught me that if I follow the rules of mathematics, even I could solve math problems. Once I understood this, I enjoyed math to the point of choosing it for my career. I now have the greatest job because I get to do math everyday and I have the opportunity to influence my students just as I was influenced. Authoring these texts has given me another avenue to reach even more students."

—*Molly O'Neill*

Nancy Hyde served as a full-time faculty member of the Mathematics Department at Broward College for 24 years. During this time she taught the full spectrum of courses from developmental math through differential equations. She received a bachelor of science degree in math education from Florida State University and master's degree in math education from Florida Atlantic University. She has conducted workshops and seminars for both students and teachers on the use of technology in the classroom. In addition to this textbook, she has authored a graphing calculator supplement for College Algebra.

"I grew up in Brevard County, Florida, where my father worked at Cape Canaveral. I was always excited by mathematics and physics in relation to the space program. As I studied higher levels of mathematics I became more intrigued by its abstract nature and infinite possibilities. It is enjoyable and rewarding to convey this perspective to students while helping them to understand mathematics."

—*Nancy Hyde*

Contents

How Will Miller/O'Neill/Hyde Help Your Students *Get Better Results?*

Better Clarity, Quality, and Accuracy!

Julie Miller, Molly O'Neill, and Nancy Hyde know what students need to be successful in mathematics. Better results come from clarity in their exposition, quality of step-by-step worked examples, and accuracy of their exercise sets, but it takes more than just great authors to build a textbook series to help students achieve success in mathematics. Our authors worked with a strong mathematical team of instructors from around the country to ensure clarity, quality, and accuracy.

> "The authors' writing style is very straight forward and easy to follow. The level of formality is just right for this level of math course."
>
> —Lynette King, *Gadsden State College*

Better Exercise Sets!

A comprehensive set of exercises are available for every student level. Julie Miller, Molly O'Neill, and Nancy Hyde worked with a national board of advisors from across the country to ensure the series will offer the appropriate depth and breadth of exercises for your students. New to this edition, **Problem Recognition Exercises** were created in direct response to student need and resulted in improved student performance on tests.

Our exercise sets help students progress from skill development to conceptual understanding. Student tested and instructor approved, the Miller/O'Neill/Hyde exercise sets will help your students get better results.

- ▶ **Problem Recognition Exercises**
- ▶ **Skill Practice Exercises**
- ▶ **Study Skills Exercises**
- ▶ **Mixed Exercises**
- ▶ **Expanding Your Skills Exercises**

> "I think that of all the textbooks that I have seen (or evaluated) they (MOH) have by far the most comprehensive sets of exercises at every level (skill-based, study skills, etc.)."
>
> —Juan Jimenez, *Springfield Technical Community College*

Better Step-By-Step Pedagogy!

The second edition provides enhanced step-by-step learning tools available to help students *get better results.*

- ▶ **Worked Examples** provide an "easy-to-understand" approach, clearly guiding each student through a step-by-step approach to master each practice exercise for better comprehension.
- ▶ **TIPS** offer students extra cautious direction to help improve understanding through hints and further insight.
- ▶ **Avoiding Mistakes** boxes alert students to common errors and provide practical ways to avoid them.

   These learning aids will help students get better results by learning how to work through a problem using a clearly defined step-by-step methodology that has been class-tested and student approved.

> "Miller/O'Neill/Hyde has a very good pedagogy that is student-friendly. It has a plethora of problems and a variety of them. It allows success for all students."
>
> —Mark Marino, *Erie Community College*

Formula for Student Success

Step-by-Step Worked Examples

▶ Do you get the feeling that there is a disconnect between your students' classwork and homework?

▶ Do your students have trouble finding worked examples that match the practice exercises?

▶ Wouldn't you like your students to see examples in the textbook that match the ones you use in class?

Miller/O'Neill/Hyde's worked examples offer a clear, concise methodology that replicates the mathematical processes used in the authors' classroom lectures!

> "In the year we've used this text I've noticed that students seem to be able to learn the material without difficulties. I attribute a lot of that to the fact the text contains examples that are worked out clearly and able to follow."
>
> —Rod Oberdick, *Durham Tech Comm Coll*

PROCEDURE Solving a Proportion

Step 1 Set the cross products equal to each other.
Step 2 Divide both sides of the equation by the number being multiplied by the variable.
Step 3 Check the solution in the original proportion.

Example 4 Applying a Proportion to Environmental Science

A biologist wants to estimate the number of elk in a wildlife preserve. She sedates 25 elk and clips a small radio transmitter onto the ear of each animal. The elk return to the wild, and after 6 months, the biologist studies a sample of 120 elk in the preserve. Of the 120 elk sampled, 4 have radio transmitters. Approximately how many elk are in the whole preserve?

— Skill Practice —

4. To estimate the number of fish in a lake, the park service catches 50 fish and tags them. After several months the park service catches a sample of 100 fish and finds that 6 are tagged. Approximately how many fish are in the lake?

Solution:

Let n represent the number of elk in the whole preserve.

$$
\begin{array}{cc}
\text{Sample} & \text{Population}
\end{array}
$$

number of elk in the sample with radio transmitters $\longrightarrow \dfrac{4}{120} = \dfrac{25}{n} \longleftarrow$ number of elk in the population with radio transmitters

total elk in the sample $\longrightarrow$ $\longleftarrow$ total elk in the population

$4 \cdot n = (120)(25)$ Equate the cross products.

$4n = 3000$

$\dfrac{\overset{1}{\cancel{4}}n}{\underset{1}{\cancel{4}}} = \dfrac{3000}{4}$ Divide both sides by 4.

$n = 750$ Divide $3000 \div 4 = 750$.

There are approximately 750 elk in the wildlife pre...

> "All of the worked examples are good and easy to understand. There are plenty of examples given. Also, it appears that there is at least one example for each different particular type of exercise in each section—very good"
>
> —Susan Haley, *Florence-Darlington Technical College*

> "Miller/O'Neill/Hyde presents each concept in clear language. Multiple examples covering various forms of problems are included and explained step by step."
>
> —Susan Harrison, *University of Wisconsin–Eau Claire*

Better Learning Tools

Avoiding Mistakes Boxes

Avoiding Mistakes boxes are integrated throughout the textbook to alert students to common errors and how to avoid them.

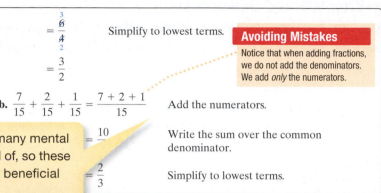

"Loving these—students make so many mental mistakes we are not always mindful of, so these were very intentionally placed and beneficial to learners."

—Sharon Morrison, *St. Petersburg College*

TIP Boxes

Teaching tips are usually only revealed in the classroom. Not anymore. Tip boxes offer students helpful hints and extra direction to help improve understanding and further insight.

TIP: To use the prefix line effectively, you must know the order of the metric prefixes. Sometimes a mnemonic (memory device) can help. Consider the following sentence. The first letter of each word represents one of the metric prefixes.

kids	have	doughnuts	until	dad	calls	mom.
kilo-	hecto-	deka-	unit	deci-	centi-	milli-

represents the main unit of measurement (meter, liter, or gram)

"I think that one of the best features of this chapter (and probably will continue throughout the text) is the tip section. I like that the students are warned in advance of common errors that can be made before they start working the problems. I also think that the tips are great to remind the instructors of the type of issues that arise when the students are working through their homework."

—Ena Salter, *Manatee Community College*

Concept Connection Boxes

Concept Connections help students understand the conceptual meaning of the problems they are solving—a vital skill in mathematics.

"This feature is one of my favorite parts in the textbook. It is useful when trying to get students to think critically about types of problems."

—Sue Duff, *Guilford Technical Community College*

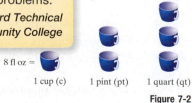

Concept Connections

7. From Figure 7-2, determine how many cups are in 1 gal.
8. From Figure 7-2, determine how many pints are in 1 gal.

8 fl oz =

1 cup (c)   1 pint (pt)   1 quart (qt)   1 gallon (gal)

Figure 7-2

New to this Edition

- ▶ Do your students have trouble with problem solving?
- ▶ Do you want to help students overcome math anxiety?
- ▶ Do you want to help your students improve performance on math assessments?

Problem Recognition Exercises!

Problem Recognition Exercises present a collection of problems that look similar to a student upon first glance, but are actually quite different in the manner of their individual solutions. Students sharpen critical thinking skills and better develop their "solution recall" to help them distinguish the method needed to solve an exercise—an essential skill in mathematics. Problem Recognition Exercises, tested in a developmental mathematics classroom, were created in direct response to student need to improve performance in testing where different problem types are mixed.

"This is a GREAT idea. This "pattern recognition" is something that I go through in my classroom, and really helps students to flesh out the idea and look at specific differences and similarities in problems."
—Matthew Robinson, *Tallahassee Community College*

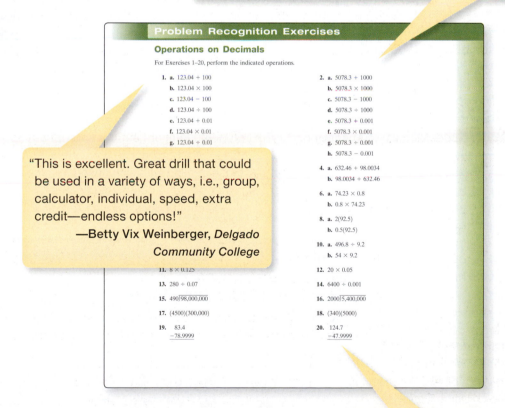

Problem Recognition Exercises

Operations on Decimals

For Exercises 1–20, perform the indicated operations.

1. a. $123.04 + 100$
   b. 123.04×100
   c. $123.04 - 100$
   d. $123.04 \div 100$
   e. $123.04 + 0.01$
   f. 123.04×0.01
   g. $123.04 \div 0.01$

2. a. $5078.3 + 1000$
   b. 5078.3×1000
   c. $5078.3 - 1000$
   d. $5078.3 \div 1000$
   e. $5078.3 + 0.001$
   f. 5078.3×0.001
   g. $5078.3 \div 0.001$
   h. $5078.3 - 0.001$

4. a. $632.46 + 98.0034$
   b. $98.0034 + 632.46$

6. a. 74.23×0.8
   b. 0.8×74.23

8. a. $2(92.5)$
   b. $0.5(92.5)$

10. a. $496.8 \div 9.2$
    b. 54×9.2

11. 8×0.125

12. 20×0.05

13. $280 \div 0.07$

14. $6400 \div 0.001$

15. $490\overline{)98,000,000}$

16. $2000\overline{)5,400,000}$

17. $(4500)(300,000)$

18. $(340)(5000)$

19. $\begin{array}{r} 83.4 \\ -78.9999 \end{array}$

20. $\begin{array}{r} 124.7 \\ -47.9999 \end{array}$

"This is excellent. Great drill that could be used in a variety of ways, i.e., group, calculator, individual, speed, extra credit—endless options!"
—Betty Vix Weinberger, *Delgado Community College*

"The MOH chapter does an excellent job giving practice with these special types of problems. I found this approach interesting and enlightening."
—Valerie Melvin, *Cape Fear Community College*

New and Improved Applications!

Class-Tested and Student Approved!

New and improved applications have been developed by an advisory team. The Miller/O'Neill/Hyde Board of Advisors Team partnered with our authors to bring you the *best applications* from every region of the country! These applications include real data and topics which are more relevant and interesting to today's student.

Objective 4: Medical Applications

87. The drug cyanocobalamin is prescribed by one doctor in the amount of 1000 mcg. How many milligrams is this?

88. An injection of naloxone is given in the amount of 800 mcg. How many milligrams is this?

89. A nurse must administer 45 mg of a drug. The drug is available in a liquid form with a concentration of 15 mg per milliliter of the solution. How many milliliters of the solution should the nurse give?

90. A patient must receive 500 mg of medication in a solution that has a strength of 250 mg per 5 milliliter of solution. How many milliliters of solution should be given?

Expanding Your Skills

91. A normal value of hemoglobin in the blood for an adult male is 18 gm/dL (that is, 18 grams per deciliter). How much hemoglobin would be expected in 20 mL of a males's blood?

92. A normal value of hemoglobin in the blood for an adult female is 15 gm/dL (that is, 15 gm per deciliter). How much hemoglobin would be expected in 40 mL of a female's blood?

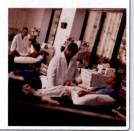

NEW Group Activities!

Each chapter concludes with a Group Activity selected by objective to promote classroom discussion and collaboration—helping students not only to solve problems but to explain their solutions for better mathematical mastery. Group Activities are great for instructors and adjuncts—bringing a more interactive approach to teaching mathematics! All required materials, activity time, and suggested group sizes are provided in the directions of the activity. Activities include: Investigating Probability, Tracking Stocks, Using Card Games with Fractions, and more!

Group Activity

Investigating Probability

Materials: Paper bags containing 10 white poker chips, 6 red poker chips, and 4 blue poker chips.

Estimated time: 15 minutes

Group Size: 3

1. Each group will receive a bag of poker chips, with 10 white, 6 red, and 4 blue chips.

2. **a.** Write the ratio of red chips in the bag to the total number of chips in the bag. _____ This value represents the *probability* of randomly selecting a red chip from the bag.

   b. Write this fraction in decimal form. _____

   c. Write the decimal from step (b) as a percent. _____
   A probability value indicates the likeliness of an event to occur. For example, to interpret this probability, one might say that there is a 30% chance of selecting a red chip at random from the bag.

Dynamic Math Animations

The Miller/O'Neill/Hyde author team has developed a series of Flash animations to illustrate difficult concepts where static images and text fall short. The animations leverage the use of on-screen movement and morphing shapes to enhance conceptual learning.

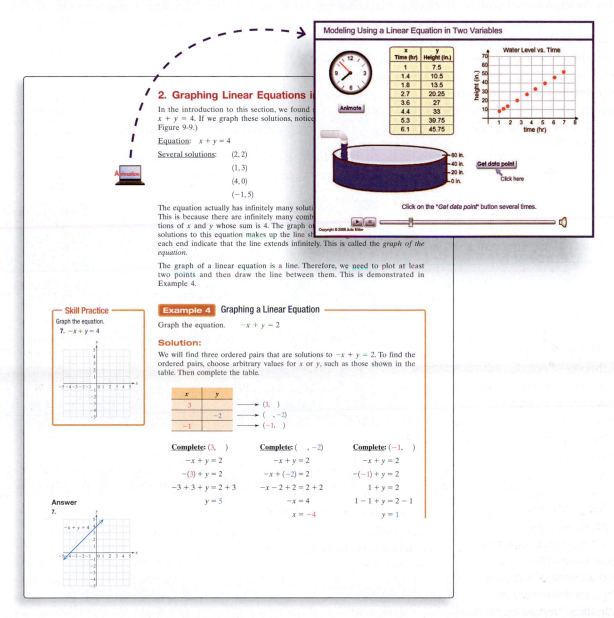

2. Graphing Linear Equations i

In the introduction to this section, we found $x + y = 4$. If we graph these solutions, notic Figure 9-9.)

Equation: $x + y = 4$

Several solutions: $(2, 2)$
                                     $(1, 3)$
                                     $(4, 0)$
                                     $(-1, 5)$

The equation actually has infinitely many soluti This is because there are infinitely many comb tions of x and y whose sum is 4. The graph o solutions to this equation makes up the line sh each end indicate that the line extends infinitely. This is called the *graph of the equation.*

The graph of a linear equation is a line. Therefore, we need to plot at least two points and then draw the line between them. This is demonstrated in Example 4.

Modeling Using a Linear Equation in Two Variables

x Time (hr)	y Height (in.)
1	7.5
1.4	10.5
1.8	13.5
2.7	20.25
3.6	27
4.4	33
5.3	39.75
6.1	45.75

Click on the "*Get data point*" button several times.

Copyright © 2008 Julie Miller

Skill Practice

Graph the equation.

7. $-x + y = 4$

Example 4   Graphing a Linear Equation

Graph the equation. $-x + y = 2$

Solution:

We will find three ordered pairs that are solutions to $-x + y = 2$. To find the ordered pairs, choose arbitrary values for x or y, such as those shown in the table. Then complete the table.

x	y
3	
	-2
-1	

→ $(3, \ )$
→ $(\ , -2)$
→ $(-1, \ )$

Complete: $(3, \ )$

$-x + y = 2$

$-(3) + y = 2$

$-3 + 3 + y = 2 + 3$

$y = 5$

Complete: $(\ , -2)$

$-x + y = 2$

$-x + (-2) = 2$

$-x - 2 + 2 = 2 + 2$

$-x = 4$

$x = -4$

Complete: $(-1, \ )$

$-x + y = 2$

$-(-1) + y = 2$

$1 + y = 2$

$1 - 1 + y = 2 - 1$

$y = 1$

Answer

7.

$-x + y = 4$

Through their classroom experience, the authors recognize that such media assets are great teaching tools for the classroom and excellent for online learning. The Miller/O'Neill/Hyde animations are interactive and quite diverse in their use. Some provide a virtual laboratory for which an application is simulated and where students can collect data points for analysis and modeling. Others provide interactive question-and-answer sessions to test conceptual learning. For word problem applications, the animations ask students to estimate answers and practice "number sense."

360° Development Process

McGraw-Hill's **360° Development Process** is an ongoing, never-ending, market-oriented approach to building accurate and innovative print and digital products. It is dedicated to continual large-scale and incremental improvement driven by multiple customer feedback loops and checkpoints. The process is initiated during the early planning stages of our new products, and is intensified during development and production. Then the process begins again upon publication in anticipation of the next edition.

A key principle in the development of any mathematics text is its ability to adapt to teaching specifications in a universal way. The only way to do so is by contacting those universal voices—and learning from their suggestions. We are confident that our book has the most current content the industry has to offer, thus pushing our desire for accuracy to the highest standard possible. In order to accomplish this, we have moved through an arduous road to production. Extensive and open-minded advice is critical in the production of a superior text.

Here is a brief overview of the initiatives included in the *Basic College Mathematics*, Second Edition, 360° Development Process:

Board of Advisors

A hand-picked group of trusted teachers active in the basic math course served as chief advisors and consultants to the authors and editorial team with regards to manuscript development. The Board of Advisors reviewed parts of the manuscript; served as a sounding board for pedagogical, media, and design concerns; consulted on organizational changes; and attended a focus group to confirm the manuscript's readiness for publication.

Basic College Mathematics

Vernon Bridges, *Durham Technical Community College*

Lynette King, *Gadsden State Community College*

Sharon Morrison, *St. Petersburg College*

Deanna Murphy, *Lane County Community College*

Rod Oberdick, *Delaware Technical and Community College*

Matthew Robinson, *Tallahassee Community College*

Pat Rome, *Delgado Community College–City Park*

Introductory Algebra

Mark Billiris, *St. Petersburg Community College*

Pauline Chow, *Harrisburg Community College*

John Close, *Salt Lake Community College*

Barbara Elzy, *Bluegrass Community College*

Lori Grady, *University of Wisconsin–Whitewater*

Patricia Roux, *Delgado Community College*

Mike Kirby, *Tidewater Community College*

Intermediate Algebra

Susan Dimick, *Spokane Community College*

Sue Duff, *Guilford Technical Community College*

Alicia Giovinazzo, *Miami Dade Community College*

Charlotte Newsome, *Tidewater Community College*

Ena Salter, *Manatee Community College*

Acknowledgments and Reviewers

The development of this textbook series would never have been possible without the creative ideas and feedback offered by many reviewers. We are especially thankful to the following instructors for their careful review of the manuscript.

Manuscript Review Panels

Over 400 teachers and academics from across the country reviewed the various drafts of the manuscript to give feedback on content, design, pedagogy, and organization. This feedback was summarized by the book team and used to guide the direction of the text.

Special "*thank you*" to our Manuscript Class-Testers

Vernon Bridges, *Durham Technical Community College*
Susan Dimick, *Spokane Community College*
Lori Grady, *University of Wisconsin–Whitewater*
Rod Oberdick, *Delaware Technical and Community College*
Matthew Robinson, *Tallahassee Community College*
Pat Rome, *Delgado Community College–City Park*

Reviewers of the Miller/O'Neill/Hyde Developmental Mathematics Series

Darla Aguilar, *Pima Community College–Desert Vista*
Ebrahim Ahmadizadeh, *Northampton Community College*
Sara Alford, *North Central Texas College*
Theresa Allen, *University of Idaho*
Sheila Anderson, *Housatonic Community College*
Lane Andrew, *Arapahoe Community College*
Jan Archibald, *Ventura College*
Yvonne Aucoin, *Tidewater Community College–Norfolk*
Eric Aurand, *Mohave Community College*
Sohrab Bakhtyari, *St. Petersburg College*
Anna Bakman, *Los Angeles Trade Technical*
Andrew Ball, *Durham Technical Community College*
Russell Banks, *Guilford Technical Community College*
Suzanne Battista, *St. Petersburg College*
Kevin Baughn, *Kirtland Community College*
Sarah Baxter, *Gloucester County College*
Lynn Beckett-Lemus, *El Camino College*
Edward Bender, *Century College*
Emilie Berglund, *Utah Valley State College*
Rebecca Berthiaume, *Edison College–Fort Myers*
John Beyers, *Miami Dade College–Hialeah*
Leila Bicksler, *Delgado Community College–West Bank*
Norma Bisulca, *University of Maine–Augusta*
Kaye Black, *Bluegrass Community and Technical College*
Deronn Bowen, *Broward College–Central*
Timmy Bremer, *Broome Community College*
Donald Bridgewater, *Broward College*
Peggy Brock, *TVI Community College*
Kelly Brooks, *Pierce College*
Susan D. Caire, *Delgado Community College–West Bank*
Peter Carlson, *Delta College*
Judy Carter, *North Shore Community College*
Veena Chadha, *University of Wisconsin–Eau Claire*
Zhixiong Chen, *New Jersey City University*
Tyrone Clinton, *Saint Petersburg College–Gibbs*
Eugenia Cox, *Palm Beach Community College*
Julane Crabtree, *Johnson Community College*
Mark Crawford, *Waubonsee Community College*
Natalie Creed, *Gaston College*
Anabel Darini, *Suffolk County Community College–Brentwood*
Antonio David, *Del Mar College*
Ron Davis, *Kennedy-King College–Chicago*
Laurie Delitsky, *Nassau Community College*
Patti D'Emidio, *Montclair State University*
Bob Denton, *Orange Coast College*
Robert Diaz, *Fullerton College*
Eileen Doran, *Palm Beach Community College*
Deborah Doucette, *Erie Community College–North Campus—Williamsville*
Thomas Drucker, *University of Wisconsin–Whitewater*
Michael Dubrowsky, *Wayne Community College*
Barbara Duncan, *Hillsborough Community College–Dale Mabry*
Jeffrey Dyess, *Bishop State Community College*
Elizabeth Eagle, *University of North Carolina–Charlotte*
Sabine Eggleston, *Edison College–Fort Myers*
Lynn Eisenberg, *Rowan-Cabarrus Community College*
Barb Elzey, *Bluegrass Community and Technical College*
Nerissa Felder, *Polk Community College*
Mark Ferguson, *Chemeketa Community College*
Diane Fisher, *Lousiana State University–Eunice*
David French, *Tidewater Community College–Chesa*

xxiii

Reviewers of the Miller/O'Neill/Hyde Developmental Mathematics Series *(continued)*

Dot French, *Community College of Philadelphia*
Deborah Fries, *Wor-Wic Community College*
Robert Frye, *Polk Community College*
Jesse M. Fuson, *Mountain State University*
Patricia Gary, *North Virginia Community College–Manassas*
Calvin Gatson, *Alabama State University*
Donna Gerken, *Miami Dade College–Kendall*
Mehrnaz Ghaffarian, *Tarrant County College South*
Mark Glucksman, *El Camino College*
Judy Godwin, *Collin County Community College*
William Graesser, *Ivy Tech Community College*
Victoria Gray, *Scott Community College*
Edna Greenwood, *Tarrant County College–Northwest*
Kimberly Gregor, *Delaware Technical Community College–Wilmington*
Vanetta Grier-Felix, *Seminole Community College*
Kathy Grigsby, *Moraine Valley Community College*
Joseph Guiciardi, *Community College of Allegheny County–Monroeville*
Susan Haley, *Florence-Darlington Technical College*
Mary Lou Hammond, *Spokane Community College*
Joseph Harris, *Gulf Coast Community College*
Lloyd Harris, *Gulf Coast Community College*
Mary Harris, *Harrisburg Area Community College–Lancaster*
Susan Harrison, *University of Wisconsin–Eau Claire*
Kristen Hathcock, *Barton County Community College*
Marie Hoover, *University of Toledo*
Linda Hoppe, *Jefferson College*
Joe Howe, *St. Charles County Community College*
Juan Jimenez, *Springfield Technical Community College*
Jennifer Johnson, *Delgado Community College*
Yolanda Johnson, *Tarrant County College South*
Shelbra Jones, *Wake Technical Community College*
Joe Jordan, *John Tyler Community College*
Cheryl Kane, *University of Nebraska–Lincoln*
Ismail Karahouni, *Lamar University*
Mike Karahouni, *Lamar University–Beaumont*
Joanne Kawczenski, *Luzerne County Community College*
Eliane Keane, *Miami Dade College–North*
Miriam Keesey, *San Diego State University*
Joe Kemble, *Lamar University–Beaumont*
Patrick Kimani, *Morrisville State College*
Sonny Kirby, *Gadsden State Community College*
Vicky Kirkpatrick, *Lane Community College*
Marcia Kleinz, *Atlantic Cape Community College*
Ron Koehn, *Southwestern Oklahoma State University*
Jeff Koleno, *Lorain County Community College*
Rosa Kontos, *Bergen Community College*
Randa Kress, *Idaho State University*

Gayle Krzemie, *Pikes Peak Community College*
Gayle Kulinsky, Carla, *Salt Lake Community College*
Linda Kuroski, *Erie Community College*
Catherine Laberta, *Erie Community College–North Campus—Williamsville*
Joyce Langguth, *University of Missouri–St. Louis*
Betty Larson, *South Dakota State University*
Katie Lathan, *Tri-County Technical College*
Kathryn Lavelle, *Westchester Community College*
Patricia Lazzarino, *North Virginia Community College–Manassas*
Julie Letellier, *University of Wisconsin–Whitewater*
Mickey Levendusky, *Pima Community College*
Barbara Little, *Central Texas College*
David Liu, *Central Oregon Community College*
Maureen Loiacano, *Montgomery College*
Wanda Long, *St. Charles County Community College*
Kerri Lookabill, *Mountain State University*
Jessica Lowenfield, *Nassau Community College*
Diane Lussier, *Pima Community College*
Mark Marino, *Erie Community College–North Campus—Williamsville*
Dorothy Marshall, *Edison College–Fort Myers*
Diane Masarik, *University of Wisconsin–Whitewater*
Louise Mataox, *Miami Dade College*
Cindy McCallum, *Tarrant County College South*
Joyce McCleod, *Florida Community College–South Campus*
Roger McCoach, *County College of Morris*
Stephen F. McCune, *Austin State University*
Ennis McKenna, Hazel, *Utah Valley State College*
Harry McLaughlin, *Montclair State University*
Valerie Melvin, *Cape Fear Community College*
Richard Moore, *St. Petersburg College–Seminole*
Elizabeth Morrison, *Valencia Community College*
Sharon Morrison, *St. Petersburg College*
Shauna Mullins, *Murray State University*
Linda Murphy, *Northern Essex Community College*
Michael Murphy, *Guilford Technical Community College*
Kathy Nabours, *Riverside Community College*
Roya Namavar, *Rogers State University*
Tony Nelson, *Tulsa Community College*
Melinda Nevels, *Utah Valley State College*
Charlotte Newsom, *Tidewater Community College–Virginia Beach*
Brenda Norman, *Tidewater Community College*
David Norwood, *Alabama State University*
Rhoda Oden, *Gadsden State Community College*
Tammy Payton, *North Idaho College*
Melissa Pedone, *Valencia Community College–Osceola*

Get Better Results

Shirley Pereira, *Grossmont College*
Pete Peterson, *John Tyler Community College*
Suzie Pickle, *St. Petersburg College*
Sheila Pisa, *Riverside Community College–Moreno Valley*
Marilyn Platt, *Gaston College*
Richard Ponticelli, *North Shore Community College*
Tammy Potter, *Gadsden State Community College*
Joel Rappaport, *Florida Community College*
Sherry Ray, *Oklahoma City Community College*
Angelia Reynolds, *Gulf Coast Community College*
Suellen Robinson, *North Shore Community College*
Jeri Rogers, *Seminole Community College–Oviedo*
Trisha Roth, *Gloucester County College*
Richard Rupp, *Del Mark College*
Dave Ruszkiewicz, *Milwaukee Area Technical College*
Nancy Sattler, *Terra Community College*
Vicki Schell, *Pensacola Junior College*
Nyeita Schult, *St. Petersburg College*
Wendiann Sethi, *Seton Hall University*
Dustin Sharp, *Pittsburg Community College*
Marvin Shubert, *Hagerstown Community College*
Plamen Simeonov, *University of Houston–Downtown*
Carolyn Smith, *Armstrong Atlantic State University*
Melanie Smith, *Bishop State Community College*
John Squires, *Cleveland State Community College*
Sharon Staver, Judith, *Florida Community College–South Campus*
Sharon Steuer, *Nassau Community College*
Trudy Streilein, *North Virginia Community College–Annandale*
Gretchen Syhre, *Hawkeye Community College*
Katalin Szucs, *Pittsburg Community College*

Mike Tiano, *Suffolk County Community College*
Stephen Toner, *Victor Valley College*
Mary Lou Townsend, *Wor-Wic Community College*
Susan Twigg, *Wor-Wic Community College*
Matthew Utz, *University of Arkansas–Fort Smith*
Joan Van Glabek, *Edison College–Fort Myers*
John Van Kleef, *Guilford Technical Community College*
Diane Veneziale, *Burlington County College–Pemberton*
Andrea Vorwark, *Metropolitan Community College–Maple Woods*
Edward Wagner, *Central Texas College*
David Wainaina, *Coastal Carolina Community College*
James Wang, *University of Alabama*
Richard Watkins, *Tidewater Community College–Virginia Beach*
Sharon Wayne, *Patrick Henry Community College*
Leben Wee, *Montgomery College*
Betty Vix Weinberger, *Delgado Community College–West Bank*
Christine Wetzel-Ulrich, *Northampton Community College*
Jackie Wing, *Angelina College*
Michelle Wolcott, *Pierce College*
Deborah Wolfson, *Suffolk County Community College–Brentwood*
Mary Wolyniak, *Broome Community College*
Rick Woodmansee, *Cosumnes River College*
Susan Working, *Grossmont College*
Karen Wyrick, *Cleveland State Community College*
Alan Yang, *Columbus State Community College*
William Young, Jr, *Century College*
Vasilis Zafiris, *University of Houston*
Vivian Zimmerman, *Prairie State College*

Special thanks go to Jon Weerts for preparing the *Instructor's Solutions Manual* and the *Student's Solutions Manual* and to Rebecca Hubiak for her work ensuring accuracy. Many thanks to Cindy Reed for her work in the video series, and to Ethel Wheland for advising us on the Instructor Notes.

Finally, we are forever grateful to the many people behind the scenes at McGraw-Hill without whom we would still be on page 1. To our developmental editor (and math instructor extraordinaire), Emilie Berglund, thanks for your day-to-day support and understanding of the world of developmental mathematics. To David Millage, our sponsoring editor and overall team captain, thanks for keeping the train on the track. Where did you find enough hours in the day? To Torie Anderson and Sabina Navsariwala, we greatly appreciate your countless hours of support and creative ideas promoting all of our efforts. To our director of development and champion,

Kris Tibbetts, thanks for being there in our time of need. To Pat Steele, where would we be without your watchful eye over our manuscript? To our editorial director, Stewart Mattson, we're grateful for your experience and energizing new ideas. Thanks for believing in us. To Jeff Huettman and Amber Bettcher, we give our greatest appreciation for the exciting technology so critical to student success. To Peggy Selle, thanks for keeping watch over the whole team as the project came together. Thank you to our wonderful designer Laurie Janssen—not only did Laurie help develop a better textbook series by delivering a clean, clear design framework for the mathematics content, Laurie also designed the best covers of the Miller/O'Neill/Hyde series to date.

Most importantly, we give special thanks to all the students and instructors who use *Basic College Mathematics* in their classes.

Supplements

For the Instructor

Instructor's Resource Manual

The *Instructor's Resource Manual* (*IRM*), written by the authors, is a printable electronic supplement available through Connect Math Hosted by ALEKS Corp. The *IRM* includes discovery-based classroom activities, worksheets for drill and practice, materials for a student portfolio, and tips for implementing successful cooperative learning. Numerous classroom activities are available for each section of text and can be used as a complement to the lectures or can be assigned for work outside of class. The activities are designed for group or individual work and take about 5–10 minutes each. With increasing demands on faculty schedules, these ready-made lessons offer a convenient means for both full-time and adjunct faculty to promote active learning in the classroom.

Instructor's Test Bank

Among the supplements is a **computerized test bank** utilizing Brownstone Diploma® algorithm-based testing software to create customized exams quickly. This user-friendly program enables instructors to search for questions by topic, format, or difficulty level; to edit existing questions or to add new ones; and to scramble questions and answer keys for multiple versions of a single test. Hundreds of text-specific, open-ended, and multiple-choice questions are included in the question bank. Sample chapter tests are also provided.

Annotated Instructor's Edition

In the *Annotated Instructor's Edition* (*AIE*), **answers to all exercises and tests appear adjacent to each exercise,** in a color used *only* for annotations. The *AIE* also contains **Instructor Notes** that appear in the margin. The notes may assist with lecture preparation. Also found in the *AIE* are icons within the Practice Exercises that serve to guide instructors in their preparation of homework assignments and lessons.

Another significant feature new to this edition is the inclusion of ***Classroom Examples*** for the instructor. In the Annotated Instructor's Edition of the text, we include references to even-numbered exercises at the end of the section for instructors to use as *Classroom Examples*. These exercises mirror the examples in the text. Therefore, if an instructor covers these exercises as classroom examples, then all the major objectives in that section will have been covered. This feature was added because we recognize the growing demands on faculty time, and to assist new faculty, adjunct faculty, and graduate assistants. Furthermore, because these exercises appear in the student edition of the text, students will not waste valuable class time copying down complicated examples from the board.

Instructor's Solutions Manual

The *Instructor's Solutions Manual* provides comprehensive, worked-out solutions to all exercises in the Chapter Openers; the Practice Exercises; the Problem Recognition Exercises; the end-of-chapter Review Exercises; the Chapter Tests; and the Cumulative Review Exercises.

For the Student

NEW Lecture Videos created by Julie Miller

Julie Miller began creating these lecture videos for her own students to use when they were absent and unable to attend one of her lectures. She found them to be so helpful that she decided to create the lecture videos for her entire developmental math book series. In these new videos, Julie walks students through the learning objectives using the same language and procedures outlined in the book. Students are able to learn and review right alongside the author! Students can also access the note files that accompany the videos so that they can take their notes while following the video just like a classroom lecture. These videos as well as the exercise videos are available online through Connect Math Hosted by ALEKS Corp.

The videos are closed-captioned for the hearing-impaired, subtitled in Spanish, and meet the Americans with Disabilities Act Standards for Accessible Design. Instructors may use them as resources in a learning center, for online courses, and to provide additional help for students who require extra practice.

ALEKS Prep for Developmental Mathematics

ALEKS Prep for Beginning Algebra and Prep for Intermediate Algebra focus on prerequisite and introductory material for Beginning Algebra and Intermediate Algebra. These prep products can be used during the first 3 weeks of a course to prepare students for future success in the course and to increase retention and pass rates. Backed by two decades of National Science Foundation funded research, ALEKS interacts with students much like a human tutor, with the ability to precisely asses a student's preparedness and provide instruction on the topics the student is most likely to learn.

ALEKS Prep Course Products Feature:

- Artificial Intelligence Targets Gaps in Individual possessive Student's Knowledge
- Assessment and Learning Directed Toward Individual possessive Student's Needs
- Open Response Environment with Realistic Input Tools
- Unlimited Online Access—PC and Mac Compatible

Free trial at www.aleks.com/free_trial/instructor

Student's Solutions Manual

The *Student's Solutions Manual* provides comprehensive, worked-out solutions to the odd-numbered exercises in the Practice Exercise sets; the Problem Recognition Exercises, the end-of-chapter Review Exercises, the Chapter Tests, and the Cumulative Review Exercises. Answers to the odd- and even-numbered entries to the Chapter Opener Puzzles are also provided.

Exercise Video Series

The video series is based on exercises from the textbook. Each presenter works through selected problems, following the solution methodology employed in the text. The video series is available online as part of Connect Math Hosted by ALEKS Corp. The videos are closed-captioned for the hearing impaired, are subtitled in Spanish, and meet the Americans with Disabilities Act Standards for Accessible Design.

Whole Numbers

1

Chapter 1

Chapter 1 begins with adding, subtracting, multiplying, and dividing whole numbers. We also include rounding, estimating, and applying the order of operations. As you work through the chapter, you can check your skills by filling in this puzzle.

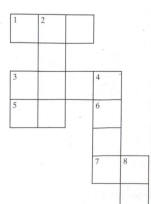

Across

1. $123 + 38$
3. $3866 - 2345$
5. $4 + 7 \times 8$
6. What digit is in the ten-thousands place?

   $4{,}532{,}891$
7. $6^2 + 18 \div 2 \times 3$

Down

2. 125×50
3. 4^2
4. $575 + 89 + 722$
8. $372 \div 12$

Section 1.1 | Introduction to Whole Numbers

Objectives

1. **Place Value**
2. **Standard Notation and Expanded Notation**
3. **Writing Numbers in Words**
4. **The Number Line and Order**

1. Place Value

Numbers provide the foundation that is used in mathematics. We begin this chapter by discussing how numbers are represented and named. All numbers in our numbering system are composed from the **digits** 0, 1, 2, 3, 4, 5, 6, 7, 8, and 9. In mathematics, the numbers 0, 1, 2, 3, 4, 5, 6, 7, 8, 9, 10, 11, 12, . . . are called the *whole numbers*. (The three dots are called *ellipses* and indicate that the list goes on indefinitely.)

For large numbers, commas are used to separate digits into groups of three called **periods**. For example, the number of live births in the United States in a recent year was 4,058,614 (*Source: The World Almanac*). Numbers written in this way are said to be in **standard form**. The position of each digit within a number determines the place value of the digit. To interpret the number of births in the United States, refer to the place value chart (Figure 1-1).

Concept Connections

1. Explain the difference between the two 3's in the number 303.

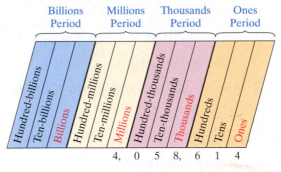

Figure 1-1

The digit 5 in the number 4,058,614 represents 5 ten-thousands because it is in the ten-thousands place. The digit 4 on the left represents 4 millions, whereas the digit 4 on the right represents 4 ones.

Skill Practice

Determine the place value of the digit 4 in each number.

2. 547,098,632
3. 1,659,984,036
4. 6,420

Example 1 **Determining Place Value**

Determine the place value of the digit 2 in each number.

a. 417,216,900   **b.** 724   **c.** 502,000,700

Solution:

a. 417,216,900   hundred-thousands

b. 724   tens

c. 502,000,700   millions

Answers

1. First 3 (on the left) represents 3 hundreds, while the second 3 (on the right) represents 3 ones.
2. Ten-millions
3. Thousands
4. Hundreds

Example 2   **Determining Place Value**

Mount Everest, the highest mountain on earth, is 29,035 feet (ft) tall. Give the place value for each digit in this number.

Solution:

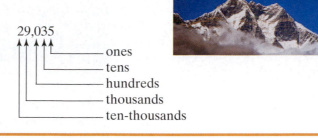

29,035

— ones
— tens
— hundreds
— thousands
— ten-thousands

Skill Practice

5. Alaska is the largest state geographically. Its land area is 571,962 square miles (mi^2). Give the place value for each digit.

2. Standard Notation and Expanded Notation

A number can also be written in an expanded form by writing each digit with its place value unit. For example, the number 287 can be written as

$$287 = 2 \text{ hundreds} + 8 \text{ tens} + 7 \text{ ones}$$

This is called **expanded form**.

Example 3   **Converting Standard Form to Expanded Form**

Convert to expanded form.

   a. 4,672   **b.** 257,016

Solution:

   a. 4,672   4 thousands + 6 hundreds + 7 tens + 2 ones

   b. 257,016   2 hundred-thousands + 5 ten-thousands + 7 thousands + 1 ten + 6 ones

Skill Practice

Convert to expanded form.
   6. 837
   7. 4,093,062

Example 4   **Converting Expanded Form to Standard Form**

Convert to standard form.

   a. 2 hundreds + 5 tens + 9 ones

   b. 1 thousand + 2 tens + 5 ones

Solution:

   a. 2 hundreds + 5 tens + 9 ones = 259

   b. Each place position from the thousands place to the ones place must contain a digit. In this problem, there is no reference to the hundreds place digit. Therefore, we assume 0 hundreds. Thus,

$$1 \text{ thousand} + 0 \text{ hundreds} + 2 \text{ tens} + 5 \text{ ones} = 1{,}025$$

Skill Practice

Convert to standard form.
   8. 8 thousands + 5 hundreds + 5 tens + 1 one
   9. 5 hundred-thousands + 4 thousands + 8 tens + 3 ones

Answers

5. 5: hundred-thousands
    7: ten-thousands
    1: thousands   9: hundreds
    6: tens   2: ones
6. 8 hundreds + 3 tens + 7 ones
7. 4 millions + 9 ten-thousands + 3 thousands + 6 tens + 2 one
8. 8,551   9. 504,083

3. Writing Numbers in Words

The word names of some two-digit numbers appear with a hyphen, while others do not. For example:

Number	Number Name
12	twelve
68	sixty-eight
40	forty
42	forty-two

To write a three-digit or larger number, begin at the leftmost group of digits. The number named in that group is followed by the period name, followed by a comma. Then the next period is named, and so on.

Example 5 Writing a Number in Words

Write the number 621,417,325 in words.

Solution:

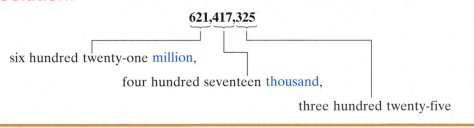

621,417,325

six hundred twenty-one million,

four hundred seventeen thousand,

three hundred twenty-five

Notice from Example 5 that when naming numbers, the name of the ones period is not attached to the last group of digits. Also note that for whole numbers, the word *and* should not appear in word names. For example, the number 405 should be written as four hundred five.

Example 6 Writing a Number in Standard Form

Write the number in standard form.

Six million, forty-six thousand, nine hundred three

Solution:

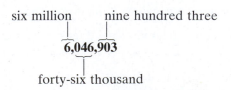

six million   nine hundred three

6,046,903

forty-six thousand

We have seen several examples of writing a number in standard form, in expanded form, and in words. Standard form is the most concise representation. Also note that when we write a four-digit number in standard form, the comma is often omitted. For example, the number 4,389 is often written as 4389.

4. The Number Line and Order

Whole numbers can be visualized as equally spaced points on a line called a *number line* (Figure 1-2).

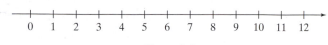

Figure 1-2

The whole numbers begin at 0 and are ordered from left to right by increasing value.

A number is graphed on a number line by placing a dot at the corresponding point. For any two numbers graphed on a number line, the number to the left is less than the number to the right. Similarly, a number to the right is greater than the number to the left. In mathematics, the symbol $<$ is used to denote "is less than," and the symbol $>$ means "is greater than." Therefore,

$3 < 5$ means   3 is less than 5

$5 > 3$ means   5 is greater than 3

| **Example 7** | **Determining Order Between Two Numbers** |

Fill in the blank with the symbol $<$ or $>$.

a. 7 ☐ 0   **b.** 30 ☐ 82

Solution:

a. 7 $>$ 0

b. 30 $<$ 82

To visualize the numbers 82 and 30 on the number line, it may be necessary to use a different scale. Rather than setting equally spaced marks in units of 1, we can use units of 10. The number 82 must be somewhere between 80 and 90 on the number line.

Skill Practice

Fill in the blank with the symbol $<$ or $>$.

12. 9 ☐ 5

13. 8 ☐ 18

Answers

12. $>$ **13.** $<$

Section 1.1 Practice Exercises

Study Skills Exercises

In this text, we provide skills for you to enhance your learning experience. Each set of practice exercises begins with an activity that focuses on one of eight areas: learning about your course, using your text, taking notes, doing homework, taking an exam (test and math anxiety), managing your time, recognizing your learning style, and studying for the final exam.

Each activity requires only a few minutes and will help you to pass this class and become a better math student. Many of these skills can be carried over to other disciplines and help you become a model college student.

1. To begin, write down the following information.

   a. Instructor's name
   b. Instructor's office number
   c. Instructor's telephone number
   d. Instructor's email address
   e. Instructor's office hours
   f. Days of the week that the class meets
   g. The room number in which the class meets
   h. Is there a lab requirement for this course? If so, where is the lab located and how often must you go?

2. Define the key terms.

   a. Digit
   b. Standard form
   c. Periods
   d. Expanded form

Objective 1: Place Value

3. Name the place values for each of the digits in the number 8,213,457.

4. Name the place values for each of the digits in the number 103,596.

For Exercises 5–24, determine the place value for each underlined digit. **(See Example 1.)**

5. 3<u>2</u>1
6. 6<u>8</u>9
7. 21<u>4</u>
8. 73<u>8</u>
9. 8,<u>7</u>10
10. 2,<u>2</u>93
11. <u>1</u>,430
12. 3,<u>1</u>01
13. 4<u>5</u>2,723
14. 655,<u>8</u>78
15. <u>1</u>,023,676,207
16. 3,<u>1</u>11,901,211
17. <u>2</u>2,422
18. 5<u>8</u>,106
19. 5<u>1</u>,033,201
20. 9<u>3</u>,971,224

21. The number of U.S. travelers abroad in a recent year was <u>1</u>0,677,881. **(See Example 2.)**

22. The area of Lake Superior is 31,<u>8</u>20 mi^2.

23. For a recent year, the total number of U.S. $1 bills in circulation was 7,653,468,440.

24. For a certain flight, the cruising altitude of a commercial jet is 31,000 ft.

Objective 2: Standard Notation and Expanded Notation

For Exercises 25–32, convert the numbers to expanded form. **(See Example 3.)**

25. 58   **26.** 71   **27.** 539   **28.** 382

29. 503   **30.** 809   **31.** 10,241   **32.** 20,873

For Exercises 33–40, convert the numbers to standard form. **(See Example 4.)**

33. 5 hundreds + 2 tens + 4 ones   **34.** 3 hundreds + 1 ten + 8 ones

35. 1 hundred + 5 tens   **36.** 6 hundreds + 2 tens

37. 1 thousand + 9 hundreds + 6 ones   **38.** 4 thousands + 2 hundreds + 1 one

39. 8 ten-thousands + 5 thousands + 7 ones   **40.** 2 ten-thousands + 6 thousands + 2 ones

41. Name the first four periods of a number (from right to left).

42. Name the first four place values of a number (from right to left).

Objective 3: Writing Numbers in Words

For Exercises 43–50, write the number in words. **(See Example 5.)**

43. 241   **44.** 327   **45.** 603   **46.** 108

47. 31,530   **48.** 52,160   **49.** 100,234   **50.** 400,199

51. The Shuowen jiezi dictionary, an ancient Chinese dictionary that dates back to the year 100, contained 9,535 characters. Write the number 9,535 in words.

52. Researchers calculate that about 590,712 stone blocks were used to construct the Great Pyramid. Write the number 590,712 in words.

53. Mt. McKinley in Alaska is 20,320 ft high. Write the number 20,320 in words.

54. There are 1,800 seats in the Regal Champlain Theater in Plattsburgh, New York. Write the number 1,800 in words.

55. Interstate I-75 is 1,377 miles (mi) long. Write the number 1,377 in words.

56. In the United States, there are approximately 60,000,000 cats living in households. Write the number 60,000,000 in words.

For Exercises 57–62, convert the number to standard form. **(See Example 6.)**

57. Six thousand, five

58. Four thousand, four

59. Six hundred seventy-two thousand

60. Two hundred forty-eight thousand

61. One million, four hundred eighty-four thousand, two hundred fifty

62. Two million, six hundred forty-seven thousand, five hundred twenty

Objective 4: The Number Line and Order

For Exercises 63–64, graph the numbers on the number line.

63. **a.** 6   **b.** 13   **c.** 8   **d.** 1

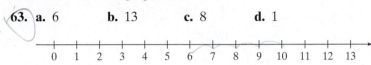

64. **a.** 5   **b.** 3   **c.** 11   **d.** 9

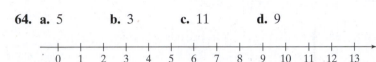

65. On a number line, what number is 4 units to the right of 6?

66. On a number line, what number is 8 units to the left of 11?

67. On a number line, what number is 3 units to the left of 7?

68. On a number line, what number is 5 units to the right of 0?

For Exercises 69–72, translate the inequality to words.

69. $8 > 2$ **70.** $6 < 11$ **71.** $3 < 7$ **72.** $14 > 12$

For Exercises 73–84, fill in the blank with the inequality symbol $<$ or $>$. **(See Example 7.)**

73. 6 ☐ 11   **74.** 14 ☐ 13   **75.** 21 ☐ 18   **76.** 5 ☐ 7

77. 3 ☐ 7   **78.** 14 ☐ 24   **79.** 95 ☐ 89   **80.** 28 ☐ 30

81. 0 ☐ 3   **82.** 8 ☐ 0   **83.** 90 ☐ 91   **84.** 48 ☐ 47

Expanding Your Skills

85. Answer true or false. The number 12 is a digit.

86. Answer true or false. The number 26 is a digit.

87. What is the greatest two-digit number?

88. What is the greatest three-digit number?

89. What is the greatest whole number?

90. What is the least whole number?

91. How many zeros are there in the number ten million?

92. How many zeros are there in the number one hundred billion?

93. What is the greatest three-digit number that can be formed from the digits 6, 9, and 4? Use each digit only once.

94. What is the greatest three-digit number that can be formed from the digits 0, 4, and 8? Use each digit only once.

Addition of Whole Numbers and Perimeter

1. Addition of Whole Numbers Using the Number Line

Objectives

1. Addition of Whole Numbers Using the Number Line
2. Addition of Whole Numbers
3. Properties of Addition
4. Translations and Applications Involving Addition
5. Perimeter

We use addition of whole numbers to represent an increase in quantity. For example, suppose Jonas typed 5 pages of a report before lunch. Later in the afternoon he typed 3 more pages. The total number of pages that he typed is found by adding 5 and 3.

$$5 \text{ pages} + 3 \text{ pages} = 8 \text{ pages}$$

The result of an addition problem is called the **sum**, and the numbers being added are called **addends**. Thus,

$$5 + 3 = 8$$

addends   sum

Concept Connections

1. Identify the addends and the sum.
   $$3 + 7 + 12 = 22$$

The number line is a useful tool to visualize the operation of addition. To add 5 and 3 on a number line, begin at 0 and move 5 units to the right. Then move an additional 3 units to the right. The final location indicates the sum.

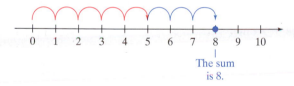

The sum is 8.

You can use a number line to find the sum of any pair of digits. The sums for all possible pairs of one-digit numbers should be memorized (see Exercise 9). Memorizing these basic addition facts will make it easier for you to add larger numbers.

2. Addition of Whole Numbers

To add whole numbers, line up the numbers vertically by place value. Then add the digits in the corresponding place positions.

Example 1   Adding Whole Numbers

Add. $24 + 61$

Solution:

$$
\begin{array}{rl}
24 = & 2 \text{ tens} + 4 \text{ ones} \\
+ 61 = & 6 \text{ tens} + 1 \text{ one} \\
\hline
85 = & 8 \text{ tens} + 5 \text{ ones}
\end{array}
$$

Skill Practice

2. Add. $\begin{array}{r} 47 \\ + 32 \\ \hline \end{array}$

Answers

1. Addends: 3, 7, and 12; sum: 22
2. 79

Example 2 Adding Whole Numbers

Add. 261 + 28

Solution:

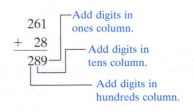

$$\begin{array}{r} 261 \\ +\ \ 28 \\ \hline 289 \end{array}$$

┌ Add digits in
ones column.
├ Add digits in
tens column.
└ Add digits in
hundreds column.

Sometimes when adding numbers, the sum of the digits in a given place position is greater than 9. If this occurs, we must do what is called *carrying* or *regrouping*. Example 3 illustrates this process.

Example 3 Adding Whole Numbers with Carrying

Add. 35 + 48

Solution:

$$\begin{array}{r} 35 = 3\text{ tens} +\ \ 5\text{ ones} \\ +\ 48 = 4\text{ tens} +\ \ 8\text{ ones} \\ \hline 7\text{ tens} + 13\text{ ones} \end{array}$$ ← The sum of the digits in the ones place exceeds 9. But 13 ones is the same as 1 ten and 3 ones. We can *carry* 1 ten to the tens column while leaving the 3 ones in the ones column. Notice that we placed the carried digit above the tens column.

$$\begin{array}{r} \overset{1}{\ }\ \ \overset{1\text{ ten}}{\ } \\ 35 = 3\text{ tens} + 5\text{ ones} \\ +\ 48 = 4\text{ tens} + 8\text{ ones} \\ \hline 83 = 8\text{ tens} + 3\text{ ones} \end{array}$$

The sum is 83.

Example 4 Adding Whole Numbers with Carrying

Add. 458 + 67

Solution:

$$\begin{array}{r} \overset{1}{4}58 \\ +\ \ 67 \\ \hline 5 \end{array}$$
Add the digits in the ones column: 8 + 7 = 15. Write 5 in the ones column, and carry the 1 to the tens column.

$$\begin{array}{r} \overset{11}{4}58 \\ +\ \ 67 \\ \hline 25 \end{array}$$
Add the digits in the tens column (including the carry): 1 + 5 + 6 = 12. Write the 2 in the tens column, and carry the 1 to the hundreds column.

$$\begin{array}{r} \overset{11}{4}58 \\ +\ \ 67 \\ \hline 525 \end{array}$$
Add the digits in the hundreds column.

The sum is 525.

Addition of numbers may include more than two addends.

<table>
<tr><td>**Example 5**</td><td>**Adding Whole Numbers**</td></tr>
</table>

Add. $21{,}076 + 84{,}158 + 2419$

Solution:

$$
\begin{array}{r}
\overset{1}{}\overset{1\,2}{2}1,076 \\
84,158 \\
+\quad 2,419 \\
\hline
107,653
\end{array}
$$

In this example, the sum of the digits in the ones column is 23. Therefore, we write the 3 and carry the 2.

Skill Practice

6. Add.
$$
\begin{array}{r}
57{,}296 \\
4{,}089 \\
+\ 9{,}762 \\
\end{array}
$$

3. Properties of Addition

We present three properties of addition that you may have already discovered.

PROPERTY **Addition Property of 0**

The sum of any number and 0 is that number.

Examples: $5 + 0 = 5$

$0 + 2 = 2$

PROPERTY **Commutative Property of Addition**

Changing the order of two addends does not affect the sum.

Example: $5 + 7$ is equivalent to $7 + 5$

In mathematics we use parentheses ( ) as grouping symbols. To add more than two numbers, we can group them and then add. For example:

$(2 + 3) + 8$ Parentheses indicate that $2 + 3$ is added first. Then 8 is added to the result.

$= 5 + 8$

$= 13$

$2 + (3 + 8)$ Parentheses indicate that $3 + 8$ is added first. Then the result is added to 2.

$= 2 + 11$

$= 13$

PROPERTY **Associative Property of Addition**

The manner in which addends are grouped does not affect the sum.

Example: $(1 + 7) + 3$ is equivalent to $1 + (7 + 3)$

Answer

6. 71,147

Example 6 Applying the Properties of Addition

a. Rewrite $9 + 6$, using the commutative property of addition.

b. Rewrite $(15 + 9) + 5$, using the associative property of addition.

Solution:

a. $9 + 6 = 6 + 9$ Change the order of the addends.

b. $(15 + 9) + 5 = 15 + (9 + 5)$ Change the grouping of the addends.

4. Translations and Applications Involving Addition

In the English language, there are many different words and phrases that imply addition. A partial list is given in Table 1-1.

Table 1-1

Word/Phrase	Example	In Symbols
Sum	The sum of 6 and 2	$6 + 2$
Added to	3 added to 8	$8 + 3$
Increased by	7 increased by 2	$7 + 2$
More than	10 more than 6	$6 + 10$
Plus	8 plus 3	$8 + 3$
Total of	The total of 9 and 6	$9 + 6$

Example 7 Translating an English Phrase to a Mathematical Statement

Translate each phrase to an equivalent mathematical statement and simplify.

a. 12 added to 109

b. The sum of 1386 and 376

Solution:

a. $109 + 12$

$$\begin{array}{r} \overset{1}{1}09 \\ +\ \ 12 \\ \hline 121 \end{array}$$

b. $1386 + 376$

$$\begin{array}{r} \overset{11}{1}386 \\ +\ \ 376 \\ \hline 1762 \end{array}$$

Addition of whole numbers is sometimes necessary to solve application problems.

Example 8 Solving an Application Problem

Carlita works as a waitress at El Pinto restaurant in Albuquerque, New Mexico. Her tips for the last five nights were $30, $18, $66, $102, and $45. Find the total amount she made in tips.

Solution:

To find the total, we add.

$$
\begin{array}{r}
\overset{1\,2}{}\$\ 30 \\
18 \\
66 \\
102 \\
+\ \ \ 45 \\
\hline
\$261
\end{array}
$$

Carlita made $261 in tips.

Skill Practice

11. Talita received test scores of 92, 100, 84, and 96 on her first four math tests. She also earned 8 points of extra credit. How many total points did she earn?

Tables and graphs are often used to summarize information in an organized manner. Examples 9 and 10 demonstrate the interpretation of these tools.

Example 9 Solving an Application Problem Involving a Table

The following table gives the number of hits for five popular websites for a recent month. Find the total number of visitors.

Website	Number of Visitors
AOL Time Warner Network	97,995
MSN-Microsoft sites	89,819
Yahoo! sites	83,433
Google sites	37,460
Terra Lycos	36,173

Solution:

$$
\begin{array}{r}
\overset{3\,3\,2\ 2\,2}{}97{,}995 \\
89{,}819 \\
83{,}433 \\
37{,}460 \\
+\ \ 36{,}173 \\
\hline
344{,}880
\end{array}
$$

There were 344,880 combined visitors to these websites.

Skill Practice

12. The table gives the number of gold, silver, and bronze medals won in the 2006 Winter Olympics for selected countries. Find the total number of medals won by Canada.

	Gold	Silver	Bronze
Germany	11	12	6
United States	9	9	7
Canada	7	10	7

Answers

11. 380 points   **12.** 24 medals

Skill Practice

13. Samira's monthly expenses are summarized in the graph. Find the sum of her expenses.

Monthly Budget

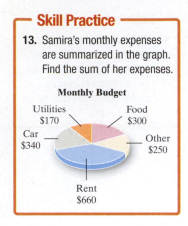

Utilities $170
Food $300
Car $340
Other $250
Rent $660

Example 10 Solving an Application Problem Involving a Graph

The graph in Figure 1-3 gives the number of new AIDS cases in the United States for the years 2003, 2004, and 2005. The red bars in the graph represent the values for the number of women (aged 13 and older). The blue bars in the graph represent the values for the number of men (aged 13 and older).

Find the total number of new AIDS cases for women in the United States in the years 2003–2005.

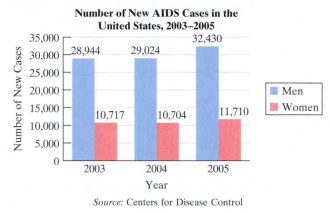

Number of New AIDS Cases in the United States, 2003–2005

Source: Centers for Disease Control

Figure 1-3

Solution:

We need to find the number of new AIDS cases for women only. Therefore, add the values corresponding to the red bars in the graph.

$$\begin{array}{r} \overset{2}{1}\overset{1}{0},717 \\ 10,704 \\ +\ 11,710 \\ \hline 33,131 \end{array}$$

There were 33,131 new AIDS cases among women in the years 2003–2005.

5. Perimeter

One special application of addition is to find the perimeter of a polygon. A **polygon** is a flat closed figure formed by line segments connected at their ends. Familiar figures such as triangles, rectangles, and squares are examples of polygons. See Figure 1-4.

Triangle   Rectangle   Square

Figure 1-4

The **perimeter** of any polygon is the distance around the outside of the figure. To find the perimeter, add the lengths of the sides.

Answer

13. $1720

Example 11   **Finding Perimeter**

Find the perimeter of the triangle.

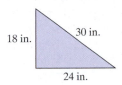

Solution:

The perimeter is the sum of the lengths of the sides.

$$\begin{array}{r} \overset{1}{1}8 \text{ in.} \\ 24 \text{ in.} \\ +\ 30 \text{ in.} \\ \hline 72 \text{ in.} \end{array}$$

The perimeter is 72 inches.

Skill Practice

14. Find the perimeter of the rectangle.

8 ft

3 ft   3 ft

8 ft

Example 12   **Finding Perimeter**

A paving company wants to edge the perimeter of a parking lot with concrete curbing. Find the perimeter of the parking lot.

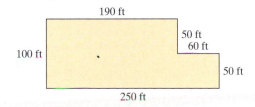

Solution:

The perimeter is the sum of the lengths of the sides.

$$\begin{array}{r} \overset{3}{1}90 \text{ ft} \\ 50 \text{ ft} \\ 60 \text{ ft} \\ 50 \text{ ft} \\ 250 \text{ ft} \\ +\ 100 \text{ ft} \\ \hline 700 \text{ ft} \end{array}$$

The distance around the parking lot (the perimeter) is 700 ft.

Skill Practice

15. Find the perimeter of the garden.

50 yd

30 yd   30 yd

15 yd   15 yd

40 yd   40 yd

20 yd

Answers

14. 22 ft

15. 240 yd

Section 1.2   Practice Exercises

Study Skills Exercises

1. Taking 12 credit-hours is the equivalent of a full-time job. Often students try to work too many hours while taking classes at school.

   a. Write down how many hours you work per week and the number of credit-hours you are taking this term.

   Number of hours worked per week _____

   Number of credit hours this term _____

   b. The table gives a recommended limit on the number of hours you should work based on the number of credit-hours you are taking at school. (Keep in mind that other responsibilities in your life such as your family might also make it necessary to limit your hours at work even more.) How do your numbers from part (a) compare to those in the table? Are you working too many hours?

Number of Credit-hours	Maximum Number of Hours of Work per Week
3	40
6	30
9	20
12	10
15	0

2. Define the key terms.

      a. Sum   **b. Addends**   **c. Polygon**   **d. Perimeter**

Review Exercises

For Exercises 3–8, write the number in the form indicated.

3. Write 351 in expanded form.

4. Write 351 in words.

5. Write 107 in expanded form.

6. Write in standard form: two thousand, four

7. Write in standard form: four thousand, twelve

8. Write in standard form:
   6 thousands + 2 hundreds + 6 ones

Objective 1: Addition of Whole Numbers Using the Number Line

9. Fill in the table. Use the number line if necessary.

+	0	1	2	3	4	5	6	7	8	9
0										
1										
2										
3										
4										
5										
6										
7										
8										
9										

For Exercises 10–15, identify the addends and the sum.

10. 5 + 9 = 14   **11.** 2 + 8 = 10   **12.** 12 + 5 = 17

13. 11 + 10 = 21   **14.** 1 + 13 + 4 = 18   **15.** 5 + 8 + 2 = 15

Objective 2: Addition of Whole Numbers

For Exercises 16–31, add. **(See Examples 1 and 2.)**

16. 42
\+ 33

17. 21
\+ 53

18. 39
\+ 20

19. 15
\+ 43

20. 12
15
\+ 32

21. 10
8
\+ 30

22. 7
21
\+ 10

23. 6
11
\+ 2

24. 341 + 225   **25.** 407 + 181   **26.** 890 + 107   **27.** 444 + 354

28. 4 + 13 + 102   **29.** 11 + 221 + 5   **30.** 31 + 7 + 430   **31.** 24 + 14 + 160

For Exercises 32–51, add the whole numbers with carrying. **(See Examples 3–5.)**

32. 76
\+ 45

33. 25
\+ 59

34. 87
\+ 24

35. 38
\+ 77

36. 658
\+ 231

37. 642
\+ 295

38. 152
\+ 549

39. 462
\+ 388

40. 15 + 5 + 9   **41.** 2 + 31 + 8   **42.** 14 + 9 + 17   **43.** 7 + 18 + 4

44. 79 + 112 + 12   **45.** 62 + 907 + 34   **46.** 331 + 422 + 76   **47.** 87 + 119 + 630

48. 4980 + 10,223   **49.** 23,112 + 892   **50.** 10,223 + 25,782 + 4980   **51.** 92,377 + 5622 + 34,659

Objective 3: Properties of Addition

For Exercises 52–55, rewrite the addition problem, using the commutative property of addition. **(See Example 6.)**

52. 12 + 6 = ☐ + ☐   **53.** 30 + 21 = ☐ + ☐   **54.** 101 + 44 = ☐ + ☐   **55.** 8 + 13 = ☐ + ☐

For Exercises 56–59, rewrite the addition problem using the associative property of addition, by inserting a pair of parentheses.

56. (4 + 8) + 13 = 4 + 8 + 13   **57.** (23 + 9) + 10 = 23 + 9 + 10

58. 7 + (12 + 8) = 7 + 12 + 8   **59.** 41 + (3 + 22) = 41 + 3 + 22

60. Explain the difference between the commutative and associative properties of addition.

61. Explain the addition property of 0. Then simplify the expressions.

a. 423 + 0   **b.** 0 + 25   **c.** 67
\+ 0

Objective 4: Translations and Applications Involving Addition

For Exercises 62–70, translate the English phrase into a mathematical statement and simplify. **(See Example 7.)**

62. The sum of 13 and 7

63. The sum of 100 and 42

64. 45 added to 7

65. 81 added to 23

66. 5 more than 18

67. 2 more than 76

68. 1523 increased by 90

69. 1320 increased by 448

70. The total of 5, 39, and 81

For Exercises 71–78, write an English phrase from the mathematical statement. Answers may vary.

71. 54 + 24

72. 33 + 15

73. 12 + 88

74. 70 + 15

75. 4 + 23 + 77

76. 11 + 41 + 53

77. 10 + 8

78. 25 + 14

79. The attendance at a high school play during one weekend was as follows: 103 on Friday, 112 on Saturday, and 61 at the Sunday matinee. What was the total attendance? **(See Example 8.)**

80. To schedule enough drivers for an upcoming week, a local pizza shop manager recorded the number of deliveries each day from the previous week: 38, 54, 44, 61, 97, 103, 124. What was the total number of deliveries for the week?

81. Three top television shows entertained the following number of viewers in one week: 26,548,000 for *American Idol*, 26,930,000 for *Grey's Anatomy*, and 20,805,000 for *House*. Find the sum of the viewers for these shows.

82. To travel from Houston to Corpus Christi, a salesperson must stop in San Antonio. If it is 195 mi from Houston to San Antonio and 228 mi from San Antonio to Corpus Christi, how far will she travel on this trip?

83. Nora earned $43,000 last year. This year her salary was increased by $2500. What is her present salary?

84. The number of participants in the Special Olympics increased by 1,205,655 since it began in 1968 with 1000 athletes. How many athletes are presently participating?

85. The table gives the number of desks and chairs delivered each quarter to an office supply store. Find the total number of desks delivered for the year. **(See Example 9.)**

86. A portion of Jonathan's checking account register is shown. What is the total amount of the four checks written?

	Chairs	Desks
March	220	115
June	185	104
September	201	93
December	198	111

Check No.	Description	Credit	Debit	Balance
1871	Electric bill		$60	$180
1872	Groceries		52	128
1873	Department store		75	53
	Payroll deposit	$1256		1309
1874	Restaurant		58	1251
	Deposit from savings	150		1401

87. The Student Career Experience Program is a program that places students in government jobs. The graph displays the number of participants in the top six agencies. Find the total number of participants in the program. **(See Example 10.)**

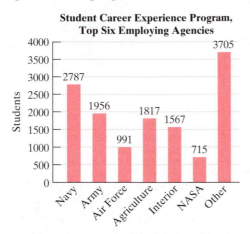

Student Career Experience Program, Top Six Employing Agencies

Source: U.S. office of Personnel Management

88. The graph displays the number of public school teachers in the United States. Find the number of elementary school, prekindergarten, and kindergarten teachers.

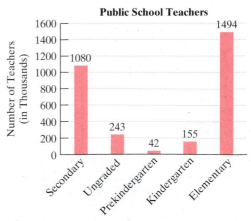

Public School Teachers

Source: National Center for Education Statistics

89. The staff for U.S. public schools is categorized in the graph. Determine the number of staff other than teachers.

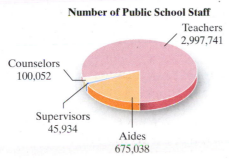

Number of Public School Staff

Teachers 2,997,741
Counselors 100,052
Supervisors 45,934
Aides 675,038

Source: National Center for Education Statistics

90. The pie graph shows the costs incurred in managing Sub-World sandwich shop for one month. From this information, determine the total cost for one month.

Sub-World Monthly Expenses

Food $7329
Labor $9560
Nonfood items $1248
Overhead $3500

Objective 5: Perimeter

For Exercises 91–98, find the perimeter. **(See Examples 11–12.)**

91.

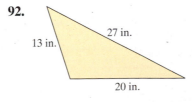

35 cm   35 cm
34 cm

92.

27 in.
13 in.
20 in.

93.

21 m   20 m
21 m   18 m
11 m   19 m

94.

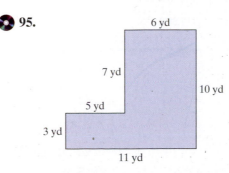

15 m
7 m   7 m
6 m

95.

6 yd
7 yd
10 yd
5 yd
3 yd
11 yd

96.

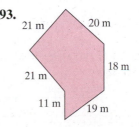

200 yd
38 yd
58 yd
98 yd   136 yd
142 yd

97. Find the perimeter of an NBA basketball court.

98. A major league baseball diamond is in the shape of a square. Find the distance a batter must run if he hits a home run.

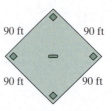

Calculator Connections

Topic: Adding Whole Numbers

The following keystrokes demonstrate the procedure to add numbers on a calculator. The **ENTER** key (or, on some calculators, the **=** key or **EXE** key) tells the calculator to complete the calculation. Notice that commas used in large numbers are not entered into the calculator.

Expression	Keystrokes	Result
92,406 + 83,168	92406 **+** 83168 **ENTER**	175574

Your calculator may use the **=** key or **EXE** key instead.

Calculator Exercises

For Exercises 99–106, add by using a calculator.

99. 9,084,037 + 452,903

100. 899,382 + 9406

101. 7,201,529 + 962,411

102.
```
   45,418
   81,990
    9,063
 + 56,309
```

103.
```
  9,300,050
  7,803,513
  3,480,009
 +  907,822
```

104.
```
  3,421,019
    822,761
  1,003,721
 +    9,678
```

105. The number of viewers for four television programs for a selected week is given in the table. (*Source:* The Nielsen Company) What is the total number of viewers?

Program	Number of Viewers
ABC premier event	17,457,000
American Idol	17,164,000
CSI	17,004,000
Law and Order	15,717,000

106. The number of votes tallied for the leading presidential candidates for the 2004 election is given in the table. (*Source:* U.S. Department of State) Find the total number of votes for these three candidates.

Candidate	Number of Votes
Nader	411,304
Kerry	59,028,109
Bush	62,040,606

Subtraction of Whole Numbers

1. Introduction to Subtraction

Jeremy bought a case of 12 sodas, and on a hot afternoon he drank 3 of the sodas. We can use the operation of subtraction to find the number of sodas remaining.

Objectives

1. **Introduction to Subtraction**
2. **Subtraction of Whole Numbers**
3. **Translations and Applications Involving Subtraction**

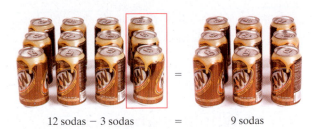

12 sodas − 3 sodas   =   9 sodas

The symbol "−" between two numbers is a subtraction sign, and the result of a subtraction is called the **difference**. The number being subtracted (in this case, 3) is called the **subtrahend**. The number 12 from which 3 is subtracted is called the **minuend**.

$$12 - 3 = 9 \quad \text{is read as} \quad \text{"12 minus 3 is equal to 9"}$$

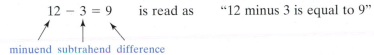

minuend subtrahend difference

Subtraction is the reverse operation of addition. To find the number of sodas that remain after Jeremy takes 3 sodas away from 12 sodas, we ask the following question:

"3 added to what number equals 12?"

That is,

$$12 - 3 = ? \quad \text{is equivalent to} \quad ? + 3 = 12$$

Subtraction can also be visualized on the number line. To evaluate $7 - 4$, start from the point on the number line corresponding to the minuend (7 in this case). Then move to the *left* 4 units. The resulting position on the number line is the difference.

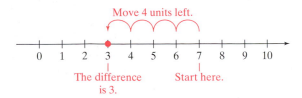

To check the result, we can use addition.

$$7 - 4 = 3 \quad \text{because} \quad 3 + 4 = 7$$

Skill Practice

Subtract. Check by using addition.

1. $11 - 5$ **2.** $8 - 0$
3. $7 - 2$ **4.** $5 - 5$

Example 1 **Subtracting Whole Numbers**

Subtract and check the answer, using addition.

a. $8 - 2$ **b.** $10 - 6$ **c.** $5 - 0$ **d.** $3 - 3$

Solution:

a. $8 - 2 = 6$ because $6 + 2 = 8$

b. $10 - 6 = 4$ because $4 + 6 = 10$

c. $5 - 0 = 5$ because $5 + 0 = 5$

d. $3 - 3 = 0$ because $0 + 3 = 3$

2. Subtraction of Whole Numbers

When subtracting large numbers, it is usually more convenient to write the numbers vertically. We write the minuend on top and the subtrahend below it. Starting from the ones column, we subtract digits having corresponding place values.

Skill Practice

Subtract. Check by using addition.

5.   472   **6.**   3947
     $-$ 261 $-$ 137

Example 2 **Subtracting Whole Numbers**

Subtract and check the answer by using addition.

a.   976   **b.**   2498
     $-$ 124 $-$ 197

Solution:

a.   976   Check: 852
     $-$ 124 $+$ 124
       852   976 ✓

└── Subtract the ones column digits.
└── Subtract the tens column digits.
└── Subtract the hundreds column digits.

b.   2498   Check: 2301
     $-$ 197 $+$ 197
       2301   2498 ✓

When a digit in the subtrahend (bottom number) is larger than the corresponding digit in the minuend (top number), we must "regroup" or borrow a value from the column to the left.

$92 = 9$ tens $+ 2$ ones
$-74 = 7$ tens $+ 4$ ones

In the ones column, we cannot take 4 away from 2. We will regroup by borrowing 1 ten from the minuend. Furthermore, 1 ten = 10 ones.

$9\!\!\!/2 = 9\!\!\!/$ tens $+ 2$ ones} We now have 12 ones in the minuend.
$-74 = 7$ tens $+ 4$ ones

$9\!\!\!/2 = 9\!\!\!/$ tens $+ 12$ ones
$-74 = 7$ tens $+ \;\;4$ ones
$\;18 = 1$ ten $+ \;\;8$ ones

TIP: The process of *borrowing* in subtraction is the reverse operation of *carrying* in addition.

Answers

1. 6   **2.** 8   **3.** 5   **4.** 0
5. 211   **6.** 3810

Example 3 **Subtracting Whole Numbers with Borrowing**

Subtract and check the result with addition.

$$134{,}616$$
$$-\ 53{,}438$$

Solution:

$$\begin{array}{r} {}^{\ 0\ 16}\\ 1\,3\,4{,}6\,\cancel{1}\,\cancel{6}\\ -\ \ 5\,3{,}4\,3\,8\\ \hline 8 \end{array}$$

In the ones place, 8 is greater than 6.
We borrow 1 ten from the tens place.

$$\begin{array}{r} {}^{\ \ 10}\\ {}^{5\ \cancel{0}\ 16}\\ 1\,3\,4{,}\cancel{6}\,\cancel{1}\,\cancel{6}\\ -\ \ 5\,3{,}4\,3\,8\\ \hline 7\,8 \end{array}$$

In the tens place, 3 is greater than 0. We
borrow 1 hundred from the hundreds place.

$$\begin{array}{r} {}^{0\ 13\ \ \ \ \ 10}\\ {}^{\ \ \ \ \ 5\ \cancel{0}\ 16}\\ \cancel{1}\,\cancel{3}\,4{,}\cancel{6}\,\cancel{1}\,\cancel{6}\\ -\ \ 5\,3{,}4\,3\,8\\ \hline 8\,1{,}1\,7\,8 \end{array}$$

In the ten-thousands place, 5 is greater than 3.
We borrow 1 hundred-thousand from the hundred-
thousands place.

Check:
$$\begin{array}{r} {}^{1\ \ \ \ 11}\\ 81{,}178\\ +\ 53{,}438\\ \hline 134{,}616\ \checkmark \end{array}$$

Example 4 **Subtracting Whole Numbers with Borrowing**

Subtract and check the result with addition. $500 - 247$

Solution:

$$\begin{array}{r} 500\\ -\ 247 \end{array}$$

In the ones place, 7 is greater than 0. We try to borrow 1 ten
from the tens place. However, the tens place digit is 0. Therefore we must first borrow from the hundreds place.

$$\begin{array}{r} {}^{4\ 10}\\ \cancel{5}\,\cancel{0}\,0\\ -\ 2\,4\,7 \end{array}$$

$$\begin{array}{r} {}^{\ \ \ 9}\\ {}^{4\ \cancel{10}\ 10}\\ \cancel{5}\,\cancel{0}\,\cancel{0}\\ -\ 2\,4\,7\\ \hline 2\,5\,3 \end{array}$$ ←Now we can borrow 1 ten to add to the ones place.

Subtract.

Check:
$$\begin{array}{r} {}^{1\ 1}\\ 253\\ +\ 247\\ \hline 500\ \checkmark \end{array}$$

3. Translations and Applications Involving Subtraction

In applications of mathematics, several words and phrases imply subtraction. A partial list is provided in Table 1-2.

Table 1-2

Word/Phrase	Example	In Symbols
Minus	15 minus 10	$15 - 10$
Difference	The difference of 10 and 2	$10 - 2$
Decreased by	9 decreased by 1	$9 - 1$
Less than	5 less than 12	$12 - 5$
Subtract . . . from	Subtract 3 from 8	$8 - 3$

In Table 1-2, make a note of the last two entries. The phrases *less than* and *subtract . . . from* imply a specific order in which the subtraction is performed. In both cases, begin with the second number listed and subtract the first number listed.

Example 5 **Translating an English Phrase to a Mathematical Statement**

Translate the English phrase to a mathematical statement and simplify.

a. The difference of 150 and 38

b. 30 subtracted from 82

Solution:

a. From Table 1-2, the *difference* of 150 and 38 implies $150 - 38$.

$$
\begin{array}{r}
\overset{4\ 10}{1\not5 0} \\
-\quad 3 8 \\
\hline
1 1 2
\end{array}
$$

b. The phrase "30 subtracted from 82" implies that 30 is taken away from 82. We have $82 - 30$.

$$
\begin{array}{r}
8 2 \\
- 3 0 \\
\hline
5 2
\end{array}
$$

In Section 1.2 we saw that the operation of addition is commutative. That is, the order in which two numbers are added does not affect the sum. This is *not* true for subtraction. For example, $82 - 30$ is not equal to $30 - 82$. The symbol $\neq$ means "is not equal to." Thus, $82 - 30 \neq 30 - 82$.

Most applications of subtraction generally fall into two categories.

1. The first type is phrased as a subtraction problem in which the minuend and subtrahend are given.

Answers

9. $12 - 8$; 4
10. $9 - 3$; 6

Example: Shawn has $52 and then spends $40. How much money does he have left? (In this problem, we subtract $40 from $52.)

$$\$52 - \$40 = \$12$$

2. The second type is phrased as an addition problem with a missing addend.

Example: Maria received 72 points on her last math test, but needed 90 points to receive an A. How many more points would she have needed to earn an A? (In this problem, the addition problem can be translated to subtraction.)

$$72 + ? = 90 \quad \text{is equivalent to} \quad 90 - 72 = ?$$

Because $90 - 72 = 18$, Maria would have needed 18 more points.

Example 6 **Solving an Application Problem**

A biology class started with 35 students. By midsemester, 7 students had dropped. How many students are still in the class?

Solution:

$$35 - 7 = 28 \qquad \text{There are } 28 \text{ students still in the class.}$$

Skill Practice

11. The temperature at 1:00 P.M. in Denver was 47°F. Three hours later, the temperature was 34°F. By how much did the temperature drop?

Example 7 **Solving an Application Problem**

A surveyor knows that the perimeter of the lot shown is 620 ft. Find the length of the missing side. See Figure 1-5.

Solution:

Recall that the perimeter of a polygon is the sum of the lengths of its sides. The sum of the three known sides in Figure 1-5 is 480 ft:

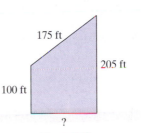
175 ft, 205 ft, 100 ft, ?

Figure 1-5

$$\begin{array}{r} \overset{1}{1}00 \\ 175 \\ + 205 \\ \hline 480 \end{array}$$

We can subtract 480 ft from the perimeter to find the length of the missing side:

$$620 \text{ ft} - 480 \text{ ft} = ?$$

$$\begin{array}{r} \overset{5}{\cancel{6}}\overset{12}{2}0 \\ - 480 \\ \hline 140 \end{array}$$

The missing side is 140 ft long.

Skill Practice

12. Teresa earned test scores of 98, 84, and 90 on her first three exams. How many points must she score on the fourth exam to earn a total of 360 points?

A third application of subtraction is to compute a change (increase or decrease) in an amount.

Answers
11. 13°F
12. 88 points

Skill Practice

Refer to the graph for Example 8.

13. a. Has the number of robberies increased or decreased between 1995 and 2005?
   b. Determine the amount of increase or decrease.

Example 8 **Solving an Application Problem**

The number of reported robberies in the United States has fluctuated each year, as shown in the graph.

Number of Robberies by Year

Source: Federal Bureau of Investigation

a. Find the increase in the number of reported robberies from the year 2000 to 2005.

b. Find the decrease in the number of reported robberies from the year 1995 to 2000.

Solution:

For the purpose of finding an amount of increase or decrease, we will subtract the smaller number from the larger number.

a. Because the number of robberies went *up* from 2000 to 2005, there was an *increase*. To find the amount of the increase, we subtract the smaller number from the larger number.

$$
\begin{array}{r}
4\,\overset{0\ 17}{1\,7},0\,0\,0 \\
-\ 4\,0\,8,0\,0\,0 \\
\hline
9,0\,0\,0
\end{array}
$$

From 2000 to 2005, there was an increase of 9,000 reported robberies in the United States.

b. Because the number of robberies went *down* from 1995 to 2000, there was a *decrease*. To find the amount of the decrease, we subtract the smaller number from the larger number.

$$
\begin{array}{r}
5\,\overset{7\ 11}{8\,1},0\,0\,0 \\
-\ 4\,0\,8,0\,0\,0 \\
\hline
1\,7\,3,0\,0\,0
\end{array}
$$

From 1995 to 2000 there was a decrease of 173,000 reported robberies in the United States.

Answers

13. a. decreased **b.** 164,000 robberies

Section 1.3   Practice Exercises

Study Skills Exercises

1. It is very important to attend class every day. Math is cumulative in nature, and you must master the material learned in the previous class to understand today's lesson. Because this is so important, many instructors tie attendance into the final grade. Write down the attendance policy for your class.

2. Define the key terms.

    a. Difference   **b. Subtrahend**   **c. Minuend**

Review Exercises

For Exercises 3–5, add.

3. $330 + 821$

4.
$$\begin{array}{r} 782 \\ 21 \\ + 1046 \\ \hline \end{array}$$

5.
$$\begin{array}{r} 46 \\ 804 \\ + 49 \\ \hline \end{array}$$

6. Circle the true statement:

$14 > 21,\ 14 < 21$

7. Circle the true statement:

$0 < 10,\ 0 > 10$

8. Write the inequality in words:

$22 < 25$

Objective 1: Introduction to Subtraction

For Exercises 9–14, identify the minuend, subtrahend, and the difference.

9. $12 - 8 = 4$

10. $6 - 1 = 5$

11. $21 - 12 = 9$

12. $32 - 2 = 30$

13.
$$\begin{array}{r} 9 \\ - 6 \\ \hline 3 \end{array}$$

14.
$$\begin{array}{r} 17 \\ - 3 \\ \hline 14 \end{array}$$

For Exercises 15–18, write the subtraction problem as a related addition problem. For example, $19 - 6 = 13$ can be written as $13 + 6 = 19$.

15. $27 - 9 = 18$

16. $20 - 8 = 12$

17. $102 - 75 = 27$

18. $211 - 45 = 166$

For Exercises 19–24, subtract, then check the answer by using addition. **(See Example 1.)**

19. $8 - 3$ Check: $\square + 3 = 8$

20. $7 - 2$ Check: $\square + 2 = 7$

21. $4 - 1$ Check: $\square + 1 = 4$

22. $9 - 1$ Check: $\square + 1 = 9$

23. $6 - 0$ Check: $\square + 0 = 6$

24. $3 - 0$ Check: $\square + 0 = 3$

Objective 2: Subtraction of Whole Numbers

For Exercises 25–36, subtract and check the answer by using addition. **(See Example 2.)**

25.
$$\begin{array}{r} 68 \\ - 23 \\ \hline \end{array}$$

26.
$$\begin{array}{r} 54 \\ - 31 \\ \hline \end{array}$$

27.
$$\begin{array}{r} 88 \\ - 27 \\ \hline \end{array}$$

28.
$$\begin{array}{r} 75 \\ - 50 \\ \hline \end{array}$$

29. 1347
      − 221

30. 4865
      − 713

31. 1525
      − 1204

32. 8843
      − 5612

33. 12,806 − 2802

34. 12,771 − 1240

35. 14,356 − 13,253

36. 34,550 − 31,450

For Exercises 37–60, subtract the whole numbers involving borrowing. **(See Examples 3–4.)**

37. 76
      − 59

38. 64
      − 48

39. 87
      − 38

40. 94
      − 75

41. 240
      − 136

42. 360
      − 225

43. 710
      − 189

44. 850
      − 303

45. 4350
      − 4327

46. 7293
      − 7255

47. 6002
      − 1238

48. 3000
      − 2356

49. 10,425
      − 9,022

50. 23,901
      − 8,064

51. 62,088
      − 59,871

52. 32,112
      − 28,334

53. 470 − 92

54. 674 − 89

55. 3700 − 2987

56. 8000 − 3788

57. 32,439 − 1498

58. 21,335 − 4123

59. 8,007,234 − 2,345,115

60. 3,045,567 − 1,871,495

Objective 3: Translations and Applications Involving Subtraction

For Exercises 61–72, translate the English phrase into a mathematical statement and simplify. **(See Example 5.)**

61. 78 minus 23

62. 45 minus 17

63. 78 decreased by 6

64. 50 decreased by 12

65. Subtract 100 from 422.

66. Subtract 42 from 89.

67. 72 less than 1090

68. 60 less than 3111

69. The difference of 50 and 13

70. The difference of 405 and 103

71. Subtract 35 from 103.

72. Subtract 14 from 91.

For Exercises 73–76, write an English phrase for the mathematical statement. (Answers will vary.)

73. 93 − 27

74. 80 − 20

75. 165 − 85

76. 171 − 42

77. Use the expression 7 − 4 to explain why subtraction is not commutative.

78. Is subtraction associative? Use the numbers 10, 6, 2 to explain.

79. A $50 bill was used to purchase $17 worth of gasoline. Find the amount of change received. **(See Example 6.)**

80. There are 55 DVDs to shelve one evening at a video rental store. If Jason puts away 39 before leaving for the day, how many are left for Patty to put away?

81. The songwriting team of John Lennon and Paul McCartney had 118 chart hits while Mick Jagger and Keith Richards had 63. How many more chart hits did Lennon and McCartney have than Jagger and Richards?

82. Due to severe drought in the state of Alabama in 2007, a local well driller said that the minimum depth to drill for water had increased from 150 ft in 2006 to 200 ft in 2007. In 2007, the driller had to dig 505 ft to find water in one rural community. How many more feet above the 2007 minimum depth did the driller have to drill?

83. In landscaping a yard, Lily would like 26 plants for a border. If she has 18 plants in her truck, how many more will she need to finish the job?

84. A collection is taken to buy flowers for a co-worker who is in the hospital. If $30 has been collected and the flower arrangement costs $43, how much more needs to be collected?

85. A recent report in *USA Today* indicated that the play, *The Lion King*, had been performed 4043 times on Broadway. At that time, the play, *Hairspray*, had been performed 2064 times on Broadway. How many more times had *The Lion King* been performed than *Hairspray*?

86. At the time of John Elway's retirement from football, his total passing yardage was 51,475 yd. Brett Favre had 42,285 yd as of 2002. How many more yards would Favre need to reach Elway's total?

For Exercises 87 and 88, for each figure find the missing length.

87. The perimeter of the triangle is 39 m.

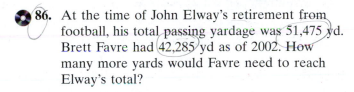

14 m   12 m   ?

88. The perimeter of the figure is 547 cm.

139 cm   87 cm   ?   201 cm

89. A homeowner knows that the perimeter of his backyard is 56 yd. Find the length of the missing side. **(See Example 7.)**

4 yd   4 yd   10 yd   14 yd   ?   14 yd

90. Barbara has 15 ft of molding to install in her bathroom, as shown in the figure. What is the missing length? *Note*: There will be no molding by the tub or door.

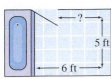

?   5 ft   6 ft

For Exercises 91–94, use the information from the graph. (**See Example 8.**)

91. What is the difference in the number of marriages between 1980 and 2000?

92. Find the decrease in the number of marriages in the United States between the years 2000 and 2005.

93. What is the difference in the number of marriages between the year having the greatest and the year having the least?

94. Between which two 5-year periods did the greatest increase in the number of marriages occur? What is the increase?

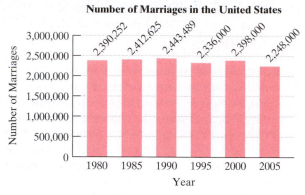

Number of Marriages in the United States

Source: National Center for Health Statistics

Figure for Exercises 91–94

Calculator Connections

Topic: Subtracting Whole Numbers

To subtract numbers on a calculator, use the subtraction key $-$. Do not confuse the subtraction key with the $(-)$ key. The $(-)$ is presented later to enter negative numbers.

Expression	Keystrokes	Result
345,899 − 43,018	345899 $-$ 43018 **ENTER**	302881

Calculator Exercises

For Exercises 95–97, subtract by using a calculator.

95. 4,905,620
    − 458,318

96. 953,400,415
    − 56,341,902

97. 82,025,160
    −79,118,705

For Exercises 98–101, refer to the table showing the land area for five states.

98. Find the difference in land area between Colorado and Wisconsin.

99. Find the difference in land area between Tennessee and West Virginia.

100. Find the difference in land area between the state with the greatest land area and the state with the least land area.

101. How much more land area does Wisconsin have than Tennessee?

State	Land Area (mi²)
Rhode Island	1,045
Tennessee	41,217
West Virginia	24,078
Wisconsin	54,310
Colorado	103,718

Rounding and Estimating

1. Rounding

Rounding a whole number is a common practice when we do not require an exact value. For example, Madagascar lost 3956 mi^2 of rainforest between 1990 and 2008. We might round this number to the nearest thousand and say that there was approximately 4000 mi^2 lost. In mathematics, we use the symbol $\approx$ to read "is approximately equal to." Therefore, 3956 mi^2 $\approx$ 4000 mi^2.

Objectives

1. **Rounding**
2. **Estimation**
3. **Using Estimation in Applications**

A number line is a helpful tool to understand rounding. For example, the number 48 is closer to 50 than it is to 40. Therefore, 48 rounded to the nearest ten is 50.

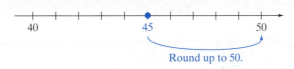

The number 43, on the other hand, is closer to 40 than to 50. Therefore, 43 rounded to the nearest ten is 40.

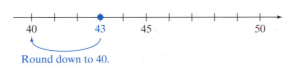

Concept Connections

1. Is the number 82 closer to 80 or to 90? Round 82 to the nearest ten.
2. Is the number 65 closer to 60 or to 70? Round the number to the nearest ten.

The number 45 is halfway between 40 and 50. In such a case, our convention will be to round *up* to the next-larger ten.

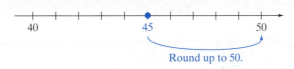

The decision to round up or down to a given place value is determined by the digit to the *right* of the given place value. The following steps outline the procedure.

PROCEDURE   Rounding Whole Numbers

Step 1   Identify the digit one position to the right of the given place value.
Step 2   If the digit in step 1 is a 5 or greater, add 1 to the digit in the given place value. Then replace each digit to the right of the given place value by 0.
Step 3   If the digit in step 1 is less than 5, replace it and each digit to its right by 0. Note that in this case, the digit in the original given place value does not change.

Answers

1. Closer to 80; 80
2. The number 65 is the same distance from 60 and 70; round up to 70.

Example 1	**Rounding a Whole Number**

Round 3741 to the nearest hundred.

Solution:

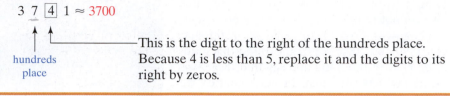

$$3\ 7\ \boxed{4}\ 1 \approx 3700$$

hundreds place

This is the digit to the right of the hundreds place. Because 4 is less than 5, replace it and the digits to its right by zeros.

Example 1 could also have been solved by drawing a number line. Use the part of a number line between 3700 and 3800.

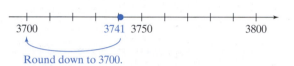

Round down to 3700.

Example 2	**Rounding a Whole Number**

Round 1,790,641 to the nearest hundred-thousand.

Solution:

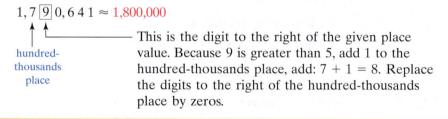

$$1,7\ \boxed{9}\ 0,6\ 4\ 1 \approx 1,800,000$$

hundred-thousands place

This is the digit to the right of the given place value. Because 9 is greater than 5, add 1 to the hundred-thousands place, add: $7 + 1 = 8$. Replace the digits to the right of the hundred-thousands place by zeros.

Example 3	**Rounding a Whole Number**

Round 1503 to the nearest thousand.

Solution:

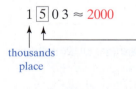

$$1\ \boxed{5}\ 0\ 3 \approx 2000$$

thousands place

This is the digit to the right of the thousands place. Because this digit is 5, we round up. We increase the thousands place digit by 1. That is, $1 + 1 = 2$. Replace the digits to its right by zeros.

Answers

3. 12,000   **4.** 150,000
5. 8,000,000

| Example 4 | **Rounding a Whole Number** |

Round the number 24,961 to the hundreds place.

Solution:

$$2\,4,9\,\boxed{6}^{+1}\,1 \approx 25,000$$

This value is greater than 5. Therefore, add 1 to the hundreds place digit. Replace the digits to the right of the hundreds place with 0.

Skill Practice
6. Round 39,823 to the nearest thousand.

2. Estimation

We use the process of rounding to estimate the result of numerical calculations. For example, to estimate the following sum, we can round each addend to the nearest ten.

31	rounds to	→	30
12	rounds to	→	10
+ 49	rounds to	→	+ 50
			90

The estimated sum is 90 (the actual sum is 92).

| Example 5 | **Estimating a Sum** |

Estimate the sum by rounding to the nearest thousand.

$$6109 + 976 + 4842 + 11,619$$

Solution:

6,109	rounds to	→	6,000
976	rounds to	→	1,000
4,842	rounds to	→	5,000
+ 11,619	rounds to	→	+ 12,000
			24,000

The estimated sum is 24,000 (the actual sum is 23,546).

Skill Practice
7. Estimate the sum by rounding each number to the nearest hundred.

3162 + 4931 + 2206

| Example 6 | **Estimating a Difference** |

Estimate the difference 4817 − 2106 by rounding each number to the nearest hundred.

Solution:

4817	rounds to	→	4800
− 2106	rounds to	→	− 2100
			2700

The estimated difference is 2700 (the actual difference is 2711).

Skill Practice
8. Estimate the difference by rounding each number to the nearest million.

35,264,000 − 21,906,210

Answers
6. 40,000   7. 10,300
8. 13,000,000

3. Using Estimation in Applications

Example 7 Estimating a Sum in an Application

A driver for a delivery service must drive from Chicago, Illinois, to Dallas, Texas, and make several stops on the way. The driver follows the route given on the map. Estimate the total mileage by rounding each distance to the nearest hundred miles.

Solution:

541	rounds to ⟶	500
418	rounds to ⟶	400
354	rounds to ⟶	400
+ 205	rounds to ⟶	+ 200
		1500

The driver traveled approximately 1500 mi.

Skill Practice

9. The two countries with the largest areas of rainforest are Brazil and the Democratic Republic of Congo. The rainforest areas are 4,776,980 square kilometers (km^2) and 1,336,100 km^2, respectively. Round the values to the nearest hundred-thousand. Estimate the total area of rainforest for these two countries.

Example 8 Estimating a Difference in an Application

In a recent year, the U.S. Census Bureau reported that the number of males over the age of 18 was 100,994,367. The same year, the number of females over 18 was 108,133,727. Round each value to the nearest million. Estimate how many more females over 18 there were than males over 18.

Solution:

The number of males was approximately 101,000,000. The number of females was approximately 108,000,000.

$$108,000,000$$
$$- 101,000,000$$
$$\overline{7,000,000}$$

There were approximately 7 million more women over age 18 in the United States than men.

Skill Practice

10. In a recent year, there were 135,073,000 persons over the age of 16 employed in the United States. During the same year, there were 6,742,000 persons over the age of 16 who were unemployed. Approximate each value to the nearest million. Use these values to approximate how many more people were employed than unemployed.

Answers

9. 6,100,000 km^2
10. Employed: 135,000,000;
    unemployed: 7,000,000;
    difference: 128,000,000

Section 1.4   Practice Exercises

Study Skills Exercises

1. Purchase a three-ring binder for your math notes and homework. Use section dividers to separate each chapter that you cover in the text. Keep your homework and notes in the appropriate section. What other course materials might you keep organized in your notebook?

2. Define the key term **rounding**.

Review Exercises

For Exercises 3–6, add or subtract as indicated.

3. $\begin{array}{r} 59 \\ -33 \\ \hline \end{array}$

4. $\begin{array}{r} 130 \\ -98 \\ \hline \end{array}$

5. $\begin{array}{r} 4009 \\ +998 \\ \hline \end{array}$

6. $\begin{array}{r} 12{,}033 \\ +23{,}441 \\ \hline \end{array}$

7. Determine the place value of the digit 6 in the number 1,860,432.

8. Determine the place value of the digit 4 in the number 1,860,432.

Objective 1: Rounding

9. Explain how to round a whole number to the hundreds place.

10. Explain how to round a whole number to the tens place.

For Exercises 11–28, round each number to the given place value. **(See Examples 1–4.)**

11. 342; tens
12. 834; tens
13. 725; tens

14. 445; tens
15. 9384; hundreds
16. 8363; hundreds

17. 8539; hundreds
18. 9817; hundreds
19. 34,992; thousands

20. 76,831; thousands
21. 2578; thousands
22. 3511; thousands

23. 9982; hundreds
24. 7974; hundreds
25. 109,337; thousands

26. 437,208; thousands
27. 489,090; ten-thousands
28. 388,725; ten-thousands

29. In the first weekend of its release, the movie *Spider-man 3* grossed $148,431,020. Round this number to the millions place.

30. The average per capita personal income in the United States in a recent year was $33,050. Round this number to the nearest thousand.

31. The average center-to-center distance from the Earth to the Moon is 238,863 mi. Round this to the thousands place.

32. A shopping center in Edmonton, Alberta, Canada, covers an area of 492,000 square meters (m^2). Round this number to the hundred-thousands place.

Objective 2: Estimation

For Exercises 33–36, estimate the sum by first rounding each number to the nearest ten. **(See Example 5.)**

33.	57	**34.**	33	**35.**	41	**36.**	29
	82		78		12		73
	+ 21		+ 41		+ 129		+ 113

For Exercises 37–40, estimate the difference by first rounding each number to the nearest hundred. **(See Example 6.)**

37.	898	**38.**	731	**39.**	3412	**40.**	9771
	− 422		− 584		− 1252		− 4544

Objective 3: Using Estimation in Applications

For Exercises 41–42, refer to the table.

Brand	Manufacturer	Sales ($)
M&Ms	Mars	97,404,576
Hershey's Milk Chocolate	Hershey Chocolate	81,296,784
Reese's Peanut Butter Cups	Hershey Chocolate	54,391,268
Snickers	Mars	53,695,428
KitKat	Hershey Chocolate	38,168,580

41. Round the sales to the nearest million to estimate the total sales brought in by the Mars company. **(See Example 7.)**

42. Round the sales to the nearest million to estimate the total sales brought in by the Hershey Chocolate Company.

43. Neil Diamond earned $71,339,710 in U.S. tours in one year while Paul McCartney earned $59,684,076. Round each value to the nearest million dollars to estimate how much more Neil Diamond earned. **(See Example 8.)**

44. The number of women in the 45–49 age group who gave birth in 1981 is 1190. By 2001 this number increased to 4844. Round each value to the nearest thousand to estimate how many more women in the 45–49 age group gave birth in 2001.

For Exercises 45–48, use the given table.

45. Round the revenue to the nearest hundred-thousand to estimate the total revenue for the years 2000 through 2002.

46. Round the revenue to the nearest hundred-thousand to estimate the total revenue for the years 2003 through 2005.

47. a. Determine the year with the greatest revenue. Round this revenue to the nearest hundred-thousand.

Beach Parking Revenue for Daytona Beach, Florida	
Year	Revenue
2000	$3,316,897
2001	3,272,028
2002	3,360,289
2003	3,470,295
2004	3,173,050
2005	1,970,380

Source: Daytona Beach News Journal
Table for Exercises 45–48

b. Determine the year with the least revenue. Round this revenue to the nearest hundred-thousand.

48. Estimate the difference between the year with the greatest revenue and the year with the least revenue.

For Exercises 49–52, use the graph provided.

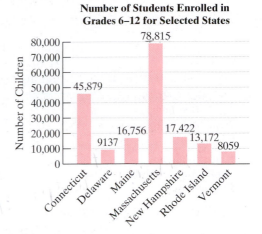

Number of Students Enrolled in Grades 6–12 for Selected States

49. Determine the state with the greatest number of students enrolled in grades 6–12. Round this number to the nearest thousand.

50. Determine the state with the least number of students enrolled in grades 6–12. Round this number to the nearest thousand.

51. Use the information in Exercises 49 and 50 to estimate the difference between the number of students in the state with the highest enrollment and that of the lowest enrollment.

52. Estimate the total number of students enrolled in grades 6–12 in the selected states by first rounding the number of students to the thousands place.

Source: National Center for Education Statistics

Figure for Exercises 49–52

53. If you were to estimate the following sum, what place value would you round to and why?

$$389,220 + 2988 + 12,824 + 101,333$$

54. Identify the place value that you would round to when estimating the answer to the following problem. Then round the values and estimate the answer.

$$4208 - 932 + 1294$$

Expanding Your Skills

For Exercises 55–58, round the numbers to estimate the perimeter of each figure. (Answers may vary.)

55.

3045 mm
1892 mm   1892 mm
3045 mm

56.
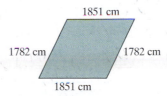

1851 cm
1782 cm   1782 cm
1851 cm

57.
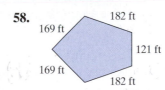

105 in.
57 in.   57 in.
57 in.   57 in.
105 in.

58.

182 ft
169 ft
121 ft
169 ft
182 ft

Section 1.5   Multiplication of Whole Numbers and Area

Objectives

1. **Introduction to Multiplication**
2. **Properties of Multiplication**
3. **Multiplying Many-Digit Whole Numbers**
4. **Estimating Products by Rounding**
5. **Translations and Applications Involving Multiplication**
6. **Area of a Rectangle**

Concept Connections

1. How can multiplication be used to compute the sum $4 + 4 + 4 + 4 + 4 + 4 + 4$?

1. Introduction to Multiplication

Suppose that Carmen buys three cartons of eggs to prepare a large family brunch. If there are 12 eggs per carton, then the total number of eggs can be found by adding three 12s.

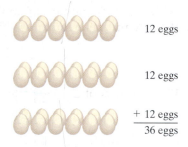

12 eggs

12 eggs

+ 12 eggs

36 eggs

When each addend in a sum is the same, we have what is called *repeated* addition. Repeated addition is also called **multiplication**. We use the multiplication sign $\times$ to express repeated addition more concisely.

$$12 + 12 + 12 \quad \text{is equal to} \quad 3 \times 12$$

The expression 3×12 is read "3 times 12" to signify that the number 12 is added 3 times. The numbers 3 and 12 are called **factors**, and the result, 36, is called the **product**.

The symbol $\cdot$ may also be used to denote multiplication such as in the expression $3 \cdot 12 = 36$. Two factors written adjacent to each other with no other operator between them also implies multiplication. The quantity $2y$, for example, is understood to be 2 times y. If we use this notation to multiply two numbers, parentheses are used to group one or both factors. For example:

$$3(12) = 36 \quad (3)12 = 36 \quad \text{and} \quad (3)(12) = 36$$

all represent the product of 3 and 12.

TIP: In the expression 3(12), the parentheses are necessary because two adjacent factors written together with no grouping symbol would look like the number 312.

The products of one-digit numbers such as $4 \times 5 = 20$ and $2 \times 7 = 14$ are basic facts. All products of one-digit numbers should be memorized (see Exercise 6).

Skill Practice

Identify the factors and the product.

2. $3 \times 11 = 33$
3. $2 \cdot 5 \cdot 8 = 80$

Example 1   Identifying Factors and Products

Identify the factors and the product.

a. $6 \times 3 = 18$ **b.** $5 \cdot 2 \cdot 7 = 70$

Solution:

a. Factors: 6, 3; product: 18   **b.** Factors: 5, 2, 7; product: 70

2. Properties of Multiplication

Recall from Section 1.2 that the order in which two numbers are added does not affect the sum. The same is true for multiplication. This is stated formally as the *commutative property of multiplication.*

Answers

1. 7×4
2. Factors: 3 and 11; product: 33
3. Factors 2, 5, and 8; product: 80

> ## PROPERTY   Commutative Property of Multiplication
>
> Changing the order of two factors does not affect the product.
>
> Example: 2×5 is equivalent to 5×2

The following rectangular arrays help us visualize the commutative property of multiplication.

$2 \times 5 = 10$ 2 rows of 5

$5 \times 2 = 10$ 5 rows of 2

Multiplication is also an associative operation.

> ## PROPERTY   Associative Property of Multiplication
>
> The manner in which factors are grouped under multiplication does not affect the product.
>
> Example: $(3 \times 5) \times 2$ is equivalent to $3 \times (5 \times 2)$

Example 2   Applying Properties of Multiplication

a. Rewrite the expression 3×9, using the commutative property of multiplication. Then find the product.

b. Rewrite the expression $(4 \times 2) \times 3$, using the associative property of multiplication. Then find the product.

Solution:

a. $3 \times 9 = 9 \times 3$. The product is 27.

b. $(4 \times 2) \times 3 = 4 \times (2 \times 3)$.

   To find the product, we have

   $$4 \times (2 \times 3)$$
   $$= 4 \times (6)$$
   $$= 24$$

The product is 24.

Skill Practice

4. Rewrite the expression 6×5, using the commutative property of multiplication. Then find the product.

5. Rewrite the expression $3 \times (1 \times 7)$, using the associative property of multiplication. Then find the product.

Two other important properties of multiplication involve factors of 0 and 1.

Answers

4. 5×6; product is 30
5. $(3 \times 1) \times 7$; product is 21

> **PROPERTY** **Multiplication Property of 0**
>
> The product of any number and 0 is 0.
>
> Examples: $5 \times 0 = 0$
>
> $0 \times 12 = 0$

The product $5 \times 0 = 0$ can easily be understood by writing the product as repeated addition.

$$\underbrace{0 + 0 + 0 + 0 + 0}_{\text{Add 0 five times.}} = 0$$

> **PROPERTY** **Multiplication Property of 1**
>
> The product of any number and 1 is that number.
>
> Examples: $1 \times 4 = 4$
>
> $3 \times 1 = 3$

The last property of multiplication involves both addition and multiplication. First consider the expression $2(4 + 3)$. By performing the operation within parentheses first, we have

$$2(4 + 3) = 2(7) = 14$$

We get the same result by multiplying 2 times each addend within the parentheses:

$$2(4 + 3) = (2 \times 4) + (2 \times 3) = 8 + 6 = 14$$

This result illustrates the distributive property of multiplication over addition (sometimes we simply say *distributive property* for short).

> **PROPERTY** **Distributive Property of Multiplication over Addition**
>
> The product of a number and a sum can be found by multiplying the number by each addend.
>
> Example: $5(7 + 3) = (5 \times 7) + (5 \times 3)$

Skill Practice

Apply the distributive property and simplify.
6. $2(6 + 4)$
7. $5(0 + 8)$

Example 3   **Applying the Distributive Property of Multiplication Over Addition**

Apply the distributive property and simplify.

a. $3(4 + 8)$ **b.** $7(3 + 0)$

Solution:

a. $3(4 + 8) = (3 \times 4) + (3 \times 8) = 12 + 24 = 36$

b. $7(3 + 0) = (7 \times 3) + (7 \times 0) = 21 + 0 = 21$

Answers

6. $(2 \times 6) + (2 \times 4)$; 20
7. $(5 \times 0) + (5 \times 8)$; 40

3. Multiplying Many-Digit Whole Numbers

When multiplying numbers with several digits, it is sometimes necessary to carry. To see why, consider the product 3×29. By writing the factors in expanded form, we can apply the distributive property. In this way, we see that 3 is multiplied by both 20 and 9.

$$3 \times 29 = 3(20 + 9) = (3 \times 20) + (3 \times 9)$$
$$= 60 + 27$$
$$= 6 \text{ tens} + 2 \text{ tens} + 7 \text{ ones}$$
$$= 8 \text{ tens} + 7 \text{ ones}$$
$$= 87$$

Now we will multiply 29×3 in vertical form.

$$
\begin{array}{r}
\overset{2}{2}9 \\
\times \quad 3 \\
\hline
7
\end{array}
$$

Multiply $3 \times 9 = 27$. Write the 7 in the ones column and carry the 2.

$$
\begin{array}{r}
\overset{2}{2}9 \\
\times \quad 3 \\
\hline
8\ 7
\end{array}
$$

Multiply 3×2 tens $= 6$ tens. Add the carry: 6 tens $+$ 2 tens $=$ 8 tens. Write the 8 in the tens place.

Example 4	Multiplying a Many-Digit Number by a One-Digit Number

Multiply.

$$
\begin{array}{r}
368 \\
\times \quad 5
\end{array}
$$

Solution:

Using the distributive property, we have

$$5(300 + 60 + 8) = 1500 + 300 + 40 = 1840$$

This can be written vertically as:

$$
\begin{array}{r}
368 \\
\times \quad 5 \\
\hline
40 \\
300 \\
+\ 1500 \\
\hline
1840
\end{array}
$$

Multiply 5×8.
Multiply 5×60.
Multiply 5×300.
Add.

The numbers 40, 300, and 1500 are called *partial sums*. The product of 386 and 5 is found by adding the partial sums. The product is 1840.

The solution to Example 4 can also be found by using a shorter form of multiplication. We outline the procedure:

$$
\begin{array}{r}
\overset{4}{3}68 \\
\times \quad 5 \\
\hline
0
\end{array}
$$

Multiply $5 \times 8 = 40$. Write the 0 in the ones place and carry the 4.

$$\overset{34}{368}$$
$$\underline{\times \quad 5}$$
$$40$$

Multiply 5×6 tens $= 300$. Add the carry. $300 + 4$ tens $= 340$. Write the 4 in the tens place and carry the 3.

$$\overset{34}{368}$$
$$\underline{\times \quad 5}$$
$$1840$$

Multiply 5×3 hundreds $= 1500$. Add the carry. $1500 + 3$ hundreds $= 1800$. Write the 8 in the hundreds place and the 1 in the thousands place.

Example 5 demonstrates the process to multiply two factors with many digits.

Skill Practice

9. Multiply.
   59
   × 26

Example 5 Multiplying a Many-Digit Number by a Many-Digit Number

Multiply.

$$72$$
$$\times 83$$

Solution:

Writing the problem vertically and computing the partial sums, we have

$$\overset{1}{72}$$
$$\underline{\times 83}$$
$$216 \qquad \text{Multiply } 3 \times 72.$$
$$\underline{+ 5760} \qquad \text{Multiply } 80 \times 72.$$
$$5976 \qquad \text{Add.}$$

The product is 5976.

Skill Practice

10. Multiply.
    274
    × 586

Example 6 Multiplying Two Multidigit Whole Numbers

Multiply. 368×497.

Solution:

$$\overset{23}{\underset{45}{\overset{67}{368}}}$$
$$\underline{\times 497}$$
$$2576$$
$$33120$$
$$\underline{+ 147200}$$
$$182{,}896$$

4. Estimating Products by Rounding

A special pattern occurs when one or more factors in a product ends in zero. Consider the following products:

$$12 \times 20 = 240 \qquad\qquad 120 \times 20 = 2400$$
$$12 \times 200 = 2400 \qquad\qquad 1200 \times 20 = 24{,}000$$
$$12 \times 2000 = 24{,}000 \qquad 12{,}000 \times 20 = 240{,}000$$

Answers
9. 1534   10. 160,564

Notice in each case the product is $12 \times 2 = 24$ followed by the total number of zeros from each factor. Consider the product 1200×20.

$$
\begin{array}{r}
12\,|\,00 \\
\times\ \ 2\,|\,0 \\
\hline
24\,|\,000
\end{array}
$$

Shift the numbers 1200 and 20 so that the zeros appear to the right of the multiplication process. Multiply $12 \times 2 = 24$. Write the product 24 followed by the total number of zeros from each factor.

Example 7 **Estimating a Product**

Estimate the product 795×4060 by rounding 795 to the nearest hundred and 4060 to the nearest thousand.

Solution:

$$
\begin{array}{llll}
795 & \text{rounds to} \longrightarrow & 800 & 8\,|\,00 \\
4060 & \text{rounds to} \longrightarrow & 4000 & \times\ 4\,|\,000 \\
& & & \overline{32\,|\,00000}
\end{array}
$$

The product is approximately 3,200,000.

Example 8 **Estimating a Product in an Application**

For a trip from Atlanta to Los Angeles, the average cost of a plane ticket was $495. If the plane carried 218 passengers, estimate the total revenue for the airline. (*Hint*: Round each number to the hundreds place and find the product.)

Solution:

$$
\begin{array}{llll}
\$495 & \text{rounds to} \longrightarrow & \$\ 5\,|\,00 \\
218 & \text{rounds to} \longrightarrow & \times 2\,|\,00 \\
& & \overline{\$10\,|\,0000}
\end{array}
$$

The airline received approximately $100,000 in revenue.

---

5. Translations and Applications Involving Multiplication

In English there are many different words that imply multiplication. A partial list is given in Table 1-3.

Table 1-3

Word/Phrase	Example	In Symbols
Product	The product of 4 and 7	4×7
Times	8 times 4	8×4
Multiply … by …	Multiply 6 by 3	6×3

Multiplication may also be warranted in applications involving unit rates. In Example 8, we multiplied the cost per customer ($495) by the number of customers (218). The value $495 is a unit rate because it gives the cost per one customer (per one unit).

Skill Practice

13. Ella can type 65 words per minute. How many words can she type in 45 minutes?

Example 9   **Solving an Application Involving Multiplication**

The average weekly income for production workers is $489. How much does a production worker make in 1 year (assume 52 weeks in 1 year).

Solution:

The value $489 per week is a unit rate. The total earnings for 1 year is given by $489 × 52.

$$
\begin{array}{r}
\overset{4\,4}{}\\[-2pt]
\overset{1\,1}{489}\\
\times\,52\\
\hline
978\\
+\,24450\\
\hline
25{,}428
\end{array}
$$

The yearly earnings are $25,428.

TIP: This product can be estimated quickly by rounding the factors.

$$
\begin{array}{rcl}
489 & \text{rounds to} \longrightarrow & 5\,|\,00\\
52 & \text{rounds to} \longrightarrow & \times\,5\,|\,0\\
\hline
& & 25\,|\,000
\end{array}
$$

The total yearly income is approximately $25,000. Estimating gives a quick approximation of a product. Furthermore, it also checks for the reasonableness of our exact product. In this case $25,000 is close to our exact value of $25,428.

6. Area of a Rectangle

Another application of multiplication of whole numbers lies in finding the area of a region. **Area** measures the amount of surface contained within the region. For example, a square that is 1 in. by 1 in. occupies an area of 1 square inch, denoted as $1\ \text{in.}^2$. Similarly, a square that is 1 centimeter (cm) by 1 cm occupies an area of 1 square centimeter. This is denoted by $1\ \text{cm}^2$.

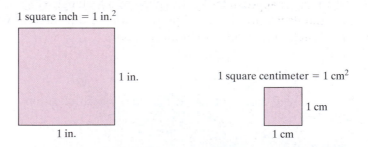

1 square inch = $1\ \text{in.}^2$

1 in.

1 in.

1 square centimeter = $1\ \text{cm}^2$

1 cm

1 cm

The units of square inches and square centimeters (in.^2 and cm^2) are called *square units*. To find the area of a region, measure the number of square units occupied in that region. For example, the region in Figure 1-6 occupies $6\ \text{cm}^2$.

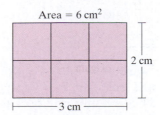

Area = $6\ \text{cm}^2$

2 cm

3 cm

Figure 1-6

Answer

13. 2925 words

The 3-cm by 2-cm region in Figure 1-6 suggests that to find the **area of a rectangle**, multiply the length by the width. If the area is represented by A, the length is represented by l, and the width is represented by w, then we have

$$\text{Area of rectangle} = (\text{length}) \times (\text{width})$$

$$A = l \times w$$

The letters A, l, and w are called **variables** because their values *vary* as they are replaced by different numbers.

Example 10   Finding the Area of a Rectangle

Find the area and perimeter of the rectangle.

Solution:

Area:

$A = l \times w$

$A = (7 \text{ yd}) \times (4 \text{ yd})$

$\quad = 28 \text{ yd}^2$

Recall from Section 1.2 that the perimeter of a polygon is the sum of the lengths of the sides. In a rectangle the opposite sides are equal in length.

Perimeter:

$P = 7 \text{ yd} + 4 \text{ yd} + 7 \text{ yd} + 4 \text{ yd}$

$\quad = 22 \text{ yd}$

The area is 28 yd^2 and the perimeter is 22 yd.

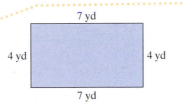

- **Skill Practice**

  14. Find the area and perimeter of the rectangle.

  12 ft

  5 ft

- **Avoiding Mistakes**

  Notice that area is measured in square units (such as yd^2) and perimeter is measured in units of length (such as yd). It is important to apply the correct units of measurement.

Example 11   Finding Area in an Application

The state of Wyoming is approximately the shape of a rectangle (Figure 1-7). Its length is 355 mi and its width is 276 mi. Approximate the total area of Wyoming by rounding the length and width to the nearest ten.

WY   276 mi

355 mi

Figure 1-7

- **Skill Practice**

  15. A house sits on a rectangular lot that is 192 ft by 96 ft. Approximate the area of the lot by rounding the length and width to the nearest hundred.

Answers

14. Area: 60 ft^2; perimeter: 34 ft

15. 20,000 ft^2

Solution:

$$
\begin{array}{r}
\overset{1}{\overset{4}{}}\\
355 \quad \text{rounds to} \longrightarrow \quad 36\,|\,0 \\
276 \quad \text{rounds to} \longrightarrow \quad \times\,28\,|\,0 \\
\hline
288 \quad\;\;\; \\
720 \quad\quad\;\; \\
\hline
1008\,|\,00 \;\;
\end{array}
$$

The area of Wyoming is approximately 100,800 mi^2.

Section 1.5   Practice Exercises

Study Tips

1. List the materials that you need to bring to class every day (for example, paper, pencil, etc.).

2. Define the key terms.

   a. **Multiplication**   b. **Factor**   c. **Product**

   d. **Area**   e. **Area of a rectangle**   f. **Variable**

Review Exercises

For Exercises 3–5, estimate the answer by rounding to the indicated place value.

3. $869{,}240 + 34{,}921 + 108{,}332$; ten-thousands

4. $907{,}801 - 413{,}560$; hundred-thousands

5. $8821 - 3401$; hundreds

Objective 1: Introduction to Multiplication

6. Fill in the table of multiplication facts.

×	0	1	2	3	4	5	6	7	8	9
0										
1										
2										
3										
4										
5										
6										
7										
8										
9										

For Exercises 7–10, write the repeated addition as multiplication and simplify.

7. $5 + 5 + 5 + 5 + 5 + 5$

8. $2 + 2 + 2 + 2 + 2 + 2 + 2 + 2 + 2$

9. $9 + 9 + 9$

10. $7 + 7 + 7 + 7$

For Exercises 11–14, identify the factors and the product. **(See Example 1.)**

11. $13 \times 42 = 546$

12. $26 \times 9 = 234$

13. $3 \cdot 5 \cdot 2 = 30$

14. $4 \cdot 3 \cdot 8 = 96$

15. Write the product of 5 and 12, using three different notations. (Answers may vary.)

16. Write the product of 23 and 14, using three different notations. (Answers may vary.)

Objective 2: Properties of Multiplication

For Exercises 17–22, match the property with the statement.

17. $8 \times 1 = 8$

18. $6 \cdot 13 = 13 \cdot 6$

19. $2(6 + 12) = 2 \cdot 6 + 2 \cdot 12$

20. $5 \cdot (3 \cdot 2) = (5 \cdot 3) \cdot 2$

21. $0 \times 4 = 0$

22. $7(14) = 14(7)$

a. Commutative property of multiplication

b. Associative property of multiplication

c. Multiplication property of 0

d. Multiplication property of 1

e. Distributive property of multiplication over addition

For Exercises 23–28, rewrite the expression, using the indicated property. **(See Examples 2–3.)**

23. 14×8; commutative property of multiplication

24. 3×9; commutative property of multiplication

25. $6 \times (2 \times 10)$; associative property of multiplication

26. $(4 \times 15) \times 5$; associative property of multiplication

27. $5(7 + 4)$; distributive property of multiplication over addition

28. $3(2 + 6)$; distributive property of multiplication over addition

Objective 3: Multiplying Many-Digit Whole Numbers

For Exercises 29–60, multiply. **(See Examples 4–6.)**

29. $\begin{array}{r} 24 \\ \times\ 6 \\ \hline \end{array}$

30. $\begin{array}{r} 18 \\ \times\ 5 \\ \hline \end{array}$

31. $\begin{array}{r} 26 \\ \times\ 2 \\ \hline \end{array}$

32. $\begin{array}{r} 71 \\ \times\ 3 \\ \hline \end{array}$

33. $\begin{array}{r} 131 \\ \times\ 5 \\ \hline \end{array}$

34. $\begin{array}{r} 725 \\ \times\ 3 \\ \hline \end{array}$

35. $\begin{array}{r} 344 \\ \times\ 4 \\ \hline \end{array}$

36. $\begin{array}{r} 105 \\ \times\ 9 \\ \hline \end{array}$

37. $\begin{array}{r} 1410 \\ \times\ 8 \\ \hline \end{array}$

38. $\begin{array}{r} 2016 \\ \times\ 6 \\ \hline \end{array}$

39. $\begin{array}{r} 3312 \\ \times\ 7 \\ \hline \end{array}$

40. $\begin{array}{r} 4801 \\ \times\ 5 \\ \hline \end{array}$

41. 42,014
     $\times\ 9$

42. 51,006
     $\times\ 8$

43.   32
     $\times\ 14$

44.   41
     $\times\ 21$

45. 68 · 24

46. 55 · 41

47. 72 · 12

48. 13 · 46

49. (143)(17)

50. (722)(28)

51. (349)(19)

52. (512)(31)

53.   151
     $\times\ 127$

54.   703
     $\times\ 146$

55.   222
     $\times\ 841$

56.   387
     $\times\ 506$

57.   3532
     $\times\ 6014$

58.   2810
     $\times\ 1039$

59.   4122
     $\times\ \ 982$

60.   7026
     $\times\ \ 528$

Objective 4: Estimating Products by Rounding

For Exercises 61–68, multiply the numbers, using the method found on page 43. **(See Example 7.)**

61.   600
     $\times\ 40$

62.   900
     $\times\ 50$

63.   3000
     $\times\ 700$

64.   4000
     $\times\ 400$

65.   8000
     $\times\ 9000$

66.   1000
     $\times\ 2000$

67. 90,000
     $\times\ \ 400$

68. 50,000
     $\times\ 6000$

For Exercises 69–72, estimate the product by first rounding the number to the indicated place value.

69. 11,784 × 5201; thousands place

70. 45,046 × 7812; thousands place

71. 82,941 × 29,740; ten-thousands place

72. 630,229 × 71,907; ten-thousands place

73. Suppose a hotel room costs $189 per night. Round this number to the nearest hundred to estimate the cost for a five-night stay. **(See Example 8.)**

74. The science department of Comstock High School must purchase a set of calculators for a class. If the cost of one calculator is $129, estimate the cost of 28 calculators by rounding the numbers to the tens place.

75. The average price for a ticket to see Kenny Chesney is $137. If a concert stadium seats 10,256 fans, estimate the amount of money received during that performance by rounding the number of seats to the nearest ten-thousand.

76. A breakfast buffet at a local restaurant serves 48 people. Estimate the maximum revenue for one week (7 days) if the price of a breakfast is $12.

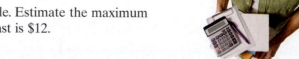

Objective 5: Translations and Applications Involving Multiplication

77. The 4-gigabyte (4-GB) iPod nano is advertised to store approximately 1000 songs. Assuming the average length of a song is 4 minutes, how many minutes of music can be stored on the iPod nano? **(See Example 9.)**

78. One CD can hold 700 megabytes (MB) of data. How many megabytes can 15 CDs hold?

79. It costs about $45 for a cat to have a medical exam. If a humane society has 37 cats, find the cost of medical exams for their cats.

80. A can of Coke contains 12 fluid ounces (fl oz). Find the number of ounces in a case of Coke containing 12 cans.

81. PaperWorld shipped 115 cases of copy paper to a business. There are 5 reams of paper in each case and 500 sheets of paper in each ream. Find the number of sheets of paper delivered to the business.

82. A dietary supplement bar has 14 grams (g) of protein. If Kathleen eats 2 bars a day for 6 days, how many grams of protein will she get from this supplement?

83. Tylee's car gets 31 miles per gallon (mpg) on the highway. How many miles can he travel if he has a full tank of gas (12 gal)?

84. Sherica manages a small business called Pizza Express. She has 23 employees who work an average of 32 hours (hr) per week. How many hours of work does Sherica have to schedule each week?

Objective 6: Area of a Rectangle

For Exercises 85–88, find the area. **(See Example 10.)**

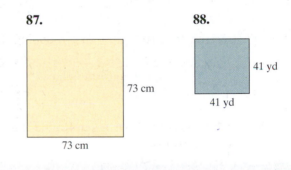

85.
12 ft
23 ft

86.
31 m
2 m

87.
73 cm
73 cm

88.
41 yd
41 yd

89. The state of Colorado is approximately the shape of a rectangle. Its length is 388 mi and its width is 269 mi. Approximate the total area of Colorado by rounding the length and width to the nearest ten. **(See Example 11.)**

90. A parcel of land has a width of 132 yd and a length of 149 yd. Approximate the total area by rounding each dimension to the nearest ten.

91. The front of a building has windows that are 44 in. by 58 in.

   a. Approximate the area of one window.

   b. If the building has three floors and each floor has 14 windows, how many windows are there?

   c. What is the approximate total area of all of the windows?

92. The length of a carport is 51 ft and its width is 29 ft. Approximate the area of the carport.

93. Mr. Slackman wants to paint his garage door that is 8 ft by 16 ft. To decide how much paint to buy, he must find the area of the door. What is the area of the door?

94. To carpet a rectangular room, Erika must find the area of the floor. If the dimensions of the room are 10 yd by 15 yd, how much carpeting does she need?

Section 1.6　Division of Whole Numbers

1. Introduction to Division

Suppose 12 pieces of pizza are to be divided evenly among 4 children (Figure 1-8). The number of pieces that each child would receive is given by $12 \div 4$, read "12 divided by 4."

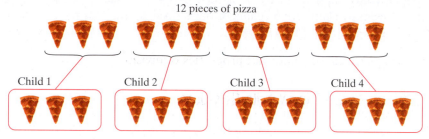

Figure 1-8

The process of separating 12 pieces of pizza evenly among 4 children is called **division**. The statement $12 \div 4 = 3$ indicates that each child receives 3 pieces of pizza. The number 12 is called the **dividend**. It represents the number to be divided. The number 4 is called the **divisor**, and it represents the number of groups. The result of the division (in this case 3) is called the **quotient**. It represents the number of items in each group.

Division can be represented in several ways. For example, the following are all equivalent statements.

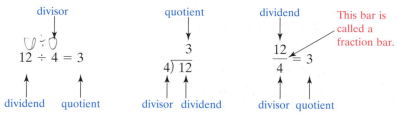

Recall that subtraction is the reverse operation of addition. In the same way, division is the reverse operation of multiplication. For example, we say $12 \div 4 = 3$ because $3 \times 4 = 12$.

| Example 1 | Identifying the Dividend, Divisor, and Quotient |

Simplify each expression. Then identify the dividend, divisor, and quotient.

a. $48 \div 6$　　b. $9\overline{)36}$　　c. $\dfrac{63}{7}$

Solution:

a. $48 \div 6 = 8$　　　because $8 \times 6 = 48$

The dividend is 48, the divisor is 6, and the quotient is 8.

b. $9\overline{)36}^{\,4}$ because $4 \times 9 = 36$

The dividend is 36, the divisor is 9, and the quotient is 4.

c. $\dfrac{63}{7} = 9$ because $9 \times 7 = 63$

The dividend is 63, the divisor is 7, and the quotient is 9.

2. Properties of Division

Example 2 illustrates the important properties of division.

Example 2 | **Dividing Whole Numbers**

Divide.

a. $8 \div 8$　　**b.** $\dfrac{6}{6}$　　　　**c.** $5 \div 1$

d. $1\overline{)7}$　　**e.** $0 \div 6$　　**f.** $\dfrac{0}{4}$　　　**g.** $6 \div 0$

Skill Practice

Divide.
4. $3\overline{)3}$　　　5. $5 \div 5$
6. $\dfrac{4}{1}$　　　　7. $8 \div 1$
8. $\dfrac{0}{7}$　　　　9. $3\overline{)0}$

Solution:

a. $8 \div 8 = 1$ because $1 \times 8 = 8$

b. $\dfrac{6}{6} = 1$ because $1 \times 6 = 6$

c. $5 \div 1 = 5$ because $5 \times 1 = 5$

d. $1\overline{)7}^{\,7}$ because $7 \times 1 = 7$

e. $0 \div 6 = 0$ because $0 \times 6 = 0$

f. $\dfrac{0}{4} = 0$ because $0 \times 4 = 0$

g. $6 \div 0$ is *undefined* because there is no number that when multiplied by 0 will produce a product of 6.

PROPERTY Properties of Division

1. Any nonzero number divided by itself is 1.
   Example: $9 \div 9 = 1$
2. Any number divided by 1 is the number itself.
   Example: $3 \div 1 = 3$
3. Zero divided by any nonzero number is zero.
   Example: $0 \div 5 = 0$
4. Any number divided by zero is undefined.
   Example: $9 \div 0$ is undefined

Answers
4. 1　5. 1　6. 4　7. 8
8. 0　9. 0

You should also note that unlike addition and multiplication, division is neither commutative nor associative. In other words, reversing the order of the dividend and divisor may produce a different quotient. Similarly, changing the manner in which numbers are grouped with division may affect the outcome. See Exercises 31 and 32.

3. Long Division

To divide larger numbers we use a process called **long division**. This process uses a series of estimates to find the quotient. We illustrate long division in Example 3.

Example 3 Using Long Division

Divide. $7)\overline{161}$

Solution:

Estimate $7)\overline{161}$ by first estimating $7)\overline{16}$ and writing the result in the tens place of the quotient. Since $7 \times 2 = 14$, there are at least 2 sevens in 16.

$$\begin{array}{r} 2 \\ 7)\overline{161} \\ -140 \\ \hline 21 \end{array}$$

The 2 in the tens place represents 20 in the quotient.
←— Multiply 7×20 and write the result under the dividend.
Subtract 140. We see that our estimate leaves 21.

Repeat the process. Now divide $7)\overline{21}$ and write the result in the ones place of the quotient.

$$\begin{array}{r} 23 \\ 7)\overline{161} \\ -140 \\ \hline 21 \\ -21 \\ \hline 0 \end{array}$$

←— Multiply 7×3.
Subtract.

The quotient is 23.

Check: $$\begin{array}{r} 23 \\ \times 7 \\ \hline 161 \checkmark \end{array}$$

We can streamline the process of long division by "bringing down" digits of the dividend one at a time.

Example 4 Using Long Division

Divide. $6138 \div 9$

Solution:

$$\begin{array}{r} 682 \\ 9)\overline{6138} \\ -54 \downarrow \\ \hline 73 \\ -72 \downarrow \\ \hline 18 \\ -18 \\ \hline 0 \end{array}$$

$9 \times 6 = 54$ and subtract.
Bring down the 3.
$9 \times 8 = 72$ and subtract.
Bring down the 8.
$9 \times 2 = 18$ and subtract.

The quotient is 682.   Check: $\overset{71}{682}$
                                    $\underline{\times\ \ 9}$
                                     6138 ✔

In many instances, quotients do not come out evenly. For example, suppose we had 13 pieces of pizza to distribute among 4 children (Figure 1-9).

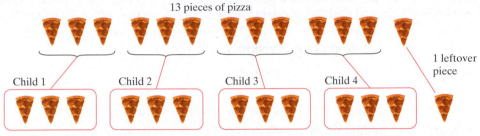

13 pieces of pizza

Child 1   Child 2   Child 3   Child 4   1 leftover piece

Figure 1-9

The mathematical term given to the "leftover" piece is called the **remainder**. The division process may be written as

$$\begin{array}{r} 3\ \text{R1} \\ 4\overline{)13} \\ -12 \\ \hline 1 \end{array}$$

The remainder is written next to the 3.

The **whole part of the quotient** is 3, and the remainder is 1. Notice that the remainder is written next to the whole part of the quotient.

We can check a division problem that has a remainder. To do so, multiply the divisor by the whole part of the quotient and then add the remainder. The result must equal the dividend. That is,

(Divisor)(whole part of quotient) + remainder = dividend

Thus,

$$(4)(3) + 1 \overset{?}{=} 13$$
$$12 + 1 \overset{?}{=} 13$$
$$13 = 13 \ ✔$$

Example 5 **Using Long Division**

Divide. $1253 \div 6$

Solution:

$$\begin{array}{r} 208\ \text{R5} \\ 6\overline{)1253} \\ -12 \\ \hline 05 \\ -00 \\ \hline 53 \\ -48 \\ \hline 5 \end{array}$$

$6 \times 2 = 12$ and subtract.

Bring down the 5.

Note that 6 does not divide into 5, so we put a 0 in the quotient.

Bring down the 3.

$6 \times 8 = 48$ and subtract.

The remainder is 5.

To check, verify that $6 \times 208 + 5 = 1253$. ✔

Skill Practice

14. Divide.

$5107 \div 5$

Answer

14. 1021 R2

4. Dividing by a Many-Digit Divisor

When the divisor has more than one digit, we still use a series of estimations to find the quotient.

Skill Practice

15. $63\overline{)4516}$

| **Example 6** | Dividing by a Two-Digit Number |

Divide. $32\overline{)1259}$

Solution:

To estimate the leading digit of the quotient, estimate the number of times 30 will go into 125. Since $30 \cdot 4 = 120$, our estimate is 4.

$$\begin{array}{r} 4 \\ 32\overline{)1259} \\ -128 \end{array}$$
$32 \times 4 = 128$ is too big. We cannot subtract 128 from 125. Revise the estimate in the quotient to 3.

$$\begin{array}{r} 3 \\ 32\overline{)1259} \\ -96 \\ \hline 299 \end{array}$$
$32 \times 3 = 96$ and subtract.
Bring down the 9.

Now estimate the number of times 30 will go into 299. Because $30 \times 9 = 270$, our estimate is 9.

$$\begin{array}{r} 39 \text{ R}11 \\ 32\overline{)1259} \\ -96 \\ \hline 299 \\ -288 \\ \hline 11 \end{array}$$
$32 \times 9 = 288$ and subtract.
The remainder is 11.

To check, verify that $32 \times 39 + 11 = 1259$. ✔

Skill Practice

16. $304\overline{)62,405}$

| **Example 7** | Dividing by a Many-Digit Number |

Divide. $\dfrac{82,705}{602}$

Solution:

$$\begin{array}{r} 137 \text{ R}231 \\ 602\overline{)82,705} \\ -602 \\ \hline 2250 \\ -1806 \\ \hline 4445 \\ -4214 \\ \hline 231 \end{array}$$

$602 \times 1 = 602$ and subtract.
Bring down the 0.
$602 \times 3 = 1806$ and subtract.
Bring down the 5.
$602 \times 7 = 4214$ and subtract.
The remainder is 231.

To check, verify that $602 \times 137 + 231 = 82,705$. ✔

Answers
15. 71 R43   **16.** 205 R85

5. Translations and Applications Involving Division

Several words and phrases imply division. A partial list is given in Table 1-4.

Table 1-4

Word/Phrase	Example	In Symbols
Divide	Divide 12 by 3	$12 \div 3$ or $\dfrac{12}{3}$ or $3\overline{)12}$
Quotient	The quotient of 20 and 2	$20 \div 2$ or $\dfrac{20}{2}$ or $2\overline{)20}$
Per	110 mi per 2 hr	$110 \div 2$ or $\dfrac{110}{2}$ or $2\overline{)110}$
Divides into	4 divides into 28	$28 \div 4$ or $\dfrac{28}{4}$ or $4\overline{)28}$
Divided, or shared equally among	64 shared equally among 4	$64 \div 4$ or $\dfrac{64}{4}$ or $4\overline{)64}$

Example 8 Solving an Application Involving Division

A painting business employs 3 painters. The business collects $1950 for painting a house. If all painters are paid equally, how much does each person make?

Solution:

This is an example where $1950 is shared equally among 3 people. Therefore, we divide.

```
      650
 3)1950
  -18↓        3 × 6 = 18 and subtract.
   15         Bring down the 5.
  -15↓        3 × 5 = 15 and subtract.
   00         Bring down the 0.
   -0         3 × 0 = 0 and subtract.
    0         The remainder is 0.
```

Each painter makes $650.

Example 9 Solving an Application Involving Division with Estimation

Elaine and Max drove from South Bend, Indiana, to Bonita Springs, Florida. The total driving distance was 1089 mi, and the driving time was approximately 20 hr. Estimate the average speed by rounding the distance to the nearest hundred.

Skill Practice

17. Four players play Hearts with a standard 52-card deck of cards. If the cards are equally distributed, how many cards does each player get?

18. A college has budgeted $4800 to buy graphing calculators. Each calculator costs $119. Estimate the number of calculators that the college can buy by rounding the cost to the nearest ten.

Answers

17. 13 cards 18. 40 calculators

Solution:

1089 mi rounds to 1100 mi. The speed is represented by 1100 mi per 20 hr, or

```
         55
    20)1100
      −100↓        20 × 5 = 100 and subtract.
       100         Bring down the 0.
      −100         20 × 5 = 100 and subtract.
         0
```

Max and Elaine averaged approximately 55 miles per hour (mph).

Skill Practice

19. The cost for four different types of pastry at a French bakery is shown in the graph.

Cost for Selected Pastries

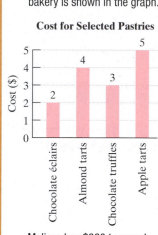

Melissa has $360 to spend on desserts.

a. If she spends all the money on chocolate éclairs, how many can she buy?

b. If she spends all the money on apple tarts, how many can she buy?

Example 10 **Solving an Application Involving Division**

The graph in Figure 1-10 depicts the number of calories burned per hour for selected activities.

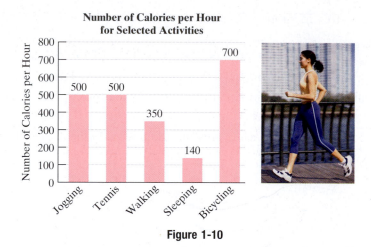

Number of Calories per Hour for Selected Activities

Figure 1-10

a. Janie wants to burn 3500 calories per week exercising. For how many hours must she jog?

b. For how many hours must Janie bicycle to burn 3500 calories?

Solution:

a. The total number of calories must be divided into 500-calorie increments. Thus, the number of hours required is given by 3500 ÷ 500.

```
        7
  500)3500
    −3500        Janie requires 7 hr of jogging to burn 3500 calories.
        0
```

b. 3500 calories must be divided into 700-calorie increments. The number of hours required is given by 3500 ÷ 700.

```
        5
  700)3500
    −3500        Janie requires 5 hr of bicycling to burn 3500 calories.
        0
```

Answers

19. a. 180 chocolate éclairs
   b. 72 apple tarts

Section 1.6 Practice Exercises

Study Skills Exercises

1. In your next math class, take notes by drawing a vertical line about three-fourths of the way across the paper, as shown. On the left side, write down what your instructor puts on the board or overhead. On the right side, make your own comments about important words, procedures, or questions that you have.

2. Define the key terms.
   a. Division   b. Dividend   c. Divisor   d. Quotient
   e. Long division   f. Remainder   g. Whole part of the quotient

Review Exercises

For Exercises 3–10, add, subtract, or multiply as indicated.

3. $48 \cdot 103$
4. $678 - 83$
5. $1008 + 245$
6. $14(220)$

7. 5230×127
8. $789(25)$
9. $4890 - 3988$
10. $38,002 + 3902$

Objective 1: Introduction to Division

For Exercises 11–16, simplify each expression. Then identify the dividend, divisor, and quotient. **(See Example 1.)**

11. $72 \div 8$
12. $32 \div 4$
13. $8\overline{)64}$

14. $5\overline{)35}$
15. $\dfrac{45}{9}$
16. $\dfrac{20}{5}$

Objective 2: Properties of Division

17. In your own words, explain the difference between dividing a number by zero and dividing zero by a number.

18. Explain what happens when a number is either divided or multiplied by 1.

For Exercises 19–30, use the properties of division to simplify the expression, if possible. **(See Example 2.)**

19. $15 \div 1$
20. $21\overline{)21}$
21. $0 \div 10$
22. $\dfrac{0}{3}$

23. $0\overline{)9}$
24. $4 \div 0$
25. $\dfrac{20}{20}$
26. $1\overline{)9}$

27. $\dfrac{16}{0}$
28. $\dfrac{5}{1}$
29. $8\overline{)0}$
30. $13 \div 13$

31. Show that $6 \div 3 = 2$ but $3 \div 6 \neq 2$ by using multiplication to check.

32. Show that division is not associative, using the numbers 36, 12, and 3.

Objective 3: Long Division

33. Explain the process for checking a division problem when there is no remainder.

34. Show how checking by multiplication can help us remember that $0 \div 5 = 0$ and that $5 \div 0$ is undefined.

For Exercises 35–46, divide and check by multiplying. (See Examples 3 and 4.)

35. $78 \div 6$ **36.** $364 \div 7$ **37.** $5\overline{)205}$ **38.** $8\overline{)152}$
Check: $6 \times \square = 78$ Check: $7 \times \square = 364$ Check: $5 \times \square = 205$ Check: $8 \times \square = 152$

39. $\dfrac{972}{2}$ **40.** $\dfrac{582}{6}$ **41.** $1227 \div 3$ **42.** $236 \div 4$

43. $5\overline{)1015}$ **44.** $5\overline{)2035}$ **45.** $\dfrac{4932}{6}$ **46.** $\dfrac{3619}{7}$

For Exercises 47–54, check the following division problems. If it does not check, find the correct answer.

47. $\dfrac{56}{4\overline{)224}}$ **48.** $\dfrac{82}{7\overline{)574}}$ **49.** $761 \div 3 = 253$ **50.** $604 \div 5 = 120$

51. $\dfrac{1021}{9} = 113 \text{ R4}$ **52.** $\dfrac{1311}{6} = 218 \text{ R3}$ **53.** $8\overline{)203}^{25 \text{ R6}}$ **54.** $7\overline{)821}^{117 \text{ R5}}$

For Exercises 55–70, divide and check the answer. (See Example 5.)

55. $61 \div 8$ **56.** $89 \div 3$ **57.** $9\overline{)92}$ **58.** $5\overline{)74}$

59. $\dfrac{55}{2}$ **60.** $\dfrac{49}{3}$ **61.** $593 \div 3$ **62.** $801 \div 4$

63. $\dfrac{382}{9}$ **64.** $\dfrac{428}{8}$ **65.** $3115 \div 2$ **66.** $4715 \div 6$

67. $6014 \div 8$ **68.** $9013 \div 7$ **69.** $6\overline{)5012}$ **70.** $2\overline{)1101}$

Objective 4: Dividing by a Many-Digit Divisor

For Exercises 71–82, divide. (See Examples 6 and 7.)

71. $9110 \div 19$ **72.** $3505 \div 13$ **73.** $24\overline{)1051}$ **74.** $41\overline{)8104}$

75. $\dfrac{8008}{26}$ **76.** $\dfrac{9180}{15}$ **77.** $68,012 \div 54$ **78.** $92,013 \div 35$

79. $69,712 \div 304$ **80.** $51,107 \div 221$ **81.** $114\overline{)34,428}$ **82.** $421\overline{)87,989}$

Objective 5: Translations and Applications Involving Division

For Exercises 83–88, for each English sentence, write a mathematical expression and simplify.

83. Find the quotient of 497 and 71.   **84.** Find the quotient of 1890 and 45.

85. Divide 877 by 14.   **86.** Divide 722 by 53.

87. Divide 6 into 42.   **88.** Divide 9 into 108.

89. There are 392 students signed up for Anatomy 101. If each classroom can hold 28 students, find the number of classrooms needed. **(See Example 8.)**

90. A wedding reception is planned to take place in the fellowship hall of a church. The bride anticipates 120 guests, and each table will seat 8 people. How many tables should be set up for the reception to accommodate all the guests?

91. A case of tomato sauce contains 32 cans. If a grocer has 168 cans, how many cases can he fill completely? How many cans will be left over?

92. Austin has $425 to spend on dining room chairs. If each chair costs $52, does he have enough to purchase 8 chairs? If so, will he have any money left over?

93. Pauline drove 312 mi in 6 hr. Find Pauline's average speed (in miles per hour).

94. A house cleaning company charges $144 to clean a 3-room apartment. At this rate, how much does it cost to clean one room?

95. If it takes 2200 lb of grapes to make 100 gal of white wine, how many pounds are needed for 1 gal?

96. There are 7280 acres of ferns in Florida that are owned by 260 farmers. Find the average size of each farm.

97. Suppose Genny can type 1234 words in 22 min. Round each number to estimate her rate in words per minute. **(See Example 9.)**

98. On a trip to California from Illinois, Lavu drove 2780 mi. The gas tank in his car allows him to travel 405 mi. Round each number to the hundreds place to estimate the number of tanks of gas needed for the trip.

99. A group of 18 people goes to a concert. Ticket prices are given in the graph. If the group has $450, can they all attend the concert? If so, which type of seats can they buy? **(See Example 10.)**

100. The graph gives the average annual income for four professions: teacher, professor, CEO, and programmer. Find the monthly income for each of the four professions.

Ticket Prices

Figure for Exercise 99

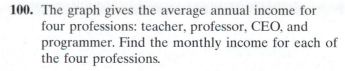
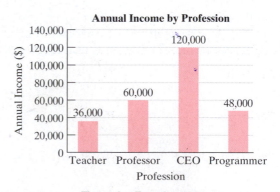

Figure for Exercise 100

101. The labels on most laundry detergents include the total number of wash loads that can be done. This makes comparison shopping easier since the amount of soap to be used can vary from one brand to the next. The label on Planet laundry detergent does not show the total number of loads. The instructions on the back label suggest using 4 fl oz of detergent per load of laundry.

   a. How many loads of laundry can be done with a 50 fl oz bottle of Planet?

   b. How many ounces of detergent are leftover?

102. At an elementary school, parents may pick up their children in a loading area in front of the school. Nine cars are allowed to pull up at one time, and it takes approximately 2 min to load and release the group of 9 cars. If it takes 26 minutes to get all cars through the line, how many cars are in line waiting to pick up children?

Calculator Connections

Topic: Multiplying and Dividing Whole Numbers

To multiply and divide numbers on a calculator, use the $\times$ and $\div$ keys, respectively.

Expression	Keystrokes	Result
38,319 × 1561	38319 ⨯ 1561 ENTER	59815959
2,449,216 ÷ 6248	2449216 ÷ 6248 ENTER	392

Calculator Exercises

For Exercises 103–106, solve the problem. Use a calculator to perform the calculations.

103. The United States consumes approximately 21,000,000 barrels (bbl) of oil per day. (*Source:* U.S. Energy Information Administration) How much does it consume in 1 year?

104. The average time to commute to work for people living in Washington state is 26 min (round trip 52 min). (*Source:* U.S. Census Bureau) How much time does a person spend commuting to and from work in 1 year if the person works 5 days a week for 50 weeks per year?

105. The budget for the U.S. federal government for 2008 is approximately $2532 billion dollars. (*Source:* U.S. Department of the Treasury) How much could the government spend each month and still stay within its budget?

106. At a weigh station, a truck carrying 96 crates weighs in at 34,080 lb. If the truck weighs 9600 lb when empty, how much does each crate weigh?

Problem Recognition Exercises

Operations on Whole Numbers

Perform the indicated operations. For each problem, estimate to check your answer.

1. 92
   + 41

2. 72
   + 17

3. 89
   − 22

4. 156
   − 40

5. 221
   × 14

6. 146
   × 25

7. 35)2240

8. 24)2088

9. 946 − 612

10. 6004 − 221

11. 2311 + 2652

12. 4299 + 1201

13. 1312 ÷ 4

14. 4376 ÷ 8

15. 82 × 3000

16. 47 × 60,000

17. 113 + 59

18. 8410 + 744

19. 621 − 539

20. 7000 − 439

21. 5064 ÷ 22

22. 6387 ÷ 15

23. 50 · 400

24. 600 · 900

25. 34,855
    − 12,137

26. 10,922
    − 10,846

27. 8231
    + 3412

28. 5912
    + 955

29. 63)34,524

30. 27)22,761

31. 548
    × 63

32. 843
    × 27

33. 418 × 10

34. 418 × 100

35. 418 × 1000

36. 418 × 10,000

37. 350,000 ÷ 10

38. 350,000 ÷ 100

39. 350,000 ÷ 1000

40. 350,000 ÷ 10,000

41. 159 + 224 + 123

42. 8091 + 225 + 426

43. 534 + 12 + 66

44. 877 + 34 + 123

| Section 1.7 | Exponents, Square Roots, and the Order of Operations |

Objectives

1. **Exponents**
2. **Square Roots**
3. **Order of Operations**
4. **Computing a Mean (Average)**

1. Exponents

Thus far in the text we have learned to add, subtract, multiply, and divide whole numbers. We now present the concept of an **exponent** to represent repeated multiplication. For example, the product

exponent

$$3 \cdot 3 \cdot 3 \cdot 3 \cdot 3 \qquad \text{can be written as} \qquad 3^5$$

base

The expression 3^5 is written in exponential form. The exponent, or **power**, is 5 and represents the number of times the **base**, 3, is multiplied. The expression 3^5 is read as "three to the fifth power." Other expressions in exponential form are shown next.

5^2 is read as   "five squared" or "five to the second power"

5^3 is read as   "five cubed" or "five to the third power"

5^4 is read as   "five to the fourth power"

5^5 is read as   "five to the fifth power"

> **TIP:** The expression $5^1 = 5$. Any number without an exponent explicitly written has a power of 1.

Exponential form is a shortcut notation for repeated multiplication. However, to simplify an expression in exponential form, we often write out the individual factors.

Skill Practice

Evaluate.
1. 8^2 2. 4^3 3. 2^5

| Example 1 | Evaluating Exponential Expressions |

Evaluate.

a. 6^2 **b.** 5^3 **c.** 2^4

Solution:

a. $6^2 = 6 \cdot 6$ The exponent, 2, indicates the number of times the base, 6, is multiplied.

$\quad\;\; = 36$

b. $5^3 = 5 \cdot 5 \cdot 5$ When three factors are multiplied, we can group the first two factors and perform the multiplication.

$\quad\;\; = (5 \cdot 5) \cdot 5$

$\quad\;\; = (25) \cdot 5$ Then multiply the product of the first two factors by the last factor.

$\quad\;\; = 125$

c. $2^4 = 2 \cdot 2 \cdot 2 \cdot 2$

$\quad\;\; = (2 \cdot 2) \cdot 2 \cdot 2$ Group the first two factors.

$\quad\;\; = 4 \cdot 2 \cdot 2$ Multiply the first two factors.

$\quad\;\; = (4 \cdot 2) \cdot 2$ Multiply the product by the next factor to the right.

$\quad\;\; = 8 \cdot 2$

$\quad\;\; = 16$

Answers

1. 64   **2.** 64   **3.** 32

One important application of exponents lies in recognizing **powers of 10**, that is, 10 raised to a whole-number power. For example, consider the following expressions.

$$10^1 = 10$$

$$10^2 = 10 \cdot 10 = 100$$

$$10^3 = 10 \cdot 10 \cdot 10 = 1000$$

$$10^4 = 10 \cdot 10 \cdot 10 \cdot 10 = 10,000$$

$$10^5 = 10 \cdot 10 \cdot 10 \cdot 10 \cdot 10 = 100,000$$

$$10^6 = 10 \cdot 10 \cdot 10 \cdot 10 \cdot 10 \cdot 10 = 1,000,000$$

From these examples, we see that a power of 10 results in a 1 followed by several zeros. The number of zeros is the same as the exponent on the base of 10.

2. Square Roots

To square a number means that we multiply the base times itself. For example, $5^2 = 5 \cdot 5 = 25$.

To find a positive **square root** of a number means that we reverse the process of squaring. For example, finding the square root of 25 is equivalent to asking, "What positive number, when squared, equals 25?" The symbol $\sqrt{}$, (called a *radical sign*) is used to denote the positive square root of a number. Therefore, $\sqrt{25}$ is the positive number, that when squared, equals 25. Thus, $\sqrt{25} = 5$ because $(5)^2 = 25$.

Example 2   Evaluating Square Roots

Find the square roots.

a. $\sqrt{9}$ **b.** $\sqrt{64}$ **c.** $\sqrt{1}$ **d.** $\sqrt{0}$

Solution:

a. $\sqrt{9} = 3$ because $(3)^2 = 3 \cdot 3 = 9$

b. $\sqrt{64} = 8$ because $(8)^2 = 8 \cdot 8 = 64$

c. $\sqrt{1} = 1$ because $(1)^2 = 1 \cdot 1 = 1$

d. $\sqrt{0} = 0$ because $(0)^2 = 0 \cdot 0 = 0$

Skill Practice

Find the square roots.
4. $\sqrt{4}$
5. $\sqrt{100}$
6. $\sqrt{400}$
7. $\sqrt{121}$

TIP: To simplify square roots, it is advisable to become familiar with the following squares and square roots.

$$0^2 = 0 \longrightarrow \sqrt{0} = 0 \qquad 7^2 = 49 \longrightarrow \sqrt{49} = 7$$
$$1^2 = 1 \longrightarrow \sqrt{1} = 1 \qquad 8^2 = 64 \longrightarrow \sqrt{64} = 8$$
$$2^2 = 4 \longrightarrow \sqrt{4} = 2 \qquad 9^2 = 81 \longrightarrow \sqrt{81} = 9$$
$$3^2 = 9 \longrightarrow \sqrt{9} = 3 \qquad 10^2 = 100 \longrightarrow \sqrt{100} = 10$$
$$4^2 = 16 \longrightarrow \sqrt{16} = 4 \qquad 11^2 = 121 \longrightarrow \sqrt{121} = 11$$
$$5^2 = 25 \longrightarrow \sqrt{25} = 5 \qquad 12^2 = 144 \longrightarrow \sqrt{144} = 12$$
$$6^2 = 36 \longrightarrow \sqrt{36} = 6 \qquad 13^2 = 169 \longrightarrow \sqrt{169} = 13$$

Answers
4. 2
5. 10
6. 20
7. 11

3. Order of Operations

A numerical expression may contain more than one operation. For example, the following expression contains both multiplication and subtraction.

$$18 - 5(2)$$

The order in which the multiplication and subtraction are performed will affect the overall outcome.

Multiplying first yields	Subtracting first yields
$18 - 5(2) = 18 - 10$	$18 - 5(2) = 13(2)$
$= 8$ (correct)	$= 26$ (incorrect)

To avoid confusion, mathematicians have outlined the proper order of operations. In particular, multiplication is performed before addition or subtraction. The guidelines for the order of operations are given next. These rules must be followed in all cases.

PROCEDURE   Order of Operations

Step 1 Perform all operations inside parentheses first.

Step 2 Simplify any expressions containing exponents or square roots.

Step 3 Perform multiplication or division in the order that they appear from left to right.

Step 4 Perform addition or subtraction in the order that they appear from left to right.

Skill Practice

Simplify.

8. $18 + 6 \div 2 - 4$
9. $(20 - 4) \div 2 + 1$
10. $2^3 - \sqrt{25}$

Example 3 Using the Order of Operations

Simplify.

a. $15 - 10 \div 2 + 3$ **b.** $(5 - 2) \cdot 7 - 1$ **c.** $\sqrt{100} - 2^3$

Solution:

a. $15 - \underbrace{10 \div 2} + 3$

$= \underbrace{15 - 5} + 3$ Perform the division $10 \div 2$ first.

$=\ \ \ \ 10 + 3$ Perform addition and subtraction from left to right.

$=\ \ \ \ \ \ 13$ Add.

b. $\underbrace{(5 - 2)} \cdot 7 - 1$

$= \underbrace{(3) \cdot 7} - 1$ Perform the operation inside parentheses first.

$=\ \ \ 21 - 1$ Perform multiplication before subtraction.

$=\ \ \ \ \ 20$ Subtract.

c. $\sqrt{100} - 2^3$

$=\ 10 - 8$ Simplify any expressions with exponents or square roots. Note that $\sqrt{100} = 10$, and $2^3 = 2 \cdot 2 \cdot 2 = 8$.

$=\ \ \ 2$ Subtract.

Answers

8. 17   **9.** 9   **10.** 3

Example 4	**Using the Order of Operations**

Simplify.

a. $300 \div (7 - 2)^2 - 2^2$ **b.** $36 + (7^2 - 3)$

Solution:

a. $300 \div (7 - 2)^2 - 2^2$

$= 300 \div (5)^2 - 2^2$ Perform the operation within parentheses first.

$= 300 \div 25 - 4$ Simplify exponents: $5^2 = 5 \cdot 5 = 25$ and $2^2 = 2 \cdot 2 = 4$.

$= \quad 12 - 4$ Perform division before subtraction.

$= \quad\quad 8$ Subtract.

b. $\quad 36 + (7^2 - 3)$

$= 36 + (49 - 3)$ Perform the operations within parentheses first. The guidelines indicate that we simplify the expression with the exponent before we subtract: $7^2 = 49$.

$= 36 + \quad 46$ Continue simplifying within parentheses.

$= 82$ Add.

4. Computing a Mean (Average)

The order of operations must be used when we compute an average. The technical term for the average of a list of numbers is the **mean** of the numbers. To find the mean of a set of numbers, first compute the sum of the values. Then divide the sum by the number of values. This is represented by the formula

$$\text{Mean} = \frac{\text{sum of the values}}{\text{number of values}}$$

Example 5	**Computing a Mean (Average)**

Ashley took 6 tests in Chemistry. Find her mean (average) score.

$$89, 91, 72, 86, 94, 96$$

Solution:

$$\text{Average score} = \frac{89 + 91 + 72 + 86 + 94 + 96}{6}$$

$$= \frac{528}{6} \quad \text{Add the values in the list first.}$$

$$= 88 \quad \text{Divide.}$$

$$\begin{array}{r} 88 \\ 6\overline{)528} \\ -48 \\ \hline 48 \\ -48 \\ \hline 0 \end{array}$$

Ashley's mean (average) score is 88.

> **TIP:** The division bar in $\dfrac{89 + 91 + 72 + 86 + 94 + 96}{6}$ is also a grouping symbol and implies parentheses:
>
> $$\dfrac{(89 + 91 + 72 + 86 + 94 + 96)}{6}$$

Section 1.7 Practice Exercises

Boost *your* GRADE at ALEKS.com!

ALEKS® version 3.0

- Practice Problems
- Self-Tests
- NetTutor
- e-Professors
- Videos

Study Skills Exercises

1. Look over the notes that you took today. Do you understand what you wrote? If there were any rules, definitions, or formulas, highlight them so that they can be easily found when studying for the test. You may want to begin by highlighting the order of operations.

2. Define the key terms.

   a. **Exponent**   b. **Power**   c. **Base**

   d. **Power of 10**   e. **Square root**   f. **Mean**

Review Exercises

For Exercises 3–8, write true or false for each statement.

3. Addition is commutative; for example, $5 + 3 = 3 + 5$.

4. Subtraction is commutative; for example, $5 - 3 = 3 - 5$.

5. $6 \cdot 0 = 6$ 6. $0 \div 8 = 0$ 7. $0 \cdot 8 = 0$ 8. $5 \div 0$ is undefined

Objective 1: Exponents

9. Write an exponential expression with 9 as the base and 4 as the exponent.

10. Write an exponential expression with 3 as the base and 8 as the exponent.

11. Write an exponential expression with 7 as the exponent and 2 as the base.

12. Write an exponential expression with 5 as the exponent and 6 as the base.

For Exercises 13–16, write the repeated multiplication in exponential form. Do not simplify.

13. $3 \cdot 3 \cdot 3 \cdot 3 \cdot 3 \cdot 3$ 14. $7 \cdot 7 \cdot 7 \cdot 7$ 15. $4 \cdot 4 \cdot 4 \cdot 4 \cdot 2 \cdot 2 \cdot 2$ 16. $5 \cdot 5 \cdot 5 \cdot 10 \cdot 10 \cdot 10$

For Exercises 17–20, expand the exponential expression as a repeated multiplication. Do not simplify.

17. 8^4 18. 2^6 19. 4^8 20. 6^2

For Exercises 21–36, evaluate the exponential expressions. **(See Example 1.)**

21. 2^3 **22.** 4^2 **23.** 3^2 **24.** 5^2

25. 3^3 **26.** 11^2 **27.** 5^3 **28.** 4^3

29. 2^5 **30.** 6^3 **31.** 3^4 **32.** 5^4

33. 1^2 **34.** 1^3 **35.** 1^4 **36.** 1^5

37. Explain what happens when the number 1 is raised to any power. **(See Exercises 33–36.)**

For Exercises 38–41, evaluate the powers of 10.

38. 10^2 **39.** 10^3 **40.** 10^4 **41.** 10^5

42. Explain how to get 10^9 *without* doing the repeated multiplication. **(See Exercises 38–41.)**

Objective 2: Square Roots

For Exercises 43–50, evaluate the square roots. **(See Example 2.)**

43. $\sqrt{4}$ **44.** $\sqrt{9}$ **45.** $\sqrt{36}$ **46.** $\sqrt{81}$

47. $\sqrt{100}$ **48.** $\sqrt{49}$ **49.** $\sqrt{0}$ **50.** $\sqrt{16}$

Objective 3: Order of Operations

51. Does the order of operations indicate that addition is always performed before subtraction? Explain.

52. Does the order of operations indicate that multiplication is always performed before division? Explain.

For Exercises 53–93, simplify using the order of operations. **(See Examples 3–4.)**

53. $6 + 10 \cdot 2$ **54.** $4 + 3 \cdot 7$ **55.** $10 - 3^2$ **56.** $11 - 2^2$

57. $(10 - 3)^2$ **58.** $(11 - 2)^2$ **59.** $36 \div 2 \div 6$ **60.** $48 \div 4 \div 2$

61. $15 - (5 + 8)$ **62.** $41 - (13 + 8)$ **63.** $(13 - 2) \cdot 5 - 2$ **64.** $(8 + 4) \cdot 6 + 8$

65. $4 + 12 \div 3$ **66.** $9 + 15 \div \sqrt{25}$ **67.** $30 \div 2 \cdot \sqrt{9}$ **68.** $55 \div 11 \cdot 5$

69. $7^2 - 5^2$ **70.** $3^3 - 2^3$ **71.** $(7 - 5)^2$ **72.** $(3 - 2)^3$

73. $100 \div 5 \cdot 2$ **74.** $60 \div 3 \cdot 2$ **75.** $90 \div 3 \cdot 3$ **76.** $80 \div 2 \cdot 2$

77. $\sqrt{81} + 2(9 - 1)$ **78.** $\sqrt{121} + 3(8 - 3)$ **79.** $36 \div (2^2 + 5)$ **80.** $42 \div (3^2 - 2)$

81. $80 - (20 \div 4) + 6$ **82.** $120 - (48 \div 8) - 40$ **83.** $(43 - 26) \cdot 2 - 4^2$ **84.** $(51 - 48) \cdot 3 + 7^2$

85. $(18 - 5) - (23 - \sqrt{100})$ **86.** $(\sqrt{36} + 11) - (31 - 16)$ **87.** $80 \div (9^2 - 7 \cdot 11)^2$

88. $108 \div (3^3 - 6 \cdot 4)^2$ **89.** $22 - 4(\sqrt{25} - 3)^2$ **90.** $17 + 3(7 - \sqrt{9})^2$

91. $96 - 3(42 \div 7 \cdot 6 - 5)$ **92.** $50 - 2(36 \div 12 \cdot 2 - 4)$ **93.** $16 + 5(20 \div 4 \cdot 8 - 3)$

Objective 4: Computing a Mean (Average)

For Exercises 94–96, find the mean (average) of each set of numbers. **(See Example 5.)**

94. 19, 21, 18, 21, 16   **95.** 105, 114, 123, 101, 100, 111   **96.** 1480, 1102, 1032, 1002

97. Neelah took 6 quizzes and received the following scores: 19, 20, 18, 19, 18, 14. Find her quiz average.

98. Shawn's scores on his last 4 tests were 83, 95, 87, and 91. What is his average for these tests?

99. At a certain grocery store, Jessie notices that the price of bananas varies from week to week. During a 3-week period she buys bananas for 33¢ per pound, 39¢ per pound, and 42¢ per pound. What does Jessie pay on average per pound?

100. On a trip, Stephen had his car washed 4 times and paid $7, $10, $8, and $7. What was the average amount spent per wash?

101. The monthly rainfall for Seattle, Washington, is given in the table. All values are in millimeters (mm).

	Jan.	Feb.	Mar.	Apr.	May	Jun.	Jul.	Aug.	Sep.	Oct.	Nov.	Dec.
Rainfall	122	94	80	52	47	40	15	21	44	90	118	123

Find the average monthly rainfall for the months of November, December, and January.

102. The monthly snowfall for Alpena, Michigan, is given in the table. All values are in inches.

	Jan.	Feb.	Mar.	Apr.	May	Jun.	Jul.	Aug.	Sep.	Oct.	Nov.	Dec.
Snowfall	22	16	13	5	1	0	0	0	0	1	9	20

Find the average monthly snowfall for the months of November, December, January, February, and March.

Expanding Your Skills

Sometimes an expression will have parentheses within parentheses. This is called *nested parentheses*. Often different shapes such as ( ), [ ], or { } are used to make it easier to match up the pairs of parentheses, for example,

$$\{300 - 4[4 + (5 + 2)^2] + 8\} - 31$$

It is important to note that the symbols ( ), [ ], or { } all represent parentheses and are used for grouping. When nested parentheses occur, simplify the innermost set first. Then work your way out. For example, simplify

$$\{300 - 4[4 + (5 + 2)^2] + 8\} - 31$$

The solution is

$$\{300 - 4[4 + (5 + 2)^2] + 8\} - 31$$

$= \{300 - 4[4 + (7)^2] + 8\} - 31$	Simplify within the innermost parentheses first ( ).
$= \{300 - 4[4 + 49] + 8\} - 31$	Simplify the exponent.
$= \{300 - 4[53] + 8\} - 31$	Simplify within the next innermost parentheses [ ].
$= \{300 - 212 + 8\} - 31$	Multiply before adding.
$= \{88 + 8\} - 31$	Subtract and add in order from left to right within the parentheses { }.
$= 96 - 31$	Simplify within the parentheses { }.
$= 65$	Simplify.

For Exercises 103–106, simplify the expressions with nested parentheses.

103. $3[4 + (6 - 3)^2] - 15$

104. $2[5(4 - 1) + 3] \div 6$

105. $5\{21 - [3^2 - (4 - 2)]\}$

106. $4\{18 - [(10 - 8) + 2^3]\}$

Calculator Connections

Topic: Evaluating Expressions with Exponents

Many calculators use the $\boxed{x^2}$ key to square a number. To raise a number to a higher power, use the $\boxed{\wedge}$ key (or on some calculators, the $\boxed{x^y}$ key or $\boxed{y^x}$ key).

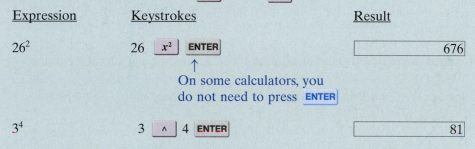

Calculator Exercises

For Exercises 107–112, use a calculator to perform the indicated operations.

107. 156^2 **108.** 418^2 **109.** 12^5 **110.** 35^4 **111.** 43^3 **112.** 71^3

For Exercises 113–118, simplify the expressions by using the order of operations. For each step use the calculator to simplify the given operation.

113. $8126 - 54{,}978 \div 561$

114. $92{,}168 + 6954 \times 29$

115. $(3548 - 3291)^2$

116. $(7500 \div 625)^3$

117. $\dfrac{89{,}880}{384 + 2184}$ *Hint:* This expression has implied grouping symbols. $\dfrac{89{,}880}{(384 + 2184)}$

118. $\dfrac{54{,}137}{3393 - 2134}$ *Hint:* This expression has implied grouping symbols. $\dfrac{54{,}137}{(3393 - 2134)}$

Section 1.8 **Problem-Solving Strategies**

1. Problem-Solving Strategies

In this section, we offer additional practice with applications of whole numbers. Keep in mind that all word problems are different and that there is no magic "trick" to solve an application problem. However, we can offer the following guidelines.

> **PROCEDURE Guidelines for Problem Solving**
>
> **Step 1** Read the problem carefully and familiarize yourself with the situation. If possible, draw a diagram or write down an appropriate formula. Sometimes you may be able to estimate a reasonable answer.
>
> **Step 2** Write down what information is given and what must be found.
>
> **Step 3** Form a strategy. Identify what mathematical operation applies (addition, subtraction, multiplication, or division). Sometimes a combination of operations is necessary.
>
> **Step 4** Perform the mathematical operations to solve for the unknown.
>
> **Step 5** Check the answer. If the answer is reasonable and checks, state the answer in words.

2. Applications Involving One Operation

We illustrate these guidelines with a variety of examples. To assist with step 3 where we must identify an appropriate mathematical operation, we summarize some of our key words and phrases. See Table 1-5.

Table 1-5

Operation	Key Word or Phrase
Addition	Sum, added to, increased by, more than, plus, total of
Subtraction	Difference, minus, decreased by, less, subtract
Multiplication	Product, times, multiply
Division	Quotient, divide, per, shared equally

Skill Practice

1. The odometer of a car read 24,316 mi last year. This year the reading is 37,134. How many miles was the car driven during the year?

Example 1   Solving a Travel Application

Kent travels from Columbus, Ohio, to Indianapolis, Indiana, and then on to Springfield, Illinois. The total distance he drives is 351 mi. The distance between Columbus and Indianapolis is 168 mi. Find the distance between Indianapolis and Springfield.

Solution:

Familiarize and draw a picture.

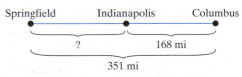

Springfield   Indianapolis   Columbus
?   168 mi
351 mi

Given: In this case, we know the total distance and one of the parts.

Find: Find the second distance (between Indianapolis and Springfield).

Answer

1. 12,818 mi

Operation: This problem can be phrased as an addition problem with a missing addend or as an equivalent subtraction problem.

$$? + 168 = 351 \quad \text{or} \quad 351 - 168 = ?$$

Subtracting yields

$$351 - 168 = 183$$

The distance between Indianapolis and Springfield is 183 mi.

Example 2 Solving a Sales Application

A used car business keeps records of vehicle sales by type of vehicle and month.

	July	August	September
Cars	23	28	32
Trucks	13	8	10
SUVs	15	18	21

Find the total number of vehicles sold in July.

Skill Practice

Refer to the table in Example 2.

2. Find the total number of trucks sold during this 3-month period.

Solution:

Given: The number of each type of vehicle sold in July (highlighted in red)

Find: The total number of vehicles sold in July

Operation: The word *total* indicates addition.

$$\text{Total} = 23 + 13 + 15$$
$$= 51$$

There were 51 vehicles sold in July.

Example 3 Solving a Business Application

A ream of paper holds 500 sheets. Kim purchases 24 reams for her office. How many sheets of paper is this?

Skill Practice

3. One page of print in a book contains 48 lines of text. How many lines of text are in one chapter containing 21 pages?

Solution:

Familiarize and draw a picture.

$$\underset{\text{sheets}}{500} + \underset{\text{sheets}}{500} + \underset{\text{sheets}}{500} + \cdots + \underset{\text{sheets}}{500}$$

Given: There are 24 reams (packages) of paper with 500 sheets per package.

Find: The total number of sheets of paper

Operation: This situation calls for repeated addition. Therefore, we will multiply.

$$24(500) = 12,000$$

There are 12,000 sheets of paper.

Answers

2. 31 trucks   3. 1008 lines

┌─ **Skill Practice** ─────┐
4. A vat of flour at a food
distributor holds 580 lb of flour.
How many 5-lb bags of flour
can be filled from one vat?
└──────────────────────┘

Example 4	**Solving a Consumer Application**

A 5-speed Jeep Cherokee gets 23 mpg (miles per gallon) on the highway. How many gallons of gas would be required for a 667-mi drive from El Paso to Dallas?

Solution:

Familiarize and draw a picture.

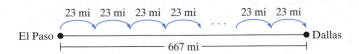

Given: The total distance, 667 mi, and the gas mileage, 23 mpg

Find: How many increments of 23 mi would be required for the trip?

Operation: This is a situation where 667 mi must be divided into 23-mi increments. Use the operation of division.

$$\begin{array}{r} 29 \\ 23\overline{)667} \\ -46 \\ \hline 207 \\ -207 \\ \hline 0 \end{array}$$

> **TIP:** The solution to Example 4 can be checked by multiplication. Twenty-nine gallons of gas at 23 mpg produces
>
> (29 gal)(23 mi/gal) = 667 mi

The drive from El Paso to Dallas in a Jeep Cherokee will require 29 gal of gas.

3. Applications Involving Multiple Operations

Sometimes more than one operation is needed to solve an application problem.

┌─ **Skill Practice** ─────┐
5. Danielle buys a new
entertainment center with a
new plasma television for
$4240. She pays $1000
down, and the rest is paid off
in equal monthly payments
over 2 years. Find Danielle's
monthly payment.
└──────────────────────┘

Example 5	**Solving a Consumer Application**

Jorge bought a car for $18,340. He paid $2500 down and then paid the rest in equal monthly payments over a 4-year period. Find the amount of Jorge's monthly payment (not including interest).

Solution:

Familiarize and draw a picture.

Given: total price: $18,340
         down payment: $2500

         payment plan: 4 years
         (48 months)

Find: monthly payment

$18,340 Original cost of car

−2,500   Minus down payment

$15,840 Amount to be paid off

Divide payments over
4 years (48 months)

Answers

4. 116 bags
5. $135 per month

Operations:

1. The amount of the loan to be paid off is equal to the original cost of the car minus the down payment. We use subtraction:

$$\begin{array}{r} \$18,340 \\ - 2,500 \\ \hline \$15,840 \end{array}$$

2. This money is distributed in equal payments over a 4-year period. Because there are 12 months in 1 year, there are $4 \times 12 = 48$ months in a 4-year period. To distribute $15,840 among 48 equal payments, we divide.

$$\begin{array}{r} 330 \\ 48)\overline{15,840} \\ -144 \\ \hline 144 \\ -144 \\ \hline 00 \end{array}$$

> **TIP:** The solution to Example 5 can be checked by multiplication. Forty-eight payments of $330 each amount to 48($330) = $15,840. This added to the down payment totals $18,340 as desired.

Jorge's monthly payments will be $330.

Example 6 **Solving a Travel Application**

Linda must drive from Clayton to Oakley. She can travel directly from Clayton to Oakley on a mountain road, but will only average 40 mph. On the route through Pearson, she travels on highways and can average 60 mph. Which route will take less time?

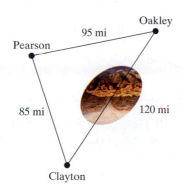

Solution:

Read and familiarize: A map is presented in the problem.

Given: The distance for each route and the speed traveled along each route

Find: Find the time required for each route. Then compare the times to determine which will take less time.

Operations:

1. First note that the total distance of the route through Pearson is found by using addition.

$$85 \text{ mi} + 95 \text{ mi} = 180 \text{ mi}$$

Skill Practice

6. Taylor makes $18 per hour for the first 40 hr worked each week. His overtime rate is $27 per hour for hours exceeding the normal 40-hr workweek. If his total salary for one week is $963, determine the number of hours of overtime worked.

Answer

6. 9 hr overtime

2. The speed of the vehicle gives us the distance traveled per hour. Therefore, the time of travel equals the total distance divided by the speed.

From Clayton to Oakley through the mountains, we divide 120 mi by 40-mph increments to determine the number of hours.

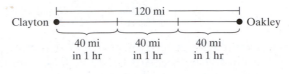

$$\text{Time} = \frac{120 \text{ mi}}{40 \text{ mph}} = 3 \text{ hr}$$

From Clayton to Oakley through Pearson, we divide 180 mi by 60-mph increments to determine the number of hours.

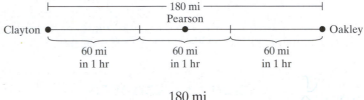

$$\text{Time} = \frac{180 \text{ mi}}{60 \text{ mph}} = 3 \text{ hr}$$

Therefore, each route takes the same amount of time, 3 hr.

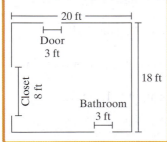

Example 7 Solving a Construction Application

A rancher must fence the corral shown in Figure 1-11. However, no fencing is required on the side adjacent to the barn. If fencing costs $4 per foot, what is the total cost?

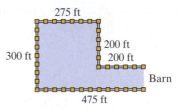

Figure 1-11

Solution:

Read and familiarize: A figure is provided.

Strategy

With some application problems, it helps to work backward from your final goal. In this case, our final goal is to find the total cost. However, to find the total cost, we must first find the total distance to be fenced. To find the total distance, we add the lengths of the sides that are being fenced.

11
275 ft
200 ft
200 ft
475 ft
+ 300 ft
————
1450 ft

Therefore,

$$\begin{pmatrix} \text{Total cost} \\ \text{of fencing} \end{pmatrix} = \begin{pmatrix} \text{total} \\ \text{distance} \\ \text{in feet} \end{pmatrix}\begin{pmatrix} \text{cost} \\ \text{per foot} \end{pmatrix}$$

$$= (1450\ \text{ft})(\$4\ \text{per ft})$$

$$= \$5800$$

The total cost of fencing is $5800.

Section 1.8 Practice Exercises

Boost your GRADE at ALEKS.com!

- Practice Problems
- Self-Tests
- NetTutor
- e-Professors
- Videos

Study Skills Exercise

1. Sometimes you may run into a problem with homework, or you may find that you are having trouble keeping up with the pace of the class. A tutor can be a good resource. Answer the following questions.
   a. Does your college offer tutoring? **b.** Is it free? **c.** Where would you go to sign up for a tutor?

Review Exercises

For Exercises 2–13, translate the English phrase into a mathematical statement and simplify.

2. 89 decreased by 66

3. 71 increased by 14

4. 16 more than 42

5. Twice 14

6. The difference of 93 and 79

7. Subtract 32 from 102

8. Divide 12 into 60

9. The product of 10 and 13

10. The total of 12, 14, and 15

11. The quotient of 24 and 6

12. 41 less than 78

13. The sum of 5, 13, and 25

Objective 1: Problem-Solving Strategies

14. In your own words, list the guidelines or strategy that you would use to solve an application problem.

For Exercises 15–18, write two or more key words or phrases that represent the given operation. Answers may vary.

15. Addition

16. Multiplication

17. Subtraction

18. Division

Objective 2: Applications Involving One Operation

19. White Mountain Peak in California is 14,246 ft high. Denali in Alaska is 20,320 ft high. How much higher is Denali than White Mountain Peak? **(See Example 1.)**

20. In a recent year, *Reader's Digest* was the best-selling U.S. magazine with 12,212,000 yearly subscriptions. *Sports Illustrated* was 15th overall and had 3,252,900 yearly subscriptions. How many more subscriptions did *Reader's Digest* have than *Sports Illustrated*?

For Exercises 21–22, refer to the table.

Country	Metric Tons of Oil Consumed in 2004	Population in 2005
United States	937,600,000	298,000,000
China	308,600,000	1,316,000,000
Japan	241,500,000	128,000,000
Russia	128,500,000	143,000,000
Germany	123,600,000	83,000,000
India	119,300,000	1,103,000,000

Source: U.S. Energy Information Administration

21. Find the total amount of oil consumed by China, Japan, Russia, Germany, and India. **(See Example 2.)**

22. Find the total population of China, Japan, Russia, Germany, and India.

23. A graphing calculator screen consists of an array of rectangular dots called *pixels*. If the screen has 96 rows of pixels and 126 pixels in each row, how many pixels are in the whole screen? **(See Example 3.)**

24. The floor of a rectangular room has 62 rows of tile with 38 tiles in each row. How many total tiles are there?

25. At one time, Tidewater Community College in Virginia had 3000 students who registered for Beginning Algebra. If the average class size is 25 students, how many Beginning Algebra classes will the college have to offer?

26. Eight people are to share equally in an inheritance of $84,480. How much money will each person receive?

27. The Honda Hybrid gets 45 miles per gallon (mpg) in stop-and-go traffic. How many gallons will it use in 405 mi of stop-and-go driving? **(See Example 4.)**

28. A couple traveled at an average speed of 52 mph for a cross-country trip. If the couple drove 1352 mi, how many hours was the trip?

29. Jeannette has two children who each attended college in Boston. Her son Ricardo attended Bunker Hill Community College where the yearly tuition and fees came to $2600. Her daughter Ricki attended M.I.T. where the yearly tuition and fees totaled $26,960. If Jeannette paid the full amount for both children to go to school, what was her total expense for tuition and fees for 1 year?

30. Clyde and Mason each leave a rest area on the Florida Turnpike. Clyde travels north and Mason travels south. After 2 hr, Clyde has gone 138 mi and Mason, who ran into heavy traffic, traveled only 96 mi. How far apart are they?

31. The Toyota Prius gets 55 mpg on the highway. How many miles can it go on 20 gal?

32. A 3 credit-hour class at a certain college meets 3 hr per week. If a semester is 16 weeks long, how many hours will the class meet during the semester?

33. A movie theater has 70 rows and 45 seats in a row. What is the maximum seating capacity?

34. A square checkerboard has 8 boxes per row and 8 rows. What is the total number of boxes?

Objective 3: Applications Involving Multiple Operations

35. Jackson purchased a car for $16,540. He paid $2500 down and paid the rest in equal monthly payments over a 36-month period. How much were his monthly payments? **(See Example 5.)**

36. Lucio purchased a refrigerator for $1170. He paid $150 at the time of purchase and then paid off the rest in equal monthly payments over 1 year. How much was his monthly payment?

37. Monika must drive from Watertown to Utica. She can travel directly from Watertown to Utica on a small county road, but will only average 40 mph. On the route through Syracuse, she travels on highways and can average 60 mph. Which route will take less time? **(See Example 6.)**

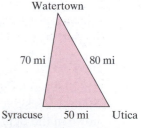

Figure for Exercise 37

38. Rex has a choice of two routes to drive from Oklahoma City to Fort Smith. On the interstate, the distance is 220 mi and he can drive 55 mph. If he takes the back roads, he can only travel 40 mph, but the distance is 200 mi. Which route will take less time?

39. If you wanted to line the outside of a garden with a decorative border, would you need to know the area of the garden or the perimeter of the garden?

40. If you wanted to know how much sod to lay down within a rectangular backyard, would you need to know the area of the yard or the perimeter of the yard?

41. Alexis wants to buy molding for a room that is 12 ft by 11 ft. No molding is needed for the doorway, which measures 3 ft. See the figure. If molding costs $2 per foot, how much money will it cost? **(See Example 7.)**

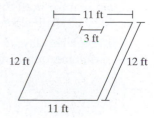

42. A homeowner wants to fence her rectangular backyard. The yard is 75 ft by 90 ft. If fencing costs $5 per foot, how much will it cost to fence the yard?

43. What is the cost to carpet the room whose dimensions are shown in the figure? Assume that carpeting costs $34 per square yard and that there is no waste.

6 yd

5 yd

44. What is the cost to tile the room whose dimensions are shown in the figure? Assume that tile costs $3 per square foot.

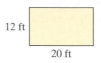

12 ft

20 ft

45. The balance in Gina's checking account is $278. If she writes checks for $82, $59, and $101, how much will be left over?

46. The balance in Jose's checking account is $3455. If he write checks for $587, $36, and $156, how much will be left over?

47. A community college bought 72 new computers and 6 new printers for a computer lab. If computers were purchased for $2118 each and the printers for $256 each, what was the total bill (not including tax)?

48. Tickets to the San Diego Zoo in California cost $22 for children aged 3–11 and $33 for adults. How much money is required to buy tickets for a class of 33 children and 6 adult chaperones?

49. A discount music store buys used CDs from its customers for $3. Furthermore, a customer can buy any used CD in the store for $8. Latayne sells 16 CDs.

a. How much money does she receive by selling the 16 CDs?

b. How many CDs can she then purchase with the money?

50. Shevona earns $8 per hour and works a 40-hr workweek. At the end of the week, she cashes her paycheck and then buys two tickets to a Janet Jackson concert.

a. How much money is her paycheck worth?

b. If the concert tickets cost $64 each, how much money does she have left over from her paycheck after buying the tickets?

51. During his 13-year career with the Chicago Bulls, Michael Jordan scored 12,192 field goals (worth 2 points each). He scored 581 three-point shots and 7327 free-throws (worth 1 point each). How many total points did he score during his career with the Bulls?

52. A matte is to be cut and placed over five small square pictures before framing. Each picture is 5 in. wide, and the matte frame is 37 in. wide, as shown in the figure. If the pictures are to be equally spaced (including the space on the left and right edges), how wide is the matte between them?

5 in.

37 in.

53. Mortimer the cat was prescribed a suspension of methimazole for hyperthyroidism. This suspension comes in a 60 milliliter bottle with instructions to give 1 milliliter twice a day. The label also shows there is one refill, but it must be called in 2 days ahead. Mortimer had his first two doses on September 1.

a. For how many days will one bottle last?

b. On what day, at the latest, should his owner order a refill to avoid running out of medicine?

54. Recently, the American Medical Association reported that there were 630,300 male doctors and 205,900 female doctors in the United States.

    a. What is the difference between the number of male doctors and the number of female doctors?

    b. What is the total number of doctors?

55. On a map, each inch represents 60 mi.

    a. If Las Vegas and Salt Lake City are approximately 6 in. apart on the map, what is the actual distance between the cities?

    b. If Madison, Wisconsin, and Dallas, Texas, are approximately 840 mi apart, how many inches would this represent on the map?

56. On a map, each inch represents 40 mi.

    a. If Wichita, Kansas, and Des Moines, Iowa, are approximately 8 in. apart on the map, what is the actual distance between the cities?

    b. If Seattle, Washington, and Sacramento, California, are approximately 600 mi apart, how many inches would this represent on the map?

57. A textbook company ships books in boxes containing a maximum of 12 books. If a bookstore orders 1250 books, how many boxes can be filled completely? How many books will be left over?

58. A farmer sells eggs in containers holding a dozen eggs. If he has 4257 eggs, how many containers will be filled completely? How many eggs will be left over?

59. Marc pays for an $84 dinner with $20 bills.

    a. How many bills must he use?

    b. How much change will he receive?

60. Shawn buys 3 CDs for a total of $54 and pays with $10 bills.

    a. How many bills must he use?

    b. How much change will he receive?

61. Ling has three jobs. He works for a lawn maintenance service 4 days a week. He also tutors math and works as a waiter on weekends. His hourly wage and the number of hours for each job are given for a 1-week period. How much money did Ling earn for the week?

	Hourly Wage	Number of Hours
Tutor	$30/hr	4
Waiter	10/hr	16
Lawn maintenance	8/hr	30

62. An electrician, a plumber, a mason, and a carpenter work at a certain construction site. The hourly wage and the number of hours each person worked are summarized in the table. What was the total amount paid for all four workers?

	Hourly Wage	Number of Hours
Electrician	$36/hr	18
Plumber	28/hr	15
Mason	26/hr	24
Carpenter	22/hr	48

Group Activity

Becoming a Successful Student

Materials: Computer with Internet access

Estimated time: 15 minutes

Group Size: 4

Good time management, good study skills, and good organization will help you be successful in this course. Answer the following questions and compare your answers with your group members.

1. For the following week, write down the times each day that you plan to study math.

Monday	Tuesday	Wednesday	Thursday	Friday	Saturday	Sunday

2. Write down the date of your next math test. _____

3. How many hours do you work each week outside of your school work? _____
Do you think that this is impacting your success in this class?

4. Look through the book in Chapter 1 and find the page number corresponding to each feature in the book. Discuss with your group members how you might use each feature.

Problem Recognition Exercises: page _____

Chapter Summary: page _____

Chapter Review Exercises: page _____

Chapter Test: page _____

5. Look at the Skill Practice exercises in the margin (for example, find Skill Practice exercises 11–13 in Section 1.7). Where are the answers to these exercises located? Discuss with your group members how you might use the Skill Practice exercises.

6. Discuss with your group members places where you can go for extra help in math. Then write down three of the suggestions.

_____

_____

_____

7. Do you keep an organized notebook for this class? _____ Can you think of any suggestions that you can share with your group members to help them keep their materials organized?

8. Some students favor different methods of learning over others. For example, you might prefer:

- Learning through listening and hearing.
- Learning through seeing images, watching demonstrations, and visualizing diagrams and charts.
- Learning by experience through a hands-on approach by doing things.
- Learning through reading and writing.

Most experts believe that the most effective learning comes when a student engages in _all_ of these activities. However, each individual is different and may benefit from one activity more than another. You can visit a number of different websites to determine your "learning style." Try doing a search on the Internet with the key words "_learning styles assessment._" Once you have found a suitable website, answer the questionnaire and the site will give you feedback on what method of learning works best for you.

Chapter 1   Summary

Section 1.1   Introduction to Whole Numbers

Key Concepts

The place value for each **digit** of a number is shown in the chart.

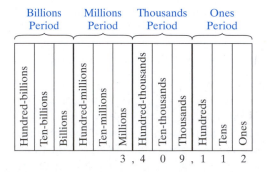

Numbers can be written in different forms, for example:

Standard Form: 3,409,112

Expanded Form: 3 millions + 4 hundred-thousands + 9 thousands + 1 hundred + 1 ten + 2 ones

Words: three million, four hundred nine thousand, one hundred twelve

The order of whole numbers can be visualized by placement on a number line.

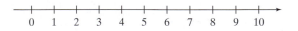

Examples

Example 1

The digit 9 in the number 24,891,321 is in the ten-thousands place.

Example 2

The standard form of the number forty-one million, three thousand, fifty-six is 41,003,056.

Example 3

The expanded form of the number 76,903 is 7 ten-thousands + 6 thousands + 9 hundreds + 3 ones.

Example 4

In words the number 2504 is two thousand, five hundred four.

Example 5

To show that $8 > 4$, note the placement on the number line: 8 is to the right of 4.

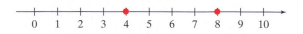

Section 1.2   Addition of Whole Numbers and Perimeter

Key Concepts

The **sum** is the result of adding numbers called **addends**.

Addition is performed with and without carrying.

Addition Property of Zero:

The sum of any number and zero is that number.

Commutative Property of Addition:

Changing the order of the addends does not affect the sum.

Associative Property of Addition:

The manner in which the addends are grouped does not affect the sum.

There are several words and phrases that indicate addition, such as *sum, added to, increased by, more than, plus,* and *total of.*

The **perimeter** of a **polygon** is the distance around the outside of the figure. To find perimeter, take the sum of the lengths of all sides of the figure.

Examples

Example 1

For $2 + 7 = 9$, the addends are 2 and 7, and the sum is 9.

Example 2

$$
\begin{array}{r}
23 \\
+\ 41 \\
\hline
64
\end{array}
\qquad
\begin{array}{r}
{\scriptstyle 1\ 1} \\
189 \\
+\ 76 \\
\hline
265
\end{array}
$$

Example 3

$16 + 0 = 16$

Example 4

$3 + 12 = 12 + 3$

Example 5

$2 + (19 + 3) = (2 + 19) + 3$

Example 6

The sum of 6 and 18 translates to $6 + 18$.

Example 7

The expression $5 + 4$ can be translated as 5 increased by 4, or 4 more than 5.

Example 8

The perimeter is found by adding the lengths of all sides.

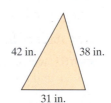

42 in. 38 in.

31 in.

Perimeter = 42 in. + 38 in. + 31 in. = 111 in.

Section 1.3   Subtraction of Whole Numbers

Key Concepts

The **difference** is the result of subtracting the **subtrahend** from the **minuend**.

Subtract numbers with and without borrowing.

There are several words and phrases that indicate subtraction, such as *minus, difference, decreased by, less than,* and *subtract from.*

Examples

Example 1

For $19 - 13 = 6$, the minuend is 19, the subtrahend is 13, and the difference is 6.

Example 2

$$
\begin{array}{r}
398 \\
-\ 227 \\
\hline
171
\end{array}
\qquad
\begin{array}{r}
{\scriptstyle 9} \\
{\scriptstyle 1\ 10\ 14} \\
2\,0\,4 \\
-\ 8\,8 \\
\hline
1\,1\,6
\end{array}
$$

Example 3

The difference of 15 and 7 translates to $15 - 7$.

Example 4

The expression $31 - 20$ can be translated to 31 decreased by 20, or subtract 20 from 31.

Section 1.4   Rounding and Estimating

Key Concepts

To **round a number**, follow these steps.

Step 1 Identify the digit one position to the right of the given place value.
Step 2 If the digit in step 1 is a 5 or greater, add 1 to the digit in the given place value. Then replace each digit to the right of the given place value by 0.
Step 3 If the digit in step 1 is less than 5, replace it and each digit to its right by 0.

Round to estimate sums and differences.

Examples

Example 1

Round each number to the indicated place.

a. 4942; hundreds place $\longrightarrow$ 4900
b. 3712; thousands place $\longrightarrow$ 4000
c. 135; tens place $\longrightarrow$ 140

Example 2

Round to the thousands place to estimate the sum: $3929 + 2528 + 5452$.

Solution: $4000 + 3000 + 5000 = 12{,}000$

The sum is approximately 12,000.

Section 1.5   Multiplication of Whole Numbers and Area

Key Concepts

Multiplication is repeated addition.

The **product** is the result of multiplying **factors**.

Properties of Multiplication

1. Commutative Property of Multiplication: Changing the order of the factors does not affect the product.
2. Associative Property of Multiplication: The manner in which the factors are grouped does not affect the product.
3. Multipliction Property of 0: The product of any number and 0 is 0.
4. Multiplication Property of 1: The product of any number and 1 is that number.
5. Distributive Property of Multiplication over Addition

Multiply whole numbers.

The **area of a rectangle** with length l and width w is given by $A = l \cdot w$.

Examples

Example 1

$16 + 16 + 16 + 16 = 4 \times 16 = 64$

Example 2

For $3 \times 13 \times 2 = 78$ the factors are 3, 13, and 2, and the product is 78.

Example 3

1. $4 \times 7 = 7 \times 4$

2. $6 \times (5 \times 7) = (6 \times 5) \times 7$

3. $43 \times 0 = 0$

4. $290 \times 1 = 290$

5. $5 \times (4 + 8) = (5 \times 4) + (5 \times 8)$

Example 4

$3 \times 14 = 42 \qquad 7(4) = 28$

$$
\begin{array}{r}
312 \\
\times\ 23 \\
\hline
936 \\
6240 \\
\hline
7176
\end{array}
$$

Example 5

Find the area of the rectangle.

23 cm [rectangle] 70 cm

$A = (23\ \text{cm}) \cdot (70\ \text{cm}) = 1610\ \text{cm}^2$

Section 1.6   Division of Whole Numbers

Key Concepts

A **quotient** is the result of dividing the **dividend** by the **divisor**.

Properties of Division:

1. Any number divided by itself is 1.
2. Any number divided by 1 is the number itself.
3. Zero divided by any nonzero number is zero.
4. A number divided by zero is undefined.

Long division, with and without a **remainder**

Examples

Example 1

For $36 \div 4 = 9$, the dividend is 36, the divisor is 4, and the quotient is 9.

Example 2

1. $13 \div 13 = 1$

2.
$$
\begin{array}{r}
37 \\
1\overline{)37}
\end{array}
$$

3. $\dfrac{0}{2} = 0$

Example 3

$\dfrac{2}{0}$ is undefined.

Example 4

$$
\begin{array}{r}
263 \\
3\overline{)789} \\
-6 \\
\hline
18 \\
-18 \\
\hline
09 \\
-9 \\
\hline
0
\end{array}
\qquad
\begin{array}{r}
41 \text{ R } 12 \\
21\overline{)873} \\
-84 \\
\hline
33 \\
-21 \\
\hline
12
\end{array}
$$

Section 1.7   Exponents, Square Roots, and the Order of Operations

Key Concepts

A number raised to an **exponent** represents repeated multiplication.

For 6^3, 6 is the **base** and 3 is the exponent or **power**.

The **square root** of 16 is 4 because $4^2 = 16$. That is, $\sqrt{16} = 4$.

Order of Operations

1. Perform all operations inside parentheses first.
2. Simplify any expressions containing exponents or square roots.
3. Perform multiplication or division in the order that they appear from left to right.
4. Perform addition or subtraction in the order that they appear from left to right.

Powers of 10

$10^1 = 10$

$10^2 = 100$

$10^3 = 1000$ and so on.

The **mean** is the average of a set of numbers. To find the mean, add all the values and divide by the number of values.

Examples

Example 1

$9^4 = 9 \cdot 9 \cdot 9 \cdot 9 = 6561$

Example 2

$\sqrt{49} = 7$

Example 3

$32 \div \sqrt{16} + (9 - 6)^2$

$= 32 \div \sqrt{16} + (3)^2$

$= 32 \div 4 + 9$

$= 8 + 9$

$= 17$

Example 4

$10^5 = 100,000$ 1 followed by 5 zeros

Example 5

Find the mean (average) of Michael's scores from his homework assignments.

40, 41, 48, 38, 42, 43

Solution:

$$\frac{40 + 41 + 48 + 38 + 42 + 43}{6} = \frac{252}{6} = 42$$

The average is 42.

Section 1.8	Problem-Solving Strategies

Key Concepts

Guidelines for Problem Solving

Step 1 Read the problem carefully. Draw a diagram or write an appropriate formula. Estimate a reasonable answer.

Step 2 Write down what information is given and what must be found.

Step 3 Form a strategy. Identify what mathematical operation or operations apply.

Step 4 Perform the mathematical operations to solve for the unknown.

Step 5 Check the answer.

Examples

Example 1

Nolan received a doctor's bill for $984. His insurance will pay $200, and the balance can be paid in 4 equal monthly payments. How much will each payment be?

Solution:

To find the amount not paid by insurance, subtract $200 from the total bill.

$$984 - 200 = 784$$

To find Nolan's 4 equal payments, divide the amount not covered by insurance by 4.

$$784 \div 4 = 196$$

Nolan must make 4 payments of $196 each.

Chapter 1   Review Exercises

Section 1.1

For Exercises 1–2, determine the place value for each underlined digit.

1. 1<u>0</u>,024

2. <u>8</u>21,811

For Exercises 3–4, convert the numbers to standard form.

3. 9 ten-thousands + 2 thousands + 4 tens + 6 ones   9,246

4. 5 hundred-thousands + 3 thousands + 1 hundred + 6 tens

For Exercises 5–6, convert the numbers to expanded form.

5. 3,400,820

6. 30,554

For Exercises 7–8, write the numbers in words.

7. 245

8. 30,861

For Exercises 9–10, write the numbers in standard form.

9. Three thousand, six-hundred two

10. Eight hundred thousand, thirty-nine

For Exercises 11–12, place the numbers on the number line.

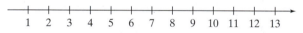

11. 2

12. 7

For Exercises 13–14, determine if the inequality is true or false.

13. $3 < 10$

14. $10 > 12$

Section 1.2

For Exercises 15–16, identify the addends and the sum.

15. $105 + 119 = 224$

16. 53
      + 21
      ────
        74

For Exercises 17–20, add.

17. $18 + 24 + 29$

18. $27 + 9 + 18$

19. 8403
      + 9007

20. 68,421
      + 2,221

21. For each of the mathematical statements, identify the property used. Choose from the commutative property or the associative property.

 a. $6 + (8 + 2) = (8 + 2) + 6$ Commute

 b. $6 + (8 + 2) = (6 + 8) + 2$ Associate

 c. $6 + (8 + 2) = 6 + (2 + 8)$ Commute

For Exercises 22–25, translate the English phrase to a mathematical statement and simplify.

22. The sum of 403 and 79

23. 92 added to 44

24. 7 more than 36

25. 23 increased by 6

26. The table gives the number of cars sold by three dealerships during one week.

	Honda	Ford	Toyota
Bob's Discount Auto	23	21	34
AA Auto	31	25	40
Car World	33	20	22

 a. What is the total number of cars sold by AA Auto?

 b. What is the total number of Fords sold by these three dealerships?

27. The bar graph represents the distribution of the U.S. population by age group for a recent year. Determine the number of seniors (aged 60 and over).

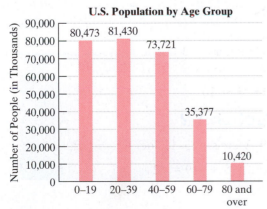

U.S. Population by Age Group

Source: U.S. Census Bureau

28. Find the perimeter of the figure.

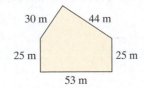

30 m   44 m
25 m   25 m
53 m

Section 1.3

For Exercises 29–30, identify the minuend, subtrahend, and difference.

29. $14 - 8 = 6$

30.   102 m
        $-$ 78 s
        ──────
          24 p

For Exercises 31–32, subtract and check your answer by addition.

31.   37   Check: ☐ $+ 11 = 37$
        $-$ 11

32.   61   Check: ☐ $+ 41 = 61$
        $-$ 41

For Exercises 33–36, subtract.

33.   2005
        $-$ 1884

34. $1389 - 299$

35. $86,000 - 54,981$

36. $67,000 - 32,812$

For Exercises 37–40, translate the English phrase into a mathematical statement and simplify.

37. 38 minus 31

38. 111 decreased by 15

39. Subtract 42 from 251

40. The difference of 90 and 52

41. There were 95,191,761 tons of watermelons and 23,299,323 tons of cantaloupes produced in 2006. What is the difference between the weight of the watermelons and the weight of the cantaloupes?

42. Tiger Woods earned $57,940,144 from the PGA tour as of 2006. If Phil Mickelson earned $36,167,360, find the difference in their winnings.

43. The graph gives the estimated number of overseas visitors (in thousands) for five cities in the United States for a recent year. What is the difference between the number of visitors to New York and the number of visitors to Orlando?

Overseas Visitors in U.S. Cities

Source: U.S. Department of Commerce

Section 1.4

For Exercises 44–45, round each number to the given place value.

44. 5,234,446; millions

45. 9,332,945; ten-thousands

For Exercises 46–47, estimate the sum by rounding to the indicated place value.

46. 894,004 − 123,883; hundred-thousands

47. 330 + 489 + 123 + 571; hundreds

48. In 2004, the population of Russia was 144,112,353, and the population of Japan was 127,295,333. Estimate the difference in their populations by rounding to the nearest million.

49. The state of Missouri has two dams: Fort Peck with a volume of 96,050 cubic meters (m^3) and Oahe with a volume of 66,517 m^3. Round the numbers to the nearest thousand to estimate the total volume of these two dams.

Section 1.5

For Exercises 50–51, identify the factors and the product.

50. 32 · 12 = 384

51. 33 × 40 = 1320

52. Indicate whether the statement is equal to the product of 8 and 13.

   a. 8(13)   **b.** (8) · 13   **c.** (8) + (13)

For Exercises 53–57, for each property listed, choose an expression from the right column that demonstrates the property.

53. Associative property of multiplication   **a.** 3(4) = 4(3)

54. Distributive property of multiplication over addition   **b.** 19 × 1 = 19

55. Multiplication property of 0   **c.** (1 · 8) · 3 = 1 · (8 · 3)

56. Commutative property of multiplication   **d.** 0 · 29 = 0

57. Multiplication property of 1   **e.** 4(3 + 1) = 4 · 3 + 4 · 1

For Exercises 58–60, multiply.

58. 142
   × 43

59. (1024)(51)

60. 6000
   × 500

61. A discussion group needs to purchase books that are accompanied by a workbook. The price of the book is $26, and the workbook costs an additional $13. If there are 11 members in the group, how much will it cost the group to purchase both the text and workbook for each student?

62. Orcas, or killer whales, eat 551 pounds (lb) of food a day. If Sea World has two adult killer whales, how much food will they eat in 1 week?

Section 1.6

For Exercises 63–64, perform the division. Then identify the divisor, dividend, and quotient.

63. 42 ÷ 6

64. 4)52

For Exercises 65–68, use the properties of division to simplify the expression, if possible.

65. $3 \div 1$ **66.** $3 \div 3$

67. $3 \div 0$ **68.** $0 \div 3$

69. Explain how you check a division problem if there is no remainder.

70. Explain how you check a division problem if there is a remainder.

For Exercises 71–73, divide and check the answer.

71. $348 \div 6$ **72.** $11\overline{)458}$ **73.** $\dfrac{1043}{20}$

For Exercises 74–75, write the English phrase as a mathematical expression and simplify.

74. The quotient of 72 and 4

75. 108 divided by 9

76. Quinita has 105 photographs that she wants to divide equally among herself and three siblings. How many photos will each person receive? How many photos will be left over?

77. Ashley has $60 to spend on souvenirs at a surf shop. The prices of several souvenirs are given in the graph.

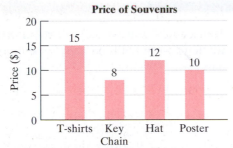

Price of Souvenirs

a. How many souvenirs can Ashley buy if she chooses all T-shirts?

b. How many souvenirs can Ashley buy if she chooses all hats?

Section 1.7

For Exercises 78–79, write the repeated multiplication in exponential form. Do not simplify.

78. $8 \cdot 8 \cdot 8 \cdot 8 \cdot 8$

79. $2 \cdot 2 \cdot 2 \cdot 2 \cdot 5 \cdot 5 \cdot 5$

For Exercises 80–83, evaluate the exponential expressions.

80. 5^3 **81.** 4^4

82. 1^7 **83.** 10^6

For Exercises 84–85, evaluate the square roots.

84. $\sqrt{64}$ **85.** $\sqrt{144}$

For Exercises 86–91, evaluate the expression using the order of operations.

86. $14 \div 7 \cdot 4 - 1$

87. $10^2 - 5^2$

88. $90 - 4 + 6 \div 3 \cdot 2$

89. $2 + 3 \cdot 12 \div 2 - \sqrt{25}$

90. $6^2 - 4^2 + (9 - 7)^3$

91. $26 - 2(10 - 1) + (3 + 4 \cdot 11)$

92. Find the mean (average) for the set of numbers 7, 6, 12, 5, 7, 6, 13.

93. Carolyn's electric bills for the past 5 months have been $80, $78, $101, $92, and $94. Find her average monthly charge.

94. The table shows the number of homes sold by a realty company in the last 6 months. Determine the average number of houses sold per month for these 6 months.

Month	Number of Houses
May	6
June	9
July	11
August	13
September	5
October	4

Section 1.8

95. The Cincinnati Zoo houses about 17,000 animals that represent 750 species. The San Diego Zoo has 4000 animals representing 800 species.

a. Which zoo has the most animals? How many more animals does it have?

b. Which zoo has the most species? How many more species does it have?

96. Doris drives her son to extracurricular activities each week. She drives 5 mi round-trip to baseball practice 3 times a week and 6 mi round-trip to piano lessons once a week.

a. How many miles does she drive in 1 week to get her child to his activities?

b. Approximately how many miles does she travel during a school year consisting of 10 months (there are approximately 4 weeks per month)?

97. At one point in his baseball career, Alex Rodriquez signed a contract for $252,000,000 for a 9-year period between 2001 and 2010. Suppose federal taxes amount to $75,600,000 for the contract. After taxes, how much will Alex receive per year?

98. Aletha wants to buy plants for a rectangular garden in her backyard that measures 12 ft by 8 ft. She wants to divide the garden into 2-square-foot (2 ft^2) areas, one for each plant.

a. How many plants should Aletha buy?

b. If the plants cost $3 each, how much will it cost Aletha for the plants?

c. If she puts a fence around the perimeter of the garden that costs $2 per foot, how much will it cost for the fence?

d. What will be Aletha's total cost for this garden?

Chapter 1   Test

1. Determine the place value for the underlined digit.

    a. 4̲92   **b.** 23,4̲41   **c.** 2̲,340,711   **d.** 340,5̲92

2. Fill in the table with either the word name for the number or the number in standard form.

State / Province	Population	
	Standard Form	Word Name
a. Kentucky		Four million, sixty-five thousand
b. Texas	21,325,000	
c. Pennsylvania	12,287,000	
d. New Brunswick, Canada		Seven hundred twenty-nine thousand
e. Ontario, Canada	11,410,000	

3. Translate the phrase by writing the numbers in standard form and inserting the appropriate inequality. Choose from < or >.

a. Fourteen is greater than six.

b. Seventy-two is less than eighty-one.

For Exercises 4–17, perform the indicated operation.

4.   51
    + 78

5.   82
    × 4

6.   154
    − 41

7. $4\overline{)908}$

8. $58 \cdot 49$

9. $149 + 298$

10. $324 \div 15$

11. $3002 − 2456$

12. $10,984 − 2881$

13. $\dfrac{840}{42}$

14. (500,000)(3000)

15. 34 + 89 + 191 + 22

16. 403(0)

17. $0\overline{)16}$

18. For each of the mathematical statements, identify the property used. Choose from the commutative property of multiplication and the associative property of multiplication. Explain your answer.

a. (11 · 6) · 3 = 11 · (6 · 3)

b. (11 · 6) · 3 = 3 · (11 · 6)

19. Round each number to the indicated place value.

a. 4850; hundreds

b. 12,493; thousands

c. 7,963,126; hundred-thousands

20. The attendance to the Van Gogh and Gauguin exhibit in Chicago was 690,951. The exhibit moved to Amsterdam, and the attendance was 739,117. Round the numbers to the ten-thousands place to estimate the total attendance for this exhibit.

For Exercises 21–24, simplify, using the order of operations.

21. $8^2 \div 2^4$

22. $26 \cdot \sqrt{4} - 4(8 - 1)$

23. $36 \div 3(14 - 10)$

24. $65 - 2(5 \cdot 3 - 11)^2$

25. Brittany and Jennifer are taking an online course in business management. Brittany has taken 6 quizzes worth 30 points each and received the following scores: 29, 28, 24, 27, 30, and 30. Jennifer has only taken 5 quizzes so far, and her scores are 30, 30, 29, 28, and 28. At this point in the course, which student has a higher average?

26. The use of the cell phone has grown every year for the past 13 years. See the graph.

a. Find the change in the number of phones used from 2003 to 2004.

b. Of the years presented in the graph, between which two years was the increase the greatest?

Cell Phone Use in the United States

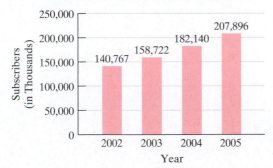

27. The table gives the number of calls to three fire departments during a selected number of weeks. Find the number of calls per week of each department to determine which department is the busiest.

	Number of Calls	Time Period (Number of Weeks)
North Side Fire Department	80	16
South Side Fire Department	72	18
East Side Fire Department	84	28

28. Find the perimeter of the figure.

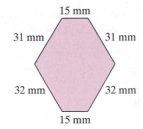

29. Find the perimeter and the area of the rectangle.

30. Round to the nearest hundred to estimate the area of the rectangle.

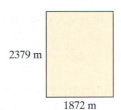

Fractions and Mixed Numbers: Multiplication and Division

2

CHAPTER OUTLINE

Chapter 2

In Chapter 2 we study the concept of a fraction and a mixed number. We learn how to simplify fractions by reducing to lowest terms. Learning the terms and vocabulary relating to fractions will help you understand this chapter. Try the crossword puzzle shown. Refer to definitions in the chapter if you need help.

Across

2. The top number of a fraction.
5. To divide two fractions, multiply the first fraction by the _____ of the second fraction.
6. A fraction whose numerator is greater than or equal to the denominator is called _____.

Down

1. The bottom number of a fraction.
3. A fraction whose numerator is less than the denominator is called _____.
4. A whole number greater than 1 that has only 1 and itself as factors is called a _____ number.

Section 2.1   Introduction to Fractions and Mixed Numbers

Objectives

1. **Definition of a Fraction**
2. **Proper and Improper Fractions**
3. **Mixed Numbers**
4. **Fractions and the Number Line**

1. Definition of a Fraction

In Chapter 1, we studied operations on whole numbers. In this chapter, we work with numbers that represent part of a whole. When a whole unit is divided into equal parts, we call the parts **fractions** of a whole. For example, the pizza in Figure 2-1 is divided into 5 equal parts. One-fifth $(\frac{1}{5})$ of the pizza has been eaten, and four-fifths $(\frac{4}{5})$ of the pizza remains.

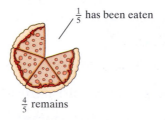

$\frac{1}{5}$ has been eaten

$\frac{4}{5}$ remains

Figure 2-1

A fraction is written in the form $\frac{a}{b}$, where a and b are whole numbers and $b \neq 0$. In the fraction $\frac{5}{8}$, the "top" number, 5, is called the **numerator**. The "bottom" number, 8, is called the **denominator**.

$$\text{numerator} \longrightarrow \frac{5}{8} \longleftarrow \text{denominator}$$

Avoiding Mistakes

The fraction $\frac{a}{b}$ can also be written as a/b. However, we discourage the use of the "slanted" fraction bar. In later applications of algebra, the slanted fraction bar can cause confusion.

Example 1   Identifying the Numerator and Denominator of a Fraction

For each fraction, identify the numerator and denominator.

a. $\frac{3}{5}$ **b.** $\frac{1}{8}$ **c.** $\frac{8}{1}$

Solution:

a. $\frac{3}{5}$ The numerator is 3. The denominator is 5.

b. $\frac{1}{8}$ The numerator is 1. The denominator is 8.

c. $\frac{8}{1}$ The numerator is 8. The denominator is 1.

Skill Practice

Identify the numerator and denominator.

1. $\frac{4}{11}$ **2.** $\frac{0}{5}$ **3.** $\frac{6}{1}$

The *denominator* of a fraction denotes the number of equal pieces into which a whole unit is divided. The *numerator* denotes the number of pieces being considered.

Answers

1. Numerator: 4, denominator: 11
2. Numerator: 0, denominator: 5
3. Numerator: 6, denominator: 1

For example, the garden in Figure 2-2 is divided into 10 equal parts. Three sections contain tomato plants. Therefore, $\frac{3}{10}$ of the garden contains tomato plants.

$\frac{3}{10}$ tomato plants

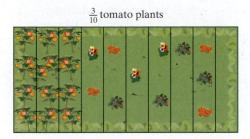

Figure 2-2

Example 2 Writing Fractions

Write a fraction for the shaded portion and a fraction for the unshaded portion of the figure.

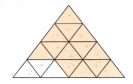

Solution:

Shaded portion:
$$\frac{13}{16}$$
← 13 pieces are shaded.
← The triangle is divided into 16 equal pieces.

Unshaded portion:
$$\frac{3}{16}$$
← 3 pieces are not shaded.
← The triangle is divided into 16 equal pieces.

Skill Practice

4. Write a fraction for the shaded portion and a fraction for the unshaded portion.

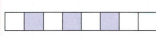

Example 3 Writing Fractions

What portion of the group of doctors shown below is female?

Solution:

The group consists of 5 members. Therefore, the denominator is 5. There are 2 women being considered. Thus, $\frac{2}{5}$ of the group is female.

Skill Practice

5. Refer to Example 3. What portion of the group of doctors is male?

In Section 1.6 we learned that fractions represent division. For example, note that the fraction $\frac{5}{1} = 5 \div 1 = 5$. In general, a fraction of the form $\frac{n}{1} = n$. This implies that any whole number can be written as a fraction by writing the whole number over 1.

Answers

4. Shaded portion: $\frac{3}{8}$; unshaded portion: $\frac{5}{8}$
5. $\frac{3}{5}$

┌─ **Concept Connections** ─┐
Simplify if possible.

6. $\dfrac{4}{0}$ **7.** $\dfrac{0}{12}$
└──────────────────────┘

Further recall that for $a \neq 0$, $0 \div a = 0$ and $a \div 0$ is undefined. Therefore, $\dfrac{0}{a} = 0$ and $\dfrac{a}{0}$ is undefined. For example:

$$\frac{0}{5} = 0 \qquad \frac{5}{0} \text{ is undefined}$$

2. Proper and Improper Fractions

If the numerator is less than the denominator in a fraction, then the fraction is called a **proper fraction.** Furthermore, a proper fraction represents a number less than 1 whole unit. The following are proper fractions.

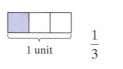

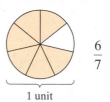

An **improper fraction** is a fraction in which the numerator is greater than or equal to the denominator. For example:

numerator greater ⟶ $\dfrac{4}{3}$ and $\dfrac{7}{7}$ ⟵ numerator equal
than denominator   to denominator

An improper fraction represents a quantity greater than 1 whole unit or equal to 1 whole unit.

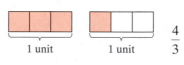

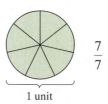

┌─ **Skill Practice** ─┐
Identify each fraction as proper or improper.

8. $\dfrac{10}{10}$ **9.** $\dfrac{7}{9}$ **10.** $\dfrac{9}{7}$
└────────────────────┘

Example 4   **Categorizing Fractions**

Identify each fraction as proper or improper.

a. $\dfrac{12}{5}$ **b.** $\dfrac{5}{12}$ **c.** $\dfrac{12}{12}$

Solution:

a. $\dfrac{12}{5}$ Improper fraction (numerator is greater than denominator)

b. $\dfrac{5}{12}$ Proper fraction (numerator is less than denominator)

c. $\dfrac{12}{12}$ Improper fraction (numerator is equal to denominator)

Answers

6. Undefined   **7.** 0
8. Improper   **9.** Proper
10. Improper

| Example 5 | Writing Improper Fractions |

Write an improper fraction to represent the fractional part of an inch for the screw shown in the figure.

Avoiding Mistakes

Each whole unit is divided into 8 pieces. Therefore the screw is $\frac{11}{8}$ in., not $\frac{11}{16}$ in.

Solution:

Each 1-in. unit is divided into 8 parts, and the screw extends for 11 parts. Therefore, the screw is $\frac{11}{8}$ in.

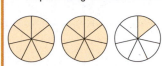
3. Mixed Numbers

Sometimes a mixed number is used instead of an improper fraction to denote a quantity greater than one whole. For example, suppose a typist typed $\frac{9}{4}$ pages of a report. We would be more likely to say that the typist typed $2\frac{1}{4}$ pages (read as "two and one-fourth pages"). The number $2\frac{1}{4}$ is called a *mixed number* and represents 2 wholes plus $\frac{1}{4}$ of a whole.

$$\frac{9}{4} = 2\frac{1}{4}$$

In general, a **mixed number** is a sum of a whole number and a fractional part of a whole. However, by convention the plus sign is left out.

$$3\frac{1}{2} \quad \text{means} \quad 3 + \frac{1}{2}$$

Suppose we want to change a mixed number to an improper fraction. From Figure 2-3, we see that the mixed number $3\frac{1}{2}$ is the same as $\frac{7}{2}$.

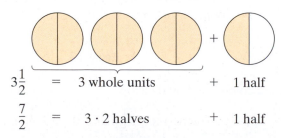

$3\frac{1}{2}$ =   3 whole units   +   1 half

$\frac{7}{2}$ =   3 · 2 halves   +   1 half

Figure 2-3

This process to convert a mixed number to an improper fraction can be summarized as follows.

PROCEDURE   Changing a Mixed Number to an Improper Fraction

   Step 1   Multiply the whole number by the denominator.

   Step 2   Add the result to the numerator.

   Step 3   Write the result from step 2 over the denominator.

For example,

$$3\frac{1}{2} = \frac{3 \times 2 + 1}{2} = \frac{7}{2}$$

Multiply the whole number by the denominator.

Add the numerator.

Write the result over the denominator.

Example 6   **Converting Mixed Numbers to Improper Fractions**

Convert the mixed number to an improper fraction.

a. $7\frac{1}{4}$ **b.** $8\frac{2}{5}$

Solution:

a. $7\frac{1}{4} = \frac{7 \times 4 + 1}{4}$ **b.** $8\frac{2}{5} = \frac{8 \times 5 + 2}{5}$

$\qquad\quad = \frac{28 + 1}{4}$ $\qquad\quad = \frac{40 + 2}{5}$

$\qquad\quad = \frac{29}{4}$ $\qquad\quad = \frac{42}{5}$

Now suppose we want to convert an improper fraction to a mixed number. In Figure 2-4, the improper fraction $\frac{13}{5}$ represents 13 slices of pizza where each slice is $\frac{1}{5}$ of a whole pizza. If we divide the 13 pieces into groups of 5, we make 2 whole pizzas with 3 pieces left over. Thus,

$$\frac{13}{5} = 2\frac{3}{5}$$

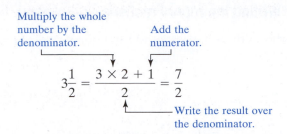

13 pieces = 2 groups of 5 + 3 left over

$\quad\frac{13}{5} = \qquad\quad 2 \qquad + \qquad \frac{3}{5}$

Figure 2-4

This process can be accomplished by division.

$$\frac{13}{5} \longrightarrow \begin{array}{r} 2 \\ 5\overline{)13} \\ -10 \\ \hline 3 \end{array} \qquad 2\frac{3}{5}$$

remainder

divisor

PROCEDURE   **Changing an Improper Fraction to a Mixed Number**

Step 1   Divide the numerator by the denominator to obtain the quotient and remainder.

Step 2   The mixed number is then given by

$$\text{Quotient} + \frac{\text{remainder}}{\text{divisor}}$$

Example 7 Converting Improper Fractions to Mixed Numbers

Convert to a mixed number.

a. $\dfrac{25}{6}$ b. $\dfrac{162}{41}$

Solution:

a. $\dfrac{25}{6} \longrightarrow$ $\begin{array}{r} 4 \\ 6\overline{)25} \\ -24 \\ \hline 1 \end{array}$ $4\dfrac{1}{6}$ ← remainder ← divisor

b. $\dfrac{162}{41} \longrightarrow$ $\begin{array}{r} 3 \\ 41\overline{)162} \\ -123 \\ \hline 39 \end{array}$ $3\dfrac{39}{41}$ ← remainder ← divisor

The process to convert an improper fraction to a mixed number indicates that the result of a division problem can be written as a mixed number.

Example 8 Writing a Quotient as a Mixed Number

Divide. Write the quotient as a mixed number.

$$28\overline{)4217}$$

Solution:

$\begin{array}{r} 150 \\ 28\overline{)4217} \\ -28 \\ \hline 141 \\ -140 \\ \hline 17 \\ -0 \\ \hline 17 \end{array}$ $150\dfrac{17}{28}$ ← remainder ← divisor

4. Fractions and the Number Line

Fractions can be visualized on a number line. For example, to graph the fraction $\frac{3}{4}$, divide the distance between 0 and 1 into 4 equal parts. To plot the number $\frac{3}{4}$, start at 0 and count over 3 parts.

Answers

14. $2\dfrac{4}{5}$ 15. $4\dfrac{7}{22}$ 16. $145\dfrac{22}{41}$

Skill Practice

Plot the numbers on a number line.

17. $\frac{4}{5}$ **18.** $\frac{1}{3}$ **19.** $\frac{13}{4}$

Example 9 Plotting Fractions on a Number Line

Plot the point on the number line corresponding to each fraction.

a. $\frac{1}{2}$ **b.** $\frac{5}{6}$ **c.** $\frac{21}{5}$

Solution:

a. $\frac{1}{2}$ Divide the distance between 0 and 1 into 2 equal parts.

b. $\frac{5}{6}$ Divide the distance between 0 and 1 into 6 equal parts.

Answers

17.

18.

19.

c. $\frac{21}{5} = 4\frac{1}{5}$ Write $\frac{21}{5}$ as a mixed number.

Thus, $\frac{21}{5} = 4\frac{1}{5}$ is located one-fifth of the way between 4 and 5 on the number line.

Section 2.1 Practice Exercises

Boost your GRADE at ALEKS.com!

ALEKS version 3.0

- Practice Problems
- Self-Tests
- NetTutor
- e-Professors
- Videos

Study Skills Exercises

1. After doing a section of homework, check the odd-numbered answers in the back of the text. Choose a method to identify the exercises that gave you trouble (i.e., circle the number or put a star by the number). List some reasons why it is important to label these problems.

2. Define the key terms.
   a. Fraction **b.** Numerator **c.** Denominator
   d. Proper fraction **e.** Improper fraction **f.** Mixed number

Objective 1: Definition of a Fraction

For Exercises 3–6, identify the numerator and the denominator for each fraction. (See Example 1.)

3. $\frac{2}{3}$ 　　　　**4.** $\frac{8}{9}$ 　　　　**5.** $\frac{12}{11}$ 　　　　**6.** $\frac{1}{2}$

For Exercises 7–14, write the fraction as a division problem and simplify, if possible.

7. $\frac{6}{1}$ 　　　**8.** $\frac{9}{1}$ 　　　**9.** $\frac{2}{2}$ 　　　**10.** $\frac{8}{8}$

11. $\frac{0}{3}$ 　　　**12.** $\frac{0}{7}$ 　　　**13.** $\frac{2}{0}$ 　　　**14.** $\frac{11}{0}$

For Exercises 15–22, write a fraction that represents the shaded area. (See Example 2.)

15. 　**16.** 　**17.** 　**18.**

19. 　**20.** 　**21.** 　**22.**

23. Write a fraction to represent the portion of gas in a gas tank represented by the gauge.

24. Write a fraction that represents the portion of medicine left in the bottle.

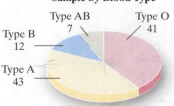

25. The scoreboard for a recent men's championship swim meet in Melbourne, Australia, shows the final standings in the men's 100-m freestyle event. What portion of the finalists are from the USA? (See Example 3.)

26. Refer to the scoreboard from Exercise 25. What portion of the finalists are from the Republic of South Africa (RSA)?

27. The graph categorizes a sample of people by blood type. What portion of the sample represents people with type O blood?

28. Refer to the graph from Exercise 27. What portion of the sample represents people with type A blood?

Name	Country	Time
Maginni, Filippo	ITA	48.43
Hayden, Brent	CAN	48.43
Sullivan, Eamon	AUS	48.47
Cielo Filho, Cesar	BRA	48.51
Lezak, Jason	USA	48.52
Van Den Hoogenband, Pieter	NED	48.63
Schoeman, Roland Mark	RSA	48.72
Neethling, Ryk	RSA	48.81

Sample by Blood Type

Type AB 7, Type O 41, Type B 12, Type A 43

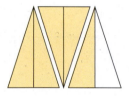 **29.** A class has 21 children—11 girls and 10 boys. What fraction of the class is made up of boys?

30. In a neighborhood in Ft. Lauderdale, 10 houses are for sale and 53 are not for sale. Write a fraction representing the portion of houses that are for sale.

Objective 2: Proper and Improper Fractions

For Exercises 31–38, label the fraction as proper or improper. **(See Example 4.)**

31. $\dfrac{7}{8}$ **32.** $\dfrac{2}{3}$ **33.** $\dfrac{10}{10}$ **34.** $\dfrac{3}{3}$

35. $\dfrac{7}{2}$ **36.** $\dfrac{21}{20}$ **37.** $\dfrac{15}{17}$ **38.** $\dfrac{13}{21}$

For Exercises 39–42, write an improper fraction for the shaded portion of each group of figures. **(See Example 5.)**

39.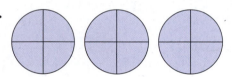

40.

41.

42.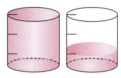

Objective 3: Mixed Numbers

For Exercises 43–44, write an improper fraction and a mixed number for the shaded portion of each group of figures.

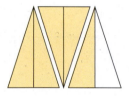 **43.**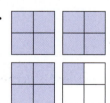

44.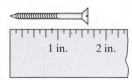

45. Write an improper fraction and a mixed number to represent the length of the nail.

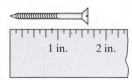
1 in.   2 in.

46. Write an improper fraction and a mixed number that represent the fraction of the amount of sugar needed for a batch of cookies, as indicated in the figure.

$1\frac{1}{2}$

For Exercises 47–58, convert the mixed number to an improper fraction. **(See Example 6.)**

47. $1\frac{3}{4}$ **48.** $6\frac{1}{3}$ **49.** $4\frac{2}{9}$ **50.** $3\frac{1}{5}$

51. $3\frac{3}{7}$ **52.** $8\frac{2}{3}$ **53.** $7\frac{1}{4}$ **54.** $10\frac{3}{5}$

55. $11\frac{5}{12}$ **56.** $12\frac{1}{6}$ **57.** $21\frac{3}{8}$ **58.** $15\frac{1}{2}$

59. How many eighths are in $2\frac{3}{8}$?   **60.** How many fifths are in $2\frac{3}{5}$?

61. How many fourths are in $1\frac{3}{4}$?   **62.** How many thirds are in $5\frac{2}{3}$?

For Exercises 63–74, convert the improper fraction to a mixed number. **(See Example 7.)**

63. $\frac{37}{8}$ **64.** $\frac{13}{7}$ **65.** $\frac{39}{5}$ **66.** $\frac{19}{4}$

67. $\frac{27}{10}$ **68.** $\frac{43}{18}$ **69.** $\frac{52}{9}$ **70.** $\frac{67}{12}$

71. $\frac{133}{11}$ **72.** $\frac{51}{10}$ **73.** $\frac{23}{6}$ **74.** $\frac{115}{7}$

For Exercises 75–82, divide. Write the quotient as a mixed number. **(See Example 8.)**

75. $7\overline{)309}$ **76.** $4\overline{)921}$ **77.** $5281 \div 5$ **78.** $7213 \div 8$

79. $8913 \div 11$ **80.** $4257 \div 23$ **81.** $15\overline{)187}$ **82.** $34\overline{)695}$

Objective 4: Fractions and the Number Line

For Exercises 83–92, plot the fraction on the number line. **(See Example 9.)**

83. $\frac{3}{4}$ 0   1   **84.** $\frac{1}{2}$ 0   1

85. $\frac{1}{3}$ 0   1   **86.** $\frac{1}{5}$ 0   1

87. $\dfrac{2}{3}$

88. $\dfrac{5}{6}$

89. $\dfrac{7}{6}$

90. $\dfrac{7}{5}$

91. $\dfrac{5}{3}$

92. $\dfrac{3}{2}$

Expanding Your Skills

93. True or false? Whole numbers can be written both as proper and improper fractions.

94. True or false? Suppose m and n are nonzero numbers, where $m > n$. Then $\dfrac{m}{n}$ is an improper fraction.

95. True or false? Suppose m and n are nonzero numbers, where $m > n$. Then $\dfrac{n}{m}$ is a proper fraction.

96. True or false? Suppose m and n are nonzero numbers, where $m > n$. Then $\dfrac{n}{3m}$ is a proper fraction.

Section 2.2   Prime Numbers and Factorization

Objectives

1. **Factors and Factorizations**
2. **Divisibility Rules**
3. **Prime and Composite Numbers**
4. **Prime Factorization**
5. **Identifying All Factors of a Whole Number**

1. Factors and Factorizations

Recall from Section 1.5 that two numbers multiplied to form a product are called factors. For example, $2 \cdot 3 = 6$ indicates that 2 and 3 are factors of 6. Likewise, because $1 \cdot 6 = 6$, the numbers 1 and 6 are factors of 6. In general, a **factor** of a number n is a nonzero whole number that divides evenly into n.

The products $2 \cdot 3$ and $1 \cdot 6$ are called factorizations of 6. In general, a **factorization** of a number n is a product of factors that equals n.

Skill Practice

1. Find four different factorizations of 18.

Example 1   Finding Factorizations of a Number

Find four different factorizations of 12.

Solution:

$$12 = \begin{cases} 1 \cdot 12 \\ 2 \cdot 6 \\ 3 \cdot 4 \\ 2 \cdot 2 \cdot 3 \end{cases}$$

TIP: Notice that a factorization may include more than two factors.

Answer

1. For example. $1 \cdot 18$
   $\qquad\quad 2 \cdot 9$
   $\qquad\quad 3 \cdot 6$
   $\qquad\quad 2 \cdot 3 \cdot 3$

2. Divisibility Rules

The number 20 is said to be divisible by 5 because 5 divides evenly into 20. To determine whether one number is divisible by another, we can perform the division and note whether the remainder is zero. However, there are several rules by which we can quickly determine whether a number is divisible by 2, 3, 5, or 10. These are called divisibility rules.

PROCEDURE Divisibility Rules for 2, 3, 5, and 10

- *Divisibility by 2.* A whole number is divisible by 2 if it is an even number. That is, the ones-place digit is 0, 2, 4, 6, or 8.
  Examples: 26 and 384
- *Divisibility by 5.* A whole number is divisible by 5 if its ones-place digit is 5 or 0.
  Examples: 45 and 260
- *Divisibility by 10.* A whole number is divisible by 10 if its ones-place digit is 0.
  Examples: 30 and 170
- *Divisibility by 3.* A whole number is divisible by 3 if the sum of its digits is divisible by 3.
  Example: 312   (sum of digits is $3 + 1 + 2 = 6$ which is divisible by 3)

We address other divisibility rules for 4, 6, 8, and 9 in the Expanding Your Skills portion of the exercises. However, these divisibility rules are harder to remember, and it is often easier simply to perform division to test for divisibility.

Example 2 Applying the Divisibility Rules

Determine whether the given number is divisible by 2, 3, 5, or 10.

a. 624   **b.** 82   **c.** 720

Solution:

		Test for Divisibility
a. 624	By 2:	Yes. The number 624 is even.
	By 3:	Yes. The sum $6 + 2 + 4 = 12$ is divisible by 3.
	By 5:	No. The ones-place digit is not 5 or 0.
	By 10:	No. The ones-place digit is not 0.
b. 82	By 2:	Yes. The number 82 is even.
	By 3:	No. The sum $8 + 2 = 10$ is not divisible by 3.
	By 5:	No. The ones-place digit is not 5 or 0.
	By 10:	No. The ones-place digit is not 0.
c. 720	By 2:	Yes. The number 720 is even.
	By 3:	Yes. The sum $7 + 2 + 0 = 9$ is divisible by 3.
	By 5:	Yes. The ones-place digit is 0.
	By 10:	Yes. The ones-place digit is 0.

Skill Practice

Determine whether the given number is divisible by 2, 3, 5, or 10.

2. 428   **3.** 75   **4.** 2100

TIP: When in doubt about divisibility, you can check by division. When we divide 624 by 3, the remainder is zero.

Answers

2. Divisible by 2
3. Divisible by 3 and 5
4. Divisible by 2, 3, 5, and 10

3. Prime and Composite Numbers

Two important classifications of whole numbers are prime numbers and composite numbers.

> **DEFINITION   Prime and Composite Numbers**
>
> - A **prime number** is a whole number greater than 1 that has only two factors (itself and 1).
> - A **composite number** is a whole number greater than 1 that is not prime. That is, a composite number will have at least one factor other than 1 and the number itself.
>
> *Note:* The whole numbers 0 and 1 are neither prime nor composite.

Skill Practice

Determine whether the number is prime, composite, or neither.

5. 39   **6.** 0   **7.** 41

Example 3   Identifying Prime and Composite Numbers

Determine whether the number is prime, composite, or neither.

a. 19   **b.** 51   **c.** 1

Solution:

a. The number 19 is prime because its only factors are 1 and 19.

b. The number 51 is composite because $3 \cdot 17 = 51$. That is, 51 has factors other than 1 and 51.

c. The number 1 is neither prime nor composite by definition.

> **TIP:** The number 2 is the only even prime number.

Prime numbers are used in a variety of ways in mathematics. We advise you to become familiar with the first several prime numbers: 2, 3, 5, 7, 11, 13, 17, 19, 23, 29, . . .

4. Prime Factorization

In Example 1 we found four factorizations of 12.

$$1 \cdot 12$$
$$2 \cdot 6$$
$$3 \cdot 4$$
$$2 \cdot 2 \cdot 3$$

The last factorization $2 \cdot 2 \cdot 3$ consists of only prime-number factors. Therefore, we say $2 \cdot 2 \cdot 3$ is the prime factorization of 12.

Concept Connections

8. Is the product $2 \cdot 3 \cdot 10$ the prime factorization of 60? Explain.

> **DEFINITION   Prime Factorization**
>
> The **prime factorization** of a number is the factorization in which every factor is a prime number.
>
> *Note:* The order in which the factors are written does not affect the product.

Answers

5. Composite   **6.** Neither   **7.** Prime
8. No. The factor 10 is not a prime number. The prime factorization of 60 is $2 \cdot 2 \cdot 3 \cdot 5$.

Prime factorizations of numbers will be particularly helpful when we add, subtract, multiply, divide, and simplify fractions.

Example 4   **Determining the Prime Factorization of a Number**

Find the prime factorization of 220.

Skill Practice

9. Find the prime factorization of 90.

Solution:

One method to factor a whole number is to make a factor tree. Begin by determining any two numbers that when multiplied equal 220. Then continue factoring each factor until the branches "end" in prime numbers.

TIP: The prime factorization from Example 4 can also be expressed by using exponents as $2^2 \cdot 5 \cdot 11$.

Therefore, the prime factorization of 220 is $2 \cdot 2 \cdot 5 \cdot 11$.

In Example 4, note that the result of a prime factorization does not depend on the original two-number factorization. Similarly, the order in which the factors are written does not affect the product, for example,

$220 = 2 \cdot 2 \cdot 5 \cdot 11$ $220 = 2 \cdot 2 \cdot 5 \cdot 11$ $220 = 11 \cdot 2 \cdot 2 \cdot 5$

Avoiding Mistakes

Make sure that the end of each branch is a prime number.

TIP: You can check the prime factorization of any number by multiplying the factors.

Another technique to find the prime factorization of a number is to divide the number by the smallest known prime factor. Then divide the quotient by its smallest known prime factor. Continue dividing in this fashion until the quotient is a prime number. The prime factorization is the product of divisors and the final quotient. For example,

2 is the smallest prime factor of 220 $\longrightarrow$ 2)220

2 is the smallest prime factor of 110 $\longrightarrow$ 2)110

5 is the smallest prime factor of 55 $\longrightarrow$ 5)55

the last quotient is prime $\longrightarrow$ 11

Therefore, the prime factorization of 220 is $2 \cdot 2 \cdot 5 \cdot 11$ or $2^2 \cdot 5 \cdot 11$.

Answer

9. $2 \cdot 3 \cdot 3 \cdot 5$

Example 5   Determining Prime Factorizations

Find the prime factorization.

a. 198   **b.** 153

Solution:

a. Since 198 is even, we ⟶ 2)198
know it is divisible by 2.   3)99 ⟵ The sum of the digits $9 + 9 = 18$ is
      3)33   divisible by 3.
       11

The prime factorization of 198 is $2 \cdot 3 \cdot 3 \cdot 11$ or $2 \cdot 3^2 \cdot 11$.

b.   3)153
      3)51
       17

The prime factorization of 153 is $3 \cdot 3 \cdot 17$ or $3^2 \cdot 17$.

5. Identifying All Factors of a Whole Number

Sometimes it is necessary to identify all factors (both prime and other) of a number. Take the number 30, for example. A list of all factors of 30 is a list of all whole numbers that divide evenly into 30.

Factors of 30:   1, 2, 3, 5, 6, 10, 15, and 30

Example 6   Listing All Factors of a Number

List all factors of 36.

Solution:

Begin by listing all the two-number factorizations of 36. This can be accomplished by systematically dividing 36 by 1, 2, 3, and so on. Notice, however, that after the product $6 \cdot 6$, the two-number factorizations are repetitious, and we can stop the process.

$1 \cdot 36$

$2 \cdot 18$

$3 \cdot 12$

$4 \cdot 9$

$6 \cdot 6$

$9 \cdot 4$

$12 \cdot 3$

$18 \cdot 2$

$36 \cdot 1$

These products repeat the factorizations above. Therefore, we can stop at $6 \cdot 6$.

TIP: When listing a set of factors, it is not necessary to write the numbers in any specified order. However, in general we list the factors in order from smallest to largest.

The list of all factors of 36 consists of the individual factors in the products. The factors are 1, 2, 3, 4, 6, 9, 12, 18, 36.

Section 2.2 Practice Exercises

Study Skills Exercises

1. In general, 2 to 3 hours of study time per week is needed for each 1 hour per week of class time. Based on the number of hours you are in class this semester, how many hours per week should you be studying?

2. Define the key terms.
   a. Factor
   b. Factorization
   c. Prime number
   d. Composite number
   e. Prime factorization

Review Exercises

For Exercises 3–5, write two fractions, one representing the shaded area and one representing the unshaded area.

3.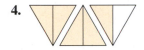

4.

5.

6. Write a fraction with numerator 6 and denominator 5. Is this fraction proper or improper?

7. Write a fraction with denominator 12 and numerator 7. Is this fraction proper or improper?

8. Write a fraction with denominator 6 and numerator 6. Is this fraction proper or improper?

9. Write the improper fraction $\frac{23}{5}$ as a mixed number.   10. Write the mixed number $6\frac{2}{7}$ as an improper fraction.

Objective 1: Factors and Factorization

For Exercises 11–14, find two different factorizations of each number. (Answers may vary.) (See Example 1.)

11. 8   12. 20   13. 24   14. 14

15. Find two factors whose product is the number in the top row and whose sum is the number in the bottom row. The first column is done for you as an example.

Product	36	42	30	15	81
Factor	12				
Factor	3				
Sum	15	13	31	16	30

16. Find two factors whose product is the number in the top row and whose difference is the number in the bottom row. The first column is done for you as an example.

Product	36	42	45	72	24
Factor	9				
Factor	4				
Difference	5	1	12	14	5

Objective 2: Divisibility Rules

17. State the divisibility rule for dividing by 2.

18. State the divisibility rule for dividing by 10.

19. State the divisibility rule for dividing by 3.

20. State the divisibility rule for dividing by 5.

For Exercises 21–28, determine if the number is divisible by   **a.** 2   **b.** 3   **c.** 5   **d.** 10

(See Example 2.)

21. 45

22. 100

23. 137

24. 241

25. 108

26. 1040

27. 3140

28. 2115

29. Ms. Berglund has 28 students in her class. Can she distribute a package of 84 candies evenly to her students?

30. Mr. Blankenship has 22 students in an algebra class. He has 110 sheets of graph paper. Can he distribute the graph paper evenly among his students?

Objective 3: Prime and Composite Numbers

For Exercises 31–46, determine whether the number is prime, composite, or neither. (See Example 3.)

31. 7

32. 17

33. 10

34. 21

35. 51

36. 57

37. 23

38. 31

39. 1

40. 0

41. 121

42. 69

43. 19

44. 29

45. 39

46. 49

47. Are there any whole numbers that are not prime or composite? If so, list them.

48. True or false? The square of any prime number is also a prime number.

49. True or false? All odd numbers are prime.

50. True or false? All even numbers are composite.

51. One method for finding prime numbers is the *sieve of Eratosthenes*. The natural numbers from 2 to 50 are shown in the table. Start at the number 2 (the smallest prime number). Leave the number 2 and cross out every second number after the number 2. This will eliminate all numbers that are multiples of 2. Then go back to the beginning of the chart and leave the number 3, but cross out every third number after the number 3 (thus eliminating the multiples of 3). Begin at the next open number and continue this process. The numbers that remain are prime numbers. Use this process to find the prime numbers less than 50.

	2	3	4	5	6	7	8	9	10
11	12	13	14	15	16	17	18	19	20
21	22	23	24	25	26	27	28	29	30
31	32	33	34	35	36	37	38	39	40
41	42	43	44	45	46	47	48	49	50

52. Use the sieve of Eratosthenes to find the prime numbers less than 80.

	2	3	4	5	6	7	8	9	10
11	12	13	14	15	16	17	18	19	20
21	22	23	24	25	26	27	28	29	30
31	32	33	34	35	36	37	38	39	40
41	42	43	44	45	46	47	48	49	50
51	52	53	54	55	56	57	58	59	60
61	62	63	64	65	66	67	68	69	70
71	72	73	74	75	76	77	78	79	80

Objective 4: Prime Factorization

For Exercises 53–56, determine whether or not the factorization represents the prime factorization. If not, explain why.

53. $36 = 2 \cdot 2 \cdot 9$ **54.** $48 = 2 \cdot 3 \cdot 8$ **55.** $210 = 5 \cdot 2 \cdot 7 \cdot 3$ **56.** $126 = 3 \cdot 7 \cdot 3 \cdot 2$

For Exercises 57–68, find the prime factorization. **(See Examples 4 and 5.)**

57. 70   **58.** 495   **59.** 260   **60.** 175

61. 147   **62.** 102   **63.** 138   **64.** 231

65. 616   **66.** 364   **67.** 47   **68.** 41

Objective 5: Identifying All Factors of a Whole Number

For Exercises 69–76, list all the factors of the number. **(See Example 6.)**

69. 12   **70.** 18   **71.** 32   **72.** 55

73. 81   **74.** 60   **75.** 48   **76.** 72

Expanding Your Skills

For Exercises 77–80, determine whether the number is divisible by 4. Use the following divisibility rule: A whole number is divisible by 4 if the number formed by its last two digits is divisible by 4.

77. 230   **78.** 1046   **79.** 4616   **80.** 10,264

For Exercises 81–84, determine whether the number is divisible by 8. Use the following divisibility rule: A whole number is divisible by 8 if the number formed by its last three digits is divisible by 8.

81. 1032   **82.** 2520   **83.** 17,126   **84.** 25,058

For Exercises 85–88, determine whether the number is divisible by 9. Use the following divisibility rule: A whole number is divisible by 9 if the sum of its digits is divisible by 9.

85. 396   **86.** 414   **87.** 8453   **88.** 1587

For Exercises 89–92, determine whether the number is divisible by 6. Use the following divisibility rule: A whole number is divisible by 6 if it is divisible by both 2 and 3 (use the divisibility rules for 2 and 3 together).

89. 522   **90.** 546   **91.** 5917   **92.** 6394

Section 2.3 **Simplifying Fractions to Lowest Terms**

Objectives

1. **Equivalent Fractions**
2. **Simplifying Fractions to Lowest Terms**
3. **Applications of Simplifying Fractions**

1. Equivalent Fractions

The fractions $\frac{3}{6}$, $\frac{2}{4}$, and $\frac{1}{2}$ all represent the same portion of a whole. See Figure 2-5. Therefore, we say that the fractions are *equivalent*.

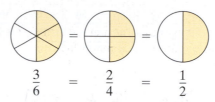

$$\frac{3}{6} \quad = \quad \frac{2}{4} \quad = \quad \frac{1}{2}$$

Figure 2-5

One method to show that two fractions are equivalent is to calculate their cross products. For example, to show that $\frac{3}{6} = \frac{2}{4}$, we have

$$\frac{3}{6} \diagdown\diagup \frac{2}{4}$$

$$3 \times 4 \overset{?}{=} 6 \times 2$$

$$12 = 12 \qquad \text{Yes. The fractions are equivalent.}$$

Avoiding Mistakes

The test to determine whether two fractions are equivalent is not the same process as multiplying fractions. Multiplying of fractions is covered in Section 2.4.

Skill Practice

Fill in the blank ☐ with = or ≠.

1. $\dfrac{13}{24} \;\square\; \dfrac{6}{11}$

2. $\dfrac{9}{4} \;\square\; \dfrac{54}{24}$

Example 1 **Determining Whether Two Fractions Are Equivalent**

Fill in the blank ☐ with = or ≠.

a. $\dfrac{18}{39} \;\square\; \dfrac{6}{13}$ b. $\dfrac{5}{7} \;\square\; \dfrac{7}{9}$

Solution:

a. $\dfrac{18}{39} \diagdown\diagup \dfrac{6}{13}$

$18 \times 13 \overset{?}{=} 39 \times 6$

$234 = 234$

Therefore, $\dfrac{18}{39} \boxed{=} \dfrac{6}{13}$.

b. $\dfrac{5}{7} \diagdown\diagup \dfrac{7}{9}$

$5 \times 9 \overset{?}{=} 7 \times 7$

$45 \neq 49$

Therefore, $\dfrac{5}{7} \boxed{\neq} \dfrac{7}{9}$.

2. Simplifying Fractions to Lowest Terms

In Figure 2-5, we see that $\frac{3}{6}$, $\frac{2}{4}$, and $\frac{1}{2}$ all represent equal quantities. However, the fraction $\frac{1}{2}$ is said to be in **lowest terms** because the numerator and denominator share no common factors other than 1.

To simplify a fraction to lowest terms, we apply the following important principle.

Answers

1. ≠   2. =

> **PROPERTY**   **Fundamental Principle of Fractions**
>
> Suppose that a number, c, is a common factor in the numerator and denominator of a fraction. Then
>
> $$\frac{a \cdot c}{b \cdot c} = \frac{a}{b} \cdot \frac{c}{c} = \frac{a}{b} \cdot 1 = \frac{a}{b}$$

To simplify a fraction, we begin by factoring the numerator and denominator into prime factors. This will help identify the common factors.

Example 2   **Simplifying a Fraction to Lowest Terms**

Simplify to lowest terms.

a. $\dfrac{6}{10}$ **b.** $\dfrac{170}{102}$ **c.** $\dfrac{20}{24}$

Skill Practice

Simplify to lowest terms.

3. $\dfrac{15}{35}$ **4.** $\dfrac{26}{195}$ **5.** $\dfrac{150}{105}$

Solution:

a. $\dfrac{6}{10} = \dfrac{3 \cdot 2}{5 \cdot 2}$ Factor the numerator and denominator. Notice that 2 is a common factor.

$= \dfrac{3}{5} \cdot \dfrac{2}{2}$ Apply the fundamental principle of fractions.

$= \dfrac{3}{5} \cdot 1$ Any nonzero number divided by itself is 1.

$= \dfrac{3}{5}$

b. $\dfrac{170}{102} = \dfrac{5 \cdot 2 \cdot 17}{3 \cdot 2 \cdot 17}$ Factor the numerator and denominator.

$= \dfrac{5}{3} \cdot \dfrac{2}{2} \cdot \dfrac{17}{17}$ Apply the fundamental principle of fractions.

$= \dfrac{5}{3} \cdot 1 \cdot 1$ Any nonzero number divided by itself is 1.

$= \dfrac{5}{3}$

c. $\dfrac{20}{24} = \dfrac{5 \cdot 2 \cdot 2}{3 \cdot 2 \cdot 2 \cdot 2}$ Factor the numerator and denominator.

$= \dfrac{5}{3 \cdot 2} \cdot \dfrac{2}{2} \cdot \dfrac{2}{2}$ Apply the fundamental principle of fractions.

$= \dfrac{5}{6} \cdot 1 \cdot 1$

$= \dfrac{5}{6}$

Answers

3. $\dfrac{3}{7}$ **4.** $\dfrac{2}{15}$ **5.** $\dfrac{10}{7}$

In Example 2, we show numerous steps to simplify fractions to lowest terms. However, the process is often made easier. For instance, we sometimes "divide out" common factors, and replace them with the new common factor of 1.

$$\frac{20}{24} = \frac{5 \cdot \overset{1}{2} \cdot \overset{1}{2}}{3 \cdot \underset{1}{2} \cdot \underset{1}{2} \cdot 2} = \frac{5}{6}$$

The largest number that divides evenly into the numerator and denominator is called their **greatest common factor**. By identifying the greatest common factor you can simplify the process even more.

$$\frac{20}{24} = \frac{5 \cdot \overset{1}{4}}{6 \cdot \underset{1}{4}} = \frac{5}{6}$$

Notice that "dividing out" the common factor of 4 has the same effect as dividing the numerator and denominator by 4. This is often done mentally.

$$\frac{\overset{5}{20}}{\underset{6}{24}} = \frac{5}{6} \quad \begin{array}{l} \leftarrow \text{ 20 divided by 4 equals 5.} \\ \leftarrow \text{ 24 divided by 4 equals 6.} \end{array}$$

> **TIP:** Simplifying a fraction is also called reducing a fraction to lowest terms. For example, the simplified (or reduced) form of $\frac{20}{24}$ is $\frac{5}{6}$.

Example 3   **Simplifying Fractions to Lowest Terms**

Simplify the fraction. Write the answer as a fraction or whole number.

a. $\frac{110}{99}$ **b.** $\frac{75}{25}$ **c.** $\frac{12}{60}$

Solution:

a. $\frac{110}{99} = \frac{10 \cdot \overset{1}{11}}{9 \cdot \underset{1}{11}} = \frac{10}{9}$ The greatest common factor in the numerator and denominator is 11.

Or alternatively: $\dfrac{\overset{10}{110}}{\underset{9}{99}} = \dfrac{10}{9} \quad \begin{array}{l} \leftarrow \text{ 110 divided by 11 equals 10.} \\ \leftarrow \text{ 99 divided by 11 equals 9.} \end{array}$

b. $\frac{75}{25} = \frac{3 \cdot \overset{1}{25}}{1 \cdot \underset{1}{25}} = \frac{3}{1} = 3$ The greatest common factor in the numerator and denominator is 25.

Or alternatively: $\dfrac{\overset{3}{75}}{\underset{1}{25}} = \dfrac{3}{1} \quad \begin{array}{l} \leftarrow \text{ 75 divided by 25 equals 3.} \\ \leftarrow \text{ 25 divided by 25 equals 1.} \end{array}$

$= 3$

> **TIP:** Recall that any fraction of the form $\frac{n}{1} = n$. Therefore, $\frac{3}{1} = 3$.

c. $\frac{12}{60} = \frac{1 \cdot \overset{1}{12}}{5 \cdot \underset{1}{12}} = \frac{1}{5}$ The greatest common factor in the numerator and denominator is 12.

Or alternatively: $= \dfrac{\overset{1}{12}}{\underset{5}{60}} = \dfrac{1}{5} \quad \begin{array}{l} \leftarrow \text{ 12 divided by 12 equals 1.} \\ \leftarrow \text{ 60 divided by 12 equals 5.} \end{array}$

Avoiding Mistakes

Suppose that you do not recognize the *greatest* common factor in the numerator and denominator. You can still divide by *any* common factor. However, you will have to repeat this process more than once to simplify the fraction completely. For instance, consider the fraction from Example 3(c).

$$\frac{\overset{2}{\cancel{12}}}{\underset{10}{\cancel{60}}} = \frac{2}{10}$$ Dividing by the common factor of 6 leaves a fraction that can be simplified further.

$$= \frac{\overset{1}{\cancel{2}}}{\underset{5}{\cancel{10}}} = \frac{1}{5}$$ Divide again, this time by 2. The fraction is now simplified completely because there are no other common factors in the numerator and denominator.

Example 4   **Simplifying Fractions by 10, 100, and 1000**

Simplify each fraction to lowest terms by first reducing by 10, 100, or 1000. Write the answer as a fraction.

a. $\dfrac{170}{30}$ **b.** $\dfrac{2500}{7500}$ **c.** $\dfrac{5000}{130{,}000}$

Solution:

a. $\dfrac{170}{30} = \dfrac{17\cancel{0}}{3\cancel{0}}$ Both 170 and 30 are divisible by 10. "Strike through" one zero.

$= \dfrac{17}{3}$ The fraction $\frac{17}{3}$ is simplified completely.

b. $\dfrac{2500}{7500} = \dfrac{25\cancel{00}}{75\cancel{00}}$ Both 2500 and 7500 are divisible by 100. Strike through two zeros.

$= \dfrac{\overset{1}{\cancel{25}}}{\underset{3}{\cancel{75}}}$ Simplify further.

$= \dfrac{1}{3}$

c. $\dfrac{5000}{130{,}000} = \dfrac{5\cancel{000}}{130{,}\cancel{000}}$ The numbers 5000 and 130,000 are both divisible by 1000. Strike through three zeros.

$= \dfrac{\overset{1}{\cancel{5}}}{\underset{26}{\cancel{130}}}$ Simplify further.

$= \dfrac{1}{26}$

Skill Practice

Simplify to lowest terms by first reducing by 10, 100, or 1000.

9. $\dfrac{630}{190}$ **10.** $\dfrac{1300}{52{,}000}$

11. $\dfrac{21{,}000}{35{,}000}$

Avoiding Mistakes

The "strike through" method only works for the digit 0 at the *end* of the numerator and denominator.

Concept Connections

12. How many zeros may be eliminated from the numerator and denominator of the fraction $\frac{430{,}000}{154{,}000{,}000}$?

Answers

9. $\dfrac{63}{19}$ **10.** $\dfrac{1}{40}$ **11.** $\dfrac{3}{5}$

12. Four zeros; the numerator and denominator are both divisible by 10,000.

3. Applications of Simplifying Fractions

Example 5	Simplifying Fractions in an Application

Madeleine got 28 out of 35 problems correct on an algebra exam. David got 27 out of 45 questions correct on a different algebra exam.

a. What fractional part of the exam did each student answer correctly?

b. Which student performed better?

Solution:

a. Fractional part correct for Madeleine:

$$\frac{28}{35} \quad \text{or equivalently} \quad \frac{\overset{4}{28}}{\underset{5}{35}} = \frac{4}{5}$$

Fractional part correct for David:

$$\frac{27}{45} \quad \text{or equivalently} \quad \frac{\overset{3}{27}}{\underset{5}{45}} = \frac{3}{5}$$

b. From the simplified form of each fraction, we see that Madeleine performed better because $\frac{4}{5} > \frac{3}{5}$. That is, 4 parts out of 5 is greater than 3 parts out of 5. This is also easily verified on a number line.

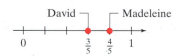

Section 2.3 Practice Exercises

Study Skills Exercises

1. Sometimes, test anxiety can be greatly reduced by adequate preparation and practice. List some places in the text where you can find extra problems for practice.

2. Define the key terms.
   a. Lowest terms
   b. Greatest common factor

Review Exercises

For Exercises 3–10, write the prime factorization for each number.

3. 145   4. 114   5. 92   6. 153

7. 85   8. 120   9. 195   10. 180

Objective 1: Equivalent Fractions

For Exercises 11–14, shade the second figure so that it expresses a fraction equivalent to the first figure.

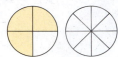 11.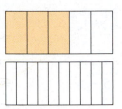

12.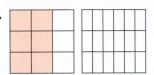

13.

14.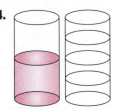

15. True or false? The fractions $\frac{4}{5}$ and $\frac{5}{4}$ are equivalent.

16. In your own words, explain the concept of equivalent fractions.

For Exercises 17–24, determine if the fractions are equivalent. Then fill in the blank with either = or ≠.
(See Example 1.)

17. $\frac{2}{3} \square \frac{3}{5}$

18. $\frac{1}{4} \square \frac{2}{9}$

19. $\frac{1}{2} \square \frac{3}{6}$

20. $\frac{6}{16} \square \frac{3}{8}$

21. $\frac{12}{16} \square \frac{3}{4}$

22. $\frac{4}{5} \square \frac{12}{15}$

23. $\frac{8}{9} \boxed{>} \frac{20}{27}$

24. $\frac{5}{6} \square \frac{12}{18}$

Objective 2: Simplifying Fractions to Lowest Terms

For Exercises 25–52, simplify the fraction to lowest terms. Write the answer as a fraction or a whole number.
(See Examples 2–3.)

25. $\frac{12}{24}$

26. $\frac{15}{18}$

27. $\frac{6}{18}$

28. $\frac{21}{24}$

29. $\frac{36}{20}$

30. $\frac{49}{42}$

31. $\frac{15}{12}$

32. $\frac{30}{25}$

33. $\frac{20}{25}$

34. $\frac{8}{16}$

35. $\frac{14}{14}$

36. $\frac{8}{8}$

37. $\frac{50}{25}$

38. $\frac{24}{6}$

39. $\frac{9}{9}$

40. $\frac{2}{2}$

41. $\frac{105}{140}$

42. $\frac{84}{126}$

43. $\frac{33}{11}$

44. $\frac{65}{5}$

45. $\dfrac{77}{110}$ **46.** $\dfrac{85}{153}$ **47.** $\dfrac{130}{150}$ **48.** $\dfrac{70}{120}$

49. $\dfrac{385}{195}$ **50.** $\dfrac{39}{130}$ **51.** $\dfrac{34}{85}$ **52.** $\dfrac{69}{92}$

For Exercises 53–60, use the order of operations to simplify.

53. $\dfrac{6-2}{10+4}$ **54.** $\dfrac{9-1}{15+3}$ **55.** $\dfrac{5-5}{7-2}$ **56.** $\dfrac{11-11}{4+7}$

57. $\dfrac{7-2}{5-5}$ **58.** $\dfrac{4+7}{11-11}$ **59.** $\dfrac{8-2}{8+2}$ **60.** $\dfrac{15+3}{15-3}$

For Exercises 61–68, simplify to lowest terms by first reducing the powers of 10. **(See Example 4.)**

61. $\dfrac{120}{160}$ **62.** $\dfrac{720}{800}$ **63.** $\dfrac{3000}{1800}$ **64.** $\dfrac{2000}{1500}$

65. $\dfrac{42,000}{22,000}$ **66.** $\dfrac{50,000}{65,000}$ **67.** $\dfrac{5100}{30,000}$ **68.** $\dfrac{9800}{28,000}$

Objective 3: Applications of Simplifying Fractions

69. André tossed a coin 48 times and heads came up 20 times. What fractional part of the tosses came up heads? What fractional part came up tails?

70. At Pizza Company, Lee made 70 pizzas one day. There were 105 pizzas sold that day. What fraction of the pizzas did Lee make?

71. a. What fraction of the alphabet is made up of vowels? (Include the letter y as a vowel, not a consonant.)

 b. What fraction of the alphabet is made up of consonants?

72. Of the 88 constellations that can be seen in the night sky, 12 are associated with astrological horoscopes. The names of as many as 36 constellations are associated with animals or mythical creatures.

 a. Of the 88 constellations, what fraction is associated with horoscopes?

 b. What fraction of the constellations have names associated with animals or mythical creatures?

73. Jonathan and Jared both sold candy bars for a fund-raiser. Jonathan sold 25 of his 35 candy bars, and Jared sold 24 of his 28 candy bars. **(See Example 5.)**

    a. What fractional part of his total number of candy bars did each boy sell?

    b. Which boy sold the greater fractional part?

75. Raymond read 720 pages of a 792-page book. His roommate, Travis, read 540 pages from a 660-page book.

    a. What fractional part of the book did each person read?

    b. Which of the roommates read a greater fraction of his book?

74. Lisa and Lynette are taking online courses. Lisa has completed 14 out of 16 assignments in her course while Lynette has completed 15 out of 24 assignments.

    a. What fractional part of her total number of assignments did each woman complete?

    b. Which woman has completed more of her course?

76. Mr. Bishop and Ms. Waymire both gave exams today. By mid-afternoon, Mr. Bishop had finished grading 16 out of 36 exams, and Ms. Waymire had finished grading 15 out of 27 exams.

    a. What fractional part of her total has Ms. Waymire completed?

    b. What fractional part of his total has Mr. Bishop completed?

77. For a recent year, the population of the United States was reported to be 296,000,000. During the same year, the population of California was 36,458,000.

    a. Round the U.S. population to the nearest hundred million.

    b. Round the population of California to the nearest million.

    c. Using the results from parts (a) and (b), write a simplified fraction showing the portion of the U.S. population represented by California.

78. For a recent year, the population of the United States was reported to be 296,000,000. During the same year, the population of Ethiopia was 75,067,000.

    a. Round the U.S. population to the nearest hundred million.

    b. Round the population of Ethiopia to the nearest million.

    c. Using the results from parts (a) and (b), write a simplified fraction comparing the population of the United States to the population of Ethiopia.

    d. Based on the result from part (c), how many times greater is the U.S. population than the population of Ethiopia?

Expanding Your Skills

79. Write three fractions equivalent to $\frac{3}{4}$.

80. Write three fractions equivalent to $\frac{1}{3}$.

81. Write three fractions equivalent to $\frac{12}{18}$.

82. Write three fractions equivalent to $\frac{80}{100}$.

Calculator Connections

Topic: Simplifying Fractions on a Calculator

Some calculators have a fraction key, $a^{b/c}$. To enter a fraction, follow this example.

Expression: $\dfrac{3}{4}$

Keystrokes: 3 $a^{b/c}$ 4 $=$

Result:

$$\boxed{\qquad 3\lrcorner 4 \qquad}$$

      ↑   ↑
   numerator denominator

To simplify a fraction to lowest terms, follow this example.

Expression: $\dfrac{22}{10}$

Keystrokes: 22 $a^{b/c}$ 10 $=$

Result:

$$\boxed{\qquad 2\lrcorner 1\lrcorner 5 \qquad} = 2\frac{1}{5}$$

     ↑   T
 whole number fraction

To convert to an improper fraction, press 2nd d/c $\boxed{\quad 11\lrcorner 5 \quad} = \dfrac{11}{5}$

Calculator Exercises

For Exercises 83–90, use a calculator to simplify the fractions. Write the answer as a proper or improper fraction.

83. $\dfrac{792}{891}$

84. $\dfrac{728}{784}$

85. $\dfrac{779}{969}$

86. $\dfrac{462}{220}$

87. $\dfrac{493}{510}$

88. $\dfrac{871}{469}$

89. $\dfrac{969}{646}$

90. $\dfrac{713}{437}$

Multiplication of Fractions and Applications

1. Multiplication of Fractions

Suppose Elija takes $\frac{1}{3}$ of a cake and then gives $\frac{1}{2}$ of this portion to his friend Max. Max gets $\frac{1}{2}$ of $\frac{1}{3}$ of the cake. This is equivalent to the expression $\frac{1}{2} \cdot \frac{1}{3}$. See Figure 2-6.

Elija takes $\frac{1}{3}$

Max gets
$\frac{1}{2}$ of $\frac{1}{3} = \frac{1}{6}$

Figure 2-6

From the illustration, the product $\frac{1}{2} \cdot \frac{1}{3} = \frac{1}{6}$. Notice that the product $\frac{1}{6}$ is found by multiplying the numerators and multiplying the denominators. This is true in general to multiply fractions.

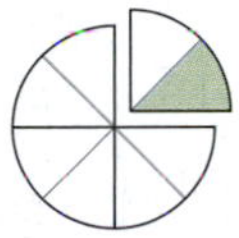

> **PROCEDURE Multiplying Fractions**
>
> To multiply fractions, write the product of the numerators over the product of the denominators. Then simplify the resulting fraction, if possible.
>
> $$\frac{a}{b} \cdot \frac{c}{d} = \frac{a \cdot c}{b \cdot d} \qquad \text{provided } b \text{ and } d \text{ are not equal to } 0$$

Objectives

1. Multiplication of Fractions
2. Fractions and the Order of Operations
3. Area of a Triangle
4. Applications of Multiplying Fractions

Concept Connections

1. What fraction is $\frac{1}{2}$ of $\frac{1}{4}$ of a whole?

Example 1   Multiplying Fractions

Multiply.

a. $\frac{2}{5} \cdot \frac{4}{7}$ **b.** $\frac{8}{3} \times 5$

Skill Practice

Multiply. Write the answer as a fraction.

2. $\frac{2}{3} \cdot \frac{5}{9}$ **3.** $\frac{7}{12} \times 11$

Solution:

a. $\frac{2}{5} \cdot \frac{4}{7} = \frac{2 \cdot 4}{5 \cdot 7} = \frac{8}{35}$ ← Multiply the numerators.
← Multiply the denominators.

Notice that the product $\frac{8}{35}$ is simplified completely because there are no common factors shared by 8 and 35.

b. $\frac{8}{3} \times 5 = \frac{8}{3} \times \frac{5}{1}$ First write the whole number as a fraction.

$= \frac{8 \times 5}{3 \times 1}$ Multiply the numerators. Multiply the denominators.

$= \frac{40}{3}$ The product is not reducible because there are no common factors shared by 40 and 3.

Answers

1. $\frac{1}{8}$ **2.** $\frac{10}{27}$ **3.** $\frac{77}{12}$

Example 2 illustrates a case where the product of fractions must be simplified.

Example 2 **Multiplying and Simplifying Fractions**

Multiply the fraction and simplify if possible.

$$\frac{4}{30} \cdot \frac{5}{14}$$

Solution:

$$\frac{4}{30} \cdot \frac{5}{14} = \frac{4 \cdot 5}{30 \cdot 14} \qquad \text{Multiply the numerators. Multiply the denominators.}$$

$$= \frac{\cancel{20}}{\cancel{420}} \qquad \text{Simplify by first dividing 20 and 420 by 10.}$$

$$= \frac{\overset{1}{\cancel{2}}}{\underset{21}{\cancel{42}}} \qquad \text{Simplify further by dividing 2 and 42 by 2.}$$

$$= \frac{1}{21}$$

It is often easier to simplify *before* multiplying. Consider the product from Example 2.

$$\frac{4}{30} \cdot \frac{5}{14} = \frac{\overset{2}{\cancel{4}}}{\underset{6}{\cancel{30}}} \cdot \frac{\overset{1}{\cancel{5}}}{\underset{7}{\cancel{14}}} \qquad \begin{array}{l}\text{4 and 14 share a common factor of 2.}\\ \text{30 and 5 share a common factor of 5.}\end{array}$$

$$= \frac{\overset{\overset{1}{\cancel{2}}}{\cancel{4}}}{\underset{\underset{3}{\cancel{6}}}{\cancel{30}}} \cdot \frac{\overset{1}{\cancel{5}}}{\underset{7}{\cancel{14}}} \qquad \text{2 and 6 share a common factor of 2.}$$

$$= \frac{1}{21}$$

As a general rule, this method is used most often in the text.

Example 3 **Multiplying and Simplifying Fractions**

Multiply and simplify.

$$\frac{10}{18} \times \frac{21}{55}$$

Solution:

$$\frac{10}{18} \times \frac{21}{55} = \frac{\overset{2}{\cancel{10}}}{\underset{6}{\cancel{18}}} \times \frac{\overset{7}{\cancel{21}}}{\underset{11}{\cancel{55}}} \qquad \begin{array}{l}\text{10 and 55 share a common factor of 5.}\\ \text{18 and 21 share a common factor of 3.}\end{array}$$

$$= \frac{\overset{\overset{1}{\cancel{2}}}{\cancel{10}}}{\underset{\underset{3}{\cancel{6}}}{\cancel{18}}} \times \frac{\overset{7}{\cancel{21}}}{\underset{11}{\cancel{55}}} \qquad \begin{array}{l}\text{We can simplify further because 2}\\ \text{and 6 share a common factor of 2.}\end{array}$$

$$= \frac{7}{33}$$

Example 4 **Multiplying and Simplifying Fractions**

Multiply and simplify. Write the answers as fractions.

a. $6\left(\dfrac{3}{8}\right)$ **b.** $\dfrac{21}{25} \cdot \dfrac{15}{39} \cdot \dfrac{65}{24}$

Solution:

a. $6\left(\dfrac{3}{8}\right) = \dfrac{6}{1} \cdot \dfrac{3}{8}$ Write the whole number as a fraction.

$= \dfrac{\overset{3}{\cancel{6}}}{1} \cdot \dfrac{3}{\underset{4}{\cancel{8}}}$ Reduce before multiplying.

$= \dfrac{9}{4}$ Multiply.

b. $\dfrac{21}{25} \cdot \dfrac{15}{39} \cdot \dfrac{65}{24} = \left(\dfrac{21}{\underset{5}{\cancel{25}}} \cdot \dfrac{\overset{3}{\cancel{15}}}{\underset{13}{\cancel{39}}}\right) \cdot \dfrac{65}{24}$ Multiply the first two fractions.

$= \left(\dfrac{21}{65}\right) \cdot \dfrac{65}{24}$

$= \dfrac{\overset{7}{\cancel{21}}}{\underset{1}{\cancel{65}}} \cdot \dfrac{\overset{1}{\cancel{65}}}{\underset{8}{\cancel{24}}}$ Simplify.

$= \dfrac{7}{8}$

2. Fractions and the Order of Operations

For problems with more than one operation, recall the order of operations.

PROCEDURE Using the Order of Operations

Step 1 Perform all operations inside parentheses first.

Step 2 Simplify any expressions containing exponents or square roots.

Step 3 Perform multiplication or division in the order that they appear from left to right.

Step 4 Perform addition or subtraction in the order that they appear from left to right.

Skill Practice

Simplify. Write the answer as a fraction.

8. $\left(\dfrac{4}{3}\right)^2$ **9.** $\left(\dfrac{6}{5}\cdot\dfrac{1}{12}\right)^3$

Example 5   **Simplifying Expressions**

Simplify.

a. $\left(\dfrac{2}{5}\right)^3$ **b.** $\left(\dfrac{2}{15}\cdot\dfrac{3}{4}\right)^2$

Solution:

a. $\left(\dfrac{2}{5}\right)^3 = \dfrac{2}{5}\cdot\dfrac{2}{5}\cdot\dfrac{2}{5}$ With an exponent of 3, multiply 3 factors of the base.

$= \dfrac{2\cdot2\cdot2}{5\cdot5\cdot5}$ Multiply the numerators. Multiply the denominators.

$= \dfrac{8}{125}$

b. $\left(\dfrac{2}{15}\cdot\dfrac{3}{4}\right)^2 = \left(\dfrac{\overset{1}{2}}{\underset{5}{15}}\cdot\dfrac{\overset{1}{3}}{\underset{2}{4}}\right)^2$ Perform the multiplication within the parentheses. Simplify.

$= \left(\dfrac{1}{10}\right)^2$ Multiply fractions within parentheses.

$= \dfrac{1}{10}\cdot\dfrac{1}{10}$ Square $\frac{1}{10}$ by multiplying $\frac{1}{10}$ times itself.

$= \dfrac{1}{100}$

In Section 1.7 we learned to recognize powers of 10. These are $10^1 = 10$, $10^2 = 100$, and so on. In this section, we learn to recognize the **powers of one-tenth**, that is, $\frac{1}{10}$ raised to a whole-number power. For example, consider the following expressions.

$$\left(\frac{1}{10}\right)^1 = \frac{1}{10}$$

$$\left(\frac{1}{10}\right)^2 = \frac{1}{10}\cdot\frac{1}{10} = \frac{1}{100}$$

$$\left(\frac{1}{10}\right)^3 = \frac{1}{10}\cdot\frac{1}{10}\cdot\frac{1}{10} = \frac{1}{1000}$$

$$\left(\frac{1}{10}\right)^4 = \frac{1}{10}\cdot\frac{1}{10}\cdot\frac{1}{10}\cdot\frac{1}{10} = \frac{1}{10,000}$$

From these examples, we see that a power of one-tenth results in a fraction with a 1 in the numerator. The denominator has a 1 followed by the same number of zeros as the exponent on the base of $\frac{1}{10}$.

3. Area of a Triangle

Recall that the area of a rectangle with length l and width w is given by

$$A = l \times w$$

Answers

8. $\dfrac{16}{9}$ **9.** $\dfrac{1}{1000}$

FORMULA Area of a Triangle

The formula for the area of a triangle is given by $A = \frac{1}{2}bh$, read "one-half base times height."

The value of b is the measure of the base of the triangle. The value of h is the measure of the height of the triangle. The base b can be chosen as the length of any of the sides of the triangle. However, once you have chosen the base, the height must be measured as the shortest distance from the base to the opposite vertex (or point) of the triangle.

Figure 2-7 shows the same triangle with different choices for the base. Figure 2-8 shows a situation in which the height must be drawn "outside" the triangle. In such a case, notice that the height is drawn down to an imaginary extension of the base line.

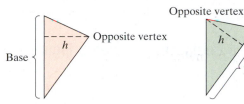

Figure 2-7

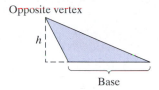

Figure 2-8

Example 6 **Finding the Area of a Triangle**

Find the area of the triangle.

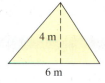

Solution:

$b = 6\,\text{m}$ and $h = 4\,\text{m}$ Identify the measure of the base and the height.

$A = \dfrac{1}{2}\,bh$

$= \dfrac{1}{2}(6\,\text{m})(4\,\text{m})$ Apply the formula for the area of a triangle.

$= \dfrac{1}{2}\left(\dfrac{6}{1}\,\text{m}\right)\left(\dfrac{4}{1}\,\text{m}\right)$ Write the whole numbers as fractions.

$= \dfrac{1}{\overset{}{2}_{1}}\left(\dfrac{\overset{3}{\cancel{6}}}{1}\,\text{m}\right)\left(\dfrac{4}{1}\,\text{m}\right)$ Simplify.

$= \dfrac{12}{1}\,\text{m}^2$ Multiply numerators. Multiply denominators.

$= 12\,\text{m}^2$ The area of the triangle is 12 square meters (m^2).

Skill Practice

Find the area of the triangle.

10.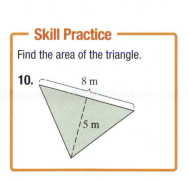

Answer

10. $20\,\text{m}^2$

Skill Practice

11. Find the area of the triangle.

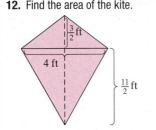

2 ft

$\frac{4}{3}$ ft

Example 7   **Finding the Area of a Triangle**

Find the area of the triangle.

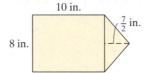

$\frac{3}{4}$ ft

$\frac{5}{3}$ ft

Solution:

$$b = \frac{5}{3} \text{ ft} \quad \text{and} \quad h = \frac{3}{4} \text{ ft} \qquad \text{Identify the measure of the base and the height.}$$

$$A = \frac{1}{2} bh$$

$$= \frac{1}{2} \left(\frac{5}{3} \text{ ft} \right) \left(\frac{3}{4} \text{ ft} \right) \qquad \text{Apply the formula for the area of a triangle.}$$

$$= \frac{1}{2} \left(\frac{5}{\overset{}{\underset{1}{3}}} \text{ ft} \right) \left(\frac{\overset{1}{3}}{4} \text{ ft} \right) \qquad \text{Simplify.}$$

$$= \frac{5}{8} \text{ ft}^2 \qquad \text{The area of the triangle is } \frac{5}{8} \text{ square feet (ft}^2\text{).}$$

Skill Practice

12. Find the area of the kite.

$\frac{3}{2}$ ft

4 ft

$\frac{11}{2}$ ft

Example 8   **Find the Area of a Composite Geometric Figure**

Find the area.

10 in.

$\frac{7}{2}$ in.

8 in.

Solution:

The total area is the sum of the areas of the rectangular region and the triangular region. That is,

Total area =

10 in.

8 in.

$+$

8 in.

$\frac{7}{2}$ in.

(area of rectangle) + (area of triangle)
$(l \times w)$ $(\frac{1}{2}bh)$

Area of rectangle: (10 in.)(8 in.)   Area of triangle: $\frac{1}{2}$ (8 in.) $\left(\frac{7}{2} \text{ in.} \right)$

$= 80 \text{ in.}^2$

$$= \frac{1}{\underset{1}{2}} \left(\frac{\overset{4}{8}}{1} \text{ in.} \right) \left(\frac{7}{2} \text{ in.} \right)$$

$$= \left(\frac{\overset{2}{4}}{1} \text{ in.} \right) \left(\frac{7}{\underset{1}{2}} \text{ in.} \right)$$

$$= \frac{14}{1} \text{ in.}^2$$

$$= 14 \text{ in.}^2$$

Total area = 80 in.^{2} + 14 in.^{2} = 94 in.^{2}

The total area of the region is 94 square inches (in.^{2}).

Answers

11. $\frac{4}{3}$ or $1\frac{1}{3}$ ft^2 **12.** 14 ft^2

4. Applications of Multiplying Fractions

Example 9 Multiplying Fractions in an Application

The population of Texas comprises roughly $\frac{3}{40}$ of the population of the United States. If the U.S. population is approximately 296,000,000, approximate the population of Texas.

Solution:

We must find $\frac{3}{40}$ of the U.S. population. This translates to

$$\frac{3}{40} \cdot 296,000,000 = \frac{3}{40} \cdot \frac{296,000,000}{1}$$

$$= \frac{888,000,000}{40} \quad \longleftarrow \text{ Multiply the numerators.}$$
$$\quad \longleftarrow \text{ Multiply the denominators.}$$

$$= \frac{888,000,0\cancel{0}}{4\cancel{0}} \quad \text{Simplify by a factor of 10.}$$

$$= \frac{88,800,000}{4} \quad \begin{array}{l}\text{Simplify by dividing:}\\ 88,800,000 \div 4.\end{array}$$

$$= 22,200,000 \quad \text{Divide.}$$

The population of Texas is approximately 22,200,000.

Skill Practice

13. Find the population of Tennessee if it is approximately $\frac{1}{50}$ of the U.S. population. Assume that the U.S. population is approximately 296,000,000.

TIP: To take a fractional portion of a quantity, *multiply* the quantity by the fraction.

Answer
13. 5,920,000

Section 2.4 Practice Exercises

Boost your GRADE at ALEKS.com!

ALEKS version 3.0

• Practice Problems
• Self-Tests
• NetTutor
• e-Professors
• Videos

Study Skills Exercises

1. Write down the page number(s) for the Chapter Summary for this chapter. _____
   Describe one way in which you can use the summary found at the end of each chapter.

2. Define the key term **powers of one-tenth**.

Review Exercises

For Exercises 3–6, identify the numerator and the denominator. Then simplify the fraction to lowest terms.

3. $\frac{10}{14}$ 4. $\frac{32}{36}$ 5. $\frac{25}{15}$ 6. $\frac{2100}{7000}$

Objective 1: Multiplication of Fractions

7. Shade the portion of the figure that represents $\frac{1}{6}$ of $\frac{1}{2}$.

8. Shade the portion of the figure that represents $\frac{1}{2}$ of $\frac{1}{2}$.

9. Shade the portion of the figure that represents $\frac{1}{4}$ of $\frac{1}{4}$.

10. Shade the portion of the figure that represents $\frac{1}{3}$ of $\frac{1}{4}$.

11. Find $\frac{1}{2}$ of $\frac{1}{4}$.

12. Find $\frac{2}{3}$ of $\frac{1}{5}$.

13. Find $\frac{3}{4}$ of 8.

14. Find $\frac{2}{5}$ of 20.

For Exercises 15–26, multiply the fractions. Write the answer as a fraction. **(See Example 1.)**

15. $\frac{1}{2} \times \frac{3}{8}$

16. $\frac{2}{3} \times \frac{1}{3}$

17. $\frac{14}{9} \cdot \frac{1}{9}$

18. $\frac{1}{8} \cdot \frac{9}{8}$

19. $\left(\frac{12}{7}\right)\left(\frac{2}{5}\right)$

20. $\left(\frac{9}{10}\right)\left(\frac{7}{4}\right)$

21. $8 \cdot \left(\frac{1}{11}\right)$

22. $3 \cdot \left(\frac{2}{7}\right)$

23. $\frac{4}{5} \cdot 6$

24. $\frac{5}{8} \cdot 5$

25. $\frac{13}{9} \times \frac{5}{4}$

26. $\frac{6}{5} \times \frac{7}{5}$

For Exercises 27–50, multiply the fractions and simplify to lowest terms. Write the answer as a fraction or whole number. **(See Examples 2–4.)**

27. $\frac{2}{9} \times \frac{3}{5}$

28. $\frac{1}{8} \times \frac{4}{7}$

29. $\frac{5}{6} \times \frac{3}{4}$

30. $\frac{7}{12} \times \frac{18}{5}$

31. $\frac{21}{5} \cdot \frac{25}{12}$

32. $\frac{16}{25} \cdot \frac{15}{32}$

33. $\frac{24}{15} \cdot \frac{5}{3}$

34. $\frac{49}{24} \cdot \frac{6}{7}$

35. $\left(\frac{6}{11}\right)\left(\frac{22}{15}\right)$

36. $\left(\frac{12}{45}\right)\left(\frac{5}{4}\right)$

37. $\left(\frac{17}{9}\right)\left(\frac{72}{17}\right)$

38. $\left(\frac{39}{11}\right)\left(\frac{11}{13}\right)$

39. $\frac{21}{4} \cdot \frac{16}{7}$

40. $\frac{85}{6} \cdot \frac{12}{10}$

41. $12 \times \frac{15}{42}$

42. $4 \times \frac{8}{92}$

43. $\frac{9}{15} \times \frac{16}{3} \times \frac{25}{8}$

44. $\frac{49}{8} \times \frac{4}{5} \times \frac{20}{7}$

45. $\frac{5}{2} \times \frac{10}{21} \times \frac{7}{5}$

46. $\frac{55}{9} \times \frac{18}{32} \times \frac{24}{11}$

47. $\frac{7}{10} \cdot \frac{3}{28} \cdot 5$

48. $\frac{11}{18} \cdot \frac{2}{20} \cdot 15$

49. $\frac{100}{49} \times 21 \times \frac{14}{25}$

50. $\frac{38}{22} \times 11 \times \frac{5}{19}$

Objective 2: Fractions and the Order of Operations

For Exercises 51–54, simplify the powers of $\frac{1}{10}$.

51. $\left(\frac{1}{10}\right)^3$

52. $\left(\frac{1}{10}\right)^4$

53. $\left(\frac{1}{10}\right)^6$

54. $\left(\frac{1}{10}\right)^9$

For Exercises 55–66, simplify. Write the answer as a fraction or whole number. **(See Example 5.)**

55. $\left(\dfrac{1}{9}\right)^2$ **56.** $\left(\dfrac{1}{4}\right)^2$ **57.** $\left(\dfrac{3}{2}\right)^3$ **58.** $\left(\dfrac{4}{3}\right)^3$

59. $\left(4 \cdot \dfrac{3}{4}\right)^3$ **60.** $\left(5 \cdot \dfrac{2}{5}\right)^3$ **61.** $\left(\dfrac{1}{9} \cdot \dfrac{3}{5}\right)^2$ **62.** $\left(\dfrac{10}{3} \cdot \dfrac{1}{100}\right)^2$

63. $\dfrac{1}{3} \cdot \left(\dfrac{21}{4} \cdot \dfrac{8}{7}\right)$ **64.** $\dfrac{1}{6} \cdot \left(\dfrac{24}{5} \cdot \dfrac{30}{8}\right)$ **65.** $\dfrac{16}{9} \cdot \left(\dfrac{1}{2}\right)^3$ **66.** $\dfrac{28}{6} \cdot \left(\dfrac{3}{2}\right)^2$

Objective 3: Area of a Triangle

For Exercises 67–70, label the height with h and the base with b, as shown in the figure.

67.   **68.**   **69.**   **70.**

For Exercises 71–80, find the area of the figure. **(See Examples 6–7.)**

71.
8 cm
11 cm

72.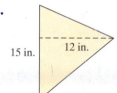
15 in.   12 in.

73.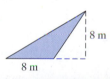
8 m
8 m

74.
1 ft
$\frac{7}{4}$ ft

75.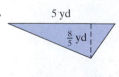
5 yd
$\frac{8}{5}$ yd

76.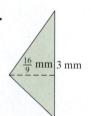
$\frac{16}{9}$ mm 3 mm

77.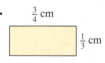
$\frac{3}{4}$ cm
$\frac{1}{3}$ cm

78.
3 m
$\frac{8}{3}$ m

79.
$\frac{15}{16}$ in.
$\frac{13}{16}$ in.

80.
$\frac{3}{4}$ ft
$\frac{23}{24}$ ft

For Exercises 81–84, find the area of the shaded region. **(See Example 8.)**

81.

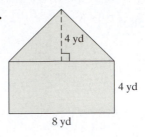

4 yd

4 yd

8 yd

82.

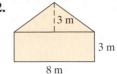

3 m

3 m

8 m

83.

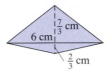

$\frac{7}{3}$ cm

6 cm

$\frac{2}{3}$ cm

84.

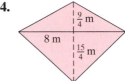

$\frac{9}{4}$ m

8 m

$\frac{15}{4}$ m

Objective 4: Applications of Multiplying Fractions

85. Ms. Robbins' car holds 16 gallons (gal) of gas. If the fuel gauge indicates that there is $\frac{5}{8}$ of a tank left, how many gallons of gas are left in the tank? **(See Example 9.)**

86. Land in a rural part of Bowie County, Texas, sells for $11,000 per acre. If Ms. Anderson purchased $\frac{3}{4}$ acre, how much did it cost?

87. Jim has half a pizza left over from dinner. If he eats $\frac{1}{4}$ of this for breakfast, how much pizza did he eat for breakfast?

88. In a certain sample of individuals, $\frac{2}{5}$ are known to have blood type O. Of the individuals with blood type O, $\frac{1}{4}$ are Rh-negative. What fraction of the individuals in the sample have O negative blood?

89. One week, Nielsen Media Research reported that a new episode of *C.S.I.* was the second most watched television show with 9,825,000 total viewers. The following week, a repeat episode was shown, but only $\frac{2}{3}$ of the viewers returned to watch. How many viewers tuned in to watch the repeat episode?

90. Nancy spends $\frac{3}{4}$ hour 3 times a day walking and playing with her dog. What is the total time she spends walking and playing with the dog each day?

91. The Bishop Gaming Center hosts a football pool. There is $1200 in prize money. The first-place winner receives $\frac{2}{3}$ of the prize money. The second-place winner receives $\frac{1}{4}$ of the prize money, and the third-place winner receives $\frac{1}{12}$ of the prize money. How much money does each person get?

92. Frankie's lawn measures 40 yd by 36 yd. In the morning he mowed $\frac{2}{3}$ of the lawn. How many square yards of lawn did he already mow? How much is left to be mowed?

Expanding Your Skills

93. Evaluate.

    a. $\left(\dfrac{1}{6}\right)^2$ **b.** $\sqrt{\dfrac{1}{36}}$

94. Evaluate.

    a. $\left(\dfrac{2}{7}\right)^2$ **b.** $\sqrt{\dfrac{4}{49}}$

For Exercises 95–98, evaluate the square roots.

95. $\sqrt{\dfrac{1}{25}}$ **96.** $\sqrt{\dfrac{1}{100}}$ **97.** $\sqrt{\dfrac{64}{81}}$ **98.** $\sqrt{\dfrac{9}{4}}$

99. Find the next number in the sequence: $\frac{1}{2}, \frac{1}{4}, \frac{1}{8}, \frac{1}{16},$ _____

100. Find the next number in the sequence: $\frac{2}{3}, \frac{2}{9}, \frac{2}{27},$ _____

101. Which is greater, $\frac{1}{2}$ of $\frac{1}{8}$ or $\frac{1}{8}$ of $\frac{1}{2}$?   **102.** Which is greater, $\frac{2}{3}$ of $\frac{1}{4}$ or $\frac{1}{4}$ of $\frac{2}{3}$?

Division of Fractions and Applications

Section 2.5

1. Reciprocal of a Fraction

Objectives

Two numbers whose product is 1 are said to be *reciprocals* of each other. For example, consider the product of $\frac{3}{8}$ and $\frac{8}{3}$.

1. Reciprocal of a Fraction
2. Division of Fractions
3. Order of Operations
4. Applications of Multiplication and Division of Fractions

$$\frac{3}{8} \cdot \frac{8}{3} = \frac{\overset{1}{\cancel{3}}}{\underset{1}{\cancel{8}}} \cdot \frac{\overset{1}{\cancel{8}}}{\underset{1}{\cancel{3}}} = 1$$

Because the product equals 1, we say that $\frac{3}{8}$ is the reciprocal of $\frac{8}{3}$ and vice versa.

    To divide fractions, first we need to learn how to find the reciprocal of a fraction.

Concept Connections

Fill in the blank.

    1. The product of a number and its reciprocal is _____.

> **PROCEDURE** Finding the Reciprocal of a Fraction
>
> To find the **reciprocal** of a nonzero fraction, interchange the numerator and denominator of the fraction. Thus, the reciprocal of $\frac{a}{b}$ is $\frac{b}{a}$ (provided $a \neq 0$ and $b \neq 0$).

Answer

1. 1

Example 1   **Finding Reciprocals**

Find the reciprocal.

a. $\dfrac{2}{5}$ **b.** $\dfrac{1}{9}$ **c.** 5   **d.** 0

Solution:

a. The reciprocal of $\frac{2}{5}$ is $\frac{5}{2}$.

b. The reciprocal of $\frac{1}{9}$ is $\frac{9}{1}$, or 9.

c. First write the whole number 5 as the improper fraction $\frac{5}{1}$. The reciprocal of $\frac{5}{1}$ is $\frac{1}{5}$.

d. The number 0 has no reciprocal because $\frac{1}{0}$ is undefined.

2. Division of Fractions

To understand the division of fractions, we compare it to the division of whole numbers. The statement $6 \div 2$ asks, "How many groups of 2 can be found among 6 wholes?" The answer is 3.

$$6 \div 2 = 3$$

In fractional form, the statement $6 \div 2 = 3$ can be written as $\frac{6}{2} = 3$. This result can also be found by multiplying.

$$6 \cdot \frac{1}{2} = \frac{6}{1} \cdot \frac{1}{2} = \frac{6}{2} = 3$$

That is, to divide by 2 is equivalent to multiplying by the reciprocal $\frac{1}{2}$.

In general, to divide two nonzero numbers we can multiply the dividend by the reciprocal of the divisor. This is how we divide by a fraction.

PROCEDURE   **Dividing Fractions**

To divide two fractions, multiply the dividend (the "first" fraction) by the reciprocal of the divisor (the "second" fraction).

The process to divide fractions can be written symbolically as

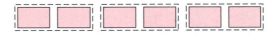

$$\frac{a}{b} \div \frac{c}{d} = \frac{a}{b} \cdot \frac{d}{c} \qquad \text{provided } b, c, \text{ and } d \text{ are not 0.}$$

Change division to multiplication.

Take the reciprocal of the divisor.

Example 2 Dividing Fractions

Divide and simplify, if possible. Write the answer as a fraction.

a. $\dfrac{2}{5} \div \dfrac{7}{4}$ **b.** $\dfrac{2}{27} \div \dfrac{8}{15}$ **c.** $\dfrac{35}{14} \div 7$ **d.** $12 \div \dfrac{8}{3}$

Skill Practice

Divide and simplify. Write the answer as a fraction.

8. $\dfrac{1}{4} \div \dfrac{2}{5}$ **9.** $\dfrac{3}{8} \div \dfrac{9}{10}$

10. $\dfrac{15}{4} \div 10$ **11.** $20 \div \dfrac{12}{5}$

Solution:

a. $\dfrac{2}{5} \div \dfrac{7}{4} = \dfrac{2}{5} \cdot \dfrac{4}{7}$ Multiply by the reciprocal of the divisor ("second" fraction).

$= \dfrac{2 \cdot 4}{5 \cdot 7}$ Multiply numerators. Multiply denominators.

$= \dfrac{8}{35}$

b. $\dfrac{2}{27} \div \dfrac{8}{15} = \dfrac{2}{27} \cdot \dfrac{15}{8}$ Multiply by the reciprocal of the divisor ("second" fraction).

$= \dfrac{\overset{1}{2}}{\underset{9}{27}} \cdot \dfrac{\overset{5}{15}}{\underset{4}{8}}$ Simplify.

$= \dfrac{5}{36}$ Multiply.

Avoiding Mistakes

Do not try to simplify until after taking the reciprocal of the divisor. In Example 2(a) it would be incorrect to "cancel" the 2 and the 4 in the expression $\frac{2}{5} \div \frac{7}{4}$.

c. $\dfrac{35}{14} \div 7 = \dfrac{35}{14} \div \dfrac{7}{1}$ Write the whole number 7 as an improper fraction *before* multiplying by the reciprocal.

$= \dfrac{35}{14} \cdot \dfrac{1}{7}$ Multiply by the reciprocal of the divisor.

$= \dfrac{\overset{5}{35}}{14} \cdot \dfrac{1}{\underset{1}{7}}$ Simplify.

$= \dfrac{5}{14}$ Multiply.

d. $12 \div \dfrac{8}{3} = \dfrac{12}{1} \div \dfrac{8}{3}$ Write the whole number 12 as an improper fraction.

$= \dfrac{12}{1} \cdot \dfrac{3}{8}$ Multiply by the reciprocal of the divisor.

$= \dfrac{\overset{3}{12}}{1} \cdot \dfrac{3}{\underset{2}{8}}$ Simplify.

$= \dfrac{9}{2}$ Multiply.

3. Order of Operations

When simplifying fractional expressions with more than one operation, be sure to follow the order of operations. Simplify within parentheses first. Then simplify expressions with exponents, followed by multiplication or division in the order of appearance from left to right.

Answers

8. $\dfrac{5}{8}$ **9.** $\dfrac{5}{12}$ **10.** $\dfrac{3}{8}$ **11.** $\dfrac{25}{3}$

TIP: In Example 3 we could also have written each division as multiplication of the reciprocal right from the start.

$$\frac{2}{3} \div \frac{4}{9} \div 6 = \left(\frac{2}{3} \cdot \frac{9}{4}\right) \cdot \frac{1}{6}$$

$$= \frac{\overset{1}{2}}{\underset{1}{3}} \cdot \frac{\overset{\overset{1}{\cancel{3}}}{\cancel{9}}}{\underset{2}{4}} \cdot \frac{1}{\underset{2}{6}} \quad \text{Simplify.}$$

$$= \frac{1}{4} \qquad\qquad \text{Multiply.}$$

Example 3 Applying the Order of Operations

Simplify. Write the answer as a fraction.

$$\frac{2}{3} \div \frac{4}{9} \div 6$$

Solution:

$$\frac{2}{3} \div \frac{4}{9} \div 6$$

$$= \left(\frac{2}{3} \div \frac{4}{9}\right) \div 6 \qquad \text{We will divide from left to right. To emphasize this order, we can insert parentheses around the first two fractions.}$$

$$= \left(\frac{\overset{1}{2}}{\underset{1}{3}} \cdot \frac{\overset{3}{9}}{\underset{2}{4}}\right) \div 6 \qquad \text{Simplify within parentheses.}$$

$$= \left(\frac{3}{2}\right) \div \frac{6}{1} \qquad \text{Simplify within parentheses and write the whole number as an improper fraction.}$$

$$= \left(\frac{3}{2}\right) \cdot \frac{1}{6} \qquad \text{Multiply by the reciprocal of the divisor.}$$

$$= \frac{\overset{1}{\cancel{3}}}{2} \cdot \frac{1}{\underset{2}{\cancel{6}}} \qquad \text{Simplify.}$$

$$= \frac{1}{4} \qquad\qquad \text{Multiply.}$$

Example 4 Applying the Order of Operations

Simplify. Write the answer as a fraction.

$$\left(\frac{3}{5} \div \frac{2}{15}\right)^2$$

Solution:

$$\left(\frac{3}{5} \div \frac{2}{15}\right)^2 \qquad \text{Perform operations within parentheses first.}$$

$$= \left(\frac{3}{5} \cdot \frac{15}{2}\right)^2 \qquad \text{Multiply by the reciprocal of the divisor.}$$

$$= \left(\frac{3}{\underset{1}{\cancel{5}}} \cdot \frac{\overset{3}{\cancel{15}}}{2}\right)^2 \qquad \text{Simplify.}$$

$$= \left(\frac{9}{2}\right)^2 \qquad \text{Multiply within parentheses.}$$

$$= \frac{9}{2} \cdot \frac{9}{2} \qquad \text{With an exponent of 2, multiply 2 factors of the base.}$$

$$= \frac{81}{4} \qquad \text{Multiply.}$$

4. Applications of Multiplication and Division of Fractions

Sometimes it is difficult to determine whether multiplication or division is appropriate to solve an application problem. Division is generally used for a problem that requires you to separate or "split up" a quantity into pieces. Multiplication is generally used if it is necessary to take a fractional part of a quantity.

Example 5 Using Division in an Application

A road crew must mow the grassy median along a stretch of highway I-95. If they can mow $\frac{5}{8}$ mile (mi) in 1 hr, how long will it take them to mow a 15-mi stretch?

Solution:

Read and familiarize.

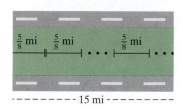

Strategy/operation: From the figure, we must separate or "split up" a 15-mi stretch of highway into pieces that are $\frac{5}{8}$ mi in length. Therefore, we must divide 15 by $\frac{5}{8}$.

$$15 \div \frac{5}{8} = \frac{15}{1} \cdot \frac{8}{5} \qquad \text{Write the whole number as a fraction. Multiply}$$
$$\text{by the reciprocal of the divisor.}$$

$$= \frac{\overset{3}{\cancel{15}}}{1} \cdot \frac{8}{\underset{1}{\cancel{5}}}$$

$$= 24$$

The 15-mi stretch of highway will take 24 hr to mow.

Example 6 Using Division in an Application

A $\frac{9}{4}$-ft length of wire must be cut into pieces of equal length that are $\frac{3}{8}$ ft long. How many pieces can be cut?

Solution:

Read and familiarize.

Operation: Here we divide the total length of wire into pieces of equal length.

$$\frac{9}{4} \div \frac{3}{8} = \frac{9}{4} \cdot \frac{8}{3} \qquad \text{Multiply by the reciprocal of the divisor.}$$

$$= \frac{\overset{3}{9}}{\underset{1}{4}} \cdot \frac{\overset{2}{8}}{\underset{1}{3}} \qquad \text{Simplify.}$$

$$= 6$$

Six pieces of wire can be cut.

Example 7   Using Multiplication in an Application

Carson estimates that his total cost for college for 1 year is $12,600. He has financial aid to pay $\frac{2}{3}$ of the cost.

a. How much money is the financial aid worth?

b. How much money will Carson have to pay?

c. If Carson's parents help him by paying $\frac{1}{3}$ of the amount not paid by financial aid, how much money will be paid by Carson's parents?

Solution:

a. Carson's financial aid will pay $\frac{2}{3}$ of $12,600. Because we are looking for a fraction of a quantity, we multiply.

$$\frac{2}{3} \cdot 12{,}600 = \frac{2}{3} \cdot \frac{12{,}600}{1}$$

$$= \frac{2}{\underset{1}{3}} \cdot \frac{\overset{4200}{12{,}600}}{1}$$

$$= 8400$$

Financial aid will pay $8400.

b. Carson will have to pay the remaining portion of the cost. This can be found by subtraction.

$$\$12{,}600 - \$8400 = \$4200$$

Carson will have to pay $4200.

TIP: The answer to Example 7(b) could also have been found by noting that financial aid paid $\frac{2}{3}$ of the cost. This means that Carson must pay $\frac{1}{3}$ of the cost, or

$$\frac{1}{3} \cdot \frac{\$12{,}600}{1} = \frac{1}{\underset{1}{3}} \cdot \frac{\overset{4200}{\$12{,}600}}{1}$$

$$= \$4200$$

Answer

16. a. $12,000,000
   b. $8,000,000
   c. $6,400,000

c. Carson's parents will pay $\frac{1}{3}$ of $4200.

$$\frac{1}{\underset{1}{\cancel{3}}} \cdot \frac{\overset{1400}{\cancel{4200}}}{1}$$

Carson's parents will pay $1400.

Example 8 **Using Multiplication in an Application**

Three-fifths of the students in the freshman class are female. Of these students, $\frac{5}{9}$ are over the age of 25. What fraction of the freshman class is female over the age of 25?

Solution:

Read and familiarize.
Strategy/operation: We must find $\frac{5}{9}$ of $\frac{3}{5}$ of one whole freshman class. This implies multiplication.

$$\frac{5}{9} \times \frac{3}{5} = \frac{\overset{1}{\cancel{5}}}{\underset{3}{\cancel{9}}} \times \frac{\overset{1}{\cancel{3}}}{\underset{1}{\cancel{5}}} = \frac{1}{3}$$

One-third of the freshman class consists of female students over the age of 25.

Skill Practice

17. In a certain police department, $\frac{1}{3}$ of the department is female. Of the female officers, $\frac{2}{7}$ were promoted within the last year. What fraction of the police department consists of females who were promoted within the last year?

Answer

17. $\frac{2}{21}$

Section 2.5   Practice Exercises

Boost your GRADE at ALEKS.com!

ALEKS version 3.0

- Practice Problems
- Self-Tests
- NetTutor
- e-Professors
- Videos

Study Skills Exercises

1. Write down the page number(s) for the Problem Recognition Exercises for this chapter. How do you think that the Problem Recognition Exercises can help you?

2. Define the key term **reciprocal**.

Review Exercises

For Exercises 3–11, multiply and simplify to lowest terms. Write the answer as a fraction or whole number.

3. $\frac{9}{11} \times \frac{22}{5}$

4. $\frac{24}{7} \cdot \frac{7}{8}$

5. $\frac{34}{5} \cdot \frac{5}{17}$

6. $3 \cdot \left(\dfrac{7}{6}\right)$ **7.** $8 \cdot \left(\dfrac{5}{24}\right)$ **8.** $\left(\dfrac{2}{7}\right)\left(\dfrac{7}{2}\right)$

9. $\left(\dfrac{9}{5}\right)\left(\dfrac{5}{9}\right)$ **10.** $\dfrac{1}{10} \times 10$ **11.** $\dfrac{1}{3} \times 3$

Objective 1: Reciprocal of a Fraction

12. For each number, determine whether the number has a reciprocal.

   a. $\dfrac{1}{2}$ **b.** $\dfrac{5}{3}$ **c.** 6   **d.** 0

For Exercises 13–20, find the reciprocal of the number, if it exists. **(See Example 1.)**

13. $\dfrac{7}{8}$ **14.** $\dfrac{5}{6}$ **15.** $\dfrac{10}{9}$ **16.** $\dfrac{14}{5}$

17. 4   **18.** 9   **19.** 0   **20.** $\dfrac{0}{4}$

Objective 2: Division of Fractions

For Exercises 21–24, fill in the blank.

21. Dividing by 3 is the same as multiplying by _____.   **22.** Dividing by 5 is the same as multiplying by _____.

23. Dividing by 8 is the same as _____ by $\dfrac{1}{8}$.   **24.** Dividing by 12 is the same as _____ by $\dfrac{1}{12}$.

For Exercises 25–48, divide and simplify the answer to lowest terms. Write the answer as a fraction or whole number. **(See Example 2.)**

25. $\dfrac{2}{15} \div \dfrac{5}{12}$ **26.** $\dfrac{11}{3} \div \dfrac{6}{5}$ **27.** $\dfrac{7}{13} \div \dfrac{2}{5}$ **28.** $\dfrac{8}{7} \div \dfrac{3}{10}$

29. $\dfrac{14}{3} \div \dfrac{6}{5}$ **30.** $\dfrac{11}{2} \div \dfrac{3}{4}$ **31.** $\dfrac{15}{2} \div \dfrac{3}{2}$ **32.** $\dfrac{9}{10} \div \dfrac{9}{2}$

33. $\dfrac{3}{4} \div \dfrac{3}{4}$ **34.** $\dfrac{6}{5} \div \dfrac{6}{5}$ **35.** $7 \div \dfrac{2}{3}$ **36.** $4 \div \dfrac{3}{5}$

37. $\dfrac{10}{9} \div \dfrac{1}{18}$ **38.** $\dfrac{4}{3} \div \dfrac{1}{3}$ **39.** $12 \div \dfrac{3}{4}$ **40.** $24 \div \dfrac{8}{5}$

41. $\dfrac{12}{5} \div 4$ **42.** $\dfrac{20}{6} \div 5$ **43.** $\dfrac{9}{50} \div \dfrac{18}{25}$ **44.** $\dfrac{30}{40} \div \dfrac{15}{8}$

45. $\dfrac{9}{100} \div \dfrac{13}{1000}$ **46.** $\dfrac{1000}{17} \div \dfrac{10}{3}$ **47.** $\dfrac{36}{5} \div \dfrac{9}{25}$ **48.** $\dfrac{13}{5} \div \dfrac{17}{10}$

Mixed Exercises

For Exercises 49–64, multiply or divide as indicated. Write the answer as a fraction or whole number.

49. $\dfrac{7}{8} \div \dfrac{1}{4}$

50. $\dfrac{7}{12} \div \dfrac{5}{3}$

51. $\dfrac{5}{8} \cdot \dfrac{2}{9}$

52. $\dfrac{1}{16} \cdot \dfrac{4}{3}$

53. $6 \cdot \dfrac{4}{3}$

54. $12 \cdot \dfrac{5}{6}$

55. $\dfrac{16}{5} \div 8$

56. $\dfrac{42}{11} \div 7$

57. $\dfrac{16}{3} \div \dfrac{2}{5}$

58. $\dfrac{17}{8} \div \dfrac{1}{4}$

59. $\dfrac{1}{8} \cdot 16$

60. $\dfrac{2}{3} \cdot 9$

61. $\dfrac{22}{7} \cdot \dfrac{5}{16}$

62. $\dfrac{40}{21} \cdot \dfrac{18}{25}$

63. $8 \div \dfrac{16}{3}$

64. $5 \div \dfrac{15}{4}$

Objective 3: Order of Operations

65. Explain the difference in the process to evaluate $\frac{2}{3} \cdot 6$ versus $\frac{2}{3} \div 6$. Then evaluate each expression.

66. Explain the difference in the process to evaluate $8 \cdot \frac{2}{3}$ versus $8 \div \frac{2}{3}$. Then evaluate each expression.

For Exercises 67–78, simplify by using the order of operations. Write the answer as a fraction or whole number. **(See Examples 3–4.)**

67. $\dfrac{54}{21} \div \dfrac{2}{3} \div 9$

68. $\dfrac{48}{56} \div \dfrac{3}{8} \div 8$

69. $\dfrac{3}{5} \div \dfrac{6}{7} \cdot \dfrac{5}{3}$

70. $\dfrac{5}{8} \div \dfrac{35}{16} \cdot \dfrac{1}{4}$

71. $\left(\dfrac{3}{8}\right)^2 \div \dfrac{9}{14}$

72. $\dfrac{7}{8} \div \left(\dfrac{1}{2}\right)^2$

73. $\left(\dfrac{2}{5} \div \dfrac{8}{3}\right)^2$

74. $\left(\dfrac{5}{12} \div \dfrac{2}{3}\right)^2$

75. $\left(\dfrac{63}{8} \div \dfrac{9}{4}\right)^2 \cdot 4$

76. $\left(\dfrac{25}{3} \div \dfrac{50}{9}\right)^2 \cdot 8$

77. $\dfrac{15}{16} \cdot \left(\dfrac{2}{3}\right)^2 \div \dfrac{20}{21}$

78. $\dfrac{8}{27} \cdot \left(\dfrac{3}{4}\right)^2 \div \dfrac{13}{18}$

Objective 4: Applications of Multiplication and Division of Fractions

79. How many eighths are in $\frac{9}{4}$?

80. How many sixths are in $\frac{4}{3}$?

81. During the month of December, a department store wraps packages free of charge. Each package requires $\frac{2}{3}$ yd of ribbon. If Li used up a 36-yd roll of ribbon, how many packages were wrapped? **(See Example 5.)**

82. A developer sells lots of land in increments of $\frac{3}{4}$ acre. If the developer has 60 acres, how many lots can be sold?

83. If one cup is $\frac{1}{16}$ gal, how many cups of orange juice can be filled from $\frac{3}{2}$ gal? **(See Example 6.)**

84. If 1 centimeter (cm) is $\frac{1}{100}$ meter (m), how many centimeters are in a $\frac{5}{4}$-m piece of rope?

85. Dorci buys 16 sheets of plywood, each $\frac{3}{4}$ in. thick, to cover her windows in the event of a hurricane. She stacks the wood in the garage. How high will the stack be?

86. Davey built a bookshelf 36 in. long. Can the shelf hold a set of encyclopedias if there are 24 books and each book averages $\frac{5}{4}$ in. thick? Explain your answer.

87. A radio station allows 18 minutes (min) of advertising each hour. How many 40-second ($\frac{2}{3}$-min) commercials can be run in

    a. 1 hr   **b.** 1 day

88. A television station has 20 min of advertising each hour. How many 30-second ($\frac{1}{2}$-min) commercials can be run in

    a. 1 hr   **b.** 1 day

89. Ricardo wants to buy a new house for $240,000. The bank requires $\frac{1}{10}$ of the cost of the house as a down payment. As a gift, Ricardo's mother will pay $\frac{2}{3}$ of the down payment. (See Example 7.)

    a. How much money will Ricardo's mother pay toward the down payment?

    b. How much money will Ricardo have to pay toward the down payment?

    c. How much is left over for Ricardo to finance?

90. Althea wants to buy a Toyota Camry for a total cost of $18,000. The dealer requires $\frac{1}{12}$ of the money as a down payment. Althea's parents have agreed to pay one-half of the down payment for her.

    a. How much money will Althea's parents pay toward the down payment?

    b. How much will Althea pay toward the down payment?

    c. How much will Althea have to finance?

91. A landowner has $\frac{9}{4}$ acres of land. She plans to sell $\frac{1}{3}$ of the land. (See Example 8.)

    a. How much land will she sell?   **b.** How much land will she retain?

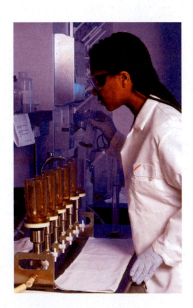

92. Josh must read 24 pages for his English class and 18 pages for psychology. He has read $\frac{1}{6}$ of the pages.

    a. How many pages has he read?

    b. How many pages does he still have to read?

93. A lab technician has $\frac{7}{4}$ liters (L) of alcohol. If she needs samples of $\frac{1}{8}$ L, how many samples can she prepare?

94. Troy has a $\frac{7}{8}$-in. nail that he must hammer into a board. Each strike of the hammer moves the nail $\frac{1}{16}$ in. into the board. How many strikes of the hammer must he make?

Expanding Your Skills

95. The rectangle shown here has an area of 30 ft². Find the length.

    $\frac{5}{2}$ ft   ?

96. The rectangle shown here has an area of 8 m². Find the width.

    ?

    14 m

Problem Recognition Exercises

Multiplication and Division of Fractions

Multiply or divide as indicated. Write the answer as a whole number or a fraction.

1. a. $\dfrac{8}{3} \cdot \dfrac{6}{5}$ b. $\dfrac{6}{5} \cdot \dfrac{8}{3}$

   c. $\dfrac{8}{3} \div \dfrac{6}{5}$ d. $\dfrac{6}{5} \div \dfrac{8}{3}$

2. a. $\dfrac{10}{3} \cdot \dfrac{12}{7}$ b. $\dfrac{12}{7} \cdot \dfrac{10}{3}$

   c. $\dfrac{10}{3} \div \dfrac{12}{7}$ d. $\dfrac{12}{7} \div \dfrac{10}{3}$

3. a. $12 \cdot \dfrac{9}{8}$ b. $\dfrac{9}{8} \cdot 12$

   c. $12 \div \dfrac{9}{8}$ d. $\dfrac{9}{8} \div 12$

4. a. $15 \cdot \dfrac{3}{5}$ b. $\dfrac{3}{5} \cdot 15$

   c. $15 \div \dfrac{3}{5}$ d. $\dfrac{3}{5} \div 15$

5. a. $\dfrac{5}{6} \cdot \dfrac{5}{6}$ b. $\dfrac{5}{6} \cdot \dfrac{6}{5}$

   c. $\dfrac{5}{6} \div \dfrac{5}{6}$ d. $\dfrac{5}{6} \div \dfrac{6}{5}$

6. a. $\dfrac{9}{8} \cdot 0$ b. $0 \cdot \dfrac{9}{8}$

   c. $\dfrac{9}{8} \div 0$ d. $0 \div \dfrac{9}{8}$

7. a. $\dfrac{1}{12} \cdot \dfrac{2}{3} \cdot \dfrac{16}{21}$ b. $\dfrac{1}{12} \cdot \dfrac{2}{3} \div \dfrac{16}{21}$

   c. $\dfrac{1}{12} \div \dfrac{2}{3} \cdot \dfrac{16}{21}$ d. $\dfrac{1}{12} \div \dfrac{2}{3} \div \dfrac{16}{21}$

8. a. $\dfrac{1}{2} \cdot \dfrac{7}{9} \cdot \dfrac{2}{3}$ b. $\dfrac{1}{2} \cdot \dfrac{7}{9} \div \dfrac{2}{3}$

   c. $\dfrac{1}{2} \div \dfrac{7}{9} \cdot \dfrac{2}{3}$ d. $\dfrac{1}{2} \div \dfrac{7}{9} \div \dfrac{2}{3}$

9. a. $\dfrac{9}{10} \cdot 6 \cdot \dfrac{1}{4}$ b. $\dfrac{9}{10} \cdot 6 \div \dfrac{1}{4}$

   c. $\dfrac{9}{10} \div 6 \cdot \dfrac{1}{4}$ d. $\dfrac{9}{10} \div 6 \div \dfrac{1}{4}$

10. a. $\dfrac{4}{5} \cdot \dfrac{1}{20} \cdot 10$ b. $\dfrac{4}{5} \cdot \dfrac{1}{20} \div 10$

    c. $\dfrac{4}{5} \div \dfrac{1}{20} \cdot 10$ d. $\dfrac{4}{5} \div \dfrac{1}{20} \div 10$

11. a. $\dfrac{2}{3} \cdot 1$ b. $1 \cdot \dfrac{2}{3}$

    c. $\dfrac{2}{3} \div 1$ d. $1 \div \dfrac{2}{3}$

12. a. $6 \div 10$ b. $10 \div 6$
    c. $6 \cdot 10$ d. $10 \cdot 6$

13. a. $8 \div \dfrac{1}{4}$ b. $8 \cdot \dfrac{1}{4}$

    c. $8 \div 4$ d. $8 \cdot 4$

14. a. $\dfrac{1}{7} \div 2$ b. $\dfrac{1}{7} \cdot 2$

    c. $\dfrac{1}{7} \cdot \dfrac{1}{2}$ d. $\dfrac{1}{7} \div \dfrac{1}{2}$

15. a. $4^2 \cdot \dfrac{1}{6}$ b. $4^2 \div \dfrac{1}{6}$

    c. $4 \cdot \left(\dfrac{1}{6}\right)^2$ d. $4 \div \left(\dfrac{1}{6}\right)^2$

16. a. $\left(\dfrac{1}{2}\right)^2 \cdot \dfrac{2}{3}$ b. $\left(\dfrac{1}{2}\right)^2 \div \dfrac{2}{3}$

    c. $\dfrac{1}{2} \cdot \left(\dfrac{2}{3}\right)^2$ d. $\dfrac{1}{2} \div \left(\dfrac{2}{3}\right)^2$

Section 2.6   Multiplication and Division of Mixed Numbers

1. Multiplication of Mixed Numbers

To multiply mixed numbers, we follow these steps.

> **PROCEDURE**   **Multiplying Mixed Numbers**
>
> **Step 1**   Change each mixed number to an improper fraction.
> **Step 2**   Multiply the improper fractions and simplify to lowest terms, if possible (see Section 2.4).
>
> Answers greater than 1 may be written as an improper fraction or as a mixed number, depending on the directions of the problem.

Example 1 demonstrates this process.

Skill Practice

1. Multiply and write the answer as a mixed number.

$$\left(4\frac{3}{5}\right)\left(5\frac{5}{6}\right)$$

Example 1   Multiplying Mixed Numbers

Multiply and write the answer as a mixed number.

$$\left(3\frac{1}{5}\right)\left(4\frac{3}{4}\right)$$

Solution:

$$\left(3\frac{1}{5}\right)\left(4\frac{3}{4}\right) = \frac{16}{5} \cdot \frac{19}{4}$$ Write each mixed number as an improper fraction.

$$= \frac{\overset{4}{\cancel{16}}}{5} \cdot \frac{19}{\underset{1}{\cancel{4}}}$$ Simplify.

$$= \frac{76}{5}$$ Multiply.

$$= 15\frac{1}{5}$$ Write the improper fraction as a mixed number.

$$
\begin{array}{r}
15 \\
5\overline{)76} \\
\underline{-5} \\
26 \\
\underline{-25} \\
1
\end{array}
$$

> **TIP:** To check whether the answer from Example 1 is reasonable, we can round each factor and estimate the product.
>
> $3\frac{1}{5}$ rounds to 3.
> $4\frac{3}{4}$ rounds to 5.
>
> Thus, $\left(3\frac{1}{5}\right)\left(4\frac{3}{4}\right) \approx (3)(5) = 15$, which is close to $15\frac{1}{5}$.

Answer

1. $26\frac{5}{6}$

Multiplying Mixed Numbers

Multiply and write the answer as a mixed number or whole number.

a. $25\frac{1}{2} \cdot 4\frac{2}{3}$ **b.** $12 \cdot \left(8\frac{7}{9}\right)$

Skill Practice

Multiply and write the answer as a mixed number or whole number.

2. $16\frac{1}{2} \cdot 3\frac{7}{11}$

3. $7\frac{1}{6} \cdot 10$

Solution:

a. $25\frac{1}{2} \cdot 4\frac{2}{3} = \frac{51}{2} \cdot \frac{14}{3}$ Write each mixed number as an improper fraction.

$= \frac{\overset{17}{\cancel{51}}}{\underset{1}{\cancel{2}}} \cdot \frac{\overset{7}{\cancel{14}}}{\underset{1}{\cancel{3}}}$ Simplify.

$= \frac{119}{1}$ Multiply.

$= 119$

Avoiding Mistakes

Do not try to multiply mixed numbers by multiplying the whole-number parts and multiplying the fractional parts. You will not get the correct answer.

For the expression $25\frac{1}{2} \cdot 4\frac{2}{3}$, it would be incorrect to multiply $(25)(4)$ and $\frac{1}{2} \cdot \frac{2}{3}$. Notice that these values do not equal 119.

b. $12 \cdot \left(8\frac{7}{9}\right) = \frac{12}{1} \cdot \frac{79}{9}$ Write the whole number and mixed number as improper fractions.

$= \frac{\overset{4}{\cancel{12}}}{1} \cdot \frac{79}{\underset{3}{\cancel{9}}}$ Simplify.

$= \frac{316}{3}$ Multiply.

$= 105\frac{1}{3}$ Write the improper fraction as a mixed number.

```
  105
3)316
 -3
  16
 -15
   1
```

2. Division of Mixed Numbers

To divide mixed numbers, we use the following steps.

PROCEDURE Dividing Mixed Numbers

Step 1 Change each mixed number to an improper fraction.

Step 2 Divide the improper fractions and simplify to lowest terms, if possible. Recall that to divide fractions, we multiply the dividend by the reciprocal of the divisor (see Section 2.5).

Answers greater than 1 may be written as an improper fraction or as a mixed number, depending on the directions of the problem.

Answers

2. 60   **3.** $71\frac{2}{3}$

Avoiding Mistakes

Be sure to take the reciprocal of the divisor *after* the mixed number is changed to an improper fraction.

Example 3   **Dividing Mixed Numbers**

Divide and write the answer as a mixed number or whole number.

a. $7\frac{1}{2} \div 4\frac{2}{3}$ b. $6 \div 5\frac{1}{7}$ c. $13\frac{5}{6} \div 7$

Solution:

a. $7\frac{1}{2} \div 4\frac{2}{3} = \frac{15}{2} \div \frac{14}{3}$ Write the mixed numbers as improper fractions.

$= \frac{15}{2} \cdot \frac{3}{14}$ Multiply by the reciprocal of the divisor.

$= \frac{45}{28}$ Multiply.

$= 1\frac{17}{28}$ Write the improper fraction as a mixed number.

b. $6 \div 5\frac{1}{7} = \frac{6}{1} \div \frac{36}{7}$ Write the whole number and mixed number as improper fractions.

$= \frac{6}{1} \cdot \frac{7}{36}$ Multiply by the reciprocal of the divisor.

$= \frac{\overset{1}{6}}{1} \cdot \frac{7}{\underset{6}{36}}$ Simplify.

$= \frac{7}{6}$ Multiply.

$= 1\frac{1}{6}$ Write the improper fraction as a mixed number.

c. $13\frac{5}{6} \div 7 = \frac{83}{6} \div \frac{7}{1}$ Write the whole number and mixed number as improper fractions.

$= \frac{83}{6} \cdot \frac{1}{7}$ Multiply by the reciprocal of the divisor.

$= \frac{83}{42}$ Multiply.

$= 1\frac{41}{42}$ Write the improper fraction as a mixed number.

3. Applications of Multiplication and Division of Mixed Numbers

Examples 4 and 5 demonstrate multiplication and division of mixed numbers in day-to-day applications.

Answers

4. $3\frac{11}{17}$ 5. $1\frac{2}{3}$ 6. $1\frac{5}{9}$

Example 4 Applying Multiplication of Mixed Numbers

Antonio has a painting company, and he recently won a contract to paint eight large classrooms at a university. Each room requires $5\frac{4}{5}$ gal of paint. How much paint is needed for this job?

Solution:

Amount needed for 8 rooms:

$$8 \cdot \left(5\frac{4}{5} \text{ gal}\right) = 8 \cdot \left(\frac{29}{5}\right) \text{ gal}$$

$$= \frac{8}{1} \cdot \frac{29}{5} \text{ gal}$$

$$= \frac{232}{5} \text{ gal}$$

$$= 46\frac{2}{5} \text{ gal}$$

The painter requires $46\frac{2}{5}$ gal of paint.

Skill Practice

7. A recipe calls for $2\frac{3}{4}$ cups of flour. How much flour is required for $2\frac{1}{2}$ times the recipe?

TIP: The solution to Example 4 can be estimated. In this case, the painter might want to over-estimate the answer to be sure that he doesn't run short of paint. We can round $5\frac{4}{5}$ gal to 6 gal. Therefore, our estimate is 8(6 gal) = 48 gal of paint.

Example 5 Applying Division of Mixed Numbers

A construction site brings in $6\frac{2}{3}$ tons of soil. Each truck holds $\frac{2}{3}$ ton. How many truckloads are necessary?

Solution:

The $6\frac{2}{3}$ tons of soil must be distributed in $\frac{2}{3}$-ton increments. This will require division.

Skill Practice

8. A department store wraps packages for $2 each. Ribbon $2\frac{5}{8}$ ft long is used to wrap each package. How many packages can be wrapped from a roll of ribbon 168 ft long?

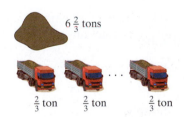

$6\frac{2}{3}$ tons

$\frac{2}{3}$ ton $\frac{2}{3}$ ton $\frac{2}{3}$ ton

$$6\frac{2}{3} \div \frac{2}{3} = \frac{20}{3} \div \frac{2}{3} \qquad \text{Write the mixed number as an improper fraction.}$$

$$= \frac{20}{3} \cdot \frac{3}{2} \qquad \text{Multiply by the reciprocal of the divisor.}$$

$$= \frac{\overset{10}{\cancel{20}}}{\underset{1}{\cancel{3}}} \cdot \frac{\overset{1}{\cancel{3}}}{\underset{1}{2}} \qquad \text{Simplify.}$$

$$= \frac{10}{1} \qquad \text{Multiply.}$$

$$= 10$$

A total of 10 truckloads of soil will be required.

Answers
7. $6\frac{7}{8}$ cups of flour   8. 64 packages

Section 2.6   Practice Exercises

Boost *your* GRADE at
ALEKS.com!

ALEKS®
version 3.0

- Practice Problems
- Self-Tests
- NetTutor

- e-Professors
- Videos

Study Skills Exercise

1. Find the page numbers for the Chapter Review Exercises, Chapter Test, and Cumulative Review Exercises for this chapter.

   Chapter Review Exercises, page(s) ——————.

   Chapter Test, page(s) ——————.

   Cumulative Review Exercises, page(s) ——————.

   Compare these features and state the advantages of each.

Review Exercises

For Exercises 2–7, multiply or divide the fractions. Write your answers as fractions.

2. $\dfrac{5}{6} \cdot \dfrac{2}{9}$

3. $\dfrac{13}{5} \cdot \dfrac{10}{9}$

4. $\dfrac{20}{9} \div \dfrac{10}{3}$

5. $\dfrac{42}{11} \div \dfrac{7}{2}$

6. $\dfrac{32}{15} \div 4$

7. $\dfrac{52}{18} \div 13$

8. Explain the process to change a mixed number to an improper fraction.

For Exercises 9–12, write the mixed number as an improper fraction.

9. $3\dfrac{2}{5}$

10. $2\dfrac{7}{10}$

11. $1\dfrac{4}{7}$

12. $4\dfrac{1}{8}$

For Exercises 13–16, write the improper fraction as a mixed number.

13. $\dfrac{77}{6}$

14. $\dfrac{57}{11}$

15. $\dfrac{39}{4}$

16. $\dfrac{31}{2}$

Objective 1: Multiplication of Mixed Numbers

For Exercises 17–32, multiply the mixed numbers. Write the answer as a mixed number or whole number.
(See Examples 1–2.)

 17. $\left(2\dfrac{2}{5}\right)\left(3\dfrac{1}{12}\right)$

18. $\left(5\dfrac{1}{5}\right)\left(3\dfrac{3}{4}\right)$

19. $2\dfrac{1}{3} \cdot \dfrac{5}{7}$

20. $6\dfrac{1}{8} \cdot \dfrac{4}{7}$

21. $4\dfrac{2}{9} \cdot 9$

22. $3\dfrac{1}{3} \cdot 6$

23. $\left(5\dfrac{3}{16}\right)\left(5\dfrac{1}{3}\right)$

24. $\left(8\dfrac{2}{3}\right)\left(2\dfrac{1}{13}\right)$

25. $\left(7\dfrac{1}{4}\right) \cdot 10$

26. $\left(2\dfrac{2}{3}\right) \cdot 3$

27. $4\dfrac{5}{8} \cdot 0$

28. $0 \cdot 6\dfrac{1}{10}$

29. $\left(3\dfrac{1}{2}\right)\left(2\dfrac{1}{7}\right)$

30. $\left(1\dfrac{3}{10}\right)\left(1\dfrac{1}{4}\right)$

31. $\left(5\dfrac{2}{5}\right)\left(\dfrac{2}{9}\right)\left(1\dfrac{4}{5}\right)$

32. $\left(6\dfrac{1}{8}\right)\left(2\dfrac{3}{4}\right)\left(\dfrac{8}{7}\right)$

Objective 2: Division of Mixed Numbers

For Exercises 33–48, divide the mixed numbers. Write the answer as a mixed number, proper fraction, or whole number. **(See Example 3.)**

33. $1\frac{7}{10} \div 2\frac{3}{4}$

34. $5\frac{1}{10} \div \frac{3}{4}$

35. $5\frac{8}{9} \div 1\frac{1}{3}$

36. $12\frac{4}{5} \div 2\frac{3}{5}$

37. $2\frac{1}{2} \div 1\frac{1}{16}$

38. $7\frac{3}{5} \div 1\frac{7}{12}$

39. $4\frac{1}{2} \div 2\frac{1}{4}$

40. $5\frac{5}{6} \div 2\frac{1}{3}$

41. $0 \div 6\frac{7}{12}$

42. $0 \div 1\frac{9}{11}$

43. $2\frac{5}{6} \div \frac{1}{6}$

44. $6\frac{1}{2} \div \frac{1}{2}$

45. $1\frac{1}{3} \div \frac{2}{7}$

46. $2\frac{1}{7} \div \frac{5}{13}$

47. $3\frac{1}{2} \div 2$

48. $4\frac{2}{3} \div 3$

Objective 3: Applications of Multiplication and Division of Mixed Numbers

49. Tabitha charges $8 per hour for baby sitting. If she works for $4\frac{3}{4}$ hr, how much should she be paid? **(See Example 4.)**

50. Kurt bought $2\frac{2}{3}$ acres of land. If land costs $10,500 per acre, how much will the land cost him?

51. According to the U.S. Census Bureau's Valentine's Day Press Release, the average American consumes $25\frac{7}{10}$ lb of chocolate in a year. Over the course of 25 years, how many pounds of chocolate would the average American consume?

52. Florida voters approved an amendment to the state constitution to set Florida's minimum wage at $6\frac{2}{3}$ per hour. Kayla earns minimum wage while working at the campus bookstore. If she works for $15\frac{3}{4}$ hr, how much should she be paid?

53. The age of a small kitten can be approximated by the following rule. The kitten's age is given as 1 week for every quarter pound of weight. **(See Example 5.)**
   a. Approximately how old is a $1\frac{3}{4}$-lb kitten?
   b. Approximately how old is a $2\frac{1}{8}$-lb kitten?

54. Richard's estate is to be split equally among his three children. If his estate is worth $1\frac{3}{4}$ million, how much will each child inherit?

55. Lucy earns $14 per hour and Ricky earns $10 per hour. Suppose Lucy worked $35\frac{1}{2}$ hr last week and Ricky worked $42\frac{1}{2}$ hr.
   a. Who earned more money and by how much?
   b. How much did they earn altogether?

56. A roll of wallpaper covers an area of 28 ft². If the roll is $1\frac{17}{24}$ ft wide, how long is the roll?

Mixed Exercises

For Exercises 57–76, perform the indicated operation. Write the answer as a mixed number, proper fraction, or whole number.

57. $2\frac{1}{5} \div 1\frac{1}{10}$

58. $3\frac{3}{4} \cdot 1\frac{5}{6}$

59. $6 \div 1\frac{1}{8}$

60. $8 \div 2\frac{1}{3}$

61. $\frac{2}{3} \cdot 2\frac{7}{10}$

62. $\frac{4}{3} \cdot 5\frac{1}{8}$

63. $4\frac{1}{12} \cdot 0$

64. $5\frac{1}{3} \cdot 6$

65. $10\frac{1}{2} \div 9$

66. $\frac{2}{7} \cdot 1\frac{8}{9}$

67. $0 \div 9\frac{2}{3}$

68. $\frac{3}{8} \div 2\frac{1}{2}$

69. $12 \cdot \frac{1}{8}$

70. $20 \cdot \frac{2}{15}$

71. $6\frac{8}{9} \div 0$

72. $0 \cdot 2\frac{1}{8}$

73. $\left(3\frac{2}{5}\right)\left(\frac{7}{34}\right)\left(3\frac{3}{4}\right)$

74. $\left(5\frac{1}{6}\right)\left(1\frac{4}{7}\right)\left(\frac{14}{33}\right)$

75. $7\frac{1}{8} \div 1\frac{1}{3} \div 2\frac{1}{4}$

76. $3\frac{1}{8} \div 5\frac{5}{7} \div 1\frac{5}{16}$

Expanding Your Skills

77. A landscaper will use a decorative concrete border around the cactus garden shown. Each concrete brick is $1\frac{1}{4}$ ft long and costs \$3. What is the total cost of the border?

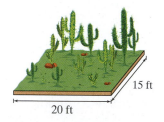

15 ft

20 ft

78. Sara drives a total of $64\frac{1}{2}$ mi to work and back. Her car gets $21\frac{1}{2}$ miles per gallon (mpg) of gas. If gas is \$5 per gallon, how much does it cost her to commute to and from work each day?

Calculator Connections

Topic: Multiplying and Dividing Fractions and Mixed Numbers

Expression	Keystrokes	Result
$\frac{8}{15} \cdot \frac{25}{28}$	8 [a^b/c] 15 × 25 [a^b/c] 28 [=]	$\boxed{10 \lrcorner 21} = \frac{10}{21}$
$2\frac{3}{4} \div \frac{1}{6}$	2 [a^b/c] 3 [a^b/c] 4 ÷ 1 [a^b/c] 6 [=] This is how you enter the mixed number $2\frac{3}{4}$.	$\boxed{16 \_ 1 \lrcorner 2} = 16\frac{1}{2}$

To convert the result to an improper fraction, press [2nd] [d/c] $\boxed{33 \lrcorner 2} = \frac{33}{2}$

Calculator Exercises

For Exercises 79–86, use a calculator to multiply or divide and simplify. Write the answer as a mixed number.

79. $12\frac{2}{3} \cdot 25\frac{1}{8}$

80. $38\frac{1}{3} \div 12\frac{1}{2}$

81. $56\frac{5}{6} \div 3\frac{1}{6}$

82. $25\frac{1}{5} \cdot 18\frac{1}{2}$

83. $32\frac{7}{12} \div 12\frac{1}{6}$

84. $106\frac{1}{9} \div 41\frac{5}{6}$

85. $11\frac{1}{2} \cdot 41\frac{3}{4}$

86. $9\frac{8}{9} \cdot 28\frac{1}{3}$

Group Activity

Cooking for Company

Estimated time: 15 minutes

Group Size: 3

This recipe for chili serves 6 people.

CHILI

1 tablespoon oil

2 onions

$1\frac{1}{2}$ lb ground beef

14 oz canned tomatoes

$1\frac{1}{4}$ teaspoons chili powder

$\frac{3}{8}$ teaspoon cumin

$1\frac{1}{8}$ teaspoon salt

$\frac{1}{8}$ teaspoon cayenne pepper

4 whole cloves

15 oz canned kidney beans

1. Suppose that this recipe is to be used for a Super Bowl party for 30 people. Each student in the group should select three ingredients and determine the amount needed to make this recipe for 30 people. Then write the revised recipe.

2. Suppose that this recipe is to be used for a dinner party for 9 guests. Each student in the group should select three ingredients and determine the amount needed to make this recipe for 9 people. Then write the revised recipe.

Chapter 2 Summary

Section 2.1 Introduction to Fractions and Mixed Numbers

Key Concepts

A **fraction** represents a part of a whole unit. For example, $\frac{1}{3}$ represents one part of a whole unit that is divided into 3 equal pieces.

In the fraction $\frac{1}{3}$, the "top" number, 1, is the **numerator**, and the "bottom" number, 3, is the **denominator**.

A fraction in which the numerator is less than the denominator is called a **proper fraction**. A fraction in which the numerator is greater than or equal to the denominator is called an **improper fraction**.

An improper fraction can be written as a **mixed number** by dividing the numerator by the denominator. Write the quotient as a whole number, and write the remainder over the divisor.

A mixed number can be written as an improper fraction by multiplying the whole number by the denominator and adding the numerator. Then write that total over the denominator.

Fractions can be represented on a number line. For example,

Examples

Example 1

$\frac{1}{3}$ of the
pie is shaded.

Example 2

For the fraction $\frac{7}{9}$, the numerator is 7 and the denominator is 9.

Example 3

$\frac{5}{3}$ is an improper fraction, $\frac{3}{5}$ is a proper fraction, and $\frac{3}{3}$ is an improper fraction.

Example 4

$\frac{10}{3}$ can be written as $3\frac{1}{3}$ because $3\overline{)10}$
$$\begin{array}{r} 3 \\ 3\overline{)10} \\ -9 \\ \hline 1 \end{array}$$

Example 5

$2\frac{4}{5}$ can be written as $\frac{14}{5}$ because $\dfrac{2 \cdot 5 + 4}{5} = \dfrac{14}{5}$.

Example 6

$\dfrac{8}{10}$

represents $\dfrac{3}{10}$.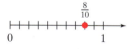

Section 2.2   Prime Numbers and Factorization

Key Concepts

A **factorization** of a number is a product of factors that equals the number.

A number is divisible by another if their quotient leaves no remainder.

Divisibility Rules for 2, 3, 5, and 10

- A whole number is divisible by 2 if the ones-place digit is 0, 2, 4, 6, or 8.
- A whole number is divisible by 3 if the sum of the digits of the number is divisible by 3.
- A whole number is divisible by 5 if the ones-place digit is 0 or 5.
- A whole number is divisible by 10 if the ones-place digit is 0.

A **prime number** is a whole number greater than 1 that has exactly two factors, 1 and itself.

Composite numbers are whole numbers that have more than two factors. The numbers 0 and 1 are neither prime nor composite.

Prime Factorization

The **prime factorization** of a number is the factorization in which every factor is a prime number.

A factor of a number n is any number that divides evenly into n. For example, the factors of 80 are 1, 2, 4, 5, 8, 10, 16, 20, 40, and 80.

Examples

Example 1

$4 \cdot 4$ and $8 \cdot 2$ are two factorizations of 16.

Example 2

15 is divisible by 5 because 5 divides into 15 evenly.

Example 3

382 is divisible by 2.
640 is divisible by 2, 5, and 10.
735 is divisible by 3 and 5.

Example 4

9 is a composite number.
2 is a prime number.
1 is neither prime nor composite.

Example 5

$$
\begin{array}{r}
2\,\overline{)756} \\
2\,\overline{)378} \\
3\,\overline{)189} \\
3\,\overline{)63} \\
3\,\overline{)21} \\
7
\end{array}
$$

The prime factorization of 756 is

$$2 \cdot 2 \cdot 3 \cdot 3 \cdot 3 \cdot 7 \quad \text{or} \quad 2^2 \cdot 3^3 \cdot 7$$

Section 2.3   Simplifying Fractions to Lowest Terms

Key Concepts

Equivalent fractions are fractions that represent the same portion of a whole unit.

Examples

Example 1

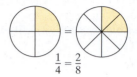

$$\frac{1}{4} = \frac{2}{8}$$

To determine if two fractions are equivalent, calculate the cross products. If the cross products are equal, then the fractions are equivalent.

Example 2

a. Compare $\frac{5}{3}$ and $\frac{6}{4}$.

$$\frac{5}{3} \overset{?}{\times} \frac{6}{4}$$

$20 \neq 18$

Fractions are
not equivalent

b. Compare $\frac{4}{5}$ and $\frac{8}{10}$.

$$\frac{4}{5} \overset{?}{\times} \frac{8}{10}$$

$40 = 40$

Fractions are
equivalent

To simplify fractions to **lowest terms**, use the fundamental principle of fractions:

Consider the fraction $\frac{a}{b}$ and the nonzero number c. Then

$$\frac{a \cdot c}{b \cdot c} = \frac{a}{b} \cdot \frac{c}{c} = \frac{a}{b} \cdot 1 = \frac{a}{b}$$

Example 3

$$\frac{25}{15} = \frac{5 \cdot 5}{3 \cdot 5} = \frac{5}{3} \cdot \frac{5}{5} = \frac{5}{3} \cdot 1 = \frac{5}{3}$$

To simplify fractions with common powers of 10, "strike through" the common zeros first.

Example 4

$$\frac{3,\cancel{000}}{12,\cancel{000}} = \frac{3}{12} = \frac{\overset{1}{\cancel{3}}}{\underset{4}{\cancel{12}}} = \frac{1}{4}$$

Section 2.4   Multiplication of Fractions and Applications

Key Concepts

Multiplication of Fractions

To multiply fractions, write the product of the numerators over the product of the denominators. Then simplify the resulting fraction, if possible.

When multiplying a whole number and a fraction, first write the whole number as a fraction by writing the whole number over 1.

Powers of one-tenth can be expressed as a fraction with 1 in the numerator. The denominator has a 1 followed by the same number of zeros as the exponent of the base of $\frac{1}{10}$.

The formula for the area of a triangle is given by $A = \frac{1}{2}bh$.

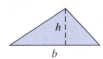

Area is expressed in square units such as ft^2, $in.^2$, yd^2, m^2, and cm^2.

Examples

Example 1

$$\frac{4}{7} \times \frac{6}{5} = \frac{24}{35}$$

$$\left(\frac{15}{16}\right)\left(\frac{4}{5}\right) = \frac{\overset{3}{\cancel{15}}}{\underset{4}{\cancel{16}}} \cdot \frac{\overset{1}{\cancel{4}}}{\underset{1}{\cancel{5}}} = \frac{3}{4}$$

Example 2

$$8 \cdot \left(\frac{5}{6}\right) = \frac{\overset{4}{\cancel{8}}}{1} \cdot \frac{5}{\underset{3}{\cancel{6}}} = \frac{20}{3}$$

Example 3

$$\left(\frac{1}{10}\right)^4 = \frac{1}{10,000}$$

Example 4

The area of the triangle is

$$A = \frac{1}{2}bh$$

$$= \left(\frac{1}{2} \cdot \frac{7}{2}\right) \cdot \frac{16}{7}$$

$$= \frac{\overset{1}{\cancel{7}}}{\underset{1}{\cancel{4}}} \cdot \frac{\overset{4}{\cancel{16}}}{\underset{1}{\cancel{7}}}$$

$$= 4$$

The area is 4 m^2.

| **Section 2.5** | **Division of Fractions and Applications** |

Key Concepts

The **reciprocal** of $\frac{a}{b}$ is $\frac{b}{a}$ for $a, b \neq 0$. The product of a fraction and its reciprocal is 1. For example,

$$\frac{6}{11} \cdot \frac{11}{6} = 1.$$

Dividing Fractions

To divide two fractions, multiply the dividend (the "first" fraction) by the reciprocal of the divisor (the "second" fraction).

When dividing by a whole number, first write the whole number as a fraction by writing the whole number over 1. Then multiply by its reciprocal.

When simplifying expressions with more than one operation, follow the order of operations.

Examples

Example 1

The reciprocal of $\frac{5}{8}$ is $\frac{8}{5}$.
The reciprocal of 4 is $\frac{1}{4}$.

The number 0 does not have a reciprocal because $\frac{1}{0}$ is undefined.

Example 2

$$\frac{18}{25} \div \frac{30}{35} = \frac{\overset{3}{18}}{\underset{5}{25}} \cdot \frac{\overset{7}{35}}{\underset{5}{30}} = \frac{21}{25}$$

Example 3

$$\frac{9}{8} \div 4 = \frac{9}{8} \div \frac{4}{1} = \frac{9}{8} \cdot \frac{1}{4} = \frac{9}{32}$$

Example 4

$$\frac{5}{42} \div \left[\left(\frac{1}{6} \right)^2 \cdot 3 \right] = \frac{5}{42} \div \left[\frac{1}{\underset{12}{36}} \cdot \frac{\overset{1}{3}}{1} \right]$$

$$= \frac{5}{42} \div \frac{1}{12}$$

$$= \frac{5}{\underset{7}{42}} \cdot \frac{\overset{2}{12}}{1} = \frac{10}{7}$$

| **Section 2.6** | **Multiplication and Division of Mixed Numbers** |

Key Concepts

Multiply Mixed Numbers

Step 1 Change each mixed number to an improper fraction.
Step 2 Multiply the improper fractions and reduce to lowest terms, if possible.

Divide Mixed Numbers

Step 1 Change each mixed number to an improper fraction.
Step 2 Divide the improper fractions and reduce to lowest terms, if possible. Recall that to divide fractions, multiply the dividend by the reciprocal of the divisor.

Examples

Example 1

$$4\frac{4}{5} \cdot 2\frac{1}{2} = \frac{\overset{12}{24}}{\underset{1}{5}} \cdot \frac{\overset{1}{5}}{\underset{1}{2}} = \frac{12}{1} = 12$$

Example 2

$$6\frac{2}{3} \div 2\frac{7}{9} = \frac{20}{3} \div \frac{25}{9} = \frac{\overset{4}{20}}{\underset{1}{3}} \cdot \frac{\overset{3}{9}}{\underset{5}{25}} = \frac{12}{5} = 2\frac{2}{5}$$

Chapter 2   Review Exercises

Section 2.1

For Exercises 1–2, write a fraction that represents the shaded area.

1.

2.

3. a. Write a fraction that has denominator 3 and numerator 5.

  b. Label this fraction as proper or improper.

4. a. Write a fraction that has numerator 1 and denominator 6.

  b. Label this fraction as proper or improper.

5. In an office supply store, 7 of the 15 computers displayed are laptops. Write a fraction representing the computers that are laptops.

For Exercises 6–7, write a fraction and a mixed number that represent the shaded area.

6.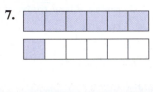

7.

For Exercises 8–9, convert the mixed number to a fraction.

8. $6\frac{1}{7}$

9. $11\frac{2}{5}$

10. How many fourths are in $4\frac{1}{4}$?

For Exercises 11–12, convert the improper fraction to a mixed number.

11. $\frac{47}{9}$

12. $\frac{23}{21}$

For Exercises 13–15, locate the numbers on the number line.

13. $\frac{10}{5}$ **14.** $\frac{7}{8}$ **15.** $\frac{13}{8}$

For Exercises 16–17, divide. Write the answer as a mixed number.

16. $7\overline{)941}$ **17.** $26\overline{)1582}$

Section 2.2

For Exercises 18–20, refer to this list of numbers: 21, 43, 51, 55, 58, 124, 140, 260, 1200.

18. List all the numbers that are divisible by 3.

19. List all the numbers that are divisible by 5.

20. List all the numbers that are divisible by 2.

For Exercises 21–24, determine if the number is prime, composite, or neither.

21. 61   **22.** 44

23. 1   **24.** 0

For Exercises 25–27, find the prime factorization.

25. 64   **26.** 330

27. 900

For Exercises 28–29, list all the factors of the number.

28. 48   **29.** 80

Section 2.3

For Exercises 30–31, determine if the fractions are equivalent. Fill in the blank with = or ≠.

30. $\frac{3}{6}\ \square\ \frac{5}{9}$ **31.** $\frac{15}{21}\ \square\ \frac{10}{14}$

For Exercises 32–39, simplify the fraction to lowest terms. Write the answer as a fraction.

32. $\dfrac{5}{20}$ **33.** $\dfrac{14}{49}$ **34.** $\dfrac{24}{16}$

35. $\dfrac{63}{27}$ **36.** $\dfrac{17}{17}$ **37.** $\dfrac{42}{21}$

38. $\dfrac{120}{150}$ **39.** $\dfrac{1400}{2000}$

40. On his final exam, Gareth got 42 out of 45 questions correct. What fraction of the test represents correct answers? What fraction represents incorrect answers?

41. Isaac proofread 6 pages of his 10-page term paper. Yulisa proofread 6 pages of her 15-page term paper.

  a. What fraction of his paper did Isaac proofread?

  b. What fraction of her paper did Yulisa proofread?

Section 2.4

For Exercises 42–47, multiply the fractions and simplify to lowest terms. Write the answer as a fraction or whole number.

42. $\dfrac{3}{5} \times \dfrac{2}{7}$ **43.** $\dfrac{4}{3} \times \dfrac{8}{3}$ **44.** $14 \cdot \dfrac{9}{2}$

45. $33 \cdot \dfrac{5}{11}$ **46.** $\dfrac{2}{9} \cdot \dfrac{5}{8} \cdot \dfrac{36}{25}$ **47.** $\dfrac{45}{7} \cdot \dfrac{6}{10} \cdot \dfrac{28}{63}$

For Exercises 48–51, evaluate by using the order of operations.

48. $\left(\dfrac{1}{10}\right)^4$ **49.** $\left(\dfrac{2}{5}\right)^2 \cdot \left(\dfrac{1}{10}\right)^2$

50. $\left(\dfrac{3}{20} \cdot \dfrac{2}{3}\right)^3$ **51.** $\left(\dfrac{1}{10}\right)^3\left(\dfrac{1000}{17}\right)$

52. Write the formula for the area of a triangle.

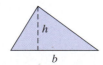

53. Write the formula for the area of a rectangle.

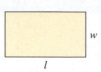

For Exercises 54–56, find the area of the shaded region.

54.

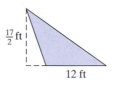

55.

56.

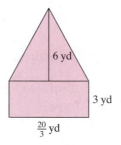

57. Maximus wants to build a workbench for his garage. He needs four boards, each cut into $\frac{7}{8}$-yd pieces. How many yards of lumber does he require?

For Exercises 58–61, refer to the graph. The graph represents the distribution of the students at a college by race/ethnicity.

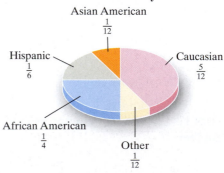

Distribution of Student Body by Race/Ethnicity

58. If the college has 3600 students, how many are African American?

59. If the college has 3600 students, how many are Asian American?

60. If the college has 3600 students, how many are Hispanic females (assume that one-half of the Hispanic student population is female)?

61. If the college has 3600 students, how many are Caucasian males (assume that one-half of the Caucasian student population is male)?

Section 2.5

For Exercises 62–63, multiply.

62. $\dfrac{3}{4} \cdot \dfrac{4}{3}$

63. $\dfrac{1}{12} \cdot 12$

For Exercises 64–67, find the reciprocal of the number, if it exists.

64. $\dfrac{7}{2}$

65. 7

66. 0

67. $\dfrac{1}{6}$

68. Dividing by 5 is the same as multiplying by _____.

69. Dividing by $\frac{2}{9}$ is the same as _____ by $\frac{9}{2}$.

For Exercises 70–75, divide and simplify the answer to lowest terms. Write the answer as a fraction or whole number.

70. $\dfrac{28}{15} \div \dfrac{21}{20}$

71. $\dfrac{7}{9} \div \dfrac{35}{63}$

72. $\dfrac{6}{7} \div 18$

73. $\dfrac{3}{10} \div \dfrac{9}{5}$

74. $\dfrac{200}{51} \div \dfrac{25}{17}$

75. $12 \div \dfrac{6}{7}$

For Exercises 76–79, simplify by using the order of operations. Write the answer as a fraction.

76. $\left(\dfrac{2}{19} \div \dfrac{8}{19}\right)^3$

77. $\left(\dfrac{12}{5}\right)^2 \div \dfrac{36}{5}$

78. $\dfrac{81}{55} \div \dfrac{3}{11} \div \dfrac{3}{2}$

79. $\dfrac{4}{13} \cdot \left(\dfrac{1}{2}\right)^3 \div 2$

For Exercises 80–81, translate to a mathematical statement. Then simplify.

80. How much is $\frac{4}{5}$ of 20?

81. How many $\frac{2}{3}$'s are in 18?

82. How many $\frac{2}{3}$-lb bags of candy can be filled from a 24-lb sack of candy?

83. Amelia worked only $\frac{4}{5}$ of her normal 40-hr workweek. If she makes $18 per hour, how much money did she earn for the week?

84. A small patio floor will be made from square pieces of tile that are $\frac{4}{3}$ ft on a side. See the figure. Find the area of the patio (in square feet) if its dimensions are 10 tiles by 12 tiles.

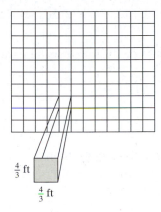

$\frac{4}{3}$ ft

$\frac{4}{3}$ ft

85. Chuck is an elementary school teacher and needs 22 pieces of wood, $\frac{3}{8}$ ft long, for a class project. If he has a 9-ft board from which to cut the pieces, will he have enough $\frac{3}{8}$-ft pieces for his class? Explain.

Section 2.6

For Exercises 86–96, multiply or divide as indicated.

86. $\left(3\dfrac{2}{3}\right)\left(6\dfrac{2}{5}\right)$

87. $\left(11\dfrac{1}{3}\right)\left(2\dfrac{3}{34}\right)$

88. $6\dfrac{1}{2} \cdot 1\dfrac{3}{13}$

89. $4 \cdot \left(5\dfrac{5}{8}\right)$

90. $45\dfrac{5}{13} \cdot 0$

91. $4\dfrac{5}{16} \div 2\dfrac{7}{8}$

92. $3\dfrac{5}{11} \div 3\dfrac{4}{5}$

93. $7 \div 1\dfrac{5}{9}$

94. $4\dfrac{6}{11} \div 2$

95. $10\dfrac{1}{5} \div 17$

96. $0 \div 3\dfrac{5}{12}$

97. It takes $1\frac{1}{4}$ gal of paint for Neva to paint her living room. If her great room (including the dining area) is $2\frac{1}{2}$ times larger than the living room, how many gallons will it take to paint that room?

98. A roll of ribbon contains $12\frac{1}{2}$ yd. How many pieces of length $1\frac{1}{4}$ yd can be cut from this roll?

Chapter 2   Test

1. a. Write a fraction that represents the shaded portion of the figure.

   b. Is the fraction proper or improper?

2. a. Write a fraction that represents the total shaded portion of the three figures.

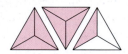

   b. Is the fraction proper or improper?

3. Write an improper fraction and a mixed number that represent the shaded region.

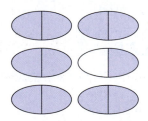

4. Is the fraction $\frac{7}{7}$ a proper or an improper fraction? Explain.

5. a. Write $\frac{11}{3}$ as a mixed number.

   b. Write $3\frac{7}{9}$ as an improper fraction.

For Exercises 6–9, plot the fraction on the number line.

6. $\dfrac{1}{2}$

7. $\dfrac{3}{4}$

8. $\dfrac{7}{12}$

9. $\dfrac{13}{5}$

10. Label the following numbers as prime, composite, or neither.

   a. 15   **b.** 0

   c. 53   **d.** 1

   e. 29   **f.** 39

11. a. List all the factors of 45.

   b. Write the prime factorization of 45.

12. a. What is the divisibility rule for 3?

   b. Is 1,981,011 divisible by 3?

13. Determine whether 1155 is divisible by

   a. 2   **b.** 3

   c. 5   **d.** 10

For Exercises 14–15, determine if the fractions are equivalent. Then fill in the blank with either = or ≠ .

14. $\frac{15}{12} \square \frac{5}{4}$

15. $\frac{2}{5} \square \frac{4}{25}$

For Exercises 16–17, simplify the fractions to lowest terms.

16. $\frac{150}{105}$

17. $\frac{1,200,000}{1,400,000}$

18. Christine and Brad are putting their photographs in scrapbooks. Christine has placed 15 of her 25 photos and Brad has placed 16 of his 20 photos.

   a. What fractional part of the total photos has each person placed?

   b. Which person has a greater fractional part completed?

For Exercises 19–26, multiply or divide as indicated. Simplify the fraction to lowest terms.

19. $\frac{2}{9} \times \frac{57}{46}$

20. $\left(\frac{75}{24}\right) \cdot 4$

21. $\frac{28}{24} \div \frac{21}{8}$

22. $\frac{105}{42} \div 5$

23. $\frac{2}{18} \times \frac{9}{25} \times \frac{40}{6}$

24. $\frac{600}{1200} \div \frac{50}{65} \div \frac{13}{15}$

25. $\frac{10}{21} \div 4\frac{1}{6}$

26. $4\frac{4}{17} \cdot 2\frac{4}{15}$

27. Perform the order of operations. Simplify the fraction to lowest terms.

$$\frac{52}{72} \div \left[\left(\frac{1}{2}\right)^2 \cdot \frac{8}{3}\right]$$

28. Find the area of the triangle.

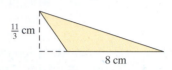

$\frac{11}{3}$ cm

8 cm

29. Which is greater, $20 \cdot \frac{1}{4}$ or $20 \div \frac{1}{4}$?

30. How many "quarter-pounders" can be made from 12 lb of ground beef?

31. The Humane Society has 120 dogs. Of the 120, $\frac{5}{8}$ are female. Among the female dogs, $\frac{1}{15}$ are pure breeds. How many of the dogs are female and pure breeds?

32. A zoning requirement indicates that a house built on less than 1 acre of land may take up no more than one-half of the land. If Liz and George purchased a $\frac{4}{5}$-acre lot of land, what is the maximum land area that they can use to build the house?

Chapters 1–2 Cumulative Review Exercises

1. Fill in the table with either the word name for the number or the number in standard form.

Mountain	Height (ft)	
	Standard Form	Words
Mt. Foraker (Alaska)		Seventeen thousand, four hundred
Mt. Kilimanjaro (Tanzania)	19,340	
El Libertador (Argentina)		Twenty-two thousand, forty-seven
Mont Blanc (France-Italy)	15,771	

For Exercises 2–13, perform the indicated operation.

2. $432 + 998$

3. $572 - 433$

4. 4122×52

5. $384 \div 16$

6. $23(81)$

7. $4\overline{)74}$

8. $\begin{array}{r} 3,000,000 \\ \times\ 40,000 \end{array}$

9. $\begin{array}{r} 1007 \\ -\ 823 \end{array}$

10. $\frac{48}{8}$

11. $6 + 2 \cdot 8$

12. $5^2 - 3^2$

13. $(5 - 3)^2$

For Exercises 14–18, match the algebraic statement with the property that it demonstrates.

14. $5 \cdot 8 = 8 \cdot 5$

    a. Commutative property of addition

15. $4(3 + 2)$
    $= 4 \cdot 3 + 4 \cdot 2$

    b. Associative property of addition

16. $(12 + 3) + 5$
    $= 12 + (3 + 5)$

    c. Distributive property of multiplication over addition

17. $8 \cdot (7 \cdot 2)$
    $= (8 \cdot 7) \cdot 2$

    d. Commutative property of multiplication

18. $32 + 9$
    $= 9 + 32$

    e. Associative property of multiplication

19. Write a fraction that represents the shaded area.

    a.

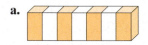

    b.

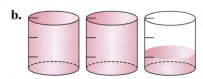

20. Identify each fraction as proper or improper.

    a. $\dfrac{7}{8}$ **b.** $\dfrac{8}{7}$ **c.** $\dfrac{8}{8}$

21. **a.** List all the factors of 30.

    b. Write the prime factorization of 30.

22. Simplify the fractions to lowest terms.

    a. $\dfrac{144}{84}$ **b.** $\dfrac{60{,}000}{150{,}000}$

23. Multiply and simplify to lowest terms. $\dfrac{35}{27} \cdot \dfrac{51}{95}$

24. Divide and simplify to lowest terms. $5\frac{2}{3} \div 6\frac{4}{5}$

25. Is multiplication of fractions a commutative operation? Explain, using the fractions $\frac{8}{13}$ and $\frac{5}{16}$.

26. Is multiplication of fractions an associative operation? Explain, using the fractions $\frac{1}{2}, \frac{2}{9},$ and $\frac{5}{3}$.

27. Simplify. $\left(\dfrac{5}{6} \cdot \dfrac{12}{25} \right)^2 \div \dfrac{2}{3}$

28. Find the area of the rectangle.

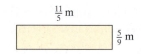

29. Find the area of the triangle.

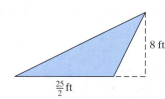

30. At one college $\frac{3}{4}$ of the students are male, and of the males, $\frac{1}{10}$ are from out of state. What fraction of the students are males who are from out of state?

Fractions and Mixed Numbers: Addition and Subtraction

3

Chapter 3

In this chapter we learn about addition and subtraction of fractions as well as operations on mixed numbers. Match each statement in the left column with the appropriate choice from the right column. Record the letter of the corresponding choice in the spaces below to complete the sentence.

Matching

1. Given $\frac{5}{7}$, select a like fraction.

h. $\frac{7}{8}$

2. The LCD of $\frac{3}{4}$ and $\frac{1}{3}$

t. $\frac{1}{7}$

3. A fraction that is not in lowest terms

y. $\frac{7}{9}$

4. The sum of $\frac{3}{8}$ and $\frac{4}{8}$

p. $\frac{15}{20}$

5. The LCD of $\frac{3}{8}$, $\frac{1}{5}$, and $\frac{1}{2}$

a. $1\frac{5}{12}$

6. The difference of $2\frac{19}{24}$ and $1\frac{3}{8}$

v. $1\frac{7}{12}$

7. The sum of $\frac{3}{4}$ and $\frac{5}{6}$

e. 40

8. Which is smaller? $\frac{7}{8}$ or $\frac{7}{9}$

o. 12

An improper fraction is $\underline{}\ \underline{}\ \underline{}\ \ \underline{}\ \underline{}\ \underline{}\ \underline{}\ \underline{}$.

1 2 3   4 5 6 7 8

163

Section 3.1	Addition and Subtraction of Like Fractions

1. Addition and Subtraction of Like Fractions

In Chapter 2 we learned how to multiply and divide fractions. The main focus of this chapter is to add and subtract fractions. The operation of addition can be thought of as combining like groups of objects. For example:

$$3 \text{ apples} + 1 \text{ apple} = 4 \text{ apples}$$

$$\text{three-fifths} + \text{one-fifth} = \text{four-fifths}$$

$$\frac{3}{5} + \frac{1}{5} = \frac{4}{5}$$

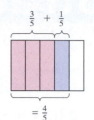

The fractions $\frac{3}{5}$ and $\frac{1}{5}$ are said to be **like fractions** because their denominators are the same. That is, the fractions have a **common denominator**. In general, two or more like fractions may be added according to the following procedure.

> **PROCEDURE Adding and Subtracting Like Fractions**
>
> **Step 1** Add or subtract the numerators.
> **Step 2** Write the sum or difference over the common denominator.
> **Step 3** Simplify the fraction to lowest terms, if possible.

Example 1 Adding Like Fractions

Add. Write the answer as a fraction or whole number.

a. $\dfrac{1}{4} + \dfrac{5}{4}$ **b.** $\dfrac{7}{15} + \dfrac{2}{15} + \dfrac{1}{15}$

Solution:

a. $\dfrac{1}{4} + \dfrac{5}{4} = \dfrac{1+5}{4}$ Add the numerators.

$\phantom{\dfrac{1}{4} + \dfrac{5}{4}} = \dfrac{6}{4}$ Write the sum over the common denominator.

$\phantom{\dfrac{1}{4} + \dfrac{5}{4}} = \dfrac{\overset{3}{\cancel{6}}}{\underset{2}{\cancel{4}}}$ Simplify to lowest terms.

$\phantom{\dfrac{1}{4} + \dfrac{5}{4}} = \dfrac{3}{2}$

> **Avoiding Mistakes**
>
> Notice that when adding fractions, we do not add the denominators. We add *only* the numerators.

b. $\dfrac{7}{15} + \dfrac{2}{15} + \dfrac{1}{15} = \dfrac{7+2+1}{15}$ Add the numerators.

$\phantom{\dfrac{7}{15} + \dfrac{2}{15} + \dfrac{1}{15}} = \dfrac{10}{15}$ Write the sum over the common denominator.

$\phantom{\dfrac{7}{15} + \dfrac{2}{15} + \dfrac{1}{15}} = \dfrac{2}{3}$ Simplify to lowest terms.

Avoiding Mistakes

Note that the process to add fractions is different from the process to multiply fractions.

$$\frac{2}{7} \times \frac{3}{7} = \frac{6}{49} \quad \text{but} \quad \frac{2}{7} + \frac{3}{7} = \frac{5}{7}$$

Concept Connections

3. Which is the correct sum for $\frac{2}{3} + \frac{5}{3}$?

   $\dfrac{7}{6}$ or $\dfrac{7}{3}$

4. Which is the correct product for $\frac{2}{3} \cdot \frac{5}{3}$?

   $\dfrac{10}{9}$ or $\dfrac{10}{3}$

Example 2 Subtracting Like Fractions

Subtract. Write the answer as a fraction or whole number.

a. $\dfrac{13}{9} - \dfrac{2}{9}$ **b.** $\dfrac{4}{3} - \dfrac{1}{3}$

Solution:

a. $\dfrac{13}{9} - \dfrac{2}{9} = \dfrac{13 - 2}{9}$ Subtract the numerators.

$\qquad = \dfrac{11}{9}$ Write the difference over the common denominator. The fraction is already in lowest terms because 11 and 9 share no common factors.

b. $\dfrac{4}{3} - \dfrac{1}{3} = \dfrac{4 - 1}{3}$ Subtract the numerators.

$\qquad = \dfrac{3}{3}$ Write the difference over the common denominator.

$\qquad = 1$ Simplify.

Skill Practice

Subtract. Write the answer as a fraction or whole number.

5. $\dfrac{14}{11} - \dfrac{8}{11}$ **6.** $\dfrac{5}{2} - \dfrac{3}{2}$

2. Order of Operations

Example 3 reviews the order of operations.

PROCEDURE Order of Operations

Step 1 Perform all operations inside parentheses first.

Step 2 Simplify expressions containing exponents or square roots.

Step 3 Perform multiplication or division in the order that they appear from left to right.

Step 4 Perform addition or subtraction in the order that they appear from left to right.

Example 3 Applying the Order of Operations

Simplify.

a. $\left(\dfrac{2}{7} + \dfrac{1}{7} \right)^2$ **b.** $\dfrac{1}{3} + \dfrac{3}{5} \div \dfrac{9}{10}$

Answers

3. $\dfrac{7}{3}$ **4.** $\dfrac{10}{9}$ **5.** $\dfrac{6}{11}$ **6.** 1

Solution:

a. $\left(\dfrac{2}{7} + \dfrac{1}{7}\right)^2 = \left(\dfrac{2+1}{7}\right)^2$ — Add fractions within parentheses first.

$= \left(\dfrac{3}{7}\right)^2$

$= \dfrac{3}{7} \cdot \dfrac{3}{7}$ — Square the fraction $\dfrac{3}{7}$.

$= \dfrac{9}{49}$ — The fraction is in lowest terms.

b. $\dfrac{1}{3} + \dfrac{3}{5} \div \dfrac{9}{10} = \dfrac{1}{3} + \left(\dfrac{3}{5} \div \dfrac{9}{10}\right)$ — We can insert parentheses to emphasize that division is performed before addition.

$= \dfrac{1}{3} + \left(\dfrac{3}{5} \cdot \dfrac{10}{9}\right)$ — Multiply by the reciprocal of the divisor.

$= \dfrac{1}{3} + \left(\dfrac{\overset{1}{\cancel{3}}}{\underset{1}{\cancel{5}}} \cdot \dfrac{\overset{2}{\cancel{10}}}{\underset{3}{\cancel{9}}}\right)$ — Multiply the fractions.

$= \dfrac{1}{3} + \left(\dfrac{2}{3}\right)$

$= \dfrac{3}{3}$ — Add the fractions.

$= 1$ — Simplify.

3. Applications of Addition and Subtraction of Fractions

Recall that the perimeter of a polygon is found by adding the lengths of the sides.

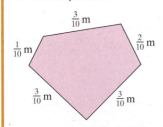

Example 4 Finding Perimeter

Find the perimeter.

Solution:

$\text{Perimeter} = \dfrac{10}{12} + \dfrac{5}{12} + \dfrac{15}{12} + \dfrac{5}{12}$

$= \dfrac{10+5+15+5}{12}$

$= \dfrac{35}{12}$ or $2\dfrac{11}{12}$

The perimeter is $2\dfrac{11}{12}$ yd.

| Example 5 | Applying Addition and Subtraction of Fractions |

Pam mixed $\frac{12}{16}$ gal of water with $\frac{1}{16}$ gal of liquid fertilizer. Then she used $\frac{7}{16}$ gal of the mixture to fertilizer her blueberry bushes. How much mixture was leftover?

Solution:

The net amount of liquid remaining is given by

Amount Pam started with · Amount Pam used

$$\overbrace{\frac{12}{16} + \frac{1}{16}} - \overbrace{\frac{7}{16}}$$

$$= \frac{12 + 1 - 7}{16}$$

$$= \frac{\overset{3}{\cancel{6}}}{\underset{8}{\cancel{16}}}$$

$$= \frac{3}{8}$$

There was $\frac{3}{8}$ gallon of mixture leftover.

Skill Practice

10. Jamie mixed $\frac{5}{8}$ gal of green paint with $\frac{7}{8}$ gal of white paint. Then she used $\frac{3}{8}$ gal of the mixture to paint a mural. How much paint is left over?

Answer

10. $\frac{9}{8}$ or $1\frac{1}{8}$ gal

Section 3.1 Practice Exercises

Boost your GRADE at ALEKS.com!

- Practice Problems
- Self-Tests
- NetTutor
- e-Professors
- Videos

Study Skills Exercises

1. How can you use the concept connections and skill practice exercises in the margins of this text?

2. Define the key terms.

  a. **Like fractions** b. **Common denominator**

Objective 1: Addition and Subtraction of Like Fractions

For Exercises 3–7, add the like units.

3. 3 ft + 5 ft

4. 7 chairs + 2 chairs

5. 7 m + 13 m

6. 8 thirds + 2 thirds

7. 1 fourth + 6 fourths

For Exercises 8–9, shade in the portion of the third figure that represents the addition of the first two figures.

8.

9.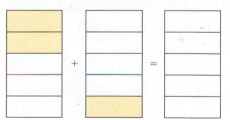

10. Explain the difference between evaluating the two expressions $\frac{2}{5} \times \frac{7}{5}$ and $\frac{2}{5} + \frac{7}{5}$.

For Exercises 11–22, add the like fractions. Write the answer as a fraction or whole number. **(See Example 1.)**

11. $\dfrac{6}{11} + \dfrac{7}{11}$ **12.** $\dfrac{5}{3} + \dfrac{2}{3}$ **13.** $\dfrac{6}{5} + \dfrac{3}{5}$ **14.** $\dfrac{3}{10} + \dfrac{4}{10}$

15. $\dfrac{1}{4} + \dfrac{3}{4}$ **16.** $\dfrac{1}{8} + \dfrac{3}{8}$ **17.** $\dfrac{2}{9} + \dfrac{4}{9}$ **18.** $\dfrac{3}{2} + \dfrac{5}{2}$

19. $\dfrac{3}{20} + \dfrac{8}{20} + \dfrac{15}{20}$ **20.** $\dfrac{5}{8} + \dfrac{4}{8} + \dfrac{9}{8}$ **21.** $\dfrac{18}{14} + \dfrac{11}{14} + \dfrac{6}{14}$ **22.** $\dfrac{7}{18} + \dfrac{22}{18} + \dfrac{10}{18}$

23. Bethany pours $\frac{1}{4}$ cup of bleach into a container and then adds $\frac{9}{4}$ cups of water. How many cups of bleach and water mixture does she have?

24. Austin rode his bike $\frac{7}{6}$ mi before he got a flat tire. He then had to walk another $\frac{1}{6}$ mi. How far did Austin travel?

For Exercises 25–28, subtract the like units.

25. 15 baskets − 4 baskets   **26.** 52 cards − 13 cards

27. 7 fifths − 1 fifth   **28.** 18 tenths − 11 tenths

For Exercises 29–30, shade in the portion of the third figure that represents the subtraction of the first two figures.

29.   **30.**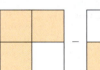

For Exercises 31–42, subtract the like fractions. Write the answer as a fraction or whole number. **(See Example 2.)**

31. $\dfrac{9}{8} - \dfrac{6}{8}$ **32.** $\dfrac{7}{9} - \dfrac{6}{9}$ **33.** $\dfrac{9}{2} - \dfrac{6}{2}$ **34.** $\dfrac{10}{4} - \dfrac{5}{4}$

35. $\dfrac{13}{3} - \dfrac{7}{3}$ **36.** $\dfrac{13}{10} - \dfrac{3}{10}$ **37.** $\dfrac{23}{12} - \dfrac{15}{12}$ **38.** $\dfrac{13}{6} - \dfrac{5}{6}$

39. $\dfrac{28}{25} - \dfrac{14}{25} - \dfrac{4}{25}$ **40.** $\dfrac{34}{15} - \dfrac{6}{15} - \dfrac{3}{15}$ **41.** $\dfrac{10}{16} - \dfrac{1}{16} - \dfrac{5}{16}$ **42.** $\dfrac{31}{40} - \dfrac{14}{40} - \dfrac{12}{40}$

43. A chemist has $\frac{5}{8}$ grams (g) of NaCl (salt). If he uses $\frac{3}{8}$ g, how much is left?

44. Jason bought $\frac{11}{4}$ acres of land and then sold $\frac{3}{4}$ acre. How much land does he have left?

Mixed Exercises

For Exercises 45–56, add or subtract as indicated. Write the answer as a fraction or whole number.

45. $\dfrac{7}{8} + \dfrac{5}{8}$

46. $\dfrac{1}{21} + \dfrac{13}{21}$

47. $\dfrac{14}{5} - \dfrac{2}{5}$

48. $\dfrac{5}{3} - \dfrac{2}{3}$

49. $\dfrac{6}{13} + \dfrac{7}{13}$

50. $\dfrac{20}{35} + \dfrac{12}{35}$

51. $\dfrac{14}{15} + \dfrac{2}{15} - \dfrac{4}{15}$

52. $\dfrac{19}{6} - \dfrac{11}{6} + \dfrac{5}{6}$

53. $\dfrac{7}{2} - \dfrac{3}{2} + \dfrac{1}{2}$

54. $\dfrac{8}{3} + \dfrac{2}{3} - \dfrac{1}{3}$

55. $\dfrac{19}{12} - \dfrac{5}{12} + \dfrac{7}{12}$

56. $\dfrac{7}{18} + \dfrac{13}{18} - \dfrac{5}{18}$

Objective 2: Order of Operations

For Exercises 57–68, simplify the expression by using the order of operations. Write the answer as a fraction or whole number. **(See Example 3.)**

57. $\left(\dfrac{11}{10} - \dfrac{2}{10}\right)^2$

58. $\left(\dfrac{7}{3} - \dfrac{5}{3}\right)^3$

59. $\dfrac{5}{4} \div \dfrac{3}{2} + \dfrac{5}{6}$

60. $\dfrac{1}{7} \div \dfrac{2}{21} + \dfrac{5}{2}$

61. $\dfrac{6}{5} + \dfrac{7}{5} - \dfrac{4}{5}$

62. $\dfrac{10}{3} - \dfrac{2}{3} + \dfrac{5}{3}$

63. $\dfrac{3}{7} + \dfrac{13}{14} \cdot 2$

64. $\dfrac{13}{6} - \dfrac{5}{18} \cdot 3$

65. $\left(\dfrac{2}{21} + \dfrac{11}{21}\right) \div \dfrac{1}{7}$

66. $\left(\dfrac{17}{30} - \dfrac{12}{30}\right) \div \dfrac{5}{6}$

67. $\dfrac{17}{30} - \dfrac{1}{2} \cdot \dfrac{7}{15}$

68. $\dfrac{5}{12} - \dfrac{1}{2} \cdot \dfrac{1}{6}$

Objective 3: Applications of Addition and Subtraction of Fractions

For Exercises 69–70, find the perimeter. **(See Example 4.)**

69.

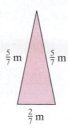

70.

71. Find the perimeter of the stamp.

72. Find the perimeter of the top of the table.

73. Gabby mixed $\frac{7}{10}$ gal of red paint with $\frac{7}{10}$ gal of white paint. She used $\frac{3}{10}$ gal of the mixture to paint a room. How much mixture was left over? **(See Example 5.)**

74. Emeril mixed $\frac{3}{8}$ cup balsamic vinegar with $\frac{4}{8}$ cup oil to make a salad dressing. Then he used $\frac{1}{8}$ cup of the mixture for a large salad. How much oil and vinegar mixture was left over?

75. A chemist mixed $\frac{5}{8}$ liter (L) of water with $\frac{7}{8}$ L of alcohol. Then he used one-quarter of the mixture in an experiment. How much mixture did he use?

76. Malcom planted tomatoes in $\frac{2}{7}$ of his garden. He planted cucumbers in $\frac{3}{7}$ of the garden and cabbage in $\frac{1}{7}$. The remaining $\frac{1}{7}$ still has not yet been planted. A deer came in one night and ate $\frac{1}{3}$ of the plants in the planted area. What fraction of the garden did the deer eat?

77. Thilan has taken up a new exercise program. He walks 6 days per week. One week he walked the distances given in the table.

    a. Find the total distance he walked for the week.

    b. Find the average distance walked per day.

Day	Distance
Monday	$\frac{4}{10}$ mi
Tuesday	$\frac{7}{10}$ mi
Wednesday	$\frac{9}{10}$ mi
Thursday	$\frac{5}{10}$ mi
Friday	$\frac{13}{10}$ mi
Saturday	$\frac{17}{10}$ mi

78. Denzel recorded the weekly rainfall for his town for 4 weeks of summer.

    a. Find the total amount of rainfall for this 4-week period.

    b. Find the average rainfall per week.

Week	Amount of Rainfall
1	$\frac{2}{10}$ in.
2	$\frac{7}{10}$ in.
3	$\frac{9}{10}$ in.
4	$\frac{17}{10}$ in.

For Exercises 79–82, find the perimeter and the area.

79.

$\frac{3}{8}$ ft

$\frac{5}{8}$ ft

80.

$\frac{7}{8}$ m

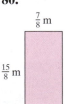

$\frac{15}{8}$ m

81.

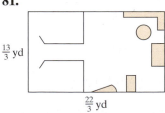

$\frac{13}{3}$ yd

$\frac{22}{3}$ yd

82.

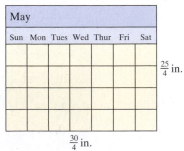

$\frac{25}{4}$ in.

$\frac{30}{4}$ in.

For Exercises 83–86, translate the phrase to a mathematical expression, then simplify.

83. The sum of three-fifths and two-fifths

84. Seven-ninths more than five-ninths

85. The difference of eleven-fifteenths and eight-fifteenths

86. Two-sevenths subtracted from five-sevenths

Least Common Multiple

1. Least Common Multiple

In Section 3.1 we learned how to add and subtract like fractions. To add or subtract fractions with different denominators, we must learn how to convert unlike fractions into like fractions. An essential concept in this process is the idea of a least common multiple of two or more numbers.

When we multiply a number by the whole numbers 1, 2, 3, and so on, we form the **multiples** of the number. For example, some of the multiples of 6 and 9 are shown below.

Objectives

1. Least Common Multiple
2. Finding the LCM by Using Prime Factors
3. Finding the LCM by Using Division by Primes (Optional)
4. Applications of the LCM
5. Equivalent Fractions and Ordering Fractions

Multiples of 6	Multiples of 9
$6 \times 1 = 6$	$9 \times 1 = 9$
$6 \times 2 = 12$	$9 \times 2 = 18$
$6 \times 3 = 18$	$9 \times 3 = 27$
$6 \times 4 = 24$	$9 \times 4 = 36$
$6 \times 5 = 30$	$9 \times 5 = 45$
$6 \times 6 = 36$	$9 \times 6 = 54$
$6 \times 7 = 42$	$9 \times 7 = 63$
$6 \times 8 = 48$	$9 \times 8 = 72$
$6 \times 9 = 54$	$9 \times 9 = 81$

In red, we have indicated several multiples that are common to both 6 and 9.

The **least common multiple (LCM)** of two given numbers is the smallest whole number that is a multiple of each given number. For example, the LCM of 6 and 9 is 18.

Multiples of 6:   6, 12, 18, 24, 30, 36, 42, . . .

Multiples of 9:   9, 18, 27, 36, 45, 54, 63, . . .

Concept Connections

1. Explain the difference between a multiple of a number and a factor of a number.

TIP: There are infinitely many numbers that are common multiples of both 6 and 9. These include 18, 36, 54, 72, and so on. However, 18 is the smallest, and is therefore the *least* common multiple.

If one number is a multiple of another number, then the LCM is the larger of the two numbers. For example, the LCM of 4 and 8 is 8.

Multiples of 4:   4, 8, 12, 16, . . .

Multiples of 8:   8, 16, 24, 32, . . .

Example 1 Finding the LCM by Listing Multiples

Find the LCM of the given numbers by listing several multiples of each number.

a. 15 and 12   **b.** 10, 15, and 8

Skill Practice

Find the LCM by listing several multiples of each number.
2. 15 and 25   3. 4, 6, and 10

Solution:

a. Multiples of 15:   15, 30, 45, 60
Multiples of 12:   12, 24, 36, 48, 60

The LCM of 15 and 12 is 60.

Answers

1. A multiple of a number is the product of the number and a whole number 1 or greater. A factor of a number is a value that divides evenly into the number.
2. 75   3. 60

b. Multiples of 10:   10, 20, 30, 40, 50, 60, 70, 80, 90, 100, 110, 120
Multiples of 15:   15, 30, 45, 60, 75, 90, 105, 120
Multiples of 8:   8, 16, 24, 32, 40, 48, 56, 64, 72, 80, 88, 96, 104, 112, 120

The LCM of 10, 15, and 8 is 120.

2. Finding the LCM by Using Prime Factors

In Example 1 we used the method of listing multiples to find the LCM of two or more numbers. As you can see, the solution to Example 1(b) required several long lists of multiples. Here we offer another method to find the LCM of two given numbers by using their prime factors.

> **PROCEDURE   Using Prime Factors to Find the LCM of Two Numbers**
> **Step 1** Write each number as a product of prime factors.
> **Step 2** The LCM is the product of unique prime factors from both numbers. Use repeated factors the maximum number of times they appear in either factorization.

This process is demonstrated in Example 2.

Skill Practice

Find the LCM by using prime factors.

4. 9 and 24
5. 16 and 9
6. 36, 42, and 30

TIP: The product $2 \cdot 2 \cdot 3 \cdot 7$ can also be written as $2^2 \cdot 3 \cdot 7$.

Example 2 Finding the LCM by Using Prime Factors

Find the LCM.

a. 14 and 12   **b.** 50 and 24   **c.** 45, 54, and 50

Solution:

a. Find the prime factorization for 14 and 12.

	2's	3's	7's
14 =	2 ·		(7)
12 =	(2 · 2 ·)	(3)	

For the factors of 2, 3, and 7, we circle the greatest number of times each occurs. The LCM is the product.

LCM = 2 · 2 · 3 · 7 = 84

b. Find the prime factorization for 50 and 24.

	2's	3's	5's
50 =	2 ·		(5 · 5)
24 =	(2 · 2 · 2 ·)	(3)	

The factor 5 is repeated twice. The factor 2 is repeated 3 times. The factor 3 is used only once.

LCM = 2 · 2 · 2 · 3 · 5 · 5 = 600   (The LCM can also be written as $2^3 \cdot 3 \cdot 5^2$.)

c. Find the prime factorization for 45, 54, and 50.

	2's	3's	5's
45 =		3 · 3 ·	5
54 =	2 ·	(3 · 3 · 3)	
50 =	(2 ·)		(5 · 5)

LCM = 2 · 3 · 3 · 3 · 5 · 5 = 1350   (The LCM can also be written as $2 \cdot 3^3 \cdot 5^2$.)

Answers
4. 72   **5.** 144   **6.** 1260

3. Finding the LCM by Using Division by Primes (Optional)

We present a third method for finding least common multiples. We systematically divide by prime numbers to determine which will be a factor of the LCM. This method is particularly helpful if three or more numbers are involved.

Example 3 Finding the LCM by Using Division by Primes

Find the LCM of 32, 48, and 30 by using division of prime factors.

Solution:

To begin this process, find any prime number that divides evenly into any of the numbers. Then divide and write the quotient as shown. We begin by dividing by the smallest prime number, 2.

$$2)\overline{32\ 48\ 30}$$
$$16\ 24\ 15$$

Repeat this process and bring down any number that is not divisible by the chosen prime.

$$2)\overline{32\ 48\ 30}$$
$$2)\overline{16\ 24\ 15}$$
$$)8\ 12\ 15$$

Bring down the 15.

Continue until all quotients are 1. The LCM is the product of the prime factors at the left.

$$2)\overline{32\ 48\ 30}$$
$$2)\overline{16\ 24\ 15}$$
$$2)\overline{8\ 12\ 15}$$
$$2)\overline{4\ 6\ 15}$$
$$2)\overline{2\ 3\ 15}$$
$$3)\overline{1\ 3\ 15}$$
$$5)\overline{1\ 1\ 5}$$
$$1\ 1\ 1$$

At this point, the prime number 2 does not divide evenly into any of the quotients. We try the next-greater prime number, 3.

The LCM is $2 \cdot 2 \cdot 2 \cdot 2 \cdot 2 \cdot 3 \cdot 5 = 480$.

4. Applications of the LCM

Example 4 Using the LCM in an Application

A tile wall is to be made from 6-in., 8-in., and 12-in. square tiles. A design is made by alternating rows with different-size tiles. The first row uses only 6-in. tiles, the second row uses only 8-in. tiles, and the third row uses only 12-in. tiles. Neglecting the grout seams, what is the shortest length of wall space that can be covered using only whole tiles?

Skill Practice

Find the LCM by using division by prime factors.

7. 20, 36, and 15

TIP: We have presented three methods to find the LCM. Try each method. Then you and your instructor can decide which method works best for you.

Skill Practice

8. Three runners run on an oval track. One runner takes 60 sec to complete the loop. The second runner requires 75 sec, and the third runner requires 90 sec. Suppose the runners begin "lined up" at the same point on the track. Find the minimum amount of time required for all three runners to be lined up again.

Answers
7. 180
8. After 900 sec (15 min) the runners will again be "lined up."

Solution:

The length of the first row must be a multiple of 6 in., the length of the second row must be a multiple of 8 in., and the length of the third row must be a multiple of 12 in. Therefore, the shortest-length wall that can be covered is given by the LCM of 6, 8, and 12.

$$6 = 2 \cdot \boxed{3}$$
$$8 = \boxed{2 \cdot 2 \cdot 2}$$
$$12 = 2 \cdot 2 \cdot 3$$

The LCM is $2 \cdot 2 \cdot 2 \cdot 3 = 24$. The shortest-length wall is 24 in.

This means that four 6-in. tiles can be placed on the first row, three 8-in. tiles can be placed on the second row, and two 12-in. tiles can be placed in the third row. See Figure 3-1.

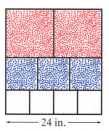

← 24 in. →

Figure 3-1

5. Equivalent Fractions and Ordering Fractions

Suppose we want to determine which of two fractions is larger. Comparing fractions with the same denominator, such as $\frac{3}{5}$ and $\frac{2}{5}$ is relatively easy. Clearly 3 parts out of 5 is greater than 2 parts out of 5.

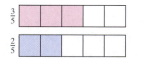

Thus, $\frac{3}{5} > \frac{2}{5}$.

So how would we compare the relative size of two fractions with *different* denominators such as $\frac{3}{5}$ and $\frac{4}{7}$? Our first step is to write the fractions as equivalent fractions with the same denominator, called a common denominator. The **least common denominator (LCD)** of two fractions is the LCM of the denominators of the fractions. The LCD of $\frac{3}{5}$ and $\frac{4}{7}$ is 35, because this is the least common multiple of 5 and 7. In Example 5, we convert the fractions $\frac{3}{5}$ and $\frac{4}{7}$ to equivalent fractions having 35 as the denominator.

Skill Practice

Write the fractions with the indicated denominator.

9. $\frac{2}{3} = \frac{}{12}$ **10.** $\frac{5}{6} = \frac{}{54}$

Answers

9. $\frac{8}{12}$ **10.** $\frac{45}{54}$

Example 5   **Writing Equivalent Fractions**

Write the fractions with the indicated denominator.

a. $\frac{3}{5} = \frac{}{35}$ **b.** $\frac{4}{7} = \frac{}{35}$

Solution:

a. $\frac{3}{5} = \frac{}{35}$ $\frac{3 \cdot 7}{5 \cdot 7} = \frac{21}{35}$ So, $\frac{21}{35}$ is equivalent to $\frac{3}{5}$.

What number must we multiply 5 by to get 35?   Multiply numerator and denominator by 7.

b. $\dfrac{4}{7} = \dfrac{}{35}$

What number must we
multiply 7 by to get 35?

$\dfrac{4 \cdot 5}{7 \cdot 5} = \dfrac{20}{35}$ So, $\dfrac{20}{35}$ is equivalent to $\dfrac{4}{7}$.

Multiply numerator and
denominator by 5.

TIP: Writing a fraction as an equivalent fraction is simply an application of the fundamental principle of fractions (See Section 2.3). In Example 5(a), we multiplied numerator and denominator of the fraction by 7. This is the same as multiplying the fraction by a convenient form of 1.

$$\frac{3}{5} = \frac{3}{5} \cdot 1 = \frac{3}{5} \cdot \frac{7}{7} = \frac{3 \cdot 7}{5 \cdot 7} = \frac{21}{35}$$

This is the same as multiplying
numerator and denominator by 7.

From Example 5, we know that $\dfrac{3}{5} = \dfrac{21}{35}$ and $\dfrac{4}{7} = \dfrac{20}{35}$.

Furthermore, because $\dfrac{21}{35} > \dfrac{20}{35}$, then we know that $\dfrac{3}{5} > \dfrac{4}{7}$.

Example 6   Comparing Two Fractions

Skill Practice

Fill in the blank with $<$, $>$,
or $=$.

11. $\dfrac{4}{7} \square \dfrac{5}{9}$

Fill in the blank with $<$, $>$, or $=$.

$$\frac{9}{8} \square \frac{7}{6}$$

Solution:

The fractions have different denominators and cannot be compared by inspection. The LCD is 24. We need to convert each fraction to an equivalent fraction with a denominator of 24.

$\dfrac{9}{8} = \dfrac{9 \cdot 3}{8 \cdot 3} = \dfrac{27}{24}$ Multiply numerator and denominator by 3,
because $8 \cdot 3 = 24$.

$\dfrac{7}{6} = \dfrac{7 \cdot 4}{6 \cdot 4} = \dfrac{28}{24}$ Multiply numerator and denominator by 4,
because $6 \cdot 4 = 24$.

Because $\dfrac{27}{24} < \dfrac{28}{24}$, then $\dfrac{9}{8} \boxed{<} \dfrac{7}{6}$.

The relationship between $\frac{9}{8}$ and $\frac{7}{6}$ is shown on the number line in Figure 3-2.

Figure 3-2

Answer

11. $>$

Skill Practice

12. Rank the fractions from least to greatest.

$$\frac{5}{9}, \frac{8}{15}, \text{ and } \frac{3}{5}$$

Example 7   Ranking Fractions in Order from Least to Greatest

Rank the fractions from least to greatest.

$$\frac{9}{20}, \frac{7}{15}, \frac{4}{9}$$

Solution:

We want to convert each fraction to an equivalent fraction with a common denominator. The least common denominator is the LCM of 20, 15, and 9.

$$20 = \textcircled{2 \cdot 2} \cdot \textcircled{5}$$
$$15 = 3 \cdot 5$$
$$9 = \textcircled{3 \cdot 3}$$

The least common denominator is $2 \cdot 2 \cdot 3 \cdot 3 \cdot 5 = 180$.

Now convert each fraction to an equivalent fraction with a denominator of 180.

$$\frac{9}{20} = \frac{9 \cdot 9}{20 \cdot 9} = \frac{81}{180}$$
Multiply numerator and denominator by 9 because $20 \cdot 9 = 180$.

$$\frac{7}{15} = \frac{7 \cdot 12}{15 \cdot 12} = \frac{84}{180}$$
Multiply numerator and denominator by 12 because $15 \cdot 12 = 180$.

$$\frac{4}{9} = \frac{4 \cdot 20}{9 \cdot 20} = \frac{80}{180}$$
Multiply numerator and denominator by 20 because $9 \cdot 20 = 180$.

Answer

12. $\frac{8}{15}, \frac{5}{9}, \frac{3}{5}$

Ranking the fractions from least to greatest, we have $\frac{80}{180}, \frac{81}{180}, \frac{84}{180}$. This is equivalent to $\frac{4}{9}, \frac{9}{20}, \frac{7}{15}$.

Section 3.2 Practice Exercises

Boost your GRADE at ALEKS.com!

ALEKS version 3.0

- Practice Problems
- Self-Tests
- NetTutor
- e-Professors
- Videos

Study Skills Exercises

1. Where do you usually do your homework? Is this the best place for you to concentrate? Explain.

2. Define the key terms.

   a. Multiple   **b. Least common multiple (LCM)**   **c. Least common denominator (LCD)**

Review Exercises

For Exercises 3–8, add and subtract as indicated. Write the answer as a whole number or fraction simplified to lowest terms.

3. $\frac{19}{6} - \frac{16}{6}$

4. $\frac{28}{4} - \frac{22}{4}$

5. $\frac{31}{15} + \frac{2}{15} - \frac{8}{15}$

6. $\frac{8}{5} + \frac{12}{5}$

7. $\frac{11}{3} + \frac{7}{3}$

8. $\frac{5}{19} - \frac{2}{19}$

Objective 1: Least Common Multiple

9. a. Circle the multiples of 24: 4, 8, 48, 72, 12, 240

   b. Circle the factors of 24: 4, 8, 48, 72, 12, 240

10. a. Circle the multiples of 30: 15, 90, 120, 3, 5, 60

   b. Circle the factors of 30: 15, 90, 120, 3, 5, 60

11. a. Circle the multiples of 36: 72, 6, 360, 12, 9, 108

   b. Circle the factors of 36: 72, 6, 360, 12, 9, 108

12. a. Circle the multiples of 28: 7, 4, 2, 56, 140, 280

   b. Circle the factors of 28: 7, 4, 2, 56, 140, 280

For Exercises 13–18, find the LCM by listing several multiples of each number. **(See Examples 1.)**

13. 10 and 25   **14.** 21 and 14   **15.** 16 and 12

16. 20 and 12   **17.** 8, 10, and 12   **18.** 4, 6, and 14

Objectives 2–3: Finding the LCM

For Exercises 19–38, find the LCM. **(See Examples 2–3.)**

19. 18 and 24   **20.** 9 and 30   **21.** 12 and 15   **22.** 27 and 45

23. 15 and 25   **24.** 16 and 24   **25.** 24 and 30   **26.** 14 and 35

27. 42 and 70   **28.** 6 and 21   **29.** 20, 18, and 27   **30.** 9, 15, and 42

31. 12, 15, and 20   **32.** 20, 30, and 40   **33.** 16, 24, and 30   **34.** 20, 42, and 35

35. 6, 12, 18, and 20   **36.** 21, 35, 50, and 75   **37.** 5, 15, 18, and 20   **38.** 28, 10, 21, and 35

Objective 4: Applications of the LCM

39. A tile floor is to be made from 10-in., 12-in., and 15-in. square tiles. A design is made by alternating rows with different-size tiles. The first row uses only 10-in. tiles, the second row uses only 12-in. tiles, and the third row uses only 15-in. tiles. Neglecting the grout seams, what is the shortest length of floor space that can be covered evenly by each row? **(See Example 4.)**

40. A patient admitted to the hospital was prescribed a pain medication to be given every 4 hr and an antibiotic to be given every 5 hr. Bandages applied to the patient's external injuries needed changing every 12 hr. The nurse changed the bandages and gave the patient both medications at 6:00 A.M. Monday morning.

   a. How many hours will pass before the patient is given both medications and has his bandages changed at the same time?

   b. What day and time will this be?

41. Four satellites revolve around the earth once every 6, 8, 10, and 15 hr, respectively. If the satellites are initially "lined up," how many hours must pass before they will again be lined up?

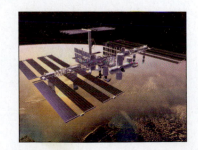

42. Mercury, Venus, and Earth revolve around the Sun approximately once every 3 months, 7 months, and 12 months, respectively (see the figure). If the planets begin "lined up," what is the minimum number of months required for them to be aligned again? (Assume that the planets lie roughly in the same plane.)

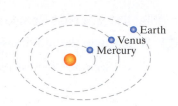

Objective 5: Equivalent Fractions and Ordering Fractions

For Exercises 43–60, rewrite each fraction with the indicated denominators. **(See Example 5.)**

43. $\dfrac{2}{3} = \dfrac{}{21}$

44. $\dfrac{7}{4} = \dfrac{}{32}$

45. $\dfrac{5}{8} = \dfrac{}{16}$

46. $\dfrac{2}{9} = \dfrac{}{27}$

47. $\dfrac{3}{4} = \dfrac{}{16}$

48. $\dfrac{3}{10} = \dfrac{}{50}$

49. $\dfrac{4}{5} = \dfrac{}{15}$

50. $\dfrac{3}{7} = \dfrac{}{70}$

51. $\dfrac{7}{6} = \dfrac{}{42}$

52. $\dfrac{10}{3} = \dfrac{}{18}$

53. $\dfrac{11}{9} = \dfrac{}{99}$

54. $\dfrac{7}{5} = \dfrac{}{35}$

55. $\dfrac{5}{13} = \dfrac{}{39}$

56. $\dfrac{6}{17} = \dfrac{}{34}$

57. $\dfrac{11}{4} = \dfrac{}{4000}$

58. $\dfrac{18}{7} = \dfrac{}{700}$

59. $\dfrac{3}{14} = \dfrac{}{70}$

60. $\dfrac{5}{66} = \dfrac{}{198}$

For Exercises 61–68, fill in the blanks with $<$, $>$, or $=$. **(See Example 6.)**

61. $\dfrac{7}{8} \square \dfrac{3}{4}$

62. $\dfrac{7}{15} \square \dfrac{11}{20}$

63. $\dfrac{13}{10} \square \dfrac{22}{15}$

64. $\dfrac{15}{4} \square \dfrac{21}{6}$

65. $\dfrac{3}{12} \square \dfrac{2}{8}$

66. $\dfrac{5}{20} \square \dfrac{4}{16}$

67. $\dfrac{5}{18} \square \dfrac{8}{27}$

68. $\dfrac{9}{24} \square \dfrac{8}{21}$

69. Which of the following fractions has the greatest value? $\dfrac{2}{3}, \dfrac{7}{8}, \dfrac{5}{6}, \dfrac{1}{2}$

70. Which of the following fractions has the least value? $\dfrac{1}{6}, \dfrac{1}{4}, \dfrac{2}{15}, \dfrac{2}{9}$

For Exercises 71–76, rank the fractions from least to greatest. **(See Example 7.)**

71. $\dfrac{7}{8}, \dfrac{2}{3}, \dfrac{3}{4}$

72. $\dfrac{5}{12}, \dfrac{3}{8}, \dfrac{2}{3}$

73. $\dfrac{5}{16}, \dfrac{3}{8}, \dfrac{1}{4}$

74. $\dfrac{2}{5}, \dfrac{3}{10}, \dfrac{5}{6}$

75. $\dfrac{4}{3}, \dfrac{13}{12}, \dfrac{17}{15}$

76. $\dfrac{5}{7}, \dfrac{11}{21}, \dfrac{18}{35}$

77. A patient had three cuts that needed stitches. A nurse recorded the lengths of the cuts. Where did the patient have the longest cut? Where did the patient have the shortest cut?

upper right arm $\frac{3}{4}$ *in.*
Right hand $\frac{11}{16}$ *in.*
above left eye $\frac{7}{8}$ *in.*

78. Three screws have lengths equal to $\frac{3}{4}$ in., $\frac{5}{8}$ in., and $\frac{11}{16}$ in. Which screw is the longest? Which is the shortest?

79. Susan buys $\frac{2}{3}$ lb of smoked turkey, $\frac{3}{5}$ lb of ham, and $\frac{5}{8}$ lb of roast beef. Which type of meat did she buy in the greatest amount? Which type did she buy in the least amount?

80. For a party, Aman had $\frac{3}{4}$ lb of cheddar cheese, $\frac{7}{8}$ lb of Swiss cheese, and $\frac{4}{5}$ lb of pepper jack cheese. Which type of cheese is in the least amount? Which type is in the greatest amount?

Expanding Your Skills

81. Which of the following fractions is between $\frac{1}{4}$ and $\frac{5}{6}$? Identify all that apply.

   a. $\dfrac{5}{12}$ **b.** $\dfrac{2}{3}$ **c.** $\dfrac{1}{8}$

82. Which of the following fractions is between $\frac{1}{3}$ and $\frac{11}{15}$? Identify all that apply.

   a. $\dfrac{2}{3}$ **b.** $\dfrac{4}{5}$ **c.** $\dfrac{2}{5}$

Addition and Subtraction of Unlike Fractions | Section 3.3

1. Addition and Subtraction of Unlike Fractions

In this section, we use the concept of the LCD to help us add and subtract unlike fractions. The first step in adding or subtracting unlike fractions is to identify the LCD. Then we change the unlike fractions to like fractions having the LCD as the denominator.

Objectives

1. Addition and Subtraction of Unlike Fractions
2. Order of Operations
3. Applications Involving Unlike Fractions

Example 1 — Adding Unlike Fractions

Add. $\dfrac{1}{6} + \dfrac{3}{4}$

Solution:

The LCD of $\frac{1}{6}$ and $\frac{3}{4}$ is 12. We can convert each individual fraction to an equivalent fraction with 12 as the denominator.

$$\frac{1}{6} = \frac{1 \cdot 2}{6 \cdot 2} = \frac{2}{12}$$ Multiply numerator and denominator by 2 because $6 \cdot 2 = 12$.

$$\frac{3}{4} = \frac{3 \cdot 3}{4 \cdot 3} = \frac{9}{12}$$ Multiply numerator and denominator by 3 because $4 \cdot 3 = 12$.

Thus, $\dfrac{1}{6} + \dfrac{3}{4}$ becomes $\dfrac{2}{12} + \dfrac{9}{12} = \dfrac{11}{12}$.

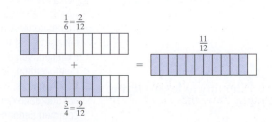

Skill Practice

Add.

1. $\dfrac{1}{10} + \dfrac{1}{15}$

Answer

1. $\dfrac{1}{6}$

> **TIP:** In Example 1, we multiplied the fraction $\frac{1}{6}$ by $\frac{2}{2}$. This is equivalent to multiplying $\frac{1}{6}$ by 1 and does not change the value.
>
> $$\frac{1}{6} = \frac{1}{6} \cdot 1 = \frac{1}{6} \cdot \frac{2}{2} = \frac{2}{12}$$
>
> The fraction $\frac{2}{12}$ is equivalent to $\frac{1}{6}$.

Adding fractions can be visualized by using a diagram. For example, the sum $\frac{1}{2} + \frac{1}{3}$ is illustrated in Figure 3-3.

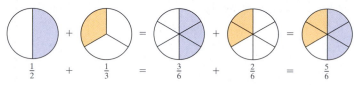

$$\frac{1}{2} \quad + \quad \frac{1}{3} \quad = \quad \frac{3}{6} \quad + \quad \frac{2}{6} \quad = \quad \frac{5}{6}$$

Figure 3-3

The general procedure to add or subtract unlike fractions is outlined as follows.

> **PROCEDURE** **Adding and Subtracting Unlike Fractions**
>
> **Step 1** Identify the LCD.
> **Step 2** Write each individual fraction as an equivalent fraction with the LCD.
> **Step 3** Add or subtract the resulting fractions as indicated.
> **Step 4** Simplify to lowest terms, if possible.

| **Example 2** | **Adding Unlike Fractions** |

Add. $\quad \dfrac{3}{10} + \dfrac{1}{5}$

Solution:

$$\frac{3}{10} + \frac{1}{5}$$

The LCD is 10. We must convert $\frac{1}{5}$ to an equivalent fraction with 10 as the denominator.

$$= \frac{3}{10} + \frac{1 \cdot 2}{5 \cdot 2}$$

Multiply numerator and denominator by 2 because $5 \cdot 2 = 10$.

$$= \frac{3}{10} + \frac{2}{10}$$

The fractions are now like.

$$= \frac{3 + 2}{10}$$

Add the like fractions.

$$= \frac{5}{10}$$

$$= \frac{\overset{1}{\cancel{5}}}{\underset{2}{\cancel{10}}}$$

Simplify to lowest terms.

$$= \frac{1}{2}$$

> **Avoiding Mistakes**
>
> Do not confuse addition and subtraction of fractions with multiplication of fractions. In multiplication, we multiply denominators. In addition we do not add denominators. We get a common denominator and then add only the numerators.

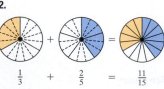

Example 3 **Subtracting Unlike Fractions**

Subtract. $\dfrac{4}{15} - \dfrac{1}{10}$

Skill Practice

Subtract.

4. $\dfrac{7}{12} - \dfrac{1}{8}$

Solution:

$\dfrac{4}{15} - \dfrac{1}{10}$ The LCD is 30.

$= \dfrac{4 \cdot 2}{15 \cdot 2} - \dfrac{1 \cdot 3}{10 \cdot 3}$ Write the fractions as equivalent fractions with the LCD.

$= \dfrac{8}{30} - \dfrac{3}{30}$

$= \dfrac{8 - 3}{30}$

$= \dfrac{5}{30}$ Subtract.

$= \dfrac{\overset{1}{\cancel{5}}}{\underset{6}{\cancel{30}}}$ Simplify to lowest terms.

$= \dfrac{1}{6}$

Sometimes when denominators are large, it is helpful to write the denominators as a product of prime factors. This is demonstrated in Example 4.

Example 4 **Adding and Subtracting Unlike Fractions**

Add or subtract as indicated.

$$\dfrac{7}{12} - \dfrac{2}{15} + \dfrac{5}{48}$$

Skill Practice

Add or subtract as indicated.

5. $\dfrac{7}{18} + \dfrac{4}{15} - \dfrac{17}{30}$

Solution:

$\dfrac{7}{12} - \dfrac{2}{15} + \dfrac{5}{48}$ To find the LCD, factor each denominator.

$= \dfrac{7}{2 \cdot 2 \cdot 3} - \dfrac{2}{3 \cdot 5} + \dfrac{5}{2 \cdot 2 \cdot 2 \cdot 2 \cdot 3}$

$\left. \begin{array}{l} 12 = 2 \cdot 2 \cdot ③ \\ 15 = 3 \cdot ⑤ \\ 48 = ②\cdot②\cdot②\cdot② \cdot 3 \end{array} \right\}$ The LCD is $2 \cdot 2 \cdot 2 \cdot 2 \cdot 3 \cdot 5 = 240$.

Answers

4. $\dfrac{11}{24}$ **5.** $\dfrac{4}{45}$

We want to convert each fraction to an equivalent fraction having a denominator of $2 \cdot 2 \cdot 2 \cdot 2 \cdot 3 \cdot 5 = 240$. Multiply numerator and denominator of each original fraction by the factors missing from the denominator.

$$= \frac{7 \cdot (2 \cdot 2 \cdot 5)}{2 \cdot 2 \cdot 3 \cdot (2 \cdot 2 \cdot 5)} - \frac{2 \cdot (2 \cdot 2 \cdot 2 \cdot 2)}{3 \cdot 5 \cdot (2 \cdot 2 \cdot 2 \cdot 2)} + \frac{5 \cdot (5)}{2 \cdot 2 \cdot 2 \cdot 2 \cdot 3 \cdot (5)}$$

$$= \frac{140}{240} - \frac{32}{240} + \frac{25}{240} \qquad \text{The fractions are now like fractions.}$$

$$= \frac{140 - 32 + 25}{240} \qquad \text{Add and subtract as indicated.}$$

$$= \frac{133}{240} \qquad \text{The fraction is in lowest terms.}$$

2. Order of Operations

In Examples 5 and 6, we must apply the order of operations to simplify the expressions.

Skill Practice

Simplify.

6. $\left(\frac{2}{3} - \frac{1}{7}\right)^2$

Example 5 Applying the Order of Operations

Simplify. $\left(\frac{1}{4} + \frac{2}{3}\right)^2$

Solution:

$$\left(\frac{1}{4} + \frac{2}{3}\right)^2 \qquad \text{Perform the operation within parentheses first.}$$

$$= \left(\frac{1 \cdot 3}{4 \cdot 3} + \frac{2 \cdot 4}{3 \cdot 4}\right)^2 \qquad \text{The common denominator is 12.}$$

$$= \left(\frac{3}{12} + \frac{8}{12}\right)^2 \qquad \text{The fractions within parentheses are now like fractions.}$$

$$= \left(\frac{11}{12}\right)^2 \qquad \text{Add fractions within parentheses.}$$

$$= \frac{11}{12} \cdot \frac{11}{12} \qquad \text{To square a number, multiply the number by itself.}$$

$$= \frac{121}{144} \qquad \text{The fraction is in lowest terms.}$$

Skill Practice

Simplify.

7. $\frac{4}{15} \div \frac{2}{5} - \frac{1}{6}$

Example 6 Applying the Order of Operations

Simplify. $\frac{5}{12} - \frac{1}{4} \div \frac{3}{2}$

Answers

6. $\frac{121}{441}$ 7. $\frac{1}{2}$

Solution:

$$\frac{5}{12} - \frac{1}{4} \div \frac{3}{2}$$ Perform the division before the subtraction.

$$= \frac{5}{12} - \frac{1}{\overset{2}{\cancel{4}}} \cdot \frac{\overset{1}{\cancel{2}}}{3}$$ To divide fractions, multiply by the reciprocal of the divisor.

$$= \frac{5}{12} - \frac{1}{6}$$ To subtract, we need the LCD of 12.

$$= \frac{5}{12} - \frac{1 \cdot 2}{6 \cdot 2}$$ Multiply numerator and denominator by 2 because $6 \cdot 2 = 12$.

$$= \frac{5}{12} - \frac{2}{12}$$ The fractions are now like fractions.

$$= \frac{3}{12}$$ Subtract.

$$= \frac{1}{4}$$ Simplify to lowest terms.

3. Applications Involving Unlike Fractions

Example 7   **Applying Operations on Unlike Fractions**

A new Kelly Safari SUV tire has $\frac{7}{16}$-in. tread. After being driven 50,000 mi, the tread depth has worn down to $\frac{7}{32}$ in. By how much has the tread depth worn away?

Tread

Solution:

In this case, we are looking for the difference in the tread depth.

$$\begin{pmatrix} \text{Difference in} \\ \text{tread depth} \end{pmatrix} = \begin{pmatrix} \text{original} \\ \text{tread depth} \end{pmatrix} - \begin{pmatrix} \text{final} \\ \text{tread depth} \end{pmatrix}$$

$$= \frac{7}{16} - \frac{7}{32}$$ The LCD is 32.

$$= \frac{7 \cdot 2}{16 \cdot 2} - \frac{7}{32}$$ Multiply numerator and denominator by 2 because $16 \cdot 2 = 32$.

$$= \frac{14}{32} - \frac{7}{32}$$ The fractions are now like.

$$= \frac{7}{32}$$ Subtract.

The tire lost $\frac{7}{32}$ in. in tread depth after 50,000 mi of driving.

> **TIP:** You can check your result by adding the final tread depth to the difference in tread depth to get the original tread depth.
>
> $$\frac{7}{32} + \frac{7}{32} = \frac{14}{32} = \frac{7}{16}$$

Skill Practice

8. On Monday, $\frac{2}{3}$ in. of rain fell on a certain town. On Tuesday, $\frac{1}{5}$ in. of rain fell. How much more rain fell on Monday than on Tuesday?

Answer

8. $\frac{7}{15}$ in.

Example 8 **Applying Operations on Unlike Fractions**

An oil tank contains 2 liters (L) of oil. A slow leak has occurred, and oil leaks out at a rate of $\frac{1}{16}$ L per day. After 7 days a mechanic notices the leak and pours $\frac{3}{8}$ L of oil back into the tank. How much oil is now in the tank?

Solution:

To find the current amount in the tank, we must determine the amount lost and the amount added.

The amount lost is given by the amount lost per day times 7 days.

$$\left(\frac{1}{16}\right)(7) = \frac{1}{16} \cdot \frac{7}{1} = \frac{7}{16} \qquad \text{The amount lost in 7 days is } \tfrac{7}{16} \text{ L.}$$

Therefore, the current amount in the tank is given by

$$\begin{array}{l} \text{Current} \\ \text{amount} \end{array} = \begin{pmatrix} \text{original} \\ \text{amount} \end{pmatrix} - \begin{pmatrix} \text{amount} \\ \text{lost} \end{pmatrix} + \begin{pmatrix} \text{amount} \\ \text{replaced} \end{pmatrix}$$

$$= 2 - \frac{7}{16} + \frac{3}{8}$$

$$= \frac{2}{1} - \frac{7}{16} + \frac{3}{8} \qquad \text{Write the whole number as a fraction.}$$

$$= \frac{2 \cdot 16}{1 \cdot 16} - \frac{7}{16} + \frac{3 \cdot 2}{8 \cdot 2} \qquad \text{The LCD is 16.}$$

$$= \frac{32}{16} - \frac{7}{16} + \frac{6}{16} \qquad \text{The fractions are now like fractions.}$$

$$= \frac{32 - 7 + 6}{16} \qquad \text{Add and subtract as indicated.}$$

$$= \frac{31}{16} \qquad \text{The fraction is in lowest terms.}$$

The tank now contains $\frac{31}{16}$ L or equivalently $1\frac{15}{16}$ L.

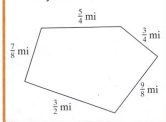
Example 9 **Finding Perimeter**

A parcel of land has the following dimensions. Find the perimeter.

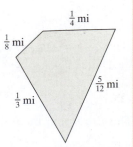

Solution:

To find the perimeter, add the lengths of the sides.

$$\frac{1}{8} + \frac{1}{4} + \frac{5}{12} + \frac{1}{3} \qquad \text{The LCD is 24.}$$

$$= \frac{1 \cdot 3}{8 \cdot 3} + \frac{1 \cdot 6}{4 \cdot 6} + \frac{5 \cdot 2}{12 \cdot 2} + \frac{1 \cdot 8}{3 \cdot 8} \qquad \begin{array}{l}\text{Convert to like} \\ \text{fractions.}\end{array}$$

$$= \frac{3}{24} + \frac{6}{24} + \frac{10}{24} + \frac{8}{24}$$

$$= \frac{27}{24} \qquad \text{Add the fractions.}$$

$$= \frac{9}{8} \qquad \text{Simplify to lowest terms.}$$

The perimeter is $\frac{9}{8}$ mi or equivalently $1\frac{1}{8}$ mi.

Section 3.3 Practice Exercises

Boost your GRADE at ALEKS.com!

- Practice Problems
- Self-Tests
- NetTutor
- e-Professors
- Videos

Study Skills Exercise

1. Do you need complete silence, or do you listen to music while you do your homework?

Try something different today so that you can compare and choose the best working environment for you.

Review Exercises

For Exercises 2–13, rewrite the fraction with the given denominator.

2. $\frac{3}{5} = \frac{}{15}$

3. $\frac{6}{7} = \frac{}{14}$

4. $\frac{4}{9} = \frac{}{36}$

5. $\frac{2}{3} = \frac{}{21}$

6. $\frac{3}{1} = \frac{}{10}$

7. $\frac{5}{1} = \frac{}{5}$

8. $\frac{4}{1} = \frac{}{12}$

9. $\frac{2}{1} = \frac{}{4}$

10. $\frac{3}{4} = \frac{}{12}$

11. $\frac{4}{5} = \frac{}{100}$

12. $\frac{3}{2} = \frac{}{18}$

13. $\frac{1}{8} = \frac{}{40}$

Objective 1: Addition and Subtraction of Unlike Fractions

14. Explain the difference between the procedures to add fractions and to multiply fractions.

For Exercises 15–50, add or subtract. Write the answer as a fraction simplified to lowest terms. **(See Examples 1–4.)**

15. $\frac{7}{8} + \frac{5}{16}$

16. $\frac{2}{9} + \frac{1}{18}$

17. $\frac{1}{15} + \frac{1}{10}$

18. $\frac{5}{6} + \frac{3}{8}$

19. $\frac{1}{10} + \frac{3}{20}$

20. $\frac{4}{15} + \frac{2}{5}$

21. $\frac{5}{6} + \frac{8}{7}$

22. $\frac{2}{11} + \frac{4}{5}$

23. $\frac{7}{8} - \frac{1}{2}$

24. $\frac{9}{10} - \frac{4}{5}$

25. $\frac{13}{12} - \frac{3}{4}$

26. $\frac{29}{30} - \frac{7}{10}$

27. $\frac{10}{9} - \frac{5}{12}$

28. $\frac{7}{6} - \frac{1}{15}$

29. $\frac{5}{8} - \frac{0}{11}$

30. $\frac{7}{12} - \frac{0}{5}$

31. $2 + \frac{9}{8}$

32. $3 + \frac{11}{9}$

33. $4 - \frac{4}{3}$

34. $2 - \frac{3}{8}$

35. $\dfrac{14}{3} + 1$ **36.** $\dfrac{12}{5} + 2$ **37.** $\dfrac{16}{7} - 2$ **38.** $\dfrac{15}{4} - 3$

39. $\dfrac{7}{10} + \dfrac{19}{100}$ **40.** $\dfrac{3}{10} + \dfrac{27}{100}$ **41.** $\dfrac{1}{10} - \dfrac{9}{100}$ **42.** $\dfrac{3}{100} - \dfrac{21}{1000}$

43. $\dfrac{3}{10} + \dfrac{9}{100} + \dfrac{1}{1000}$ **44.** $\dfrac{1}{10} + \dfrac{3}{100} + \dfrac{7}{1000}$ **45.** $\dfrac{5}{3} - \dfrac{7}{6} + \dfrac{5}{8}$ **46.** $\dfrac{7}{12} - \dfrac{2}{15} + \dfrac{5}{18}$

47. $\dfrac{1}{20} + \dfrac{5}{8} - \dfrac{7}{24}$ **48.** $\dfrac{5}{8} + \dfrac{3}{10} - \dfrac{1}{12}$ **49.** $\dfrac{1}{2} + \dfrac{1}{4} - \dfrac{1}{8} - \dfrac{1}{16}$ **50.** $\dfrac{1}{3} - \dfrac{1}{9} + \dfrac{1}{27} - \dfrac{1}{81}$

Objective 2: Order of Operations

For Exercises 51–64, simplify by applying the order of operations. Write the answer as a fraction. **(See Examples 5–6.)**

51. $\left(\dfrac{1}{2} - \dfrac{1}{3}\right)^2$ **52.** $\left(\dfrac{2}{3} + \dfrac{1}{6}\right)^2$ **53.** $\dfrac{2}{3} \div \dfrac{1}{2} - \dfrac{3}{4}$ **54.** $\dfrac{3}{5} \div \dfrac{6}{7} - \dfrac{2}{5}$

55. $\dfrac{5}{6} + \dfrac{3}{8} \div \dfrac{1}{4}$ **56.** $\dfrac{11}{12} + \dfrac{1}{9} \div \dfrac{7}{9}$ **57.** $\left(\dfrac{7}{10} - \dfrac{1}{5}\right) \cdot \dfrac{8}{3}$ **58.** $\left(\dfrac{2}{5} + \dfrac{9}{10}\right) \cdot \dfrac{5}{6}$

59. $\dfrac{4}{5} + \dfrac{5}{8} \cdot \dfrac{16}{35}$ **60.** $\dfrac{1}{6} + \dfrac{3}{7} \cdot \dfrac{14}{15}$ **61.** $\left(\dfrac{2}{5}\right)^3 + \dfrac{1}{25}$ **62.** $\left(\dfrac{3}{2}\right)^3 - \dfrac{5}{4}$

63. $\left(\dfrac{1}{4}\right)^2 \div \left(\dfrac{5}{6} - \dfrac{2}{3}\right) + \dfrac{7}{12}$ **64.** $\left(\dfrac{1}{2} + \dfrac{1}{3}\right) \cdot \left(\dfrac{2}{5}\right)^2 + \dfrac{3}{10}$

Objective 3: Applications Involving Unlike Fractions

65. When doing her laundry, Inez added $\frac{3}{4}$ cup of bleach to $\frac{3}{8}$ cup of liquid detergent. How much total liquid is added to her wash?

66. What is the smallest possible length of screw needed to pass through two pieces of wood, one that is $\frac{7}{8}$ in. thick and one that is $\frac{1}{2}$ in. thick?

67. Before a storm, a rain gauge has $\frac{1}{8}$ in. of water. After the storm, the gauge has $\frac{9}{32}$ in. How many inches of rain did the storm deliver? **(See Example 7.)**

68. In one week it rained $\frac{5}{16}$ in. If a garden needs $\frac{9}{8}$ in. of water per week, how much more water does it need?

69. A watering trough holds 5 gal of water. In the summer, water evaporates at a rate of approximately $\frac{3}{8}$ gal per day. After 4 days, Jeff pours in another gallon and a half of water ($\frac{3}{2}$ gal). How much water is in the trough? **(See Example 8.)**

70. A contractor hired two electricians to do a job. One did $\frac{3}{5}$ of the job and the other did $\frac{3}{8}$ of the job. Did the job get completed? If not, what fraction of the job is left?

71. The information in the graph shows the distribution of a college student body by class.

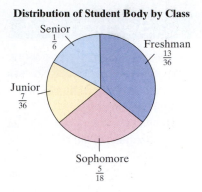

Distribution of Student Body by Class

a. What fraction of the student body consists of upper classmen (juniors and seniors)?

b. What fraction of the student body consists of freshmen and sophomores?

72. A group of college students took part in a survey. One of the survey questions read:

"Do you think the government should spend more money on research to produce alternative forms of fuel?"

The results of the survey are shown in the figure.

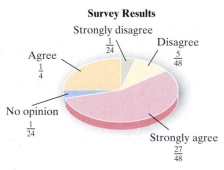

Survey Results

a. What fraction of the survey participants chose to strongly agree or agree?

b. What fraction of the survey participants chose to strongly disagree or disagree?

For Exercises 73–74, find the perimeter. **(See Example 9.)**

73.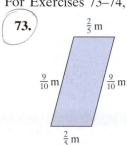

$\frac{2}{5}$ m

$\frac{9}{10}$ m $\frac{9}{10}$ m

$\frac{2}{5}$ m

74.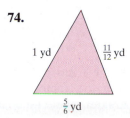

1 yd $\frac{11}{12}$ yd

$\frac{5}{6}$ yd

For Exercises 75–76, find the missing dimensions. Then calculate the perimeter.

75.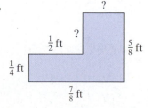

?

?

$\frac{1}{2}$ ft $\frac{5}{8}$ ft

$\frac{1}{4}$ ft

$\frac{7}{8}$ ft

76.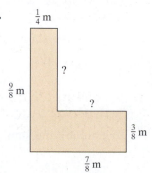

$\frac{1}{4}$ m

?

$\frac{9}{8}$ m   ?

$\frac{3}{8}$ m

$\frac{7}{8}$ m

Expanding Your Skills

77. Which fraction is closest to $\frac{1}{2}$?

a. $\frac{3}{4}$ b. $\frac{7}{10}$ c. $\frac{5}{6}$

78. Which fraction is closest to $\frac{3}{4}$?

a. $\frac{5}{8}$ b. $\frac{7}{12}$ c. $\frac{5}{6}$

Section 3.4   Addition and Subtraction of Mixed Numbers

1. Addition of Mixed Numbers

In this section, we learn to add and subtract mixed numbers. To find the sum of two or more mixed numbers, add the whole-number parts and add the fractional parts.

Skill Practice

Add.

1. $7\dfrac{2}{15} + 2\dfrac{1}{15}$

Example 1   Adding Mixed Numbers

Add. $1\dfrac{5}{9} + 2\dfrac{1}{9}$

Solution:

$$
\begin{array}{r}
1\dfrac{5}{9} \\[4pt]
+\,2\dfrac{1}{9} \\[4pt]
\hline
3\dfrac{6}{9}
\end{array}
$$

Add the whole ⎯⎯⎯⎯⎯⎯ Add the
numbers.   fractional parts.

The sum is $3\dfrac{6}{9}$ which simplifies to $3\dfrac{2}{3}$.

TIP: To understand why mixed numbers can be added in this way, recall that $1\dfrac{5}{9} = 1 + \dfrac{5}{9}$ and $2\dfrac{1}{9} = 2 + \dfrac{1}{9}$. Therefore,

$$1\dfrac{5}{9} + 2\dfrac{1}{9} = 1 + \dfrac{5}{9} + 2 + \dfrac{1}{9}$$

$$= 3 + \dfrac{6}{9}$$

$$= 3\dfrac{6}{9}$$

$$= 3\dfrac{2}{3}$$

When we perform operations on mixed numbers, it is often desirable to estimate the answer first. When rounding a mixed number, we offer the following convention.

1. If the fractional part of a mixed number is greater than or equal to $\frac{1}{2}$ (that is, if the numerator is half the denominator or greater), round to the next-greater whole number.

2. If the fractional part of the mixed number is less than $\frac{1}{2}$ (that is, if the numerator is less than half the denominator), the mixed number rounds down to the whole number.

Skill Practice

Estimate the sum and then find the actual sum.

2. $6\dfrac{1}{11} + 3\dfrac{1}{2}$

Example 2   Adding Mixed Numbers

Estimate the sum and then find the actual sum.

$$42\dfrac{1}{12} + 17\dfrac{7}{8}$$

Solution:

To estimate the sum, first round the addends.

$$
\begin{array}{lll}
42\dfrac{1}{12} & \text{rounds to} & 42 \\[8pt]
+\,17\dfrac{7}{8} & \text{rounds to} & \dfrac{+\ 18}{60}
\end{array}
$$
   The estimated value is 60.

Answers

1. $9\dfrac{1}{5}$

2. Estimate: 10; actual sum: $9\dfrac{13}{22}$

To find the actual sum, we must first write the fractional parts as like fractions. The LCD is 24.

$$42\frac{1}{12} = 42\frac{1 \cdot 2}{12 \cdot 2} = 42\frac{2}{24}$$
$$+ 17\frac{7}{8} = + 17\frac{7 \cdot 3}{8 \cdot 3} = + 17\frac{21}{24}$$
$$\overline{\phantom{+ 17\frac{7}{8}}} \qquad \overline{\phantom{+ 17\frac{7 \cdot 3}{8 \cdot 3}}} \qquad 59\frac{23}{24}$$

The actual sum is $59\frac{23}{24}$. This is close to our estimate of 60.

Example 3 **Adding Mixed Numbers With Carrying**

Estimate the sum and then find the actual sum.

$$7\frac{5}{6} + 3\frac{3}{5}$$

Solution:

$$7\frac{5}{6} \qquad \text{rounds to} \qquad 8$$
$$+ 3\frac{3}{5} \qquad \text{rounds to} \qquad \frac{+ 4}{12} \qquad \text{The estimated value is 12.}$$

To find the actual sum, first write the fractional parts as like fractions. The LCD is 30.

$$7\frac{5}{6} = 7\frac{5 \cdot 5}{6 \cdot 5} = 7\frac{25}{30}$$
$$+ 3\frac{3}{5} = + 3\frac{3 \cdot 6}{5 \cdot 6} = + 3\frac{18}{30}$$
$$\overline{\phantom{+ 3\frac{3}{5}}} \qquad \overline{\phantom{+ 3\frac{3 \cdot 6}{5 \cdot 6}}} \qquad 10\frac{43}{30}$$

Notice that the number $\frac{43}{30}$ is an improper fraction. By convention, a mixed number is written as a whole number and a *proper* fraction. We have $\frac{43}{30} = 1\frac{13}{30}$. Therefore,

$$10\frac{43}{30} = 10 + 1\frac{13}{30} = 11\frac{13}{30}$$

The sum is $11\frac{13}{30}$. This is close to our estimate of 12.

We have shown how to add mixed numbers by writing the numbers in columns. Another approach to add or subtract mixed numbers is to write the numbers first as improper fractions. Then add or subtract the fractions, as you learned in Section 3.3. To demonstrate this process, we add the mixed numbers from Example 3.

Skill Practice

Estimate the sum and then find the actual sum.

3. $5\frac{2}{5} + 7\frac{8}{9}$

Concept Connections

4. Explain how you would rewrite $2\frac{9}{8}$ as a mixed number containing a proper fraction.

Answers

3. Estimate: 13; actual sum: $13\frac{13}{45}$
4. Write the improper fraction $\frac{9}{8}$ as $1\frac{1}{8}$, and add the result to 2. The result is $3\frac{1}{8}$.

Skill Practice

Add the mixed numbers by first converting the addends to improper fractions. Write the answer as a mixed number.

5. $12\frac{1}{3} + 4\frac{3}{4}$

Example 4 **Adding Mixed Numbers by Using Improper Fractions**

Add. $7\frac{5}{6} + 3\frac{3}{5}$

Solution:

$$7\frac{5}{6} + 3\frac{3}{5} = \frac{47}{6} + \frac{18}{5}$$

Write each mixed number as an improper fraction.

$$= \frac{47 \cdot 5}{6 \cdot 5} + \frac{18 \cdot 6}{5 \cdot 6}$$

Convert the fractions to like fractions. The LCD is 30.

$$= \frac{235}{30} + \frac{108}{30}$$

The fractions are now like fractions.

$$= \frac{343}{30}$$

Add the like fractions.

$$= 11\frac{13}{30}$$

Convert the improper fraction to a mixed number.

$$\begin{array}{r} 11 \\ 30\overline{)343} \\ -30 \\ \hline 43 \\ -30 \\ \hline 13 \end{array} \quad 11\frac{13}{30}$$

The mixed number $11\frac{13}{30}$ is the same as the value obtained in Example 3.

2. Subtraction of Mixed Numbers

To subtract mixed numbers, we subtract the fractional parts and subtract the whole-number parts.

Skill Practice

6. Subtract.

$6\frac{3}{4} - 2\frac{1}{3}$

Example 5 **Subtracting Mixed Numbers**

Subtract. $15\frac{2}{3} - 4\frac{1}{6}$

Solution:

To subtract the fractional parts, we need a common denominator. The LCD is 6.

$$15\frac{2}{3} = 15\frac{2 \cdot 2}{3 \cdot 2} = 15\frac{4}{6}$$
$$-4\frac{1}{6} = -4\frac{1}{6} = -4\frac{1}{6}$$
$$\overline{\qquad\qquad 11\frac{3}{6}}$$

Subtract the whole numbers. ── Subtract the fractional parts.

The difference is $11\frac{3}{6}$ which simplifies to $11\frac{1}{2}$.

Answers

5. $17\frac{1}{12}$ **6.** $4\frac{5}{12}$

Borrowing is sometimes necessary when subtracting mixed numbers.

Example 6 **Subtracting Mixed Numbers With Borrowing**

Subtract.

a. $17\frac{2}{7} - 11\frac{5}{7}$ b. $14\frac{2}{9} - 9\frac{3}{5}$

Skill Practice

Subtract.

7. $24\frac{2}{7} - 8\frac{5}{7}$

8. $9\frac{2}{3} - 8\frac{3}{4}$

Solution:

a. We cannot subtract $\frac{5}{7}$ from $\frac{2}{7}$. Therefore, borrow 1 from 17. The borrowed 1 is written as $\frac{7}{7}$ because the common denominator is 7.

$$17\frac{2}{7} = \quad \overset{16}{\cancel{17}}\frac{2}{7} + \frac{7}{7} = \quad 16\frac{9}{7}$$
$$-11\frac{5}{7} = \quad -11\frac{5}{7} \qquad = \quad -11\frac{5}{7}$$
$$\rule{2cm}{0.4pt} \qquad \rule{2.5cm}{0.4pt} \qquad \rule{2.5cm}{0.4pt}$$
$$5\frac{4}{7}$$

The difference is $5\frac{4}{7}$.

b. To subtract the fractional parts, we need a common denominator. The LCD is 45.

$$14\frac{2}{9} = \quad 14\frac{2 \cdot 5}{9 \cdot 5} \qquad = \quad 14\frac{10}{45}$$
$$-9\frac{3}{5} = \quad -9\frac{3 \cdot 9}{5 \cdot 9} \qquad = \quad -9\frac{27}{45}$$

We cannot subtract $\frac{27}{45}$ from $\frac{10}{45}$. Therefore, borrow 1 (or equivalently $\frac{45}{45}$) from 14.

$$= \quad \overset{13}{\cancel{14}}\frac{10}{45} + \frac{45}{45} = \quad 13\frac{55}{45}$$
$$= \quad -9\frac{27}{45} \qquad = \quad -9\frac{27}{45}$$
$$\rule{2.5cm}{0.4pt} \qquad \rule{2.5cm}{0.4pt}$$
$$4\frac{28}{45}$$

The difference is $4\frac{28}{45}$.

Answers

7. $15\frac{4}{7}$ 8. $\frac{11}{12}$

Skill Practice

Skill Practice

9. Subtract.

$$10 - 3\frac{1}{6}$$

Example 7 **Subtracting Mixed Numbers With Borrowing**

Subtract. $4 - 2\frac{5}{8}$

Solution:

$$\begin{array}{r} 4 \\ -2\frac{5}{8} \end{array}$$

In this case, we have no fractional part from which to subtract.

$$\overset{3}{\cancel{4}}\frac{8}{8}$$

We can borrow 1 or equivalently $\frac{8}{8}$ from the whole number 4.

$$\begin{array}{r} \overset{3}{\cancel{4}}\frac{8}{8} \\ -2\frac{5}{8} \\ \hline 1\frac{3}{8} \end{array}$$

> **TIP:** The borrowed 1 is written as $\frac{8}{8}$ because the common denominator is 8.

The difference is $1\frac{3}{8}$.

> **TIP:** The subtraction problem $4 - 2\frac{5}{8} = 1\frac{3}{8}$ can be checked by adding:
>
> $$1\frac{3}{8} + 2\frac{5}{8} = 3\frac{8}{8} = 3 + 1 = 4 \checkmark$$

Skill Practice

Subtract by first converting to improper fractions. Write the answer as a mixed number.

10. $8\frac{2}{9} - 3\frac{5}{6}$

Example 8 **Subtracting Mixed Numbers by Using Improper Fractions**

Subtract by first converting to improper fractions. Write the answer as a mixed number.

$$10\frac{2}{5} - 4\frac{3}{4}$$

Solution:

$$10\frac{2}{5} - 4\frac{3}{4} = \frac{52}{5} - \frac{19}{4}$$

Write each mixed number as an improper fraction.

$$= \frac{52 \cdot 4}{5 \cdot 4} - \frac{19 \cdot 5}{4 \cdot 5}$$

Convert the fractions to like fractions. The LCD is 20.

$$= \frac{208}{20} - \frac{95}{20}$$

$$= \frac{113}{20}$$

Subtract the like fractions.

$$= 5\frac{13}{20}$$

Write the result as a mixed number.

$$\begin{array}{r} 5 \\ 20\overline{)113} \\ -100 \\ \hline 13 \end{array}$$

Answers

9. $6\frac{5}{6}$ 10. $4\frac{7}{18}$

As you can see from Examples 4 and 8, when we convert mixed numbers to improper fractions, the numerators of the fractions become larger numbers. Thus, we must add (or subtract) larger numerators than if we had used the method involving columns. This is one drawback. However, an advantage to converting to improper fractions first is that there is no need for carrying or borrowing.

3. Applications of Mixed Numbers

Example 9 Subtracting Mixed Numbers in an Application

The average height of a 3-year-old girl is $38\frac{1}{3}$ in. The average height of a 4-year-old girl is $41\frac{3}{4}$ in. On average, by how much does a girl grow between the ages of 3 and 4?

Solution:

We use subtraction to find the difference in heights.

$$
\begin{array}{r}
41\dfrac{3}{4} = \quad 41\dfrac{3\cdot 3}{4\cdot 3} = \quad 41\dfrac{9}{12} \\[2mm]
-\,38\dfrac{1}{3} = -\,38\dfrac{1\cdot 4}{3\cdot 4} = -\,38\dfrac{4}{12} \\[2mm]
\hline
3\dfrac{5}{12}
\end{array}
$$

The average amount of growth is $3\frac{5}{12}$ in.

Skill Practice

11. On December 1, the snow base at the Bear Mountain Ski Resort was $4\frac{1}{3}$ ft. By January 1, the base was $6\frac{1}{2}$ ft. By how much did the base amount of snow increase?

Answer

11. $2\frac{1}{6}$ ft

Section 3.4 Practice Exercises

Study Skills Exercise

1. Write down your instructor's policies for the following:

   a. Missing a test **b.** Missing a class **c.** Doing homework

Review Exercises

For Exercises 2–8, add or subtract as indicated. Write the answer as a fraction or whole number.

2. $\dfrac{3}{16} + \dfrac{7}{12}$

3. $\dfrac{25}{8} - \dfrac{23}{24}$

4. $4 - \dfrac{15}{7}$

5. $\dfrac{9}{5} + 3$

6. $\dfrac{23}{6} + \dfrac{5}{6} - \dfrac{2}{3}$

7. $\dfrac{125}{32} - \dfrac{51}{32} - \dfrac{58}{32}$

8. $\dfrac{17}{10} - \dfrac{23}{100} + \dfrac{321}{1000}$

Objective 1: Addition of Mixed Numbers

For Exercises 9–16, add the mixed numbers. (See Example 1.)

9. $2\frac{1}{11}$
$+\,5\frac{3}{11}$

10. $5\frac{2}{7}$
$+\,4\frac{3}{7}$

11. $12\frac{1}{14}$
$+\,3\frac{5}{14}$

12. $1\frac{3}{20}$
$+\,17\frac{7}{20}$

13. $4\frac{5}{16}$
$+\,11\frac{1}{4}$

14. $21\frac{2}{9}$
$+\,10\frac{1}{3}$

15. $6\frac{2}{3}$
$+\,4\frac{1}{5}$

16. $7\frac{1}{6}$
$+\,3\frac{5}{8}$

For Exercises 17–20, round the mixed number to the nearest whole number.

17. $5\frac{1}{3}$

18. $2\frac{7}{8}$

19. $1\frac{3}{5}$

20. $6\frac{3}{7}$

For Exercises 21–24, write the mixed number in proper form (that is, as a whole number with a proper fraction that is simplified to lowest terms).

21. $2\frac{6}{5}$

22. $4\frac{8}{7}$

23. $7\frac{5}{3}$

24. $1\frac{9}{5}$

For Exercises 25–34, round the numbers to estimate the answer. Then find the exact sum. In Exercise 25, the estimate is done for you. (See Examples 2–3.)

	Estimate	Exact		Estimate	Exact		Estimate	Exact
25.	7	$6\frac{3}{4}$	**26.**		$8\frac{3}{5}$	**27.**		$14\frac{7}{8}$
	$+\,8$	$+\,7\frac{3}{4}$		$+$	$+\,13\frac{4}{5}$		$+$	$+\,8\frac{1}{4}$
	$\overline{15}$							

	Estimate	Exact		Estimate	Exact		Estimate	Exact
28.		$21\frac{3}{5}$	**29.**		$3\frac{7}{16}$	**30.**		$7\frac{7}{9}$
	$+$	$+\,24\frac{9}{10}$		$+$	$+\,15\frac{11}{12}$		$+$	$+\,8\frac{5}{6}$

31. $3 + 6\frac{7}{8}$

32. $5 + 11\frac{1}{13}$

33. $32\frac{2}{7} + 10$

34. $2\frac{18}{37} + 16$

Objective 2: Subtraction of Mixed Numbers

For Exercises 35–42, subtract the mixed numbers. (See Example 5.)

35. $21\frac{9}{10}$
$-\,10\frac{3}{10}$

36. $19\frac{2}{3}$
$-\,4\frac{1}{3}$

37. $5\frac{9}{15}$
$-\,3\frac{7}{15}$

38. $33\frac{11}{12}$
$-\,14\frac{5}{12}$

39. $18\dfrac{5}{6}$

  $-\ 6\dfrac{2}{3}$

40. $21\dfrac{17}{20}$

  $-\ 20\dfrac{1}{10}$

41. $11\dfrac{5}{7}$

  $-\ 9\dfrac{5}{14}$

42. $5\dfrac{9}{11}$

  $-\ 2\dfrac{13}{22}$

For Exercises 43–46, rewrite the number 1 as a fraction having the given denominator.

43. 3   **44.** 5   **45.** 12   **46.** 6

For Exercises 47–60, round the numbers to estimate the answer. Then find the exact difference. In Exercise 47, the estimate is done for you. **(See Examples 6–7.)**

Estimate	Exact
47. 25	$25\dfrac{1}{4}$
$-\ 14$	$-\ 13\dfrac{3}{4}$
$\overline{11}$	

Estimate	Exact
48.	$36\dfrac{1}{5}$
$-$	$-\ 12\dfrac{3}{5}$

Estimate	Exact
49.	$17\dfrac{1}{6}$
$-$	$-\ 15\dfrac{5}{12}$

Estimate	Exact
50.	$22\dfrac{5}{18}$
$-$	$-\ 10\dfrac{7}{9}$

Estimate	Exact
51.	$46\dfrac{3}{7}$
$-$	$-\ 38\dfrac{1}{2}$

Estimate	Exact
52.	$23\dfrac{1}{2}$
$-$	$-\ 18\dfrac{10}{13}$

53. $6 - 2\dfrac{5}{6}$

54. $9 - 4\dfrac{1}{2}$

55. $12 - 9\dfrac{2}{9}$

56. $10 - 9\dfrac{1}{3}$

57. $5\dfrac{3}{17} - 3$

58. $16\dfrac{4}{11} - 5$

59. $23\dfrac{5}{14} - 17$

60. $21\dfrac{3}{4} - 10$

Mixed Exercises

For Exercises 61–76, add or subtract the mixed numbers by using improper fractions. Write the answers as mixed numbers, if possible. **(See Examples 4 and 8.)**

61. $2\dfrac{2}{3} + 4\dfrac{5}{8}$

62. $5\dfrac{1}{4} - 3\dfrac{1}{2}$

63. $1\dfrac{11}{15} + 4\dfrac{2}{5}$

64. $2\dfrac{10}{11} + 2\dfrac{1}{2}$

65. $3\dfrac{7}{8} - 3\dfrac{3}{16}$

66. $3\dfrac{1}{6} - 1\dfrac{23}{24}$

67. $4\dfrac{1}{12} + 5\dfrac{1}{9}$

68. $10\dfrac{2}{25} - 7\dfrac{13}{20}$

69. $9\dfrac{5}{32} - 8\dfrac{1}{4}$

70. $4\dfrac{3}{40} - 2\dfrac{7}{8}$

71. $6\dfrac{11}{14} + 4\dfrac{1}{6}$

72. $8\dfrac{3}{22} + 4\dfrac{1}{4}$

73. $12\dfrac{1}{5} - 11\dfrac{2}{7}$

74. $5\dfrac{11}{30} + 5\dfrac{3}{4}$

75. $10\dfrac{1}{8} - 2\dfrac{17}{18}$

76. $3\dfrac{8}{21} + 6\dfrac{8}{9}$

Objective 3: Applications of Mixed Numbers

For Exercises 77–80, use the table to find the lengths of several common birds.

Bird	Length
Cuban Bee Hummingbird	$2\frac{1}{4}$ in.
Sedge Wren	$3\frac{1}{2}$ in.
Great Carolina Wren	$5\frac{1}{2}$ in.
Belted Kingfisher	$11\frac{1}{4}$ in.

77. How much longer is the Belted Kingfisher than the Sedge Wren? **(See Example 9.)**

78. How much longer is the Great Carolina Wren than the Cuban Bee Hummingbird?

79. Estimate or measure the length of your index finger. Which is longer, your index finger or a Cuban Bee Hummingbird?

80. For a recent year, the smallest living dog in the United States was Brandy, a female Chihuahua, who measures 6 in. in length. How much longer is a Belted Kingfisher than Brandy?

81. A student has three part-time jobs. She tutors, delivers newspapers, and takes notes for a blind student. During a typical week she works $8\frac{2}{3}$ hr delivering newspapers, $4\frac{1}{2}$ hr tutoring, and $3\frac{3}{4}$ hr note-taking. What is the total number of hours worked in a typical week?

82. A contractor ordered three loads of gravel. The orders were for $2\frac{1}{2}$ tons, $3\frac{1}{8}$ tons, and $4\frac{1}{3}$ tons. What is the total amount of gravel ordered?

83. A plumber fits together two pipes. Find the length of the larger piece.

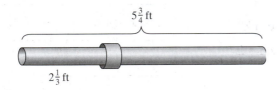

$5\frac{3}{4}$ ft

$2\frac{1}{3}$ ft

84. Find the thickness of the carpeting and pad.

$\frac{3}{8}$ in.

$\frac{7}{16}$ in.

85. A pipe has an outer diameter of $1\frac{3}{16}$ in. and an inner diameter of $1\frac{1}{8}$ in. Find the thickness of the pipe.

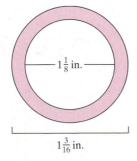

$1\frac{1}{8}$ in.

$1\frac{3}{16}$ in.

86. Find the inside diameter of the washer.

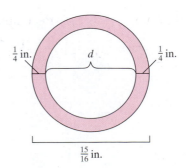

$\frac{1}{4}$ in. d $\frac{1}{4}$ in.

$\frac{15}{16}$ in.

87. When using the word processor, Microsoft Word, the default margins are $1\frac{1}{4}$ in. for the left and right margins. If an $8\frac{1}{2}$ in. by 11 in. piece of paper is used, what is the width of the printing area?

88. A water gauge in a pond measured $25\frac{7}{8}$ in. on Monday. After 2 days of rain and runoff, the gauge read $32\frac{1}{2}$ in. By how much did the water level rise?

89. A flight from Atlanta to San Diego takes $5\frac{1}{3}$ hr. After $2\frac{1}{2}$ hr, how much time remains?

90. In a triathlon, an athlete must swim $\frac{1}{4}$ mi, bike $10\frac{1}{2}$ mi, and run $3\frac{1}{5}$ mi. What is the total distance?

91. Vertical blinds were purchased for a window that is $3\frac{5}{12}$ ft high. The blinds are $3\frac{3}{4}$ ft in length. Find the distance that the blinds will hang below the window.

92. The number of hours worked per day for a plumber is given in the table. How many more hours did he work on Monday than on Saturday?

Monday	Tuesday	Wednesday	Thursday	Friday	Saturday	Sunday
$9\frac{1}{6}$ hr	$7\frac{3}{4}$ hr	$8\frac{1}{3}$ hr	$8\frac{1}{2}$ hr	$4\frac{1}{2}$ hr	$3\frac{3}{4}$ hr	0

93. A patient admitted to the hospital was dehydrated. In addition to intravenous (IV) fluids, the doctor told the patient that she must drink at least 4 L of an electrolyte solution within the next 12 hr. A nurse recorded the amounts the patient drank in the patient's chart.

   a. How many liters of electrolyte solution did the patient drink?

   b. How much more would the patient need to drink to reach 4 L?

Time	Amount
7 A.M.–10 A.M.	$1\frac{1}{4}$ L
10 A.M.–1 P.M.	$\frac{7}{8}$ L
1 P.M.–4 P.M.	$\frac{3}{4}$ L
4 P.M.–7 P.M.	$\frac{1}{2}$ L

Expanding Your Skills

For Exercises 94–97, fill in the blank to complete the pattern.

94. $1, 1\frac{1}{3}, 1\frac{2}{3}, 2, 2\frac{1}{3}, \square$

95. $\frac{1}{4}, 1, 1\frac{3}{4}, 2\frac{1}{2}, 3\frac{1}{4}, \square$

96. $\frac{5}{6}, 1\frac{1}{6}, 1\frac{1}{2}, 1\frac{5}{6}, \square$

97. $\frac{1}{2}, 1\frac{1}{4}, 2, 2\frac{3}{4}, 3\frac{1}{2}, \square$

Calculator Connections

Topic: Adding and Subtracting Fractions and Mixed Numbers on a Calculator

Expression	Keystrokes		Result
$\dfrac{7}{18} + \dfrac{1}{3}$	7 $\boxed{a^{b}\!/_{c}}$ 18 + 1 $\boxed{a^{b}\!/_{c}}$ 3 $\boxed{=}$		$\boxed{13 \rfloor 18} = \dfrac{13}{18}$
$7\dfrac{5}{8} - 4\dfrac{2}{3}$	7 $\boxed{a^{b}\!/_{c}}$ 5 $\boxed{a^{b}\!/_{c}}$ 8 $\boxed{-}$ 4 $\boxed{a^{b}\!/_{c}}$ 2 $\boxed{a^{b}\!/_{c}}$ 3 $\boxed{=}$		$\boxed{2\_23 \rfloor 24} = 2\dfrac{23}{24}$
	$\underbrace{\hspace{3cm}}_{7\frac{5}{8}}$ $\underbrace{\hspace{3cm}}_{4\frac{2}{3}}$		

To convert the result to an improper fraction, press $\boxed{2^{nd}}$ $\boxed{d/c}$ $\boxed{71 \rfloor 24} = \dfrac{71}{24}$

Calculator Exercises

98. $\dfrac{23}{42} + \dfrac{17}{24}$

99. $\dfrac{14}{75} + \dfrac{9}{50}$

100. $\dfrac{31}{44} - \dfrac{14}{33}$

101. $\dfrac{29}{68} - \dfrac{7}{92}$

102. $32\dfrac{7}{18} + 14\dfrac{2}{27}$

103. $21\dfrac{3}{28} + 4\dfrac{31}{42}$

104. $7\dfrac{11}{21} - 2\dfrac{10}{33}$

105. $5\dfrac{14}{17} - 2\dfrac{47}{68}$

Problem Recognition Exercises

Operations on Fractions and Mixed Numbers

Perform the indicated operations. Check the reasonableness of your answers by estimating.

1. $\dfrac{7}{5} + \dfrac{2}{5}$

2. $\dfrac{7}{5} \times \dfrac{2}{5}$

3. $\dfrac{7}{5} \div \dfrac{2}{5}$

4. $\dfrac{7}{5} - \dfrac{2}{5}$

5. $\dfrac{4}{3} \times \dfrac{5}{6}$

6. $\dfrac{4}{3} \div \dfrac{5}{6}$

7. $\dfrac{4}{3} + \dfrac{5}{6}$

8. $\dfrac{4}{3} - \dfrac{5}{6}$

9. $2\dfrac{3}{4} + 1\dfrac{1}{2}$

10. $2\dfrac{3}{4} - 1\dfrac{1}{2}$

11. $2\dfrac{3}{4} \div 1\dfrac{1}{2}$

12. $2\dfrac{3}{4} \times 1\dfrac{1}{2}$

13. $\left(4\dfrac{1}{3}\right) \times \left(2\dfrac{5}{6}\right)$

14. $\left(4\dfrac{1}{3}\right) \div \left(2\dfrac{5}{6}\right)$

15. $\left(4\dfrac{1}{3}\right) - \left(2\dfrac{5}{6}\right)$

16. $\left(4\dfrac{1}{3}\right) + \left(2\dfrac{5}{6}\right)$

17. $4 - \dfrac{3}{8}$

18. $4 \times \dfrac{3}{8}$

19. $4 \div \dfrac{3}{8}$

20. $4 + \dfrac{3}{8}$

21. $3\dfrac{2}{3} \div 2$

22. $3\dfrac{2}{3} - 2$

23. $3\dfrac{2}{3} + 2$

24. $3\dfrac{2}{3} \cdot 2$

25. $4\dfrac{1}{5} - \dfrac{2}{3}$

26. $4\dfrac{1}{5} + \dfrac{2}{3}$

27. $\left(4\dfrac{1}{5}\right) \cdot \dfrac{2}{3}$

28. $\left(4\dfrac{1}{5}\right) \div \dfrac{2}{3}$

29. $\dfrac{25}{9} \div 2$

30. $\dfrac{25}{9} \cdot 2$

31. $\dfrac{25}{9} - 2$

32. $\dfrac{25}{9} + 2$

33. $1\dfrac{4}{5} \cdot \dfrac{5}{9}$

34. $1\dfrac{4}{5} + \dfrac{5}{9}$

35. $1\dfrac{4}{5} \div \dfrac{5}{9}$

36. $1\dfrac{4}{5} - \dfrac{5}{9}$

37. $8 \cdot \dfrac{1}{8}$

38. $\dfrac{1}{9} \cdot 9$

39. $\dfrac{3}{7} \cdot \dfrac{7}{3}$

40. $\dfrac{5}{13} \cdot \dfrac{13}{5}$

Order of Operations and Applications of Fractions and Mixed Numbers

1. Order of Operations

At this point in the text, we have learned how to add, subtract, multiply, and divide whole numbers, fractions, and mixed numbers. In this section we practice putting all these skills to use by applying the order of operations.

To review the steps in the order of operations, refer to section 3.1, page 165.

Objectives

1. Order of Operations
2. Applications of Fractions and Mixed Numbers
3. Applications of Geometry

Example 1   Applying the Order of Operations

Simplify. $\left(3 - \dfrac{3}{4}\right)^2$

Solution:

$\left(3 - \dfrac{3}{4}\right)^2 = \left(\dfrac{3}{1} - \dfrac{3}{4}\right)^2$ First subtract the numbers within parentheses. Write the whole number as an improper fraction.

$= \left(\dfrac{3 \cdot 4}{1 \cdot 4} - \dfrac{3}{4}\right)^2$ Convert the fractions to like fractions. The LCD is 4.

$= \left(\dfrac{12}{4} - \dfrac{3}{4}\right)^2$ The fractions are now like.

$= \left(\dfrac{9}{4}\right)^2$ Subtract.

$= \dfrac{9}{4} \cdot \dfrac{9}{4}$ Square the quantity $\dfrac{9}{4}$.

$= \dfrac{81}{16}$ or $5\dfrac{1}{16}$

Skill Practice

Simplify.

1. $\left(4 - \dfrac{5}{2}\right)^2$

Answer

1. $\dfrac{9}{4}$

Skill Practice

Simplify.

2. $3\dfrac{1}{4} \div 4 \cdot \left(\dfrac{2}{5}\right)$

Example 2 Applying the Order of Operations

Simplify. $1\dfrac{2}{3} \div 6 \cdot \left(\dfrac{3}{10}\right)$

Solution:

$$1\dfrac{2}{3} \div 6 \cdot \left(\dfrac{3}{10}\right) = \dfrac{5}{3} \div \dfrac{6}{1} \cdot \dfrac{3}{10}$$

Write the mixed number and whole number as improper fractions. Multiply and divide in order from left to right.

$$= \dfrac{5}{3} \cdot \dfrac{1}{6} \cdot \dfrac{3}{10}$$

Multiply by the reciprocal of the second fraction.

$$= \dfrac{\overset{1}{5}}{\underset{1}{3}} \cdot \dfrac{1}{6} \cdot \dfrac{\overset{1}{3}}{\underset{2}{10}}$$

Simplify.

$$= \dfrac{1}{12}$$

Multiply.

Skill Practice

Simplify.

3. $3\dfrac{2}{5} - \left(\dfrac{1}{3}\right)^2 \cdot 6$

Avoiding Mistakes

Do not try to "cancel" the 4 and the 8. The fraction $\frac{4}{25}$ is being *added*, not multiplied.

Example 3 Applying the Order of Operations

Simplify. $\left(\dfrac{2}{5}\right)^2 + \left(2\dfrac{5}{8}\right) \cdot \dfrac{3}{7}$

Solution:

$$\left(\dfrac{2}{5}\right)^2 + \left(2\dfrac{5}{8}\right) \cdot \dfrac{3}{7}$$

Perform the exponent operation first.

$$= \dfrac{2}{5} \cdot \dfrac{2}{5} + \left(2\dfrac{5}{8}\right) \cdot \dfrac{3}{7}$$

Square the quantity $\frac{2}{5}$.

$$= \dfrac{4}{25} + \dfrac{21}{8} \cdot \dfrac{3}{7}$$

Write the mixed number as an improper fraction.

$$= \dfrac{4}{25} + \dfrac{\overset{3}{21}}{8} \cdot \dfrac{3}{\underset{1}{7}}$$

Multiply before adding. Simplify common factors within the second two fractions.

$$= \dfrac{4}{25} + \dfrac{9}{8}$$

$$= \dfrac{4 \cdot 8}{25 \cdot 8} + \dfrac{9 \cdot 25}{8 \cdot 25}$$

Add the fractions. The LCD is $25 \cdot 8 = 200$.

$$= \dfrac{32}{200} + \dfrac{225}{200}$$

The fractions are now like.

$$= \dfrac{257}{200} \text{ or } 1\dfrac{57}{200}$$

The answer can be written as either an improper fraction or a mixed number.

Answers

2. $\dfrac{13}{40}$ **3.** $\dfrac{41}{15}$ or $2\dfrac{11}{15}$

2. Applications of Fractions and Mixed Numbers

Examples 4 and 5 use operations on fractions and mixed numbers in real-world applications.

| Example 4 | Using Mixed Numbers in a Sports Application |

The graph in Figure 3-4 gives the winning height for the men's high jump for selected Olympic games.

  a. What is the difference between the winning high jump in 1992 versus 1948?

  b. What is the average height from the 1948, 1960, 1968, and 1992 results?

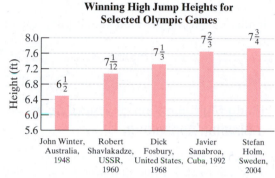

Winning High Jump Heights for Selected Olympic Games

Source: International Association of Athletics Federations

Figure 3-4

Skill Practice

4. The graph gives the winning distance for men's and women's discus throw for selected Olympic games.

Winning Discus Throw Results for Selected Olympic Games

$180\frac{3}{4}$ $194\frac{1}{6}$ $224\frac{5}{12}$ $227\frac{1}{3}$

■ Women   ■ Men

Source: International Association of Athletics Federations

  a. How much farther was the men's throw than the women's throw in 2000?
  b. What was the average throw for women for these two years?

Solution:

  a. The word "difference" implies subtraction. We subtract the 1948 height from the 1992 height.

$$1992 \text{ height} \longrightarrow \quad 7\frac{2}{3} = \quad 7\frac{2\cdot 2}{3\cdot 2} = \quad 7\frac{4}{6}$$

$$1948 \text{ height} \longrightarrow \quad -6\frac{1}{2} = \quad -6\frac{1\cdot 3}{2\cdot 3} = \quad -6\frac{3}{6}$$

$$1\frac{1}{6}$$

The difference between the winning heights in 1992 and 1948 is $1\frac{1}{6}$ ft.

  b. The average height is found by taking the sum of the four heights and dividing by 4. The sum of the heights is given by

$$6\frac{1}{2} = 6\frac{6}{12}$$
$$7\frac{1}{12} = 7\frac{1}{12}$$
$$7\frac{1}{3} = 7\frac{4}{12}$$
$$+ 7\frac{2}{3} = + 7\frac{8}{12}$$

The LCD of all four fractions is 12.

$$27\frac{19}{12} = 27 + 1\frac{7}{12} = 28\frac{7}{12}$$

Answers

4. **a.** $2\frac{11}{12}$ ft

   b. $202\frac{7}{12}$ ft

Now divide by 4 to find the average:

$$28\frac{7}{12} \div 4 = \frac{343}{12} \div \frac{4}{1}$$ Write the mixed number and whole number as improper fractions.

$$= \frac{343}{12} \cdot \frac{1}{4}$$ Multiply by the reciprocal of the divisor.

$$= \frac{343}{48}$$ Multiply.

$$= 7\frac{7}{48}$$ Write the result as a mixed number. $48\overline{)343}$
$$-336$$
$$7$$

The average winning height for the men's high jump for the selected years is $7\frac{7}{48}$ ft.

Skill Practice

5. Sylvia made $20,000 working part time at a tutoring center. On her federal income tax return, she claimed the standard deduction of $5150. If $\frac{1}{5}$ of her adjusted income goes toward taxes, how much did she pay in taxes?

Example 5   **Using Multiplication of Fractions in a Finance Application**

Sheila makes $75,000 as a self-employed graphics artist. However, she pays approximately $\frac{2}{5}$ of her income in income tax, Medicare tax, and social security. If she has business deductions of $15,000 per year, how much total tax does she pay?

Solution:

Sheila will be taxed on her income, minus her business deductions. Therefore, her total tax is given by

$$\text{Total tax} = \frac{2}{5}(75{,}000 - 15{,}000)$$

$$= \frac{2}{5}(60{,}000)$$

$$= \frac{2}{\cancel{5}}\left(\frac{\cancel{60{,}000}^{12{,}000}}{1}\right)$$

$$= 24{,}000$$

Sheila will have to pay $24,000 in taxes.

3. Applications of Geometry

In Example 6, we review the concepts of perimeter and area in applications.

Answer

5. Sylvia paid $2970 in taxes.

Example 6   **Using Fractions and Mixed Numbers in a Geometry Application**

Jason and Sara plan to paint a side of their house (Figure 3-5).

a. How much area will they have to paint?

b. They want to string Christmas lights around the triangular portion of the house. What length is required for the string of lights?

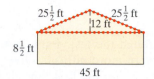

Figure 3-5

Skill Practice

6. A homeowner wants to sod the yard and fence the perimeter as shown in the figure.

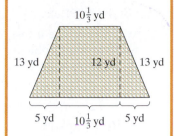

a. How much sod (area) is required?

b. How much fencing is required?

Solution:

a. The area of the side of the house is given by the sum of the rectangular area and the triangular area.

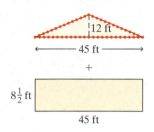

$$\text{Area of triangle} = \frac{1}{2}bh$$

$$= \frac{1}{2}(45\text{ ft})(12\text{ ft})$$

$$= \frac{1}{2}\left(\frac{45}{1}\text{ ft}\right)\left(\frac{\overset{6}{\cancel{12}}}{1}\text{ ft}\right)$$

$$= 270\text{ ft}^2$$

$$\text{Area of rectangle} = l \cdot w$$

$$= (45\text{ ft})\left(8\frac{1}{2}\text{ ft}\right)$$

$$= \left(\frac{45}{1}\text{ ft}\right)\left(\frac{17}{2}\text{ ft}\right)$$

$$= \frac{765}{2}\text{ft}^2 \quad \text{or} \quad 382\frac{1}{2}\text{ ft}^2$$

The total area is given by $270\text{ ft}^2 + 382\frac{1}{2}\text{ ft}^2 = 652\frac{1}{2}\text{ ft}^2$.

The total area to paint is $652\frac{1}{2}\text{ ft}^2$.

b. The perimeter of the triangle is found by adding the lengths of the sides.

$$25\frac{1}{2}\text{ ft}$$
$$25\frac{1}{2}\text{ ft}$$
$$+\ 45\ \ \text{ft}$$
$$\overline{95\frac{2}{2}\text{ ft} = 96\text{ ft}}$$

Jason and Sara will need 96 ft of lights.

Avoiding Mistakes

In reality, Jason and Sara might want to overestimate the amount of paint needed so that they don't run out of supplies during the job. For example:

Triangle area = 270 ft²
Rectangle area ≈ (45 ft)(10 ft)
≈ 450 ft²
Total area ≈ 270 ft² + 450 ft²
≈ 720 ft²

Answers

6. a. 184 yd² of sod is needed.
  b. $56\frac{2}{3}$ yd of fencing is needed.

Study Skills Exercise

1. When you take a test, go through the test, doing all the problems that you know first. Then go back and work on the problems that were more difficult. Give yourself a time limit for how much time you spend on each problem (maybe 3 to 5 min the first time through). Circle the importance of each statement.

	Not important	Somewhat important	Very important
a. Read through the entire test first.	1	2	3
b. If time allows, go back and check each problem.	1	2	3
c. Write out all steps instead of doing the work in your head.	1	2	3

Review Exercises

For Exercises 2–8, perform the indicated operation.

2. $7\frac{3}{10} + 2\frac{14}{15}$

3. $16 - 3\frac{7}{9}$

4. $5\frac{5}{8} \cdot 2\frac{1}{9}$

5. $7\frac{1}{9} \div 2\frac{2}{3}$

6. $24\frac{3}{5} - 14\frac{3}{4}$

7. $\left(1\frac{5}{6}\right)^2$

8. $13\frac{1}{14} + 4\frac{5}{7}$

For Exercises 9–12, convert the mixed number to an improper fraction.

9. $5\frac{2}{13}$

10. $2\frac{7}{11}$

11. $3\frac{9}{10}$

12. $1\frac{15}{16}$

For Exercises 13–16, convert the improper fraction to a mixed number.

13. $\frac{29}{5}$

14. $\frac{50}{7}$

15. $\frac{30}{19}$

16. $\frac{25}{8}$

Objective 1: Order of Operations

For Exercises 17–34, simplify, using the order of operations. Write the answer as a mixed number, if possible. (See Examples 1–3.)

17. $\left(2 - \frac{1}{2}\right)^2$

18. $\left(3 - \frac{2}{5}\right)^2$

19. $1\frac{5}{6} \cdot 2\frac{1}{2} \div 1\frac{1}{4}$

20. $2\frac{1}{7} \div 1\frac{1}{3} \cdot \frac{7}{10}$

21. $6\frac{1}{6} + 2\frac{1}{3} \div 1\frac{3}{4}$

22. $8\frac{7}{9} + 2\frac{1}{6} \cdot 3\frac{1}{3}$

23. $6 - 5\frac{1}{7} \cdot \frac{1}{3}$

24. $11 - 6\frac{1}{3} \div 1\frac{1}{6}$

25. $\left(3\frac{1}{4} + 1\frac{5}{8}\right) \cdot 2\frac{2}{3}$

26. $\left(1\frac{3}{5} + 2\frac{4}{7}\right) \cdot 5\frac{5}{6}$

27. $\left(1\frac{1}{5}\right)^2 \cdot \left(1\frac{7}{9} - 1\frac{5}{12}\right)$

28. $\left(1\frac{1}{3}\right)^3 \div \left(2\frac{7}{9} + 1\frac{2}{3}\right)$

29. $\left(6\frac{3}{4} - 2\frac{1}{8}\right) \div \left(1\frac{1}{2}\right)^3$

30. $\left(2\frac{1}{2} + 1\frac{7}{8}\right) \cdot \left(1\frac{1}{7}\right)^2$

31. $\left(\frac{1}{2}\right)^2 + \left(1\frac{1}{3}\right) \cdot \frac{9}{4}$

32. $\left(\frac{2}{3}\right)^2 + \left(2\frac{1}{2}\right) \cdot \frac{2}{9}$

33. $\left(5 - 1\frac{7}{8}\right) \div \left(3 - \frac{13}{16}\right)$

34. $\left(4 + 2\frac{1}{9}\right) \div \left(2 - 1\frac{11}{36}\right)$

Objective 2: Applications of Fractions and Mixed Numbers

35. Acceleration on cars is often compared by the time it takes to go from 0 to 60 miles per hour (mph). The graph gives the times for four cars. **(See Example 4.)**

   a. What is the difference between the times for the Lamborghini and the Caterham?

   b. Find the average time for these four cars.

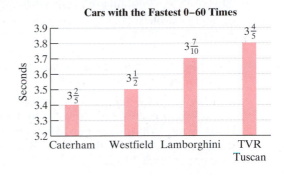

Cars with the Fastest 0–60 Times

36. Luis keeps a portfolio and tracks the number of hours he spends working on math each day.

   a. How much time did Luis spend last week working on math?

   b. Find the average amount of time spent per day working on math.

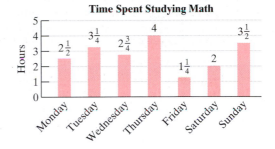

Time Spent Studying Math

37. Six women went on a weight loss program for 4 weeks. The amount of weight lost for each woman is given in the graph.

   a. Find the total weight loss.

   b. Find the average weight loss per person.

   c. Find the difference between the maximum weight loss and the minimum weight loss.

Weight Loss

38. Geoff ran $4\frac{1}{4}$ mi on Monday, $2\frac{1}{2}$ mi on Tuesday, 3 mi on Wednesday, and $8\frac{3}{4}$ mi on Thursday.

   a. What is the total distance he ran?

   b. What is the average distance he ran per day?

39. A financial analyst followed a certain stock. On Monday, the stock was at $15\frac{5}{8}$. By Friday, it had fallen to $11\frac{3}{4}$. By how much did the stock drop?

40. On Monday a stock closed at $12\frac{3}{4}$. On Tuesday, the stock rose $1\frac{1}{2}$. What was the closing price of the stock on Tuesday?

41. George will get $\frac{1}{3}$ of an $80,250 inheritance. How much money will he receive? **(See Example 5.)**

42. Aaron pays about $\frac{7}{25}$ of his annual salary in federal income tax. If he makes $45,000 per year and claims the standard deduction of $5150, how much tax must he pay?

43. A $15\frac{1}{4}$-ft cable is cut into 4 pieces of equal length. How long is each piece?

44. Twenty-seven pounds of candy is distributed in $\frac{3}{4}$-lb bags. How many bags can be filled?

45. A cheese plate advertises a total of 3 lb of assorted cheeses. At the end of a party, there was $\frac{1}{4}$ lb of Swiss cheese, $\frac{1}{3}$ lb of cheddar, and $\frac{1}{6}$ lb of Jack cheese left over. How many pounds of cheese were eaten?

46. Meade gave $\frac{1}{4}$ of a candy bar to Max and then ate $\frac{1}{3}$ of the candy bar himself. What fraction of the candy bar is left?

47. A bread recipe calls for $3\frac{1}{4}$ cups of flour. If the Daily Bread bakery has large bags of flour containing 65 cups each, how many loaves of bread can be made?

48. A carpenter worked $37\frac{1}{4}$ hr last week and earned $894. What is his hourly rate?

49. If interest rates average $6\frac{1}{2}$ points and go up $\frac{3}{4}$ point, what is the new rate?

50. The annual consumption of tea in Hong Kong is $1\frac{9}{25}$ kg per capita. The per capita tea consumption in the United Kingdom is $\frac{21}{25}$ kg more than that in Hong Kong. What is the per capita amount of tea consumed in the United Kingdom?

51. Stephanie is planning to sew her bridesmaids' dresses. The pattern calls for $2\frac{1}{2}$ yd of material for one dress with an additional $1\frac{1}{4}$ yd for the matching jacket. If she has three bridesmaids, how many yards of material will she need to buy?

52. Grace travels $1\frac{1}{4}$ hr to work each day. If she works 5 days a week, how much time is spent traveling to and from work?

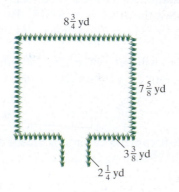

53. Wilma has $6\frac{1}{2}$ lb of mixed nuts. If she gives Fred one-half of the mixture and Barney one-third of the mixture, how many pounds of nuts does she have left?

54. Jeremy mowed $\frac{2}{3}$ of his front lawn in the morning and then $\frac{1}{4}$ of the lawn in the evening. What portion of the lawn still needs to be mowed?

55. Joan waters her plants each day with $22\frac{3}{4}$ gal of water. With a new irrigation system in place, she uses only $17\frac{2}{3}$ gal of water. How many gallons of water does she save in a 30-day period with the new irrigation system?

56. A school fund-raiser began a bake sale with $36\frac{1}{2}$ lb of cookies. At the end of the day $5\frac{3}{4}$ lb remained. How many pounds of cookies were sold?

Objective 3: Applications of Geometry

57. A decorator has $14\frac{2}{3}$ ft of wallpaper border. How much more does she need to place wallpaper border around the walls of the bathroom shown in the figure? (*Note:* No border is needed on the door.)

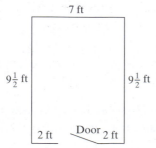

7 ft

$9\frac{1}{2}$ ft $9\frac{1}{2}$ ft

2 ft   Door 2 ft

58. Leonie wants to put a border of shrubs around her garden, as seen in the figure. Find the distance that must be lined with shrubs.

$8\frac{3}{4}$ yd

$7\frac{5}{8}$ yd

$3\frac{3}{8}$ yd

$2\frac{1}{4}$ yd

59. A stop sign is in the shape of an 8-sided polygon in which each of the sides is $12\frac{1}{2}$ in. Find the perimeter.

$12\frac{1}{2}$ in.

60. The shutters on the front of the house need to be painted. Determine the area of the four shutters.

$3\frac{2}{3}$ ft

$1\frac{5}{6}$ ft

61. Matt needs to replace gutters on his townhouse. If the front and back of the townhouse are each $20\frac{1}{2}$ ft long and the side of the house is $35\frac{1}{3}$ ft long, how much gutter should he buy to go around the three sides of the house?

62. Find the area of the triangular portion of the roof.

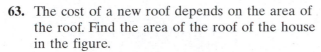

$8\frac{1}{2}$ ft

50 ft

63. The cost of a new roof depends on the area of the roof. Find the area of the roof of the house in the figure.

$14\frac{1}{4}$ ft

$35\frac{7}{8}$ ft

64. Find the area of the calculator screen.

$6\frac{1}{5}$ cm

$3\frac{7}{10}$ cm

65. The Krajewskis want to improve the backyard by fertilizing the grass and putting up a decorative fence. To know how much fertilizer to use, they must know the area of the yard. **(See Example 6.)**

   a. Using the figure, determine the area of the yard.

   b. How many meters of fencing will they need?

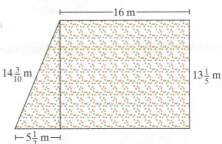

16 m

$14\frac{3}{10}$ m

$13\frac{1}{5}$ m

$5\frac{1}{2}$ m

66. A homeowner plans to put an addition onto her house. The dimensions of the front face of the addition are shown in the figure.

   a. If the homeowner wants to paint the front face of the house, how many square feet of paint will be needed?

   b. If the homeowner strings lights along the slanted edge and left edge, how long a string of lights will be needed?

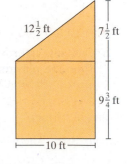

$12\frac{1}{2}$ ft

$7\frac{1}{2}$ ft

$9\frac{3}{4}$ ft

10 ft

Expanding Your Skills

67. Richard wants to paint his garage. Determine the area that needs to be painted. (He will paint the garage door, front, back, and sides of the garage, but not the roof.)

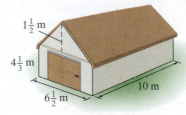

$1\frac{1}{2}$ m

$4\frac{1}{3}$ m

10 m

$6\frac{1}{2}$ m

Group Activity

Card Games with Fractions

Materials: A deck of fraction cards for each group. These can be made from index cards where one side of the card is blank, and the other side has a fraction written on it. The deck should consist of several cards of each of the following fractions.

$$\frac{1}{4}, \frac{1}{2}, \frac{3}{4}, \frac{1}{6}, \frac{1}{3}, \frac{2}{3}, \frac{5}{6}, \frac{1}{8}, \frac{3}{8}, \frac{5}{8}, \frac{7}{8}, \frac{1}{5}, \frac{2}{5}, \frac{3}{5}, \frac{4}{5}, \frac{1}{10}, \frac{3}{10}, \frac{4}{9}, \frac{2}{9}, \frac{3}{7}$$

Estimated time: Instructor discretion

In this activity, we outline three different games for students to play in their groups as a fun way to reinforce skills of adding fractions, recognizing equivalent fractions, and ordering fractions.

Game 1 "Blackjack"

Group Size: 3

1. In this game, one student in the group will be the dealer, and the other two will be players. The dealer will deal each player one card face down and one card face up. Then the players individually may elect to have more cards given to them (face up). The goal is to have the sum of the fractions get as close to "2" without going over.

2. Once the players have taken all the cards that they want, they will display their cards face up for the group to see. The player who has a sum closest to "2" without going over wins. The dealer with resolve any "disputes."

3. The members of the group should rotate after several games so that each person has the opportunity to be a player and to be the dealer.

Game 2 "War"

Group Size: 2

1. In this game, each player should start with half of the deck of cards. The players should shuffle the cards and then stack them neatly face down on the table. Then each player will select the top card from the deck, turn it over, and place it on the table. The player who has the fraction with the greatest value "wins" that round and takes both cards.

2. Continue overturning cards and deciding who "wins" each round until all of the cards have been overturned. Then the players will count the number of cards they each collected. The player with the most cards wins.

Game 3 "Bingo"

Group Size: The whole class

1. Each student gets five fraction cards. The instructor will call out fractions that are not in lowest terms. The students must identify whether the fraction that was called is the same as one of the fractions on their cards. For example, if the instructor calls out "three-ninths," then students with the fraction card $\frac{1}{3}$ would have a match.

2. The student who first matches all five cards, wins.

Chapter 3   Summary

Section 3.1   Addition and Subtraction of Like Fractions

Key Concepts	Examples

Key Concepts

Adding Like Fractions

1. Add the numerators.
2. Write the sum over the common denominator.
3. Simplify the fraction to lowest terms, if possible.

Subtracting Like Fractions

1. Subtract the numerators.
2. Write the difference over the common denominator.
3. Simplify the fraction to lowest terms, if possible.

Example 3 is an application involving subtraction of **like fractions.**

Examples

Example 1

$$\frac{5}{8} + \frac{7}{8} = \frac{12}{8} = \frac{\overset{3}{\cancel{12}}}{\underset{2}{\cancel{8}}} = \frac{3}{2}$$

Example 2

$$\frac{25}{10} - \frac{7}{10} = \frac{18}{10} = \frac{\overset{9}{\cancel{18}}}{\underset{5}{\cancel{10}}} = \frac{9}{5}$$

Example 3

A nail that is $\frac{13}{8}$ in. long is driven through a board that is $\frac{11}{8}$ in. thick. How much of the nail extends beyond the board?

$$\frac{13}{8} - \frac{11}{8} = \frac{2}{8} = \frac{1}{4}$$

The nail will extend $\frac{1}{4}$ in.

Section 3.2 Least Common Multiple

Key Concepts

The numbers obtained by multiplying a number by the whole numbers 1, 2, 3, and so on are called the **multiples** of the number.

The **least common multiple (LCM)** of two given numbers is the smallest whole number that is a multiple of each given number.

Using Prime Factors to Find the LCM of Two Numbers

1. Write each number as a product of prime factors.
2. The LCM is the product of unique prime factors from both numbers. Use repeated factors the maximum number of times they appear in either factorization.

Writing Equivalent Fractions

Use the fundamental principle of fractions to convert a fraction to an equivalent fraction with a given denominator.

Ordering Fractions

Write the fractions with a common denominator. Then compare the numerators.

Examples

Example 1

The numbers 5, 10, 15, 20, 25, 30, 35, and 40 are several multiples of 5.

Example 2

Find the LCM of 8 and 10.
Some multiples of 8 are 8, 16, 24, 32, 40.
Some multiples of 10 are 10, 20, 30, 40.

40 is the least common multiple.

Example 3

Find the LCM for the numbers 24 and 16.

$24 = 2 \cdot 2 \cdot 2 \cdot 3$

$16 = 2 \cdot 2 \cdot 2 \cdot 2$

$LCM = 2 \cdot 2 \cdot 2 \cdot 2 \cdot 3 = 48$

Example 4

Write the fraction with the indicated denominator.

$$\frac{3}{4} = \frac{}{36}$$

$$\frac{3 \cdot 9}{4 \cdot 9} = \frac{27}{36}$$

The fraction $\frac{27}{36}$ is equivalent to $\frac{3}{4}$.

Example 5

Fill in the blank with the appropriate symbol, $<$ or $>$.

$\frac{5}{9} \square \frac{7}{12}$ The LCD is 36.

$$\frac{5 \cdot 4}{9 \cdot 4} \square \frac{7 \cdot 3}{12 \cdot 3}$$

$$\frac{20}{36} < \frac{21}{36}$$

Section 3.3   Addition and Subtraction of Unlike Fractions

Key Concepts

To add or subtract unlike fractions, first we must write each fraction as an equivalent fraction with a common denominator.

Steps to Add or Subtract Unlike Fractions

1. Identify the LCD.
2. Write each individual fraction as an equivalent fraction with the LCD.
3. Add or subtract the resulting fractions as indicated.
4. Simplify to lowest terms, if possible.

Examples

Example 1

Simplify. $\dfrac{7}{5} - \dfrac{3}{10} + \dfrac{13}{15}$

$\dfrac{7 \cdot 6}{5 \cdot 6} - \dfrac{3 \cdot 3}{10 \cdot 3} + \dfrac{13 \cdot 2}{15 \cdot 2}$ The LCD = 30.

$= \dfrac{42}{30} - \dfrac{9}{30} + \dfrac{26}{30}$

$= \dfrac{42 - 9 + 26}{30}$

$= \dfrac{59}{30}$

Section 3.4   Addition and Subtraction of Mixed Numbers

Key Concepts

Addition of Mixed Numbers

To find the sum of two or more mixed numbers, add the whole-number parts and add the fractional parts.

Examples

Example 1

$$\begin{array}{r} 3\dfrac{5}{8} = 3\dfrac{10}{16} \\[2mm] + 1\dfrac{1}{16} = 1\dfrac{1}{16} \\[1mm] \hline 4\dfrac{11}{16} \end{array}$$

Example 2

$$\begin{array}{r} 2\dfrac{9}{10} = 2\dfrac{27}{30} \\[2mm] + 6\dfrac{5}{6} = 6\dfrac{25}{30} \\[1mm] \hline 8\dfrac{52}{30} = 8 + 1\dfrac{22}{30} \\[2mm] = 9\dfrac{11}{15} \end{array}$$

Subtraction of Mixed Numbers

To subtract mixed numbers, subtract the fractional parts and subtract the whole-number parts.
When the fractional part in the subtrahend is larger than the fractional part in the minuend, we borrow from the whole number part of the minuend.

Example 3

$$\begin{array}{r} 5\dfrac{3}{4} = 5\dfrac{9}{12} \\[2mm] - 2\dfrac{2}{3} = 2\dfrac{8}{12} \\[1mm] \hline 3\dfrac{1}{12} \end{array}$$

Example 4

$$\begin{array}{r} 7\dfrac{1}{2} = \overset{6}{\cancel{7}}\dfrac{\overset{5+10}{5}}{10} = 6\dfrac{15}{10} \\[2mm] - 3\dfrac{4}{5} = 3\dfrac{8}{10} \;\; = 3\dfrac{8}{10} \\[1mm] \hline 3\dfrac{7}{10} \end{array}$$

We can also add or subtract mixed numbers by writing the numbers as improper fractions. Then add or subtract the fractions.

Example 5

$$4\dfrac{7}{8} + 2\dfrac{1}{16} - 3\dfrac{1}{4} = \dfrac{39}{8} + \dfrac{33}{16} - \dfrac{13}{4}$$

$$= \dfrac{78}{16} + \dfrac{33}{16} - \dfrac{52}{16} = \dfrac{59}{16} = 3\dfrac{11}{16}$$

| **Section 3.5** | **Order of Operations and Applications of Fractions and Mixed Numbers** |

Key Concepts

Order of Operations

1. Perform all operations inside parentheses first.
2. Simplify expressions containing exponents or square roots.
3. Perform multiplication or division in the order that they appear from left to right.
4. Perform addition or subtraction in the order that they appear from left to right.

Example 2 is an example of an application involving mixed numbers and fractions.

Examples

Example 1

$$4\frac{2}{9} + 2\frac{1}{12} \cdot 5\frac{1}{3} = 4\frac{2}{9} + \frac{25}{\overset{3}{\cancel{12}}} \cdot \frac{\overset{4}{\cancel{16}}}{3}$$

$$= 4\frac{2}{9} + \frac{100}{9}$$

$$= \frac{38}{9} + \frac{100}{9}$$

$$= \frac{138}{9} = 15\frac{3}{9} = 15\frac{1}{3}$$

Example 2

Wallace has a budget of $750 for putting a curb around an area in his front yard. Curbing costs $3\frac{1}{2}$ per foot. To stay within his budget, how many feet of curbing can he afford?

Solution:

$$750 \div 3\frac{1}{2} = \frac{750}{1} \div \frac{7}{2}$$

$$= \frac{750}{1} \cdot \frac{2}{7}$$

$$= \frac{1500}{7} = 214\frac{2}{7}$$

Wallace can purchase up to $214\frac{2}{7}$ ft of curbing.

Chapter 3   Review Exercises

Section 3.1

For Exercises 1–4, add or subtract the like units.

1. 5 books + 3 books
2. 12 cm + 6 cm

3. 25 mi – 13 mi
4. 13 CDs – 2 CDs

5. Explain what is meant by the term like fractions.

6. Give an example of two like fractions and two unlike fractions. Answers may vary.

For Exercises 7–12, add or subtract the like fractions. Simplify the answer to lowest terms.

7. $\frac{5}{6} + \frac{4}{6}$

8. $\frac{4}{15} + \frac{6}{15}$

9. $\frac{5}{12} + \frac{1}{12}$

10. $\frac{2}{9} + \frac{7}{9}$

11. $\frac{15}{7} - \frac{6}{7}$

12. $\frac{21}{5} - \frac{6}{5}$

For Exercises 13–16, simplify the expression by using the order of operations.

13. $\dfrac{3}{8} \cdot \dfrac{3}{2} + \dfrac{3}{16}$ **14.** $\dfrac{4}{9} + \left(\dfrac{4}{3}\right)^2$

15. $\dfrac{21}{13} - \dfrac{5}{2} \div \dfrac{13}{4}$ **16.** $\left(\dfrac{7}{10} - \dfrac{2}{10}\right)^3 \cdot \dfrac{8}{7}$

17. Find the perimeter of the picture.

$\frac{13}{4}$ in.

$\frac{11}{4}$ in.

18. Refer to the table that gives the average snowfall for selected cities.

	January	February	March
Spartanburg, SC	$\frac{25}{10}$ in.	$\frac{19}{10}$ in.	$\frac{12}{10}$ in.
Amarillo, TX	$\frac{39}{10}$ in.	$\frac{36}{10}$ in.	$\frac{28}{10}$ in.
Portland, OR	$\frac{33}{10}$ in.	$\frac{10}{10}$ in.	$\frac{4}{10}$ in.

a. What was the total snowfall for the months of January, February, and March for Spartanburg?

b. How much more snow did Amarillo get than Portland in the month of March?

Section 3.2

19. List the first four multiples for each number.

   a. 7   **b.** 13   **c.** 22

20. Explain the difference between a common multiple and the *least* common multiple of a pair of numbers. Use the numbers 6 and 8 in your explanation.

21. List all factors for each number.

   a. 100   **b.** 65   **c.** 70

22. Find the prime factorization.

   a. 100   **b.** 65   **c.** 70

For Exercises 23–26, find the LCM by using any method.

23. 30 and 25   **24.** 22 and 144

25. 105 and 28   **26.** 16, 24, and 32

27. Sharon and Tonya signed up at a gym on the same day. Sharon will be able to go to the gym every third day and Tonya will go to the gym every fourth day. In how many days will they meet again at the gym?

For Exercises 28–31, rewrite each fraction with the indicated denominator.

28. $\dfrac{5}{16} = \dfrac{}{48}$ **29.** $\dfrac{9}{5} = \dfrac{}{35}$

30. $\dfrac{7}{12} = \dfrac{}{60}$ **31.** $\dfrac{17}{15} = \dfrac{}{150}$

For Exercises 32–34, fill in the blanks with $<$, $>$, or $=$.

32. $\dfrac{11}{24} \square \dfrac{7}{12}$ **33.** $\dfrac{5}{6} \square \dfrac{7}{9}$ **34.** $\dfrac{5}{6} \square \dfrac{15}{18}$

35. Rank the following numbers from least to greatest: $\dfrac{7}{10}, \dfrac{72}{105}, \dfrac{8}{15}, \dfrac{27}{35}$

Section 3.3

For Exercises 36–46, add or subtract. Write the answer as a fraction simplified to lowest terms.

36. $\dfrac{1}{8} + \dfrac{7}{12}$ **37.** $\dfrac{9}{10} - \dfrac{61}{100}$ **38.** $\dfrac{11}{25} - \dfrac{2}{5}$

39. $\dfrac{3}{26} + \dfrac{5}{13}$ **40.** $\dfrac{25}{11} + 2$ **41.** $4 - \dfrac{37}{20}$

42. $\dfrac{4}{15} - \dfrac{0}{3}$ **43.** $\dfrac{0}{17} + \dfrac{1}{34}$ **44.** $\dfrac{7}{100} - \dfrac{33}{1000}$

45. $\dfrac{2}{15} + \dfrac{5}{8} - \dfrac{1}{3}$ **46.** $\dfrac{11}{14} - \dfrac{4}{7} + \dfrac{3}{2}$

For Exercises 47–48, simplify by applying the order of operations.

47. $\left(\dfrac{2}{5} + \dfrac{1}{40}\right) \div \dfrac{15}{8} - \dfrac{4}{25}$

48. $\dfrac{20}{7} \cdot \left(\dfrac{11}{15} - \dfrac{1}{3}\right)^2 + \dfrac{1}{7}$

For Exercises 49–50, find (a) the perimeter and (b) the area.

49.

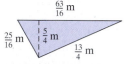

$\frac{63}{16}$ m

$\frac{25}{16}$ m $\frac{5}{4}$ m

$\frac{13}{4}$ m

50.

$\frac{3}{2}$ yd

$\frac{7}{3}$ yd

Section 3.4

For Exercises 51–62, add or subtract the mixed numbers.

51. $\begin{array}{r} 9\frac{8}{9} \\ + 1\frac{2}{7} \\ \hline \end{array}$

52. $\begin{array}{r} 10\frac{1}{2} \\ + 3\frac{15}{16} \\ \hline \end{array}$

53. $\begin{array}{r} 7\frac{5}{24} \\ - 4\frac{7}{12} \\ \hline \end{array}$

54. $\begin{array}{r} 5\frac{1}{6} \\ - 3\frac{1}{4} \\ \hline \end{array}$

55. $\begin{array}{r} 5\frac{3}{8} \\ - 2\frac{1}{3} \\ \hline \end{array}$

56. $\begin{array}{r} 3\frac{4}{5} \\ - 1\frac{4}{15} \\ \hline \end{array}$

57. $\begin{array}{r} 6\frac{4}{7} \\ + 5\frac{11}{14} \\ \hline \end{array}$

58. $\begin{array}{r} 3\frac{3}{8} \\ + 2\frac{13}{16} \\ \hline \end{array}$

59. $\begin{array}{r} 6 \\ - 2\frac{3}{5} \\ \hline \end{array}$

60. $\begin{array}{r} 8 \\ - 4\frac{11}{14} \\ \hline \end{array}$

61. $\begin{array}{r} 42\frac{1}{8} \\ + 21\frac{13}{16} \\ \hline \end{array}$

62. $\begin{array}{r} 38\frac{9}{10} \\ + 11\frac{3}{5} \\ \hline \end{array}$

For Exercises 63–66, round the numbers to estimate the answer. Then find the exact sum or difference.

63. $2\frac{1}{4} + 4\frac{2}{9} + 1\frac{29}{36}$

Estimate: _____

Exact: _____

64. $5\frac{2}{5} + 1\frac{9}{10} + 3\frac{19}{30}$

Estimate: _____

Exact: _____

65. $65\frac{1}{8} - 14\frac{9}{10}$

Estimate: _____

Exact: _____

66. $43\frac{13}{15} - 20\frac{23}{25}$

Estimate: _____

Exact: _____

67. Corry drove for $4\frac{1}{2}$ hr in the morning and $3\frac{2}{3}$ hr in the afternoon. Find the total number of hours he drove.

68. Denise owned $2\frac{1}{8}$ acres of land. If she sells $1\frac{1}{4}$ acres, how much will she have left?

Section 3.5

For Exercises 69–74, simplify by using the order of operations. Write the answer as a mixed number, if possible.

69. $1\frac{1}{5} + 4\frac{9}{10} \cdot 2\frac{2}{7}$

70. $5\frac{3}{4} - 23\frac{1}{2} \div 5\frac{2}{9}$

71. $\left(8\frac{1}{9} - 6\frac{2}{3}\right) \div 9\frac{3}{4}$

72. $\left(5\frac{1}{8} + 1\frac{1}{16}\right) \cdot 2\frac{10}{11}$

73. $\left(1\frac{1}{5}\right)^2 \cdot \left(4\frac{1}{2} + 3\frac{5}{6}\right)$

74. $\left(1\frac{5}{16}\right) \div \left(11\frac{1}{8} - 10\frac{3}{4}\right)^2$

75. In a certain region, the appraised value of a house is $\frac{9}{10}$ of its market value. If the market value of Owen's house is \$160,000, what is the appraised value?

76. Nuts 'N Things makes a nut mixture from $2\frac{1}{4}$ lb of cashews, $7\frac{3}{4}$ lb of peanuts, and $2\frac{1}{2}$ lb of pecans. The mixture is then divided into 10 bags. How many pounds are in each bag?

Chapter 3 Test

For Exercises 1–2, add or subtract the like fractions.

1. $\dfrac{4}{5} + \dfrac{3}{5}$

2. $\dfrac{23}{16} - \dfrac{15}{16}$

3. Explain the difference between evaluating these two expressions:

$$\dfrac{5}{11} - \dfrac{3}{11} \quad \text{and} \quad \dfrac{5}{11} \times \dfrac{3}{11}$$

4. a. List the first four multiples of 24.

    b. List all factors of 24.

    c. Write the prime factorization of 24.

5. Find the LCM for the numbers 16, 24, and 30.

For Exercises 6–8, write each fraction with the indicated denominator.

6. $\dfrac{5}{9} = \dfrac{}{63}$

7. $\dfrac{11}{21} = \dfrac{}{63}$

8. $\dfrac{4}{7} = \dfrac{}{63}$

9. Rank the fractions in Exercises 6–8 from least to greatest.

For Exercises 10–13, add or subtract as indicated. Write the answer as a fraction.

10. $\dfrac{3}{8} + \dfrac{3}{16}$

11. $\dfrac{7}{3} - 2$

12. $\dfrac{7}{12} - \dfrac{1}{4}$

13. $\dfrac{3}{5} + \dfrac{1}{15}$

For Exercises 14–17, add or subtract the mixed numbers. Write the answer as a mixed number.

14. $6\dfrac{3}{4} + 10\dfrac{5}{8}$

15. $12 - 9\dfrac{10}{11}$

16. $22\dfrac{1}{4} + 35\dfrac{1}{2} + 2\dfrac{2}{3}$

17. $15\dfrac{1}{6} - 12\dfrac{3}{8} - 1\dfrac{7}{24}$

For Exercises 18–21, perform the indicated operations.

18. $4\dfrac{5}{8} \cdot 2\dfrac{2}{3} - 8\dfrac{1}{6}$

19. $3\dfrac{1}{3} \div 2\dfrac{1}{2} + 5\dfrac{2}{3}$

20. $\left(\dfrac{2}{5}\right)^2 \div \left(1\dfrac{1}{10} + 2\dfrac{5}{6}\right)$

21. $\left(7\dfrac{1}{4} - 5\dfrac{1}{6}\right) \cdot 1\dfrac{3}{5}$

22. A fudge recipe calls for $1\dfrac{1}{2}$ lb of chocolate. How many pounds are required for $\dfrac{2}{3}$ of the recipe?

23. The towing capacity of a Ford Expedition is $4\dfrac{19}{40}$ times that of a Buick Rendezvous. If the Rendezvous can tow 1 ton (2000 lb), what is the towing capacity of the Expedition (in pounds)?

24. Find the area and perimeter of this parking area.

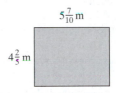

25. Justin has a budget of $14,000 to redecorate his kitchen. If he spends $\dfrac{3}{28}$ of the money on a stove and $\dfrac{1}{7}$ on a refrigerator, how much is left for new cabinets?

26. The figure gives the national records for long jump and high jump for men's and women's indoor track and field for a recent year. What is the difference for the record long jump between men and women?

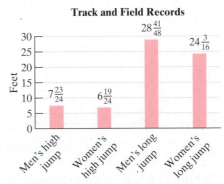

Source: International Association of Athletics Federations

Chapters 1–3   Cumulative Review Exercises

1. Write the number in words: 23,400,806

2. Find the sum of 72 and 24.

3. Find the difference of 72 and 24.

4. Find the product of 72 and 24.

5. Find the quotient of 72 and 24.

6. Round the numbers to the ten-thousands place to estimate the product: $54{,}923 \times 28{,}543$.

7. Write the expression by using exponents:
   $4 \cdot 4 \cdot 5 \cdot 5 \cdot 5 \cdot 5 \cdot 8 \cdot 8$

8. Simplify. $72 \div (4^2 - 10) \cdot 3$

9. List all the prime numbers between 15 and 35.

10. Write the prime factorization of 70.

11. Label the numerator and denominator of the fraction $\frac{21}{17}$.

12. What fraction is represented by the figure?

13. Kevin delivered 22 pizzas one evening. Of the 22 pizzas, 17 had pepperoni. What fraction of the pizzas had pepperoni? What fraction did not have pepperoni?

14. Label the fractions as proper or improper.

   a. $\dfrac{13}{5}$ **b.** $\dfrac{5}{13}$ **c.** $\dfrac{13}{13}$

15. Which of the numbers is divisible by 3 and 5?

   a. 2390   **b.** 1245   **c.** 9321

16. Label the numbers as prime, composite, or neither.

   a. 51   **b.** 52   **c.** 53

17. Find the prime factorization of 360.

18. Simplify the fraction to lowest terms: $\dfrac{180}{900}$

19. Multiply: $\dfrac{15}{16} \cdot \dfrac{2}{5}$

20. Divide: $\dfrac{20}{63} \div \dfrac{5}{9}$

21. Subtract: $\dfrac{13}{8} - \dfrac{7}{8}$

22. Add: $\dfrac{3}{16} + \dfrac{5}{16} + \dfrac{25}{16}$

23. Subtract: $4 - \dfrac{18}{5}$

24. Multiply: $6\dfrac{1}{10} \cdot 2\dfrac{3}{11}$

25. Divide: $2\dfrac{3}{5} \div 1\dfrac{7}{10}$

26. Simplify the expression.

$$\left(8\dfrac{1}{4} \div 2\dfrac{3}{4}\right)^2 \cdot \dfrac{5}{18} + \dfrac{5}{6}$$

27. To approximate the distance around a circle, multiply the diameter by the fraction $\frac{22}{7}$. Find the distance around a circle with diameter 28 cm.

28. Find the perimeter of the triangle.

29. Find the area of the triangle.

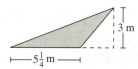

30. The figure gives the earthquake intensity on the Richter scale for four earthquakes in 2007.

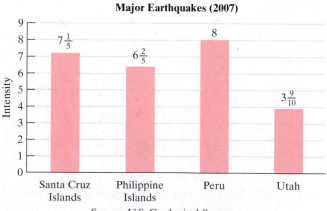

   Source: U.S. Geological Survey

   a. What was the difference in intensity between the earthquake in the Philippines and the earthquake in Utah?

   b. What was the average intensity of these earthquakes?

Decimals

4

Chapter 4

This chapter is devoted to the study of decimal numbers. We begin with a discussion of place values and then perform addition, subtraction, multiplication, and division. The applications of decimal numbers are far-reaching. Almost every day, we use decimals in transactions involving money.

   Use what you learn in this chapter to perform the operations shown in hints a–d. Then use the hints to help you complete the puzzle. Each row and each column in the grid must use the numbers 1, 2, 3, 4, and 5 exactly once.

	1			3
		4		
a			3	c
	5			
d		b		

Hints

a. Simplify $0.8 + 12.12 + 7.33$. Place the tenths place digit in box a.

b. Simplify $60.75 \div 0.3$. Place the number of digits to the left of the decimal point in box b.

c. Simplify $1.7625 - 1.56$. Place the number of digits to the right of the decimal point in box c.

d. Simplify 8.1×0.25. Place the thousandths place digit in box d.

Section 4.1 Decimal Notation and Rounding

1. Decimal Notation

In Chapters 2 and 3, we studied fraction notation to denote equal parts of a whole. In this chapter, we introduce decimal notation to denote equal parts of a whole. We first introduce the concept of a decimal fraction. A **decimal fraction** is a fraction whose denominator is a power of 10. The following are examples of decimal fractions.

$$\frac{3}{10} \text{ is read as ``three-tenths''}$$

$$\frac{7}{100} \text{ is read as ``seven-hundredths''}$$

$$\frac{9}{1000} \text{ is read as ``nine-thousandths''}$$

We now want to write these fractions in **decimal notation**. This means that we will write the numbers by using place values, as we did with whole numbers. The place value chart from Section 1.1 can be extended as shown in Figure 4-1.

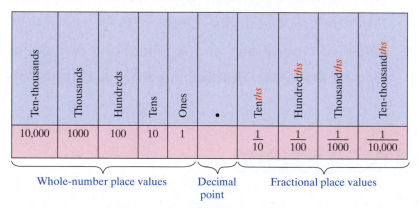

Figure 4-1

From Figure 4-1, we see that the decimal point separates the whole-number part from the fractional part. The place values for decimal fractions are located to the right of the decimal point. Their place value names are similar to those for whole numbers, but end in *ths*. Notice the correspondence between the tens place and the ten*ths* place. Similarly notice the hundreds place and the hundred*ths* place. Each place value on the left has a corresponding place value on the right, with the exception of the ones place. There is no "one*ths*" place.

Example 1	**Identify the Place Values**

Identify the place value of each underlined digit.

  a. 30,804.0<u>9</u>   **b.** 0.8469<u>2</u>0   **c.** 2<u>9</u>3.604

Solution:

  a. 30,804.0<u>9</u>   The digit 9 is in the hundredths place.

  b. 0.8469<u>2</u>0   The digit 2 is in the hundred-thousandths place.

  c. 2<u>9</u>3.604   The digit 9 is in the tens place.

For a whole number, the decimal point is understood to be after the ones place, and is usually not written. For example:

$$42. = 42$$

Using Figure 4-1, we can write the numbers $\frac{3}{10}$, $\frac{7}{100}$, and $\frac{9}{1000}$ in decimal notation.

Fraction	**Word name**	**Decimal notation**
$\dfrac{3}{10}$	Three-tenths	0.3 ↑ └─ tenths place
$\dfrac{7}{100}$	Seven-hundredths	0.07 ↑ └─ hundredths place
$\dfrac{9}{1000}$	Nine-thousandths	0.009 ↑ └─ thousandths place

TIP: The 0 to the left of the decimal point is a place-holder so that the position of the decimal point can be easily identified. It does not contribute to the value of the number. Thus, 0.3 and .3 are equal.

Now consider the number $15\frac{7}{10}$. This value represents 1 ten + 5 ones + 7 tenths. In decimal form we have 15.7.

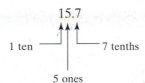

1 ten   5 ones   7 tenths

The decimal point is interpreted as the word *and*. Thus, 15.7 is read as "fifteen *and* seven tenths." The number 356.29 can be represented as

$$356 + 2 \text{ tenths} + 9 \text{ hundredths} = 356 + \frac{2}{10} + \frac{9}{100}$$

$$= 356 + \frac{20}{100} + \frac{9}{100} \qquad \text{We can use the LCD of } 100 \text{ to add the fractions.}$$

$$= 356\frac{29}{100}$$

We can read the number 356.29 as "three hundred fifty-six *and* twenty-nine hundredths."

This discussion leads to a quicker method to read decimal numbers.

> **PROCEDURE** **Reading a Decimal Number**
>
> **Step 1** The part of the number to the left of the decimal point is read as a whole number. *Note:* If there is no whole-number part, skip to step 3.
>
> **Step 2** The decimal point is read *and*.
>
> **Step 3** The part of the number to the right of the decimal point is read as a whole number but is followed by the name of the place position of the digit farthest to the right.

─ **Skill Practice** ─

Write a word name for each number.

5. 1004.6   **6.** 3.042

7. 0.0063

Example 2 **Reading Decimal Numbers**

Write the word name for each number.

a. 1028.4   **b.** 2.0736   **c.** 0.478

Solution:

a. 1028.4   is written as "one thousand, twenty-eight and four-tenths."

b. 2.0736   is written as "two and seven hundred thirty-six ten-thousandths."

c. 0.478   is written as "four hundred seventy-eight thousandths."

─ **Skill Practice** ─

Write the word name as a numeral.

8. Two hundred and two hundredths

9. Seventy-nine and sixteen thousandths

Example 3 **Writing a Numeral from a Word Name**

Write the word name as a numeral.

a. Four hundred eight and fifteen ten-thousandths

b. Five thousand eight hundred and twenty-three hundredths

Solution:

a. Four hundred eight and fifteen ten-thousandths: 408.0015

b. Five thousand eight hundred and twenty-three hundredths: 5800.23

2. Writing Decimals as Mixed Numbers or Fractions

A fractional part of a whole may be written as a fraction or as a decimal. To convert a decimal to an equivalent fraction, it is helpful to think of the decimal in words. For example:

Decimal	Word Name	Fraction	
0.3	Three tenths	$\dfrac{3}{10}$	
0.67	Sixty-seven hundredths	$\dfrac{67}{100}$	
0.048	Forty-eight thousandths	$\dfrac{48}{1000} = \dfrac{6}{125}$	(simplified)
6.8	Six and eight-tenths	$6\dfrac{8}{10} = 6\dfrac{4}{5}$	(simplified)

From the list, we notice several patterns that can be summarized as follows.

> ### PROCEDURE Converting a Decimal to a Mixed Number or Proper Fraction
>
> **Step 1** The digits to the right of the decimal point are written as the numerator of the fraction.
> **Step 2** The place value of the digit farthest to the right of the decimal point determines the denominator.
> **Step 3** The whole-number part of the number is left unchanged.
> **Step 4** Once the number is converted to a fraction or mixed number, simplify the fraction to lowest terms, if possible.

Example 4 **Writing Decimals as Proper Fractions or Mixed Numbers**

Write the decimals as proper fractions or mixed numbers and simplify.

a. 0.847   **b.** 0.0025   **c.** 4.16

Solution:

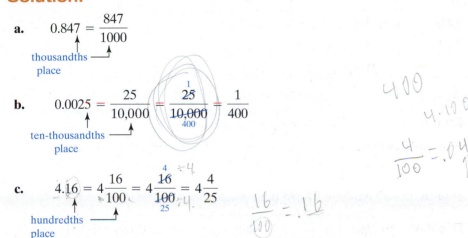

a. $0.847 = \dfrac{847}{1000}$

thousandths place

b. $0.0025 = \dfrac{25}{10{,}000} = \dfrac{25}{10{,}000} = \dfrac{1}{400}$

ten-thousandths place

c. $4.16 = 4\dfrac{16}{100} = 4\dfrac{16}{100} = 4\dfrac{4}{25}$

hundredths place

Skill Practice

Write the decimals as proper fractions or mixed numbers.
10. 0.034   **11.** 0.00086
12. 3.184

A decimal number larger than 1 can be written as a mixed number or as an improper fraction. The number 4.16 from Example 4(c) can be expressed as follows.

$$4.16 = 4\dfrac{16}{100} = 4\dfrac{4}{25} \quad \text{or} \quad \dfrac{104}{25}$$

A quick way to obtain an improper fraction for a decimal number greater than 1 is outlined here.

Concept Connections

13. Which is a correct representation of 3.17?
$$3\dfrac{17}{100} \quad \text{or} \quad \dfrac{317}{100}$$

> ### PROCEDURE Writing a Decimal Number Greater Than 1 as an Improper Fraction
>
> **Step 1** The denominator is determined by the place position of the rightmost digit to the right of the decimal point.
> **Step 2** The numerator is obtained by removing the decimal point of the original number. The resulting whole number is then written over the denominator.
> **Step 3** Simplify the improper fraction to lowest terms, if possible.

Answers

10. $\dfrac{17}{500}$ **11.** $\dfrac{43}{50{,}000}$ **12.** $3\dfrac{23}{125}$
13. They are both correct representations.

For example:

Remove decimal point.

$$4.16 = \frac{416}{100} = \frac{104}{25} \quad \text{(simplified)}$$

hundredths
place

Write the decimals as improper fractions and simplify.

14. 6.38   **15.** 15.1

Example 5 **Writing Decimals as Improper Fractions**

Write the decimals as improper fractions and simplify.

a. 40.2   **b.** 2.113

Solution:

a. $40.2 = \dfrac{402}{10} = \dfrac{\overset{201}{\cancel{402}}}{\underset{5}{\cancel{10}}} = \dfrac{201}{5}$

b. $2.113 = \dfrac{2113}{1000}$ Note that the fraction is already in lowest terms.

3. Ordering Decimal Numbers

It is often necessary to compare the values of two decimal numbers. One way of doing this is to compare the numbers in fractional form. First note that adding 0 after the rightmost digit in a decimal number does not change its value. For example,

$$0.7 = 0.70 \qquad \text{because} \qquad \frac{7}{10} = \frac{70}{100}$$

Write the numbers from least to greatest.

16. 3.7, 3.07, 3.69

Example 6 **Comparing Decimal Numbers**

Write the numbers from least to greatest.

2.1,   2.09,   2.15

Solution:

First write each number with the same number of digits to the right of the decimal point.

2.10,   2.09,   2.15   We can now write each number as a decimal fraction with the same denominator. The value

$\dfrac{210}{100}$, $\dfrac{209}{100}$, $\dfrac{215}{100}$ 209 hundredths is less than 210 hundredths, which is less than 215 hundredths.

Writing the numbers from least to greatest, we have 2.09, 2.1, and 2.15.

A quicker way to compare two decimals is outlined next.

Answers

14. $\dfrac{319}{50}$ **15.** $\dfrac{151}{10}$

16. 3.07, 3.69, 3.7

PROCEDURE   Comparing Two Positive Decimal Numbers

Step 1   Starting at the left (and moving toward the right), compare the digits in each corresponding place position.

Step 2   As we move from left to right, the first instance in which the digits differ determines the order of the numbers. The number having the greater digit is greater overall.

Example 7   Ordering Decimals

Fill in the blank with < or >.

a. 0.68 ☐ 0.7

b. 3.462 ☐ 3.4619

Solution:

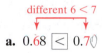

a. 0.68 < 0.70

different 6 < 7

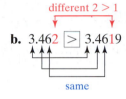

b. 3.462 > 3.4619

different 2 > 1

same

Skill Practice

Fill in the blank with < or >.

17. 4.163 ☐ 4.159

18. 218.38 ☐ 218.41

4. Rounding Decimals

The process to round the decimal part of a number is nearly the same as rounding whole numbers (see Section 1.4). The main difference is that the digits to the right of the rounding place position are dropped instead of being replaced by zeros.

PROCEDURE   Rounding Decimals to a Place Value to the Right of the Decimal Point

Step 1   Identify the digit one position to the right of the given place value.

Step 2   If the digit in step 1 is 5 or greater, add 1 to the digit in the given place value. Then discard the digits to its right.

Step 3   If the digit in step 1 is less than 5, discard it and any digits to its right.

Skill Practice

Round 187.26498 to the indicated place value.

19. Hundredths

20. Ten-thousandths

Example 8   Rounding Decimal Numbers

Round 14.795 to the indicated place value.

a. Tenths   **b.** Hundredths

Solution:

remaining digits discarded

a. 1 4 . 7 9 5 ≈ 14.8
            +1

tenths place — This digit is 5 or greater. Add 1 to the tenths place.

remaining digit discarded

b. 1 4 . 7 9 5 ≈ 14.80
              +1

hundredths place — This digit is 5 or greater. Add 1 to the hundredths place.

Answers

17. >   **18.** <   **19.** 187.26

20. 187.2650

In Example 8(b) the 0 in 14.80 indicates that the number was rounded to the hundredths place. It would be incorrect to drop the zero. Even though 14.8 has the same numerical value as 14.80, it implies a different level of accuracy. For example, when measurements are taken using some instrument such as a ruler or scale, the measured values are not exact. The place position to which a number is rounded reflects the accuracy of the measuring device. Thus, the value 14.8 lb indicates that the scale is accurate to the nearest tenth of a pound. The value 14.80 lb indicates that the scale is accurate to the nearest hundredth of a pound.

Skill Practice

21. Round 45.372 to the hundredths place.
22. Round 134.9996 to the thousandths place.

Example 9 **Rounding Decimal Numbers**

a. Round 4.81542 to the thousandths place.

b. Round 52.9999 to the hundredths place.

Solution:

remaining digits discarded

a. $4.81542 \approx 4.815$

thousandths place — This digit is less than 5. Discard it and all digits to the right.

b.  discard remaining digits

5 2 . 9 9 9 9

hundredths place — This digit is greater than 5. Add 1 to the hundredths place digit.

• Since the hundredths place digit is 9, adding 1 requires us to carry 1 to the tenths place digit.
• Since the tenths place digit is 9, adding 1 requires us to carry 1 to the ones place digit.

≈ 53.00

Answers

21. 45.37   22. 135.000

Section 4.1   Practice Exercises

Boost your GRADE at ALEKS.com!

ALEKS version 3.0

• Practice Problems   • e-Professors
• Self-Tests   • Videos
• NetTutor

Study Skills Exercises

1. After you get a test back, it is a good idea to correct the test so that you do not make the same errors again. One recommended approach is to use a clean sheet of paper and divide the paper down the middle vertically, as shown. For each problem that you missed on the test, rework the problem correctly on the left-hand side of the paper. Then write a written explanation on the right-hand side of the paper.

   Take the time this week to make corrections from your last test.

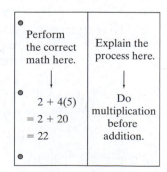

Perform the correct math here.	Explain the process here.
$2 + 4(5)$	Do multiplication before addition.
$= 2 + 20$	
$= 22$	

2. Define the key terms.

   a. Decimal fraction   **b. Decimal notation**

Objective 1: Decimal Notation

For Exercises 3–6, expand the powers of 10.

3. 10^2 4. 10^3 5. 10^4 6. 10^5

For Exercises 7–10, expand the powers of $\frac{1}{10}$.

7. $\left(\dfrac{1}{10}\right)^2$ 8. $\left(\dfrac{1}{10}\right)^3$ 9. $\left(\dfrac{1}{10}\right)^4$ 10. $\left(\dfrac{1}{10}\right)^5$

For Exercises 11–22, identify the place value of each underlined digit. **(See Example 1.)**

11. 3.9_8_3   12. 34.8_2_   13. 440.3_9_   14. 2_4_8.94

15. 4_8_9.02   16. 4.092_8_4   17. 9.283_4_5   18. 0.32_1_

19. 0.48_9_   20. 5_8_.211   21. 93._8_34   22. 5.00000_1_

For Exercises 23–30, write the word name for each decimal fraction.

23. $\dfrac{9}{10}$ 24. $\dfrac{7}{10}$ 25. $\dfrac{23}{100}$ 26. $\dfrac{19}{100}$

27. $\dfrac{33}{1000}$ 28. $\dfrac{51}{1000}$ 29. $\dfrac{407}{10,000}$ 30. $\dfrac{20}{10,000}$

For Exercises 31–38, write the word name for the decimal. **(See Example 2.)**

31. 3.24   32. 4.26   33. 5.9   34. 3.4

35. 52.3   36. 21.5   37. 6.219   38. 7.338

For Exercises 39–44, write the word name as a numeral. **(See Example 3.)**

39. Eight thousand, four hundred seventy-two and fourteen thousandths

40. Sixty thousand, twenty-five and four hundred one ten-thousandths

41. Seven hundred and seven hundredths

42. Nine thousand and nine thousandths

43. Two million, four hundred sixty-nine thousand and five hundred six thousandths

44. Eighty-two million, six hundred fourteen and ninety-seven ten-thousandths

Objective 2: Writing Decimals as Mixed Numbers or Fractions

For Exercises 45–56, write the decimal as a proper fraction or as a mixed number and simplify. **(See Example 4.)**

45. 3.7   46. 1.9   47. 2.8   48. 4.2

49. 0.25   **50.** 0.75   **51.** 0.55   **52.** 0.45

53. 20.812   **54.** 32.905   **55.** 15.0005   **56.** 4.0015

For Exercises 57–64, write the decimal as an improper fraction and simplify. **(See Example 5.)**

57. 8.4   **58.** 2.5   **59.** 3.14   **60.** 5.65

61. 23.5   **62.** 14.6   **63.** 11.91   **64.** 21.33

Objective 3: Ordering Decimal Numbers

For Exercises 65–68, arrange the numbers from least to greatest. **(See Example 6.)**

65. 34.25, 34.2, 34.3, 34.29   **66.** 12.46, 12.4, 12.5, 12.49

67. 0.42, 0.043, $\frac{4}{10}$, 0.042, 0.43   **68.** 0.04999, 0.0499, $\frac{5}{10}$, 0.4999, 0.05001

For Exercises 69–76, fill in the blank with $<$ or $>$. **(See Example 7.)**

69. 6.312 ☐ 6.321   **70.** 8.503 ☐ 8.530   **71.** 11.21 ☐ 11.2099   **72.** 10.51 ☐ 10.5098

73. 0.762 ☐ 0.76   **74.** 0.1291 ☐ 0.129   **75.** 51.72 ☐ 51.721   **76.** 49.06 ☐ 49.062

77. Which number is between 3.12 and 3.13? Circle all that apply.

     a. 3.127   **b.** 3.129   **c.** 3.134   **d.** 3.139

78. Which number is between 42.73 and 42.86? Circle all that apply.

     a. 42.81   **b.** 42.64   **c.** 42.79   **d.** 42.85

79. The batting averages for five legends are given in the table. Rank the players' batting averages from lowest to highest. (*Source:* Baseball Almanac)

Player	Average
Joe Jackson	0.3558
Ty Cobb	0.3664
Lefty O'Doul	0.3493
Ted Williams	0.3444
Rogers Hornsby	0.3585

80. The average speed, in miles per hour (mph), of the Daytona 500 for selected years is given in the table. Rank the speeds from slowest to fastest.

Year	Driver	Speed (mph)
1989	Darrell Waltrip	148.466
1991	Ernie Irvan	148.148
1997	Jeff Gordon	148.295
2007	Kevin Harvick	149.333

Source: NASCAR

Objective 4: Rounding Decimals

81. The numbers given all have equivalent value. However, suppose they represent measured values from a scale. Explain the difference in the interpretation of these numbers.

$$0.25, \quad 0.250, \quad 0.2500, \quad 0.25000$$

82. Which number properly represents 3.499999 rounded to the thousandths place?

    a. 3.500   **b.** 3.5   **c.** 3.500000   **d.** 3.499

83. Which value is rounded to the nearest tenth, 7.1 or 7.10?

84. Which value is rounded to the nearest hundredth, 34.50 or 34.5?

For Exercises 85–96, round the decimals to the indicated place values. **(See Examples 8–9.)**

85. 49.943; tenths	86. 12.7483; tenths	87. 33.416; hundredths
88. 4.359; hundredths	89. 9.0955; thousandths	90. 2.9592; thousandths
91. 21.0239; tenths	92. 16.804; hundredths	93. 6.9995; thousandths
94. 21.9997; thousandths	95. 0.0079499; ten-thousandths	96. 0.00084985; ten-thousandths

97. The average snail moves at a rate of about 0.00362005 miles per hour. Round the decimal value to the ten-thousandths place.

For Exercises 98–101, round the number to the indicated place position.

	Number	Hundreds	Tens	Tenths	Hundredths	Thousandths
98.	349.2395					
99.	971.0948					
100.	79.0046					
101.	21.9754					

Expanding Your Skills

102. What is the least number with three places to the right of the decimal that can be created with the digits 2, 9, and 7? Assume that the digits cannot be repeated.

103. What is the greatest number with three places to the right of the decimal that can be created from the digits 2, 9, and 7? Assume that the digits cannot be repeated.

Addition and Subtraction of Decimals

Objectives

1. Addition and Subtraction of Decimals
2. Applications of Addition and Subtraction of Decimals

1. Addition and Subtraction of Decimals

In this section, we learn to add and subtract decimals. To begin, consider the sum $5.67 + 3.12$.

$$5.67 = 5 + \frac{6}{10} + \frac{7}{100}$$
$$+\ 3.12 = +\ 3 + \frac{1}{10} + \frac{2}{100}$$
$$8 + \frac{7}{10} + \frac{9}{100} = 8.79$$

Notice that the decimal points and place positions are lined up to add the numbers. In this way, we can add digits with the same place values because we are effectively adding decimal fractions with like denominators. The intermediate step of using fraction notation is often skipped. We can get the same result more quickly by adding digits in like place positions.

> ### PROCEDURE   Adding and Subtracting Decimals
> **Step 1**   Write the numbers in a column with the decimal points and corresponding place values lined up. (You may insert additional zeros as placeholders after the last digit to the right of the decimal point.)
> **Step 2**   Add or subtract the digits in columns from right to left, as you would whole numbers. The decimal point in the answer should be lined up with the decimal points from the original numbers.

Skill Practice

Add.
1. $184.218 + 14.12$

Example 1   Adding Decimals

Add. $27.486 + 6.37$

Solution:

$$\begin{array}{r} 27.486 \\ +\ 6.370 \\ \end{array}$$ Line up decimal points.
Insert an extra zero as a placeholder.

$$\begin{array}{r} \overset{1}{2}7.\overset{1}{4}86 \\ +\ 6.370 \\ \hline 33.856 \end{array}$$ Add digits with common place values.

Line up the decimal point in the answer.

Concept Connections

2. Check your answer to problem 1 by estimation.

With operations on decimals it is important to locate the correct position of the decimal point. A quick estimate can help you determine whether your answer is reasonable. From Example 1, we have

$$\begin{array}{lll} 27.486 & \text{rounds to} & 27 \\ 6.370 & \text{rounds to} & +6 \\ \hline & & 33 \end{array}$$

Answers

1. 198.338
2. $\approx 184 + 14 = 198$, which is close to 198.338.

The estimated value, 33, is close to the actual value of 33.856.

Example 2 Adding Decimals

Add. $3.7026 + 43 + 816.3$

Skill Practice
Add.
3. $2.90741 + 15.13 + 3$

Solution:

```
    3.7026      Line up decimal points.
   43.0000      Insert a decimal point and four zeros after it.
+ 816.3000      Insert three zeros.

    11
    3.7026      Add digits with common place values.
   43.0000
+ 816.3000
  863.0026      Line up decimal point in the answer.
```

The sum is 863.0026.

TIP: To check that the answer is reasonable, round each addend.

```
3.7026    rounds to        4
                            1
43        rounds to        43
816.3     rounds to     + 816
                         863
```

which is close to the actual sum, 863.0026.

Example 3 Subtracting Decimals

Subtract.

a. $0.2868 - 0.056$ **b.** $139 - 28.63$ **c.** $192.4 - 89.387$

Skill Practice
Subtract.
4. $3.194 - 0.512$
5. $0.397 - 0.1584$
6. $566.4 - 414.231$

Solution:

a.
```
   0.2868      Line up the decimal points.
 - 0.0560      Insert an extra zero as a placeholder.
   0.2308      Subtract digits with common place values.
               Decimal point in the answer is lined up.
```

b.
```
  139.00       Insert extra zeros as placeholders.
 - 28.63       Line up the decimal points.

       9
     8 10 10
  139.00       Subtract digits with common place values.
 - 28.63       Borrow where necessary.
  110.37       Line up the decimal point in the answer.
```

c.
```
  192.400      Insert extra zeros as placeholders.
 - 89.387      Line up the decimal points.

            9
     8 12 3 10 10
  192.400      Subtract digits with common place values.
 - 89.387      Borrow where necessary.
  103.013      Line up the decimal point in the answer.
```

Answers
3. 21.03741   **4.** 2.682
5. 0.2386   **6.** 152.169

Example 4 Adding and Subtracting Decimals

Simplify. $\qquad$ $27.819 - 13.78 + 9.6$

Solution:

$27.819 - 13.78 + 9.6$ We must apply the order of operations.

First subtract: $27.819 - 13.78$

$$
\begin{array}{r}
{\scriptstyle 7\ 11} \\
2\,7.\,8\,\cancel{1}\,9 \\
-\ 1\,3.\,7\,8\,0 \\
\hline
1\,4.\,0\,3\,9
\end{array}
$$

$= 14.039 + 9.6$ Now add 9.6 to the result.

$$
\begin{array}{r}
{\scriptstyle 1} \\
1\,4.\,0\,3\,9 \\
+\ \ 9.\,6\,0\,0 \\
\hline
2\,3.\,6\,3\,9
\end{array}
$$

$= 23.639$

2. Applications of Addition and Subtraction of Decimals

Decimals are used often in measurements and in day-to-day applications.

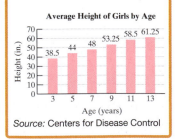
Example 5 Subtracting Decimals in an Application

A graph of the U.S. population (in millions) is given for selected years (Figure 4-2).

a. What is the difference in population between the years 1970 and 1960?

b. What is the difference in population between the years 2000 and 1990?

U.S. Population for Selected Years

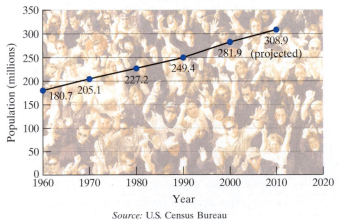

Source: U.S. Census Bureau

Figure 4-2

Answers

7. 191.48
8. a. 14.75 in.   **b.** 5.25 in.

Solution:

a. The difference in population between the years 1970 and 1960 is given by

$$\begin{array}{r} \overset{1}{2}\overset{10}{\not{0}}\overset{4}{\not{5}}.\overset{11}{\not{1}} \\ -180.7 \\ \hline 24.4 \end{array}$$

The difference in population is 24.4 million.

b. The difference in population between the years 2000 and 1990 is given by

$$\begin{array}{r} 2\overset{7}{\not{8}}\overset{11}{\not{1}}.9 \\ -249.4 \\ \hline 32.5 \end{array}$$

The difference in population is 32.5 million.

Comparing the values from parts (a) and (b), we see that the U.S. population increased during both 10-year periods. However, there was a greater increase between 1990 and 2000. This indicates that the rate of increase in population is increasing.

Example 6 **Applying Addition and Subtraction of Decimals in a Checkbook**

Fill in the balance for each line in the checkbook register, shown in Figure 4-3. What is the ending balance?

Check No.	Description	Debit	Credit	Balance
				$684.60
2409	Doctor	$ 75.50		
2410	Mechanic	215.19		
2411	Home Depot	94.56		
	Paycheck		$981.46	
2412	Veterinarian	49.90		

Figure 4-3

Solution:

We begin with $684.60 in the checking account. For each debit, we subtract. For each credit, we add.

Check No.	Description	Debit	Credit	Balance	
				$ 684.60	
2409	Doctor	$ 75.50		609.10	= $684.60 − $75.50
2410	Mechanic	215.19		393.91	= $609.10 − $215.19
2411	Home Depot	94.56		299.35	= $393.91 − $94.56
	Paycheck		$981.46	1280.81	= $299.35 + $981.46
2412	Veterinarian	49.90		1230.91	= $1280.81 − $49.90

The ending balance is $1230.91.

Skill Practice

9. Fill in the balance for each line in the checkbook register.

Check	Debit	Credit	Balance
			$437.80
1426	$82.50		
Pay		$514.02	
1427	26.04		

Answer

9.

Check	Debit	Credit	Balance
			$ 437.80
1426	$82.50		355.30
Pay		$514.02	869.32
1427	26.04		843.28

Skill Practice

10. Consider the figure.

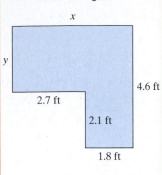

a. Find the length of side *x*.
b. Find the length of side *y*.
c. Find the perimeter.

Example 7   Applying Decimals to Perimeter

a. Find the length of the side labeled *x*.

b. Find the length of the side labeled *y*.

c. Find the perimeter of the figure.

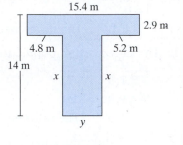

Solution:

a. If we extend the line segment labeled *x* with the dashed line as shown below, we see that the sum of side *x* and the dashed line must equal 14 m. Therefore, subtract $14 - 2.9$ to find the length of side *x*.

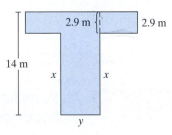

Length of side *x*:
$$\begin{array}{r} {}^{3\ 10}\\ 1\cancel{4}.\cancel{0}\\ -\ 2.9\\ \hline 11.1 \end{array}$$

Side *x* is 11.1 m long.

b. The dashed line in the figure below has the same length as side *y*. We also know that $4.8 + 5.2 + y$ must equal 15.4. Since $4.8 + 5.2 = 10.0$,

$$y = 15.4 - 10.0$$
$$= 5.4$$

The length of side *y* is 5.4 m.

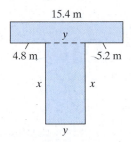

c. Now that we have the lengths of all sides, add them to get the perimeter.

$$\begin{array}{r} {}^{2\ 3}\\ 15.4\\ 2.9\\ 5.2\\ 11.1\\ 5.4\\ 11.1\\ 4.8\\ +\ 2.9\\ \hline 58.8 \end{array}$$

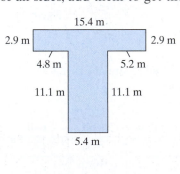

The perimeter is 58.8 m.

Answer

10. a. Side *x* is 4.5 ft.
b. Side *y* is 2.5 ft.
c. The perimeter is 18.2 ft.

Section 4.2 Practice Exercises

Boost your GRADE at ALEKS.com!

- Practice Problems
- Self-Tests
- NetTutor
- e-Professors
- Videos

Study Skills Exercise

1. Go to the online services that accompany this text. List two options that this online service offers that could help you in this course.

   a. _____

   b. _____

Review Exercises

2. Which number is equal to 5.03? Circle all that apply.

   a. 5.030   **b.** 5.30   **c.** 5.0300   **d.** 5.3

3. Which number is equal to $\frac{7}{100}$? Circle all that apply.

   a. 0.7   **b.** 0.07   **c.** 0.070   **d.** 0.007

4. Which number is equal to $\frac{9}{10}$? Circle all that apply.

   a. 0.09   **b.** 0.090   **c.** 0.90   **d.** 0.900

For Exercises 5–10, round the decimals to the indicated place values.

5. 23.489; tenths   **6.** 42.314; hundredths   **7.** 8.6025; thousandths

8. 0.981; tenths   **9.** 2.82998; ten-thousandths   **10.** 2.78999; thousandths

Objective 1: Addition and Subtraction of Decimals

For Exercises 11–16, add the decimal numbers. Then round the numbers and find the sum to determine if your answer is reasonable. The first estimate is done for you. **(See Examples 1–2.)**

Expression	Estimate		Expression	Estimate
11. 44.6 + 18.6	45 + 19 = 64		**12.** 28.2 + 23.2	
13. 5.306 + 3.645			**14.** 3.451 + 7.339	
15. 12.9 + 3.091			**16.** 4.125 + 5.9	

For Exercises 17–28, add the decimals. **(See Examples 1–2.)**

17. 78.9 + 0.9005   **18.** 44.2 + 0.7802   **19.** 23 + 8.0148   **20.** 7.9302 + 34

21. 34 + 23.0032 + 5.6   **22.** 23 + 8.01 + 1.0067   **23.** 68.394 + 32.02   **24.** 2.904 + 34.229

25. 103.94 + 24.5   **26.** 93.2 + 43.336   **27.** 54.2 + 23.993 + 3.87   **28.** 13.9001 + 72.4 + 34.13

For Exercises 29–34, subtract the decimal numbers. Then round the numbers and find the difference to determine if your answer is reasonable. The first estimate is done for you. **(See Example 3.)**

Expression	Estimate		Expression	Estimate
29. $35.36 - 21.12$	$35 - 21 = 14$		**30.** $53.9 - 22.4$	
31. $7.24 - 3.56$			**32.** $23.3 - 20.8$	
33. $45.02 - 32.7$			**34.** $66.15 - 42.9$	

For Exercises 35–46, subtract the decimals. **(See Example 3.)**

35. $14.5 - 8.823$ **36.** $33.2 - 21.932$ **37.** $2 - 0.123$

38. $4 - 0.42$ **39.** $103.4 - 45.05 - 0.982$ **40.** $98.5 - 23.21 - 0.144$

41. $55.9 - 34.2354$ **42.** $49.1 - 24.481$ **43.** $18.003 - 3.238$

44. $21.03 - 16.446$ **45.** $183.01 - 23.452$ **46.** $164.23 - 44.3893$

Mixed Exercises

For Exercises 47–58, add and subtract as indicated. **(See Example 4.)**

47. $6.007 + 12.74 - 3.4$ **48.** $3.005 + 25.127 - 13.7$ **49.** $23.37 - 21.9 + 5.111$

50. $0.78 - 0.028 + 6.1$ **51.** $8.962 + 51 - 40.05$ **52.** $11.957 + 45 - 3.55$

53. $5.3 + 5.03 + 5.003 - 5.0003$ **54.** $2.6 + 2.06 + 2.006 - 2.0006$ **55.** $5.84 + 5.084 - 5.0084$

56. $85.3 - 47.0092 + 4.06$ **57.** $10 - 0.9 - 0.09 - 0.009$ **58.** $5 - 0.9 - 0.99 - 0.999$

Objective 2: Applications of Addition and Subtraction of Decimals

59. The amount of time that it takes Mercury, Venus, Earth, and Mars to revolve about the Sun is given in the graph. **(See Example 5.)**

  a. How much longer does it take Mars to complete a revolution around the Sun than the Earth?

  b. How much longer does it take Venus than Mercury to revolve around the Sun?

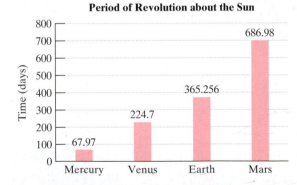

Source: National Aeronautics and Space Administration

60. The birth weights of the Dilley sextuplets are given in the graph.

  a. What is the difference between the weights of Julian and Quinn?

  b. What is the total weight of all six babies?

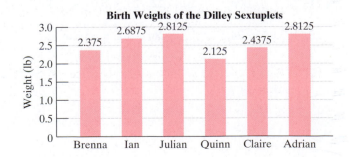

61. Water flows into a pool at a constant rate. The water level is recorded at several 1-hr intervals.

a. From the table, how many inches is the water level rising each hour?

b. At this rate, what will the water level be at 1:00 P.M.?

c. At this rate, what will the water level be at 3:00 P.M.?

Time	Water Level
9:00 A.M.	4.2 in.
10:00 A.M.	5.9 in.
11:00 A.M.	7.6 in.
12:00 P.M.	9.3 in.

62. The total gross earnings worldwide for four top animated films are given in the table.

a. What was the difference between the earnings for *Shrek 2* and *Ice Age*?

b. What was the difference between the earnings for *Finding Nemo* and *The Lion King*?

c. What was the total earnings for all four films?

Movie	Earnings ($ millions)
Shrek 2	920.7
Finding Nemo	864.6
The Lion King	783.8
Ice Age: The Meltdown	651.6

63. Fill in the balance for each line in the checkbook register shown in the figure. What was the ending balance?
(See Example 6.)

Check No.	Description	Debit	Credit	Balance
				$ 245.62
2409	Electric bill	$ 52.48		
2410	Groceries	72.44		
2411	Department store	108.34		
	Paycheck		$1084.90	
2412	Restaurant	23.87		
	Transfer from savings		200	

64. A section of a bank statement is shown in the figure. Find the mistake that was made by the bank.

Date	Action	Debit	Credit	Balance
				$1124.35
1/2/08	Check #4214	$749.32		375.03
1/3/08	Check #4215	37.29		337.74
1/4/08	Transfer from savings		$ 400.00	737.74
1/5/08	Paycheck		1451.21	2188.95
1/6/08	Cash withdrawal	150.00		688.95

65. A normal human red blood cell count is between 4.2 and 6.9 million cells per microliter (μL). A cancer patient undergoing chemotherapy has a red blood cell count of 2.85 million cells per microliter. How far below the lower normal limit is this?

66. A laptop computer was originally priced at $1299.99 and was discounted to $998.95. By how much was it marked down?

67. The table shows the widths of four U.S. coins. If you stacked three quarters and a dime in one pile and two nickels and two pennies in another pile, which pile would be higher?

Coin	Width
Quarter	1.75 mm
Dime	1.35 mm
Nickel	1.95 mm
Penny	1.55 mm

Source: U.S. Department of the Treasury

68. How much thicker is a nickel than a quarter?

For Exercises 69–72, find the length of the sides labeled *x* and *y*. Then find the perimeter. **(See Example 7.)**

69.

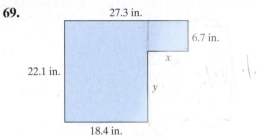

70.

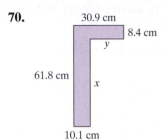

71.

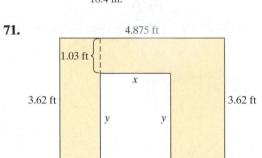

72.

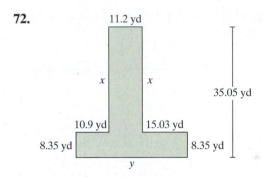

73. A city bus follows the route shown in the map. How far does it travel in one circuit?

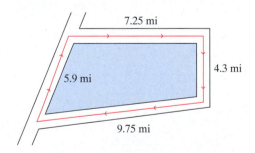

74. Santos built a new deck and needs to put a railing around the sides. He does not need railing where the deck is against the house. How much railing should he purchase?

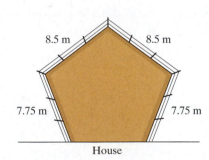

Expanding Your Skills

In a circle, the length of a line segment connecting two points on the circle and passing through the center is called a *diameter*.

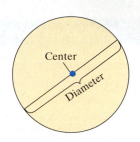

Use the definition of a diameter for Exercises 75–76.

75. The wire in a cable has a diameter of 6 mm. The insulation is 0.5 mm. What is the diameter of the cable with the insulation included?

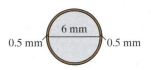

76. Find the inner diameter of the cable if the total diameter is 1.65 cm and the insulation is 0.15 cm thick.

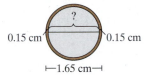

Calculator Connections

Topic: Entering decimals on a calculator

To enter decimals on a calculator, use the [.] key.

Expression	Keystrokes	Result
984.126 + 37.11	984 [.] 126 [+] 37 [.] 11 **ENTER**	1021.236

Calculator Exercises

For Exercises 77–82, refer to the table. The table gives the closing stock prices (in dollars per share) for the first day of trading for the given month.

Stock	January	February	March	April	May
IBM	$97.27	$99.00	$92.27	$95.21	$103.17
FedEx	$109.77	$111.89	$114.16	$106.07	$106.11

77. By how much did the IBM stock increase between January and May?

78. By how much did the FedEx stock decrease between January and May?

79. Between which two consecutive months did the FedEx stock increase the most? What was the amount of increase?

80. Between which two consecutive months did the IBM stock increase the most? What was the amount of increase?

81. Between which two consecutive months did the FedEx stock decrease the most? What was the amount of decrease?

82. Between which two consecutive months did the IBM stock decrease the most? What was the amount of decrease?

Section 4.3   Multiplication of Decimals

Objectives

1. Multiplication of Decimals
2. Multiplication by a Power of 10 and by a Power of 0.1
3. Applications Involving Multiplication of Decimals

1. Multiplication of Decimals

Multiplication of decimals is much like multiplication of whole numbers. However, we need to know where to place the decimal point in the product. Consider the product 0.3×0.41. One way to multiply these numbers is to write them first as decimal fractions.

$$0.3 \times 0.41 = \frac{3}{10} \times \frac{41}{100} = \frac{123}{1000} \text{ or } 0.123$$

Another method multiplies the factors vertically. First we multiply the numbers as though they were whole numbers. We temporarily disregard the decimal point in the product because it will be placed later.

$$\begin{array}{r} 0.41 \\ \times\ 0.3 \\ \hline 123 \end{array} \quad \longleftarrow \text{ decimal point not yet placed}$$

From the first method, we know that the correct answer to this problem is 0.123. Notice that 0.123 contains the same number of decimal places as the two factors combined. That is,

Concept Connections

1. How many decimal places will be in the product 2.72×1.4?
2. Explain the difference between the process to multiply 123×51 and the process to multiply 1.23×5.1.

$$\begin{array}{rll} 0.41 & \longleftarrow & \text{2 decimal places} \\ \times\ 0.3 & \longleftarrow & \text{1 decimal place} \\ \hline .123 & \longleftarrow & \text{3 decimal places} \end{array}$$

The process to multiply decimals is summarized as follows.

TIP: When multiplying decimals, it is *not* necessary to line up the decimal points as we do when we add or subtract decimals. Instead, we write the factors "right-justified."

PROCEDURE   Multiplying Two Decimals

Step 1   Multiply as you would whole numbers.

Step 2   Place the decimal point in the product so that the number of decimal places equals the combined number of decimal places of both factors.

Note: You may need to insert zeros to the left of the whole-number product to get the correct number of decimal places in the answer.

Skill Practice

3. Multiply. 19.7×4.1

In Example 1, we multiply decimals by using this process.

Example 1   Multiplying Decimals

Multiply.

$$\begin{array}{r} 11.914 \\ \times\ 0.8 \end{array}$$

Solution:

$$\begin{array}{rl} \overset{17\ 13}{11.914} & \text{3 decimal places} \\ \times\quad 0.8 & +\ \text{1 decimal place} \\ \hline 9.5312 & \text{4 decimal places} \end{array}$$

The product is 9.5312.

Answers

1. 3
2. The actual process of vertical multiplication is the same for both cases. However, for the product 1.23×5.1, the decimal point must be placed so that the product has the same number of decimal places as both factors combined (in this case, 3).
3. 80.77

Example 2   **Multiplying Decimals**

Multiply. Then use estimation to check the location of the decimal point.

$$29.3 \times 2.8$$

Solution:

Actual product:

$$\begin{array}{r} \overset{1}{}\overset{7\,2}{} \\ 29.3 \\ \times\ 2.8 \\ \hline 2344 \\ 5860 \\ \hline 82.04 \end{array}$$ 1 decimal place
+ 1 decimal place

2 decimal places   The product is 82.04.

To check the answer, we can round the factors and estimate the product. The purpose of the estimate is primarily to determine whether we have placed the decimal point correctly. Therefore, it is usually sufficient to round each factor to the left-most nonzero digit. This is called **front-end rounding.** Thus,

$$\begin{array}{ccc} 29.3 & \text{rounds to} & 30 \\ \times\ 2.8 & \text{rounds to} & \times\ 3 \\ & & \hline & & 90 \end{array}$$

The first digit for the actual product 8̲2.04 and the first digit for the estimate 9̲0 is the tens place. Therefore, we are reasonably sure that we have located the decimal point correctly. The estimate 90 is close to 82.04.

Skill Practice

4. Multiply 1.9×29.1, and check your answer using estimation.

Example 3   **Multiplying Decimals**

Multiply. Then use estimation to check the location of the decimal point.

$$2.79 \times 0.0003$$

Solution:

Actual product: Estimate:

$$\begin{array}{r} \overset{2\ 2}{} \\ 2.79 \\ \times\ 0.0003 \\ \hline .000837 \end{array}$$ 2 decimal places
+ 4 decimal places

6 decimal places
(insert 3 zeros to the left)

$$\begin{array}{ccc} 2.79 & \text{rounds to} & \sim 3 \\ \times\ 0.0003 & \text{rounds to} & \times\ 0.0003 \\ & & \hline & & 0.0009 \end{array}$$

The first digit for both the actual product and the estimate is in the ten-thousandths place. We are reasonably sure the decimal point is positioned correctly.

The product is 0.000837.

Skill Practice

5. Multiply 4.6×0.00008, and check your answer using estimation.

Answers

4. 55.29; $\approx 2 \times 30 = 60$ which is close to 55.29.
5. 0.000368; using front-end rounding, we have $\approx 5 \times 0.00008 = 0.0004$ which is close to 0.000368.

2. Multiplication by a Power of 10 and by a Power of 0.1

Consider the number 2.7 multiplied by the powers of 10; that is, 10, 100, 1000 . . .

10	100	1000
× 2.7	× 2.7	× 2.7
70	700	7000
200	2000	20000
27.0	270.0	2700.0

Multiplying 2.7 by 10 moves the decimal point 1 place to the right.
Multiplying 2.7 by 100 moves the decimal point 2 places to the right.
Multiplying 2.7 by 1000 moves the decimal point 3 places to the right.

This leads us to the following generalization.

> **PROCEDURE** Multiplying a Decimal by a Power of 10
>
> Move the decimal point to the right the same number of decimal places as the number of zeros in the power of 10.

Skill Practice

Multiply.
6. 81.6×1000
7. $0.0000085 \times 10,000$
8. $2.396 \times 10,000,000$

Example 4 Multiplying Decimals by Powers of 10

Multiply.

a. $14.78 \times 10,000$ **b.** 0.0064×100 **c.** $8.271 \times 1,000,000$

Solution:

a. $14.78 \times 10,000 = 147,800$ Move the decimal point 4 places to the right.

b. $0.0064 \times 100 = 0.64$ Move the decimal point 2 places to the right.

c. $8.271 \times 1,000,000 = 8,271,000$ Move the decimal point 6 places to the right.

Multiplying a decimal by 10, 100, 1000, and so on increases its value. Therefore, it makes sense to move the decimal point to the *right*. Now suppose we multiply a decimal by 0.1, 0.01, and 0.001. These numbers represent the decimal fractions $\frac{1}{10}$, $\frac{1}{100}$, and $\frac{1}{1000}$, respectively, and are easily recognized as powers of 0.1 (see Section 2.4). Taking one-tenth of a number or one-hundredth of a number makes the number smaller. To multiply by 0.1, 0.01, 0.001, and so on (powers of 0.1), move the decimal point to the *left*.

3.6	3.6	3.6
× 0.1	× 0.01	× 0.001
.36	.036	.0036

Answers
6. 81,600   **7.** 0.085   **8.** 23,960,000

PROCEDURE Multiplying a Decimal by Powers of 0.1

Move the decimal point to the left the same number of places as there are decimal places in the power of 0.1.

Concept Connections

9. Explain the difference between multiplying a number by 100 versus 0.01.

Example 5 Multiplying by Powers of 0.1

Multiply.

a. 62.074×0.0001 b. 7965.3×0.1 c. 0.0057×0.00001

Solution:

a. $62.074 \times 0.0001 = 0.0062074$ Move the decimal point 4 places to the left. Insert extra zeros.

b. $7965.3 \times 0.1 = 796.53$ Move the decimal point 1 place to the left.

c. $0.0057 \times 0.00001 = 0.000000057$ Move the decimal point 5 places to the left.

Skill Practice

Multiply.
10. 471.034×0.01
11. $9,437,214.5 \times 0.00001$
12. 0.0004×0.001

Sometimes people prefer to use number names to express very large numbers. For example, we might say that the U.S. population in 2004 was approximately 280 million. To write this in decimal form, we note that 1 million = 1,000,000. In this case, we have 280 of this quantity. Thus,

$$280 \text{ million} = 280 \times 1,000,000 \text{ or } 280,000,000$$

Example 6 Naming Large Numbers

Write the decimal number representing each word name.

a. The distance between the Earth and Sun is approximately 92.9 million miles.

b. The number of deaths in the United States due to heart disease in 2010 is projected to be 8 hundred thousand.

c. A recent estimate claimed that collectively Americans throw away 472 billion pounds of garbage each year.

Skill Practice

Write a decimal number representing the word name.
13. The population in Bexar County, Texas, is approximately 1.6 million.
14. Light travels approximately 5.9 trillion miles in 1 year.
15. The legislative branch of the federal government employs approximately 31 thousand employees.

Solution:

a. 92.9 million = $92.9 \times 1,000,000 = 92,900,000$

b. 8 hundred thousand = $8 \times 100,000 = 800,000$

c. 472 billion = $472 \times 1,000,000,000 = 472,000,000,000$

Answers

9. Multiplying a number by 100 increases its value. Therefore, we move the decimal point to the right two places. Multiplying a number by 0.01 decreases its value. Therefore, move the decimal point to the left two places.
10. 4.71034   11. 94.372145
12. 0.0000004   13. 1,600,000
14. 5,900,000,000,000   15. 31,000

3. Applications Involving Multiplication of Decimals

Skill Practice

16. A book club ordered 12 books on www.amazon.com for $8.99 each. The shipping cost was $4.95. What was the total bill?

Example 7 Applying Decimal Multiplication

Jane Marie bought 8 cans of tennis balls for $1.98 each. She paid $1.03 in tax. What was the total bill?

Solution:

The cost of the tennis balls before tax is

$$8(\$1.98) = \$15.84$$

$$\begin{array}{r} {}^{7\ 6}1.98 \\ \times\ \ 8 \\ \hline 15.84 \end{array}$$

Adding the tax to this value, we have

$$\left(\begin{array}{c}\text{Total} \\ \text{cost}\end{array}\right) = \left(\begin{array}{c}\text{Cost of} \\ \text{tennis balls}\end{array}\right) + (\text{Tax})$$

$$\begin{array}{r} = \$15.84 \\ +\ 1.03 \\ \hline \$16.87 \end{array}$$ The total cost is $16.87.

Skill Practice

17. The IMAX movie screen at the Museum of Science and Discovery in Ft. Lauderdale, Florida, is 18 m by 24.4 m. What is the area of the screen?

Example 8 Finding the Area of a Rectangle

The *Mona Lisa* is perhaps the most famous painting in the world. It was painted by Leonardo da Vinci somewhere between 1503 and 1506 and now hangs in the Louvre in Paris, France. The dimensions of the painting are 30 in. by 20.875 in. What is the total area?

Solution:

Recall that the area of a rectangle is given by

$$A = \ell \cdot w$$

$$A = (30 \text{ in.})(20.875 \text{ in.})$$

$$\begin{array}{r} {}^{2\ 21}20.875 \\ \times\ 30 \\ \hline 0 \\ 626250 \\ \hline 626.250 \end{array}$$

$$= 626.25 \text{ in.}^2$$

The area of the *Mona Lisa* is 626.25 in.^{2}.

Answers

16. The total bill was $112.83.

17. The screen area is 439.2 m^2.

Section 4.3 | Practice Exercises

Study Skills Exercises

1. Look through this chapter and write down page numbers in which you can find the following features.

    a. Avoiding mistakes _____   **b.** TIP box _____

    c. A key term (shown in bold) _____   **d.** A skill practice _____

2. Define the key term **front-end rounding**.

Review Exercises

For Exercises 3–6, expand the powers of 10 and 0.1.

3. 10^3 **4.** 0.1^3 **5.** 0.1^2 **6.** 10^2

Objecitve 1: Multiplication of Decimals

For Exercises 7–18, multiply the decimals. **(See Examples 1–3.)**

7. $\begin{array}{r} 0.8 \\ \times\, 0.5 \\ \hline \end{array}$ **8.** $\begin{array}{r} 0.6 \\ \times\, 0.5 \\ \hline \end{array}$ **9.** $(0.9)(4)$ **10.** $(0.2)(9)$

11. $\begin{array}{r} 0.4 \\ \times\, 20 \\ \hline \end{array}$ **12.** $\begin{array}{r} 0.9 \\ \times\, 30 \\ \hline \end{array}$ **13.** $(60)(0.003)$ **14.** $(40)(0.005)$

15. $\begin{array}{r} 22.38 \\ \times\, 0.8 \\ \hline \end{array}$ **16.** $\begin{array}{r} 31.67 \\ \times\, 0.4 \\ \hline \end{array}$ **17.** $\begin{array}{r} 14 \\ \times\, 0.002 \\ \hline \end{array}$ **18.** $\begin{array}{r} 0.25 \\ \times\, 40 \\ \hline \end{array}$

For Exercises 19–26, round each number by using front-end rounding.

19. 135   **20.** 481   **21.** 28   **22.** 52

23. 0.0672   **24.** 0.0807   **25.** 0.241   **26.** 0.339

For Exercises 27–40, multiply the decimals. Then estimate the answer by rounding. The first estimate is done for you.

	Exact	Estimate		Exact	Estimate		Exact	Estimate
27.	8.3	8	**28.**	4.3		**29.**	0.58	
	× 4.5	× 5		× 9.2			× 7.2	
		40						

30. $\begin{array}{r} 0.83 \\ \times\, 6.5 \\ \hline \end{array}$ **31.** 5.92×0.8 **32.** 9.14×0.6

33. $(0.413)(7)$ **34.** $(0.321)(6)$ **35.** 35.9×3.2 **36.** 41.7×6.1

37. 562×0.004 **38.** 984×0.009 **39.** 0.0004×3.6 **40.** 0.0008×6.5

Objective 2: Multiplication by a Power of 10 and by a Power of 0.1

41. If 417.43 is multiplied by 100, will the decimal point move to the left or to the right? By how many places?

42. If 2498.613 is multiplied by 10,000, will the decimal point move to the left or to the right? By how many places?

43. Multiply the numbers.
   a. 5.1×10 **b.** 5.1×100 **c.** 5.1×1000 **d.** $5.1 \times 10,000$

44. If 256.8 is multiplied by 0.001, will the decimal point move to the left or to the right? By how many places?

45. If 0.45 is multiplied by 0.1, will the decimal point move to the left or to the right? By how many places?

46. Multiply the numbers.

   a. 5.1×0.1 **b.** 5.1×0.01 **c.** 5.1×0.001 **d.** 5.1×0.0001

For Exercises 47–58, multiply the numbers by the powers of 10 and 0.1. (See Examples 4–5.)

47. 34.9×100 **48.** 2.163×100 **49.** 96.59×1000 **50.** 18.22×1000

51. 93.3×0.01 **52.** 80.2×0.01 **53.** 54.03×0.001 **54.** 23.11×0.001

55. 2.001×10 **56.** 5.932×10 **57.** 0.5×0.0001 **58.** 0.8×0.0001

For Exercises 59–64, write the decimal number representing each word name. (See Example 6.)

59. The number of beehives in the United States is 2.6 million. (*Source:* U.S. Department of Agriculture)

60. The people of France collectively consume 34.7 million gallons of champagne per year. (*Source:* Food and Agriculture Organization of the United Nations)

61. The most stolen make of car worldwide is Toyota. For a recent year, there were four hundred-thousand Toyota's stolen. (*Source:* Interpol)

62. The musical *Miss Saigon* ran for about 4 thousand performances in a 10-year period.

64. Coca-Cola Classic was the greatest selling brand of soft-drinks. For a recent year, over 4.8 billion gallons were sold in the United States. (*Source:* Beverage Marketing Corporation)

63. The people in the United States have spent over $20.549 billion on DVDs.

Objective 3: Applications Involving Multiplication of Decimals

65. One gallon of gasoline weighs about 6.3 lb. However, when burned, it produces 20 lb of carbon dioxide (CO_2). This is because most of the weight of the CO_2 comes from the oxygen in the air. (See Example 7.)

   a. How many pounds of gasoline does a Hummer H2 carry when its tank is full (the tank holds 32 gal).

   b. How many pounds of CO_2 does a Hummer H2 produce after burning an entire tankful of gasoline?

66. Corrugated boxes for shipping cost $2.27 each. How much will 10 boxes cost including tax of $1.59?

67. The Athletic Department at Broward Community College bought 20 pizzas for $10.95 each, 10 Greek salads for $3.95 each, and 60 soft drinks for $0.60 each. What was the total bill including a sales tax of $17.67?

68. A hotel gift shop ordered 40 T-shirts at $8.69 each, 10 hats at $3.95 each, and 20 beach towels at $4.99 each. What was the total cost of the merchandise, including the $29.21 sales tax?

69. Firestone tires cost $50.20 each. A set of four Lemans tires costs $197.99. How much can a person save by buying the set of four Lemans tires compared to four Firestone tires?

70. Certain DVD titles are on sale for 2 for $32. If they regularly sell for $19.99, how much can a person save by buying 4 DVDs?

For Exercises 71–72, find the area. (**See Example 8.**)

71.

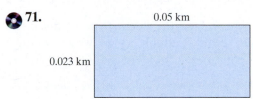

0.05 km

0.023 km

72.

4.5 yd

6.7 yd

73. Blake plans to build a rectangular patio that is 15 yd by 22.2 yd. What is the total area of the patio?

74. The front page of a newspaper is 56 cm by 31.5 cm. Find the area of the page.

For Exercises 75–78, write the amount in terms of cents.

75. $3.24   76. $21.56   77. $0.37   78. $0.75

For Exercises 79–82, write the amount in terms of dollars.

79. 347¢   80. 512¢   81. 2041¢   82. 5712¢

83. a. Round $1.499 to the nearest dollar.

   b. Round $1.499 to the nearest cent.

84. a. Round $20.599 to the nearest dollar.

   b. Round $20.599 to the nearest cent.

85. Compare the quantities $(0.2)^2$ and 0.4. Are they equal?

86. Compare the quantities $(0.5)^2$ and 2.5. Are they equal?

For Exercises 87–94, simplify the expressions.

87. $(0.4)^2$ 88. $(0.7)^2$ 89. $(1.3)^2$ 90. $(2.4)^2$

91. $(0.1)^3$ 92. $(0.2)^3$ 93. $(0.2)^4$ 94. $(0.3)^3$

Expanding Your Skills

95. Evaluate.

   a. $(0.3)^2$ b. $\sqrt{0.09}$

96. Evaluate.

   a. $(0.5)^2$ b. $\sqrt{0.25}$

For Exercises 97–100, evaluate the square roots.

97. $\sqrt{0.01}$ 98. $\sqrt{0.04}$ 99. $\sqrt{0.36}$ 100. $\sqrt{0.49}$

Section 4.4 Division of Decimals

1. Division of Decimals

Dividing decimals is much the same as dividing whole numbers. However, we must determine where to place the decimal point in the quotient.

First consider the quotient $3.5 \div 7$. We can write the numbers in fractional form and then divide.

$$3.5 \div 7 = \frac{35}{10} \div \frac{7}{1} = \frac{35}{10} \cdot \frac{1}{7} = \frac{35}{70} = \frac{5}{10} = 0.5$$

Now consider the same problem by using the efficient method of long division: $7\overline{)3.5}$.

When the divisor is a whole number, we place the decimal point directly above the decimal point in the dividend. Then we divide as we would whole numbers.

Decimal point placed above
the decimal point in the dividend.

$$7\overline{)3.5}^{.5}$$

> **PROCEDURE** Dividing a Decimal by a Whole Number
>
> To divide by a whole number:
>
> **Step 1** Place the decimal point in the quotient directly above the decimal point in the dividend.
>
> **Step 2** Divide as you would whole numbers.

Skill Practice

Divide. Check by using multiplication.

1. $502.96 \div 8$

Example 1 Dividing by a Whole Number

Divide and check the answer by multiplying.

$$30.55 \div 13$$

Solution:

Locate the decimal point in the quotient.

$$13\overline{)30.55}$$

$$
\begin{array}{r}
2.35 \\
13\overline{)30.55} \\
-26 \\
\hline
45 \\
-39 \\
\hline
65 \\
-65 \\
\hline
0
\end{array}
$$

Divide as you would whole numbers.

Check by multiplying:

$$
\begin{array}{r}
\overset{1\ 1}{2.35} \\
\times\ 13 \\
\hline
705 \\
2350 \\
\hline
30.55\ \checkmark
\end{array}
$$

Answer

1. 62.87

When dividing decimals, we do not use a remainder. Instead we insert zeros to the right of the dividend and continue dividing. This is demonstrated in Example 2.

Example 2 **Dividing by a Whole Number**

Divide and check the answer by multiplying.

$$3.5 \div 4$$

Solution:

Locate the decimal point in the quotient.

$$4\overline{)3.5}$$

$$\begin{array}{r} .8 \\ 4\overline{)3.5} \\ -32 \\ \hline 3 \end{array}$$

← Rather than using a remainder, we insert zeros in the dividend and continue dividing.

$$\begin{array}{r} .875 \\ 4\overline{)3.500} \\ -32 \\ \hline 30 \\ -28 \\ \hline 20 \\ -20 \\ \hline 0 \end{array}$$

Check by multiplying:

$$\begin{array}{r} \overset{3\;2}{0.875} \\ \times\ \ 4 \\ \hline 3.500 \ \checkmark \end{array}$$

The quotient is 0.875.

Skill Practice

Divide.

2. $6.8 \div 5$

Example 3 **Dividing by a Whole Number**

Divide and check the answer by multiplying. $40\overline{)5}$

Solution:

$$40\overline{)5.}$$

The dividend is a whole number, and the decimal point is understood to be to its right. Insert the decimal point above it in the quotient.

$$\begin{array}{r} .125 \\ 40\overline{)5.000} \\ -40 \\ \hline 100 \\ -80 \\ \hline 200 \\ -200 \\ \hline 0 \end{array}$$

Since 40 is greater than 5, we need to insert zeros to the right of the dividend.

Check by multiplying.

$$\begin{array}{r} \overset{1\;2}{.125} \\ \times\ 40 \\ \hline 000 \\ 5000 \\ \hline 5.000 \ \checkmark \end{array}$$

The quotient is 0.125.

Skill Practice

3. $20\overline{)3}$

Answers

2. 1.36 **3.** 0.15

Sometimes when dividing decimals, the quotient follows a repeated pattern. The result is called a **repeating decimal**.

Skill Practice

Divide.

4. $2.4 \div 9$

Example 4   Dividing Where the Quotient Is a Repeating Decimal

Divide. $1.7 \div 30$

Solution:

$$
\begin{array}{r}
.05666\ldots \\
30\overline{)1.70000} \\
-150 \\
\hline
200 \\
-180 \\
\hline
200 \\
-180 \\
\hline
200
\end{array}
$$

Notice that as we continue to divide, we get the same values for each successive step. This causes a pattern of repeated digits in the quotient. Therefore, the quotient is a *repeating decimal*.

The quotient is $0.05666\ldots$. To denote the repeated pattern, we often use a bar over the first occurrence of the repeat cycle to the right of the decimal point. That is,

$0.05666\ldots = 0.05\overline{6}$ ← repeat bar

Avoiding Mistakes

In Example 4, notice that the repeat bar goes over only the 6. The 5 is not being repeated.

Skill Practice

Divide.

5. $11\overline{)57}$

Example 5   Dividing Where the Quotient is a Repeating Decimal

Divide. $11\overline{)68}$

Solution:

$$
\begin{array}{r}
6.1818\ldots \\
11\overline{)68.0000} \\
-66 \\
\hline
20 \\
-11 \\
\hline
90 \\
-88 \\
\hline
20 \\
-11 \\
\hline
90 \\
-88 \\
\hline
20
\end{array}
$$

Could have stopped here

Once again, we see a repeated pattern. The quotient is a repeating decimal. Notice that we could have stopped dividing when we obtained the second value of 20.

The quotient is $6.\overline{18}$.

Avoiding Mistakes

Be sure to put the repeating bar over the entire block of numbers that is being repeated. In Example 5, the bar extends over both the 1 and the 8. We have $6.\overline{18}$.

The numbers $0.05\overline{6}$ and $6.\overline{18}$ are examples of repeating decimals. A decimal that "stops" is called a **terminating decimal**. For example, 6.18 is a terminating decimal, whereas $6.\overline{18}$ is a repeating decimal.

Answers

4. $0.2\overline{6}$ **5.** $5.\overline{18}$

In Examples 1–5, we performed division where the divisor was a whole number. Suppose now that we have a divisor that is *not* a whole number, for example, 0.56 ÷ 0.7. Because division can also be expressed in fraction notation, we have

$$0.56 \div 0.7 = \frac{0.56}{0.7}$$

If we multiply the numerator and denominator by 10, the denominator (divisor) becomes the whole number 7.

$$\frac{0.56}{0.7} = \frac{0.56 \times 10}{0.7 \times 10} = \frac{5.6}{7} \longrightarrow 7\overline{)5.6}$$

Recall that multiplying decimal numbers by 10 (or any power of 10, such as 100, 1000, etc.) moves the decimal point to the right. We use this idea to divide decimal numbers when the divisor is not a whole number.

> **PROCEDURE** Dividing When the Divisor Is Not a Whole Number
>
> **Step 1** Move the decimal point in the divisor to the right to make it a whole number.
> **Step 2** Move the decimal point in the dividend to the right the same number of places as in step 1.
> **Step 3** Place the decimal point in the quotient directly above the decimal point in the dividend.
> **Step 4** Divide as you would whole numbers.

Example 6 Dividing Decimals

Divide.

a. 0.56 ÷ 0.7   **b.** 0.005)3.1

Solution:

a. .7)̄.56   Move the decimal point in the divisor and dividend one place to the right.

7)5.6   Line up the decimal point in the quotient.

$$\begin{array}{r} 0.8 \\ 7\overline{)5.6} \\ -5\,6 \\ \hline 0 \end{array}$$

The quotient is 0.8.

b. .005)3.100   Move the decimal point in the divisor and dividend three places to the right. Insert additional zeros in the dividend if necessary. Line up the decimal point in the quotient.

$$\begin{array}{r} 620. \\ 5\overline{)3100.} \\ -30 \\ \hline 10 \\ -10 \\ \hline 00 \end{array}$$

The quotient is 620.

Concept Connections

7. Describe the pattern between the fractions $\frac{1}{9}$ and $\frac{2}{9}$ and their decimal forms.

Table 4-1

$\frac{1}{4} = 0.25$	$\frac{2}{4} = \frac{1}{2} = 0.5$	$\frac{3}{4} = 0.75$	
$\frac{1}{9} = 0.\overline{1}$	$\frac{2}{9} = 0.\overline{2}$	$\frac{3}{9} = \frac{1}{3} = 0.\overline{3}$	$\frac{4}{9} = 0.\overline{4}$
$\frac{5}{9} = 0.\overline{5}$	$\frac{6}{9} = \frac{2}{3} = 0.\overline{6}$	$\frac{7}{9} = 0.\overline{7}$	$\frac{8}{9} = 0.\overline{8}$

Skill Practice

Convert the fraction to a decimal rounded to the indicated place value.

8. $\frac{9}{7}$; tenths

9. $\frac{17}{37}$; hundredths

Example 3 **Converting Fractions to Decimals with Rounding**

Convert the fraction to a decimal rounded to the indicated place value.

a. $\frac{162}{7}$; tenths place

b. $\frac{21}{31}$; hundredths place

Solution:

a. $\frac{162}{7}$

$$\begin{array}{r} 23.14 \\ 7)\overline{162.00} \\ -14 \\ \hline 22 \\ -21 \\ \hline 10 \\ -7 \\ \hline 30 \\ -28 \\ \hline 2 \end{array}$$

← tenths place
← hundredths place

To round to the tenths place, we must determine the hundredths-place digit and use it to base our decision on rounding.

$23.14 \approx 23.1$

The fraction $\frac{162}{7}$ is approximately 23.1.

b. $\frac{21}{31}$

$$\begin{array}{r} .677 \\ 31)\overline{21.000} \\ -186 \\ \hline 240 \\ -217 \\ \hline 230 \\ -217 \\ \hline 13 \end{array}$$

← hundredths place
← thousandths place

To round to the hundredths-place, we must determine the thousandths-place digit and use it to base our decision on rounding.

$0.67\overset{1}{7} \approx 0.68$

The fraction $\frac{21}{31}$ is approximately 0.68.

2. Writing Decimals as Fractions

In Section 4.1 we converted terminating decimals to fractions. We did this by writing the decimal as a decimal fraction and then reducing the fraction to lowest terms. For example:

$$0.46 = \frac{46}{100} = \frac{\overset{23}{\cancel{46}}}{\underset{50}{\cancel{100}}} = \frac{23}{50}$$

Answers

7. The numerator of the fraction is the repeated digit in the decimal form.
8. 1.3   9. 0.46

We do not yet have the tools to convert a repeating decimal to its equivalent fraction form. However, we can make use of our knowledge of the common fractions and their repeating decimal forms from Table 4-1.

Example 4 Writing Decimals as Fractions and Fractions as Decimals

Complete the table.

	Decimal Form	Fractional Form
a.	0.475	
b.		$\frac{3}{16}$
c.		$2\frac{4}{5}$
d.	$0.\overline{6}$	
e.		$\frac{19}{11}$

Skill Practice

Complete the table.

	Decimal Form	Fractional Form
10.	0.875	
11.		$\frac{7}{20}$
12.		$2\frac{1}{3}$
13.	$0.\overline{7}$	

Solution:

a. $0.475 = \dfrac{475}{1000} = \dfrac{19 \cdot \overset{1}{\cancel{25}}}{40 \cdot \underset{1}{\cancel{25}}} = \dfrac{19}{40}$

b. $\dfrac{3}{16} = 3 \div 16$

$$16)\overline{3.0000} \quad \begin{array}{r} .1875 \\ \hline \end{array}$$

$$\begin{array}{r} .1875 \\ 16)\overline{3.0000} \\ -16 \\ \hline 140 \\ -128 \\ \hline 120 \\ -112 \\ \hline 80 \\ -80 \\ \hline 0 \end{array}$$

Therefore, $\dfrac{3}{16} = 0.1875$.

c. To convert $2\frac{4}{5}$ to decimal form, we need to convert $\frac{4}{5}$ to decimal form. This can be done by dividing. Or we can easily convert $\frac{4}{5}$ to a decimal fraction with a denominator of 10.

$$\frac{4}{5} = \frac{4 \cdot 2}{5 \cdot 2} = \frac{8}{10} = 0.8$$

Therefore, $2\dfrac{4}{5} = 2.8$.

d. From Table 4-1, the decimal $0.\overline{6} = \dfrac{2}{3}$.

e. $\dfrac{19}{11}$ means $19 \div 11$.

$$\begin{array}{r} 1.7272\ldots \\ 11)\overline{19.0000} \\ -11 \\ \hline 80 \\ -77 \\ \hline 30 \\ -22 \\ \hline 80 \end{array}$$

$$\dfrac{19}{11} = 1.\overline{72}$$

The cycle will repeat.

Answers

10. $\frac{7}{8}$ **11.** 0.35

12. $2.\overline{3}$ **13.** $\frac{7}{9}$

We can now complete the table.

	Decimal Form	Fractional Form
a.	0.475	$\frac{19}{40}$
b.	0.1875	$\frac{3}{16}$
c.	2.8	$2\frac{4}{5}$
d.	$0.\overline{6}$	$\frac{2}{3}$
e.	$1.\overline{72}$	$\frac{19}{11}$

3. Decimals and the Number Line

In Example 5, we rank the numbers from least to greatest and visualize the position of the numbers on the number line.

Example 5   Ordering Decimals and Fractions

Rank the numbers from least to greatest. Then approximate the position of the points on the number line.

$$0.\overline{45}, \quad 0.45, \quad \frac{1}{2}$$

Solution:

First note that $\frac{1}{2} = 0.5$ and that $0.\overline{45} = 0.454545\ldots$. By writing each number in decimal form, we can compare the decimals as we did in Section 4.1.

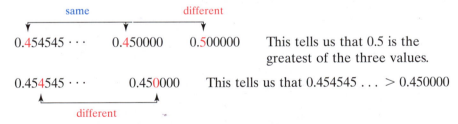

Ranking the numbers from least to greatest we have: $0.45, \quad 0.\overline{45}, \quad 0.5$

The position of these numbers can be seen on the number line. Note that we have expanded the segment of the number line between 0.4 and 0.5 to see more place values to the right of the decimal point.

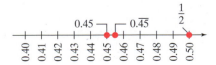

Recall that numbers that lie to the left on the number line have lesser value than numbers that lie to the right.

Answer

14. $0.16, 0.161, \frac{1}{6}$

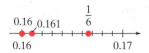

Section 4.5   Practice Exercises

Study Skills Exercise

1. In a study group, check which activities you might try to help you learn and understand the material.

☐ Quiz one another by asking each other questions.

☐ Practice teaching one another.

☐ Share and compare class notes.

☐ Support and encourage one another.

☐ Work together on exercises and sample problems.

Review Exercises

For Exercises 2–5, write the decimal fraction in decimal form.

2. $\dfrac{9}{10}$ **3.** $\dfrac{39}{100}$ **4.** $\dfrac{141}{1000}$ **5.** $\dfrac{71}{10,000}$

For Exercises 6–9, write the decimals as fractions.

6. 0.6 **7.** 0.0016 **8.** 0.35 **9.** 0.125

10. Round $0.\overline{37}$ to the thousandths place.

Objective 1: Writing Fractions as Decimals

For Exercises 11–14, write each fraction as a decimal fraction, that is, a fraction whose denominator is a power of 10. Then write the number in decimal form.

11. $\dfrac{2}{5}$ **12.** $\dfrac{4}{5}$ **13.** $\dfrac{49}{50}$ **14.** $\dfrac{3}{50}$

For Exercises 15–34, write each fraction or mixed number as a decimal. **(See Example 1.)**

15. $\dfrac{7}{25}$ **16.** $\dfrac{4}{25}$ **17.** $\dfrac{316}{500}$ **18.** $\dfrac{19}{500}$

19. $\dfrac{7}{8}$ **20.** $\dfrac{16}{64}$ **21.** $\dfrac{16}{5}$ **22.** $\dfrac{68}{25}$

23. $5\dfrac{3}{12}$ **24.** $4\dfrac{1}{16}$ **25.** $1\dfrac{1}{5}$ **26.** $6\dfrac{5}{8}$

27. $\dfrac{18}{24}$ **28.** $\dfrac{24}{40}$ **29.** $\dfrac{53}{16}$ **30.** $\dfrac{105}{56}$

31. $7\dfrac{9}{20}$ **32.** $3\dfrac{11}{25}$ **33.** $\dfrac{22}{25}$ **34.** $\dfrac{11}{20}$

For Exercises 35–46, write each fraction or mixed number as a repeating decimal. **(See Example 2.)**

35. $3\frac{8}{9}$ **36.** $4\frac{7}{9}$ **37.** $\frac{7}{15}$ **38.** $\frac{5}{18}$

39. $\frac{19}{36}$ **40.** $\frac{7}{12}$ **41.** $\frac{6}{11}$ **42.** $\frac{8}{33}$

43. $\frac{14}{111}$ **44.** $\frac{58}{111}$ **45.** $\frac{25}{22}$ **46.** $\frac{45}{22}$

For Exercises 47–56, convert the fraction to a decimal and round to the indicated place value. **(See Example 3.)**

47. $\frac{1}{7}$; thousandths   **48.** $\frac{2}{7}$; thousandths   **49.** $\frac{1}{13}$; hundredths   **50.** $\frac{9}{13}$; hundredths

51. $\frac{15}{16}$; tenths   **52.** $\frac{3}{11}$; tenths   **53.** $\frac{5}{7}$; hundredths   **54.** $\frac{1}{8}$; hundredths

55. $\frac{25}{21}$; tenths   **56.** $\frac{18}{13}$; tenths

57. Write the fractions as decimals. Explain how to memorize the decimal form for these fractions with a denominator of 9.

   a. $\frac{1}{9}$ **b.** $\frac{2}{9}$ **c.** $\frac{4}{9}$ **d.** $\frac{5}{9}$

58. Write the fractions as decimals. Explain how to memorize the decimal forms for these fractions with a denominator of 3.

   a. $\frac{1}{3}$ **b.** $\frac{2}{3}$

Objective 2: Writing Decimals as Fractions

For Exercises 59–62, complete the table. **(See Example 4.)**

59.

	Decimal Form	Fraction Form
a.	0.45	
b.		$\frac{13}{8}$ or $1\frac{5}{8}$
c.	$0.\overline{7}$	
d.		$\frac{5}{11}$

60.

	Decimal Form	Fraction Form
a.		$\frac{2}{3}$
b.	1.6	
c.		$\frac{152}{25}$
d.	$0.\overline{2}$	

61.

	Decimal Form	Fraction Form
a.	$0.\overline{3}$	
b.	2.125	
c.		$\dfrac{19}{22}$
d.		$\dfrac{42}{25}$

62.

	Decimal Form	Fraction Form
a.	0.75	
b.		$\dfrac{7}{11}$
c.	$1.\overline{8}$	
d.		$\dfrac{74}{25}$

Historically stock prices were given as fractions or mixed numbers, but are now given as decimals. For Exercises 63–64, complete the table that gives recent stock prices taken from the *Wall Street Journal*.

63.

Stock	Closing Price ($) (Decimal)	Closing Price ($) (Fraction)
McGraw-Hill	69.25	
Walgreens	44.95	
Home Depot		$38\dfrac{1}{2}$
General Electric		$37\dfrac{11}{25}$

64.

Stock	Closing Price ($) (Decimal)	Closing Price ($) (Fraction)
Dell	26.3	
StrideRite		$15\dfrac{18}{25}$
Intel	28.10	
Burger King		$24\dfrac{3}{20}$

Objective 3: Decimals and the Number Line

For Exercises 65–76, insert the appropriate symbol. Choose from $<$, $>$, or $=$.

65. $0.2 \;\square\; \dfrac{1}{5}$

66. $1.5 \;\square\; \dfrac{3}{2}$

67. $0.2 \;\square\; 0.\overline{2}$

68. $\dfrac{3}{5} \;\square\; 0.\overline{6}$

69. $\dfrac{1}{3} \;\square\; 0.3$

70. $\dfrac{2}{3} \;\square\; 0.66$

71. $4\dfrac{1}{4} \;\square\; 4.\overline{25}$

72. $2.12 \;\square\; 2.\overline{12}$

73. $0.\overline{5} \;\square\; \dfrac{5}{9}$

74. $\dfrac{7}{4} \;\square\; 1.75$

75. $0.27 \;\square\; \dfrac{3}{11}$

76. $6.4\overline{3} \;\square\; 6.43$

For Exercises 77–80, rank the numbers from least to greatest. Then approximate the position of the points on the number line. **(See Example 5.)**

77. $0.\overline{1}, \dfrac{1}{10}, \dfrac{1}{5}$

78. $3\dfrac{1}{4}, 3\dfrac{1}{3}, 3.3$

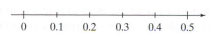

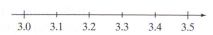

Skill Practice

3. Simplify.

$$2.6 \times \frac{3}{13} \div \left(1\frac{1}{2}\right)^2$$

Example 3   **Applying the Order of Operations**

Simplify.

$$6.4 \times 2\frac{5}{8} \div \left(\frac{3}{5}\right)^2$$

Solution:

Approach 1
Convert all numbers to fractional form.

$$6.4 \times 2\frac{5}{8} \div \left(\frac{3}{5}\right)^2 = \frac{64}{10} \times \frac{21}{8} \div \left(\frac{3}{5}\right)^2 \quad \text{Convert the decimal and mixed number to fractions.}$$

$$= \frac{64}{10} \times \frac{21}{8} \div \frac{9}{25} \quad \text{Square the quantity } \frac{3}{5}.$$

$$= \frac{64}{10} \times \frac{21}{8} \times \frac{25}{9} \quad \text{Multiply by the reciprocal of } \frac{9}{25}.$$

$$= \frac{\overset{4}{\cancel{64}}}{\underset{2}{\cancel{10}}} \times \frac{\overset{7}{\cancel{21}}}{\underset{1}{\cancel{8}}} \times \frac{\overset{5}{\cancel{25}}}{\underset{3}{\cancel{9}}} \quad \text{Simplify common factors.}$$

$$= \frac{140}{3} \text{ or } 46\frac{2}{3} \text{ or } 46.\overline{6} \quad \text{Multiply.}$$

Approach 2
Convert all numbers to decimal form.

$$6.4 \times 2\frac{5}{8} \div \left(\frac{3}{5}\right)^2 = 6.4 \times 2.625 \div (0.6)^2 \quad \text{The fraction } \frac{5}{8} = 0.625 \text{ and } \frac{3}{5} = 0.6.$$

$$= 6.4 \times 2.625 \div 0.36 \quad \text{Square the quantity 0.6. That is, } (0.6)(0.6) = 0.36.$$

Multiply 6.4 × 2.625.

```
  2.625
×   6.4
─────────
 10500
157500
─────────
16.8000
```

$$= 16.8 \div 0.36 \quad \text{Divide } 16.8 \div 0.36.$$

$$= 46.\overline{6}$$

```
.36)16.80
```

```
   46.6 . . .
36)1680
  −144
────
   240
  −216
────
   240
```

3. Calculations with Round-off Error

Example 4   **Multiplying a Fraction and a Decimal**

Multiply.

$$2.52 \cdot \left(\frac{5}{6}\right)$$

Answer

3. $\frac{4}{15}$ or $0.2\overline{6}$

Solution:

Approach 1
Convert 2.52 to fractional form and then multiply fractions.

$$\frac{252}{100} \cdot \frac{5}{6} = \frac{\overset{126}{\cancel{252}}}{\underset{20}{\cancel{100}}} \cdot \frac{\overset{1}{\cancel{5}}}{\underset{3}{\cancel{6}}} = \frac{\overset{21}{\cancel{126}}}{\underset{10}{\cancel{60}}} = \frac{21}{10} \qquad \text{Simplify common factors and multiply fractions.}$$

$$= 2.1 \qquad \text{Divide: } 21 \div 10 = 2.1.$$

Approach 2
Convert $\frac{5}{6}$ to decimal form and then multiply the decimals. However, $\frac{5}{6} = 0.8\overline{3} = 0.8333\ldots$ and we do not know how to multiply repeating decimals. We can approximate the product by rounding the value $0.8\overline{3}$ to some desired level of accuracy. Suppose we round $0.8\overline{3}$ to the thousandths place. Then $0.8\overline{3} \approx 0.833$.

$$2.52 \cdot (0.8\overline{3}) \approx 2.52 \cdot (0.833)$$

$$= 2.09916 \qquad \text{Multiply decimals.}$$

The approximated value 2.09916 is close to 2.1.

```
   2.52
×  .833
   756
  7560
201600
2.09916
```

Notice that the second approach in Example 4 was not as accurate as the first method. This is so because we used "intermediate rounding." Intermediate rounding refers to rounding a number before it is used in a calculation. We rounded the number $0.8\overline{3}$ *before* multiplying. Keep in mind that a rounded number is not exact. Any calculation performed on a rounded number compounds the error.

To minimize the effects of round-off error, keep the fraction notation as long as possible in the expression. In this way, if you do choose to convert to decimal form, you perform the division in the last step. Any rounding is done at the end.

Example 5 **Dividing a Fraction and Decimal**

Divide $\frac{4}{7} \div 3.6$. Round the answer to the nearest hundredth.

Solution:

If we attempt to write $\frac{4}{7}$ as a decimal, we find that it is the repeating decimal $0.\overline{571428}$. Therefore, we choose to change 3.6 to fractional form: $3.6 = \frac{36}{10}$.

$$\frac{4}{7} \div 3.6 = \frac{4}{7} \div \frac{36}{10} \qquad \text{Write 3.6 as a fraction.}$$

$$= \frac{\overset{1}{\cancel{4}}}{7} \cdot \frac{10}{\underset{9}{\cancel{36}}} \qquad \text{Multiply by the reciprocal of the divisor.}$$

$$= \frac{10}{63} \qquad \text{Multiply and reduce to lowest terms.}$$

We must write the answer in decimal form, rounded to the nearest hundredth.

$$\begin{array}{r} .158 \\ 63\overline{)10.00} \\ -63 \\ \hline 370 \\ -315 \\ \hline 550 \\ -504 \\ \hline 46 \end{array}$$

Divide. To round to the hundredths place, divide until we find the thousandths-place digit in the quotient. Use that digit to make a decision for rounding.

≈ 0.16 Round to the nearest hundredth.

4. Applications of Decimals and Fractions

Example 6 **Using Decimals and Fractions in a Consumer Application**

Joanne filled the gas tank in her car and noted that the odometer read 22,341.9 mi. Ten days later she filled the tank again with $11\frac{1}{2}$ gal of gas. Her odometer reading at that time was 22,622.5 mi.

a. How many miles had she driven between fill-ups?

b. How many miles per gallon did she get?

Solution:

a. To find the number of miles driven, we need to subtract the initial odometer reading from the final reading.

$$\begin{array}{r} \overset{5\ \ 12\ \ 1\ \ \ 15}{2\,2\,,6\,2\,2\,.5} \\ -\ 2\,2\,,3\,4\,1\,.9 \\ \hline 2\,8\,0\,.6 \end{array}$$

Recall that to add or subtract decimals, line up the decimal points.

Joanne had driven 280.6 mi between fill-ups.

b. To find the number of miles per gallon (mi/gal), we divide the number of miles driven by the number of gallons.

$280.6 \div 11\frac{1}{2} = 280.6 \div 11.5$ We convert to decimal form because the fraction $11\frac{1}{2}$ is recognized as 11.5.

$= 24.4$

$$11.5\overline{)280.6} \qquad \begin{array}{r} 24.4 \\ 115\overline{)2806.0} \\ -230 \\ \hline 506 \\ -460 \\ \hline 460 \\ -460 \\ \hline 0 \end{array}$$

Joanne got 24.4 mi/gal.

Example 7   **Using Decimals and Fractions to Compute a Lawsuit Settlement**

Althea won a legal settlement for $4105.20. Her lawyer received $\frac{1}{3}$ of the settlement.

a. How much money did the lawyer get?

b. How much money did Althea get?

Solution:

a. The lawyer got $\frac{1}{3}$ of $4105.20. This implies multiplication.

$\frac{1}{3} \cdot (4105.20)$ The fraction $\frac{1}{3}$ cannot be written as a terminating decimal.

$= \frac{1}{3} \cdot \frac{4105.20}{1}$ We can write $4105.20 as $\frac{4105.20}{1}$ and multiply fractions.

$= \frac{4105.20}{3}$ Multiply numerators. Multiply denominators.

$= 1368.4$ Divide: $4105.20 \div 3 = 1368.4$.

The lawyer got $1368.40.

b. Althea got the remaining portion of the money.

$$\begin{array}{r} \$4105.20 \\ -\ 1368.40 \\ \hline \$2736.80 \end{array}$$

Althea received $2736.80.

TIP: We can check the answer to Example 7 by realizing that Althea will receive $\frac{2}{3}$ of the settlement.

$\frac{2}{3}(\$4105.20) = \$2736.80$

Example 8   **Finding an Average**

Table 4-2 represents the average snowfall for 6 winter months in Syracuse, New York. What is the mean (average) amount of snowfall per winter month? Round to the nearest tenth of an inch. (*Source:* National climate data center)

Table 4-2

Month	Nov.	Dec.	Jan.	Feb.	March	April
Snowfall (in.)	9.3	26.8	29.6	26.2	17.3	4.0

Answers
8. a. He ran 15.72 mi.
    b. 10.48 mi was left.
9. The average is $484.41.

Solution:

To find the average, we must add the values. Then divide by the number of values (in this case, 6).

To find the sum, line up the addends:

$$
\begin{array}{r}
{\scriptstyle 1\,4\,2} \\
9.3 \\
26.8 \\
29.6 \\
26.2 \\
17.3 \\
+\ \ 4.0 \\
\hline
113.2
\end{array}
$$

The total snowfall for these 6 months is 113.2 in. To find the average, divide by 6.

$$
\begin{array}{r}
18.86 \\
6\overline{)113.20} \\
-6 \\
\hline
53 \\
-48 \\
\hline
52 \\
-48 \\
\hline
40 \\
-36 \\
\hline
4
\end{array}
$$

To round to the tenths place, divide until we find the hundredths-place digit. Use that digit to make a decision for rounding.

The number 18.86 rounds to 18.9.

Syracuse averages about 18.9 in. of snow per month during these 6 months.

Section 4.6   Practice Exercises

Study Skills Exercise

1. In addition to studying the material for a test, here are some other activities that people use when preparing for a test. Circle the importance of each statement.

	Not important	Somewhat important	Very important
a. Get a good night's sleep the night before the test.	1	2	3
b. Eat a good breakfast on the day of the test.	1	2	3
c. Wear comfortable clothes on the day of the test.	1	2	3
d. Arrive early to class on the day of the test.	1	2	3

Review Exercises

For Exercises 2–9, perform the indicated operation.

2. $\left(\dfrac{24}{7}\right)\left(\dfrac{35}{36}\right)$

3. 34.1×9.2

4. $790.9 + 23.91$

5. $\dfrac{34}{9} + \dfrac{5}{27}$

6. $56.7 \div 1.2$

7. $\dfrac{55}{16} \div \dfrac{11}{4}$

8. $\dfrac{9}{4} - \dfrac{7}{8}$

9. $13 - 6.04$

Objective 1: Order of Operations Involving Decimals

10. List the order of operations.

For Exercises 11–22, simplify by using the order of operations. **(See Example 1.)**

11. $(3.7 - 1.2)^2$

12. $(6.8 - 4.7)^2$

13. $16.25 - (18.2 - 15.7)^2$

14. $11.38 - (10.42 - 7.52)^2$

15. $12.46 - 3.05 - 0.8^2$

16. $15.06 - 1.92 - 0.4^2$

17. $63.75 - 9.5(4)$

18. $6.84 + (3.6)(9)$

19. $6.8 \div 2 \div 1.7$

20. $8.4 \div 2 \div 2.1$

21. $2.2 + [9.34 + (1.2)^2]$

22. $(3.1)^2 - (4.2 \div 2.1)$

Objective 2: Calculations with Decimals and Fractions

For Exercises 23–28, simplify by using the order of operations. Express the answer in decimal form. **(See Examples 2–3.)**

23. $89.8 \div 1\dfrac{1}{3}$

24. $30.12 \div 1\dfrac{3}{5}$

25. $20.04 \div \dfrac{4}{5}$

26. $(78.2 - 60.2) \div \dfrac{9}{13}$

27. $14.4 \times \left(\dfrac{7}{4} - \dfrac{1}{8}\right)$

28. $6.5 + \dfrac{1}{8} \times \left(\dfrac{1}{5}\right)^2$

Objective 3: Calculations with Round-off Error

For Exercises 29–34, perform the indicated operations. Round the answer to the nearest hundredth when necessary. **(See Examples 4–5.)**

29. $2.3 \times \dfrac{5}{9}$

30. $4.6 \times \dfrac{7}{6}$

31. $6.5 \div \dfrac{3}{5}$

32. $\dfrac{1}{12} \times 6.24 \div 2.1$

33. $(42.81 - 30.01) \div \dfrac{9}{2}$

34. $\dfrac{2}{7} \times 5.1 \times \dfrac{1}{10}$

For Exercises 35–38, perform the indicated operation. Write the answer as a repeating decimal.

35. $\dfrac{2}{9} \times 4.21$

36. $6.02 \div \dfrac{22}{23}$

37. $5.32 \div \dfrac{6}{5}$

38. $\dfrac{34}{11} \times 2.5$

Section 5.1 | Ratios

1. Writing a Ratio

Thus far we have seen two interpretations of fractions.

- The fraction $\frac{5}{8}$ represents 5 parts of a whole that has been divided evenly into 8 pieces.
- The fraction $\frac{5}{8}$ represents $5 \div 8$.

Now we consider a third interpretation.

- The fraction $\frac{5}{8}$ represents the ratio of 5 to 8.

A **ratio** is a comparison of two quantities. There are three different ways to write a ratio.

Concept Connections

1. When forming the ratio $\frac{a}{b}$, why must b not equal zero?

> **PROCEDURE   Writing a Ratio**
>
> The ratio of a to b can be written as follows, provided $b \neq 0$.
>
> **1.** a to b **2.** $a : b$ **3.** $\dfrac{a}{b}$
>
> The colon means "to."   The fraction bar means "to."

Although there are three ways to write a ratio, we primarily use the fraction form.

Skill Practice

2. For a recent flight from Atlanta to San Diego, 291 seats were occupied and 29 were unoccupied. Write the ratio of
   a. The number of occupied seats to unoccupied seats
   b. The number of unoccupied seats to occupied seats
   c. The number of occupied seats to the total number of seats

---

> **Example 1   Writing a Ratio**
>
> In an algebra class there are 15 women and 17 men.
>
> **a.** Write the ratio of women to men.   15:17
>
> **b.** Write the ratio of men to women.   17 to 15
>
> **c.** Write the ratio of women to the total number of people in the class.
>
>   15:32

Solution:

It is important to observe the *order* of the quantities mentioned in a ratio. The first quantity mentioned is the numerator. The second quantity is the denominator.

a. The ratio of women to men is

$$\frac{15}{17}$$

b. The ratio of men to women is

$$\frac{17}{15}$$

c. First find the total number of people in the class.

Total = number of women + number of men

$$= 15 + 17$$

$$= 32$$

Therefore the ratio of women to the total number of people in the class is

$$\frac{15}{32}$$

Answers

1. The value b must not be zero because division by zero is undefined.
2. **a.** $\dfrac{291}{29}$ **b.** $\dfrac{29}{291}$ **c.** $\dfrac{291}{320}$

2. Simplifying a Ratio to Lowest Terms

It is often desirable to write a ratio in lowest terms. The process is similar to simplifying fractions to lowest terms.

Example 2   Writing Ratios in Lowest Terms

Write each ratio in lowest terms.

a. 15 ft to 10 ft   **b.** $20 to $10

Skill Practice

Write the ratios in lowest terms.

3. 72 m to 16 m
4. 30 gal to 5 gal

Solution:

In part (a) we are comparing feet to feet. In part (b) we are comparing dollars to dollars. We can "cancel" the like units in the numerator and denominator as we would common factors.

a.
$$\frac{15 \text{ ft}}{10 \text{ ft}} = \frac{3 \cdot 5 \text{ ft}}{2 \cdot 5 \text{ ft}}$$

$$= \frac{3 \cdot \overset{1}{\cancel{5}} \cancel{\text{ft}}}{2 \cdot \underset{1}{\cancel{5}} \cancel{\text{ft}}} \qquad \text{Simplify common factors. "Cancel" common units.}$$

$$= \frac{3}{2}$$

Even though the number $\frac{3}{2}$ is equivalent to $1\frac{1}{2}$, we do not write the ratio as a mixed number. Remember that a ratio is a comparison of *two* quantities. If you did convert $\frac{3}{2}$ to the mixed number $1\frac{1}{2}$, you would write the ratio as $\frac{1\frac{1}{2}}{1}$. This would imply that the numerator is one and one-half times as large as the denominator.

Concept Connections

5. Which value properly represents the ratio 4 to 1?

a. $\frac{1}{4}$ **b.** $\frac{4}{1}$

b.
$$\frac{\$20}{\$10} = \frac{\overset{2}{\cancel{\$20}}}{\underset{1}{\cancel{\$10}}} \qquad \text{Simplify common factors. "Cancel" common units.}$$

$$= \frac{2}{1}$$

Although the fraction $\frac{2}{1}$ is equivalent to 2, we do not generally write ratios as whole numbers. Again, a ratio compares *two* quantities. In this case, we say that there is a 2-to-1 ratio between the original dollar amounts.

3. Writing Ratios of Mixed Numbers and Decimals

It is often desirable to express a ratio in lowest terms by using whole numbers in the numerator and denominator. This is demonstrated in Examples 3 and 4.

Example 3   Rewriting a Ratio as a Ratio of Whole Numbers

The length of a rectangular picture frame is 10.8 in., and the width is 8.64 in. Express the ratio of the length to the width. Then rewrite the ratio as a ratio of whole numbers reduced to lowest terms.

Skill Practice

Write the ratio as a ratio of whole numbers expressed in lowest terms.

6. $4.20 to $2.88

Answers

3. $\frac{9}{2}$ **4.** $\frac{6}{1}$ **5.** b   **6.** $\frac{35}{24}$

Solution:

The ratio of length to width is $\frac{10.8}{8.64}$. We now want to rewrite the ratio, using whole numbers in the numerator and denominator. Notice that if we multiply 8.64 by 100, the decimal point will move to the right 2 places. The result will be the whole number 864. Multiplying the numerator by the same number, 100, will produce a whole number in the numerator: $10.8 \times 100 = 1080$.

$$\frac{10.8}{8.64} = \frac{10.8 \times 100}{8.64 \times 100}$$ Multiply numerator and denominator by 100.

$$= \frac{1080}{864}$$ Because the numerator and denominator are large numbers, we write the prime factorization of each. The common factors are now easy to identify.

$$= \frac{\overset{1}{2} \cdot \overset{1}{2} \cdot \overset{1}{2} \cdot \overset{1}{3} \cdot \overset{1}{3} \cdot \overset{1}{3} \cdot 5}{\underset{1}{2} \cdot \underset{1}{2} \cdot \underset{1}{2} \cdot 2 \cdot 2 \cdot \underset{1}{3} \cdot \underset{1}{3} \cdot \underset{1}{3}}$$ Simplify common factors to lowest terms.

$$= \frac{5}{4}$$

The ratio of length to width is $\frac{5}{4}$.

In Example 3, we multiplied by 100 to move the decimal point *two* places to the right. Multiplying by 10 would not have been sufficient, because $8.64 \times 10 = 86.4$, which is not a whole number.

Skill Practice

7. A recipe calls for $2\frac{1}{2}$ cups of flour and $\frac{3}{4}$ cup of sugar. Write the ratio of flour to sugar. Then rewrite the ratio as a ratio of whole numbers in lowest terms.

Example 4   **Rewriting a Ratio as a Ratio of Whole Numbers**

Ling walked $2\frac{1}{4}$ mi on Monday and $3\frac{1}{2}$ mi on Tuesday. Write the ratio of miles walked Monday to miles walked Tuesday. Then rewrite the ratio as a ratio of whole numbers reduced to lowest terms.

Solution:

The ratio of miles walked on Monday to miles walked on Tuesday is $\frac{2\frac{1}{4}}{3\frac{1}{2}}$.

To convert this to a ratio of whole numbers, first we rewrite each mixed number as an improper fraction. Then we can divide the fractions and simplify.

$$\frac{2\frac{1}{4}}{3\frac{1}{2}} = \frac{\frac{9}{4}}{\frac{7}{2}}$$ Write the mixed numbers as improper fractions. Recall that a fraction bar also implies division.

$$= \frac{9}{4} \div \frac{7}{2}$$

$$= \frac{9}{4} \cdot \frac{2}{7}$$ Multiply by the reciprocal of the divisor.

$$= \frac{9}{\underset{2}{4}} \cdot \frac{\overset{1}{2}}{7}$$ Simplify common factors to lowest terms.

$$= \frac{9}{14}$$ This is a ratio of whole numbers in lowest terms.

Answer

7. $\frac{2\frac{1}{2}}{\frac{3}{4}}; \frac{10}{3}$

4. Applications of Ratios

Ratios are used in a number of applications.

Example 5   **Using Ratios to Express Population Increase**

The town of Roxbury, Connecticut, had 1825 people in the year 1990. By the year 2008, the U.S. Census Bureau projects its population to be 2441. Write a ratio depicting the increase in population to the number of people in the town in 1990.

Solution:

To write this ratio, we need to know the increase in population.

$$\text{Increase in population} = 2441 - 1825$$

$$= 616$$

The ratio of the increase in population to the number of people in 1990 is

Increase in population $\longrightarrow \dfrac{616}{1825} \longleftarrow$ number of people in 1990

Skill Practice

8. The average retail price of 1 gal of regular unleaded gas in the United States in 1990 was \$1.16. By 2007 it had increased to \$3.24. Write the ratio representing the increase in price to the price in 1990. Then rewrite the ratio as a ratio of whole numbers reduced to lowest terms.

Example 6   **Applying Ratios to Unit Conversion**

A fence is 12 yd long and $1\frac{1}{2}$ ft high.

a. Find the ratio of the length to the height with all units measured in yards.

b. Find the ratio of length to height with all units measured in feet.

$1\frac{1}{2}$ ft = $\frac{1}{2}$ yd

12 yd

Skill Practice

9. A painting is $1\frac{1}{2}$ ft long by 9 in. wide.
   a. Write the ratio of length to width with all units measured in feet.
   b. Write the ratio of length to width with all units measured in inches.

Solution:

a. Since 3 ft = 1 yd. Therefore, $1\frac{1}{2}$ ft = $\frac{1}{2}$ yd.

   Measuring in yards, we see that the ratio of length to height is

   $$\dfrac{12 \text{ yd}}{\frac{1}{2} \text{ yd}} = \dfrac{12}{1} \cdot \dfrac{2}{1} = \dfrac{24}{1}.$$

b. The length is 12 yd = 36 ft.

   Measuring in feet, we see that the ratio of length to height is

   $$\dfrac{36 \text{ ft}}{1\frac{1}{2} \text{ ft}} = \dfrac{36}{\frac{3}{2}} = \dfrac{\overset{12}{\cancel{36}}}{1} \cdot \dfrac{2}{\underset{1}{\cancel{3}}} = \dfrac{24}{1}.$$

Notice that regardless of the units used, the ratio is the same, 24 to 1. This means that the length is 24 times the height.

Answers

8. $\dfrac{2.08}{1.16}$; $\dfrac{52}{29}$ 9. a. $\dfrac{2}{1}$ b. $\dfrac{2}{1}$

Section 5.1   Practice Exercises

Study Skills Exercises

1. Does your school have a learning resource center or a tutoring center? If so, do you remember the location and hours of operation? Write them here.

   Location of learning resource center or tutoring center:

   Hours of operation:

2. Define the key term **ratio**.

Objective 1: Writing a Ratio

For Exercises 3–8, write the ratio in two other ways.

3. 5 to 6

4. 3 to 7

5. $11 : 4$

6. $8 : 13$

7. $\dfrac{1}{2}$

8. $\dfrac{1}{8}$

For Exercises 9–12, write the ratios in fraction form. **(See Example 1.)**

9. Nancy has 3 cats and 2 dogs.

   a. Write a ratio of cats to dogs.

   b. Write a ratio of dogs to cats.

   c. Write a ratio of cats to the total number of pets.

10. In a kindergarten classroom, there are 11 boys and 9 girls.

    a. Write a ratio of girls to boys.

    b. Write a ratio of boys to girls.

    c. Write a ratio of boys to the total number of children in class.

11. There are 52 cars in the parking lot, of which 21 are silver.

    a. Write a ratio of silver cars to the total number of cars.

    b. Write a ratio of silver cars to cars that are not silver.

12. On one city block, there are 21 houses and 10 of them have pools in the backyard.

    a. Write a ratio of the number of houses with a pool to the total number of houses.

    b. Write a ratio of the number of houses with a pool to the number of houses without a pool.

Objective 2: Simplifying a Ratio to Lowest Terms

For Exercises 13–24, write the ratio in lowest terms. **(See Example 2.)**

13. 4 yr to 6 yr

14. 10 lb to 14 lb

15. 5 mi to 25 mi

16. 20 ft to 12 ft

17. 8 m to 2 m

18. 14 oz to 7 oz

19. 33 cm to 15 cm

20. 21 days to 30 days

21. $60 to $50

22. 75¢ to 100¢

23. 18 in. to 36 in.

24. 3 cups to 9 cups

Objective 3: Writing Ratios of Mixed Numbers and Decimals

For Exercises 25–36, write the ratio in lowest terms with whole numbers in the numerator and denominator. **(See Examples 3 and 4.)**

25. 3.6 ft to 2.4 ft

26. 10.15 hr to 8.12 hr

27. 8 gal to $9\frac{1}{3}$ gal

28. 24 yd to $13\frac{1}{3}$ yd

29. $16\frac{4}{5}$ m to $18\frac{9}{10}$ m

30. $1\frac{1}{4}$ in. to $1\frac{3}{8}$ in.

31. $16.80 to $2.40

32. $18.50 to $3.70

33. $\frac{1}{2}$ day to 4 days

34. $\frac{1}{4}$ mi to $1\frac{1}{2}$ mi

35. 10.25 L to 8.2 L

36. 11.55 km to 6.6 km

Objective 4: Applications of Ratios

37. The temperature at 8:00 A.M. in Los Angeles was 66°F. By 2:00 P.M., the temperature had risen to 90°F. Write a ratio representing the increase in temperature to the temperature at 8:00 A.M. **(See Example 5.)**

38. The city of Goddard, Kansas, had approximately 2100 people in the year 2000. By 2008, the U.S. Census Bureau projects its population to be 4130. Write a ratio representing the increase in population to the number of people in the year 2000.

39. A plaque is 6 in. wide and $1\frac{1}{3}$ ft long ($\frac{1}{3}$ ft is 4 in.). **(See Example 6.)**

   a. Find the ratio of width to length with all units in inches.

   b. Find the ratio of width to length with all units in feet.

40. A construction company needs 2 weeks to construct a family room and 3 days to add a porch.

   a. Find the ratio of the time it takes for constructing the porch to the time constructing the family room, with all units in weeks.

   b. Find the ratio of the time it takes for constructing the porch to the time constructing the family room, with all units in days.

For Exercises 41–44, refer to the table showing Alex Rodriguez's salary (rounded to the nearest $100,000) for selected years during his career. Write each ratio in lowest terms.

Year	Team	Salary	Position
2007	New York Yankees	$22,700,000	Third baseman
2004	New York Yankees	$22,000,000	Third baseman
2000	Seattle Mariners	$4,400,000	Shortstop
1996	Seattle Mariners	$400,000	Shortstop

Source: USA TODAY

41. Write the ratio of Alex's salary for the year 1996 to the year 2000.

42. Write a ratio of Alex's salary for the year 2004 to the year 1996.

43. Write a ratio of the increase in Alex's salary between the years 1996 and the year 2000 to his salary in 1996.

44. Write a ratio of the increase in Alex's salary between the years 2004 and 2007 to his salary in 2004.

For Exercises 45–48, refer to the table that shows the average spending per person for reading (books, newspapers, magazines, etc.) by age group. Write each ratio in lowest terms.

45. Find the ratio of spending for the under-25 group to the spending for the group 75 years and over.

46. Find the ratio of spending for the group 25 to 34 years old to the spending for the group of 65 to 74 years old.

47. Find the ratio of spending for the group under 25 years old to the spending for the group of 55 to 64 years old.

48. Find the ratio of spending for the group 35 to 44 years old to the spending for the group 45 to 54 years old.

Age Group	Annual Average ($)
Under 25 years	60
25 to 34 years	111
35 to 44 years	136
45 to 54 years	172
55 to 64 years	183
65 to 74 years	159
75 years and over	128

Source: Mediamark Research Inc.

For Exercises 49–52, find the ratio of the shortest side to the longest side. Write each ratio in lowest terms with whole numbers in the numerator and denominator.

49.

$1\frac{1}{2}$ ft

$2\frac{1}{4}$ ft

2 ft

50.

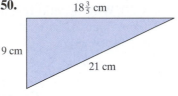

$18\frac{3}{5}$ cm

9 cm

21 cm

51.

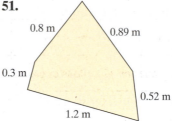

0.8 m

0.89 m

0.3 m

0.52 m

1.2 m

52.

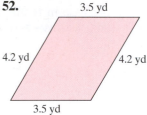

3.5 yd

4.2 yd

4.2 yd

3.5 yd

Expanding Your Skills

For Exercises 53–55, refer to the figure. The lengths of the sides for squares A, B, C, and D are given.

53. What are the lengths of the sides of Square E?

54. Find the ratio of the lengths of the sides for the given pairs of squares.

a. Square B to square A

b. Square C to square B

c. Square D to square C

d. Square E to square D

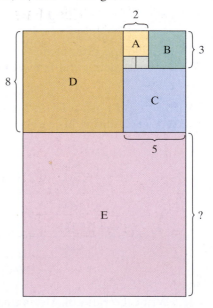

55. Write the decimal equivalents for each ratio in Exercise 54. Do these values seem to be approaching a number close to 1.618 (this is an approximation for the *golden ratio*, which is equal to $\frac{1+\sqrt{5}}{2}$)? Applications of the golden ratio are found throughout nature. In particular, as a result of the geometrically pleasing pattern, artists and architects have proportioned their work to approximate the golden ratio.

56. The ratio of a person's height to the length of the person's lower arm (from elbow to wrist) is approximately 6.5 to 1. Measure your own height and lower arm length. Is the ratio you get close to the average of 6.5 to 1?

57. The ratio of a person's height to the person's shoulder width (measured from outside shoulder to outside shoulder) is approximately 4 to 1. Measure your own height and shoulder width. Is the ratio you get close to the average of 4 to 1?

Rates

1. Definition of a Rate

A **rate** is a type of ratio used to compare different types of quantities, for example:

$$\frac{270 \text{ mi}}{13 \text{ gal}} \quad \text{and} \quad \frac{\$8.55}{1 \text{ hr}}$$

Several key words imply rates. These are given in Table 5-1.

Table 5-1

Key Word	Example	Rate
Per	117 miles per 2 hours	$\dfrac{117 \text{ mi}}{2 \text{ hr}}$
For	$12 for 3 lb	$\dfrac{\$12}{3 \text{ lb}}$
In	400 meters in 43.5 seconds	$\dfrac{400 \text{ m}}{43.5 \text{ sec}}$
On	270 miles on 12 gallons of gas	$\dfrac{270 \text{ mi}}{12 \text{ gal}}$

Because a rate compares two different quantities it is important to include the units in both the numerator and the denominator. It is also desirable to write rates in lowest terms.

Concept Connections

1. Why is it important to include units when you are expressing a rate?

Example 1   **Writing Rates in Lowest Terms**

Write each rate in lowest terms.

a. In one region, there are approximately 640 trees on 12 acres.

b. Latonya drove 138 mi on 6 gal of gas.

c. Jounne can type 625 words in 10 min.

Solution:

a. The rate of 640 trees on 12 acres can be expressed as $\dfrac{640 \text{ trees}}{12 \text{ acres}}$.

Now write this rate in lowest terms. $\dfrac{\overset{160}{\cancel{640}} \text{ trees}}{\underset{3}{\cancel{12}} \text{ acres}} = \dfrac{160 \text{ trees}}{3 \text{ acres}}$

b. The rate of 138 mi on 6 gal of gas can be expressed as $\dfrac{138 \text{ mi}}{6 \text{ gal}}$.

Now write this rate in lowest terms. $\dfrac{\overset{23}{\cancel{138}} \text{ mi}}{\underset{1}{\cancel{6}} \text{ gal}} = \dfrac{23 \text{ mi}}{1 \text{ gal}}$

Skill Practice

Write each rate in lowest terms.

2. Maria reads 15 pages in 10 min.
3. A Chevrolet Corvette Z06 gets 163.4 mi on 8.6 gal of gas.
4. At Nash Community College in Rocky Mount, North Carolina, there are 2200 students to 120 instructors.

Answers

1. The units are different in the numerator and denominator and will not "cancel."
2. $\dfrac{3 \text{ pages}}{2 \text{ min}}$
3. $\dfrac{19 \text{ mi}}{1 \text{ gal}}$ or 19 mi/gal
4. $\dfrac{55 \text{ students}}{3 \text{ instructors}}$

c. The rate of 625 words in 10 min can be represented as $\dfrac{625 \text{ words}}{10 \text{ min}}$.

Now write this rate in lowest terms. $\dfrac{\overset{125}{\cancel{625}} \text{ words}}{\underset{2}{\cancel{10}} \text{ min}} = \dfrac{125 \text{ words}}{2 \text{ min}}$

This rate indicates that 625 words in 10 min is the same speed as 125 words in 2 min.

2. Unit Rates

A rate having a denominator of 1 unit is called a **unit rate**. Furthermore the number 1 is often omitted in the denominator.

$\dfrac{23 \text{ mi}}{1 \text{ gal}} = 23 \text{ mi/gal}$ is read as "twenty-three miles per gallon."

$\dfrac{52 \text{ ft}}{1 \text{ sec}} = 52 \text{ ft/sec}$ is read as "fifty-two feet per second."

$\dfrac{\$15}{1 \text{ hr}} = \$15/\text{hr}$ is read as "fifteen dollars per hour."

PROCEDURE   Converting a Rate to a Unit Rate

To convert a rate to a unit rate, divide the numerator by the denominator and maintain the units of measurement.

Example 2   Finding Unit Rates

Write each rate as a unit rate. Round to three decimal places if necessary.

a. A health club charges $125 for 20 visits. Find the unit rate in dollars per visit.

b. In 1960, Wilma Rudolph won the women's 200-m run in 24 sec. Find her speed in meters per second.

c. During one baseball season, Barry Bonds got 149 hits in 403 at bats. Find his batting average. (*Hint:* Batting average is defined as the number of hits per the number of at bats.)

Solution:

a. The rate of $125 for 20 visits can be expressed as $\dfrac{\$125}{20 \text{ visits}}$.

To convert this to a unit rate, divide $125 by 20 visits.

$\dfrac{\$125}{20 \text{ visits}} = \dfrac{\$6.25}{1 \text{ visit}}$ or $6.25/\text{visit}$

$$\begin{array}{r} 6.25 \\ 20\overline{)125.00} \\ -120 \\ \hline 50 \\ -40 \\ \hline 100 \\ -100 \\ \hline 0 \end{array}$$

b. The rate of 200 m per 24 sec can be expressed as $\dfrac{200 \text{ m}}{24 \text{ sec}}$.

To convert this to a unit rate, divide 200 m by 24 sec.

$$\frac{200 \text{ m}}{24 \text{ sec}} \approx \frac{8.333 \text{ m}}{1 \text{ sec}} \quad \text{or approximately } 8.333 \text{ m/sec}$$

$$
\begin{array}{r}
8.\overline{3} \\
24\overline{)200.00} \\
-192 \\
\hline
80 \\
-72 \\
\hline
80
\end{array}
$$

The quotient repeats.

Wilma Rudolph's speed was approximately 8.333 m/sec.

c. The rate of 149 hits in 403 at bats can be expressed as $\dfrac{149 \text{ hits}}{403 \text{ at bats}}$.

To convert this to a unit rate, divide 149 hits by 403 at bats.

$$\frac{149 \text{ hits}}{403 \text{ at bats}} \approx \frac{0.370 \text{ hit}}{1 \text{ at bat}} \quad \text{or } 0.370 \text{ hit/at bat}$$

> **Avoiding Mistakes**
>
> Units of measurement must be included for the answer to be complete.

3. Unit Cost

A **unit cost** or unit price is the cost per 1 unit of something. At the grocery store, for example, you might purchase meat for $3.79/lb ($3.79 per 1 lb). Unit cost is useful in day-to-day life when we compare prices. Example 3 compares the prices of three different sizes of apple juice.

Example 3   **Finding Unit Costs**

Apple juice comes in a variety of sizes and packaging options. Find the unit price per ounce and determine which is the best buy.

a. $1.69

Apple Juice
48 oz

b. $2.39

Apple Juice
64 oz

c. $2.99

Apple Juice
10-pack 6 oz each

> **Skill Practice**
>
> **10.** Gatorade comes in several size packages. Compute the unit price per ounce for each option (round to the nearest thousandth of a dollar). Then determine which is the best buy.
>   **a.** $2.99 for a 64-oz bottle
>   **b.** $3.99 for four 24-oz bottles
>   **c.** $3.79 for six 12-oz bottles

Solution:

When we compute a unit cost, the cost is always placed in the numerator of the rate. Furthermore, when we divide the cost by the amount, we need to obtain enough digits in the quotient to see the variation in unit price. In this example, we have rounded to the nearest thousandth of a dollar (nearest tenth of a cent). This means that we use the ten-thousandths-place digit in the quotient on which to base our decision on rounding.

> **Answers**
>
> **10. a.** $0.047/oz
>   **b.** $0.042/oz
>   **c.** $0.053/oz
>   The 4-pack of 24-oz bottles is the best buy.

	Rate	Quotient	Unit Rate (Rounded)
a.	$\dfrac{\$1.69}{48 \text{ oz}}$	$\$1.69 \div 48 \text{ oz} \approx \$0.0352/\text{oz}$	$\$0.035/\text{oz}$ or $3.5¢/\text{oz}$
b.	$\dfrac{\$2.39}{64 \text{ oz}}$	$\$2.39 \div 64 \text{ oz} \approx \$0.0373/\text{oz}$	$\$0.037/\text{oz}$ or $3.7¢/\text{oz}$
c.	$6 \text{ oz} \times 10 = 60 \text{ oz}$ $\dfrac{\$2.99}{60 \text{ oz}}$	$\$2.99 \div 60 \text{ oz} \approx \$0.0498/\text{oz}$	$\$0.050/\text{oz}$ or $5.0¢/\text{oz}$

From the table, we see that the most economical buy is the 48-oz size because its unit rate is the least expensive.

4. Applications of Rates

Example 4 uses a unit rate for comparison in an application.

 **Example 4**   Computing Mortality Rates

Mortality rate is defined to be the total number of people who die due to some risk behavior divided by the total number of people who engage in the risk behavior. Based on the following statistics, compare the mortality rate for undergoing heart bypass surgery to the mortality rate of flying on the space shuttle.

a. Roughly 28 people will die for every 1000 who undergo heart bypass surgery. (*Source:* The Society of Thoracic Surgeons)

b. As of August 2007, there have been 14 astronauts killed in space shuttle missions out of 830 astronauts who have flown.

Solution:

a. Mortality rate for heart bypass surgery: $\frac{28}{1000} = 0.028$ death/surgery

b. Mortality rate for flying on the space shuttle: $\frac{14}{830} \approx 0.017$ deaths/flight

Comparing these rates show that it is riskier to have heart bypass surgery than to fly on the space shuttle.

Section 5.2   Practice Exercises

Study Skills Exercises

1. Budgeting enough time to do homework and to study for a class is one of the most important steps to success in a class. Use a weekly calendar to help you plan your time for your studies this week. Also write in other obligations such as the time required for your job, for your family, for sleeping, and for eating. Be realistic when you estimate the time for each activity.

2. Define the key terms.

  a. Rate   **b.** Unit rate   **c.** Unit cost

Review Exercises

3. Write the ratio 3 to 5 in two other ways.

4. Write the ratio 4:1 in two other ways.

For Exercises 5–8, write the ratio in lowest terms.

5. 36¢ to 27¢

6. $6\frac{3}{4}$ ft to $8\frac{1}{4}$ ft

7. 1.08 mi to 2.04 mi

8. $28.40 to $20.80

Objective 1: Definition of a Rate

For Exercises 9–20, write each rate in lowest terms. **(See Example 1.)**

9. A type of laminate flooring sells for $32 for 5 square feet (ft²).

10. A remote control car can go up to 44 ft in 5 sec.

11. Elaine drives 234 mi in 4 hr.

12. Travis has 14 blooms on 6 of his plants.

13. Tyler earned $58 in 8 hr.

14. Neil can type only 336 words in 15 min.

15. A printer can print 13 pages in 26 sec.

16. During a bad storm there was 2 in. of rain in 6 hr.

17. There are 130 calories in 8 snack crackers.

18. The driveway is lined with 14 plants for a length of 22 ft.

19. An advertisement states that TV trays are selling for $30 for a set of 4 trays.

20. There are 50 students assigned to 4 advisers.

Objective 2: Unit Rates

21. Of the following rates, identify those that are unit rates.

a. $\dfrac{\$0.37}{1\ oz}$

b. $\dfrac{333.2\ mi}{14\ gal}$

c. 16 ft/sec

d. $\dfrac{59\ mi}{1\ hr}$

22. Of the following rates, identify those that are unit rates.

a. $\dfrac{3\ lb}{\$1.00}$

b. $\dfrac{21\ ft}{1\ sec}$

c. 50 mi/hr

d. $\dfrac{232\ words}{2\ min}$

For Exercises 23–28, write each rate as a unit rate. **(See Example 2.)**

23. The Osborne family drove 452 mi in 4 days.

24. The book of poetry *The Prophet* by Kahlil Gibran has estimated sales of $6,000,000 over an 80-year period.

25. Philip drove 480 km in 5 hr.

26. Ian flew 1120 mi in 4 hr.

27. If Oscar bought an easy chair for $660 and plans to make 12 payments, what is the amount per payment?

28. The jockey David Gall had 7396 wins in 43 years of riding.

For Exercises 29–32, determine the unit rates and round to the nearest hundredth when necessary.

29. At the market, bananas cost $1.50 for 4 lb.

30. Ceramic tiles sell for $13.08 for a box of 12 tiles. Find the price per tile.

31. Lottery prize money of $1,792,000 is for 7 people.

32. One WeightWatchers group lost 123 lb for its 11 members.

33. A male speed skater skated 500 m in 35 sec. Find the rate in meters per second. (Round to one decimal place.)

34. A female speed skater skated 500 m in 38 sec. Find the rate in meters per second.

Objective 3: Unit Cost

For Exercises 35–44, find the unit costs (that is, dollars per unit). Round the answers to 3 decimal places when necessary. **(See Example 3.)**

35. Tide laundry detergent costs $4.99 for 100 oz.

36. Dove liquid body wash costs $3.49 for 12 oz.

37. Soda costs $1.99 for a 2-L bottle.

38. Four chairs cost $221.00.

39. A set of 4 tires costs $210.

40. A package of 3 shirts costs $64.80.

41. A package of 6 newborn bodysuits costs $32.50.

42. A package of 8 AAA batteries costs $9.84.

43. **a.** 40 oz of shampoo for $3.00

  b. 28 oz of shampoo for $2.10

  c. Which is the best buy?

44. **a.** 10 lb of potting soil for $1.70

  b. 30 lb of potting soil for $5.10

  c. Which is the best buy?

45. Corn comes in two size cans, 29 oz and 15 oz. The larger can costs $1.19 and the smaller can costs $0.77. Find the unit cost of each can. Which is the better buy? (Round to three decimal places.)

46. Napkins come in a variety of packages. A package of 400 napkins sells for $2.99, and a package of 100 napkins sells for $1.50. Find the unit cost of each package. Which is the better buy? (Round to four decimal places.)

Objective 4: Applications of Rates

47. Carbonated beverages come in different sizes and contain different amounts of sugar. Compute the amount of sugar per fluid ounce for each soda. Then determine which has the greatest amount of sugar per fluid ounce. **(See Example 4.)**

Soda	Amount	Sugar
Coca-Cola	20 fl oz	65 g
Mello Yello	12 fl oz	47 g
Canada Dry Ginger Ale	8 fl oz	24 g

48. Compute the amount of carbohydrate per fluid ounce for each soda. Then determine which has the greatest amount of per fluid ounce.

Soda	Amount	Carbohydrates
Coca-Cola	20 fl oz	65 g
Mello Yello	12 fl oz	47 g
Canada Dry Ginger Ale	8 fl oz	25 g

49. Carbonated beverages come in different sizes and have a different number of calories. Compute the number of calories per fluid ounce for each soda. Then determine which has the least number of calories per fluid ounce.

Soda	Amount	Calories
Coca-Cola	20 fl oz	240
Mello Yello	12 fl oz	170
Canada Dry Ginger Ale	8 fl oz	90

50. According to the National Institutes of Health, a platelet count below 20,000 per microliter of blood is considered a life-threatening condition. Suppose a patient's test results yield a platelet count of 13,000,000 for 100 microliters. Write this as a unit rate (number of platelets per microliter). Does the patient have a life-threatening condition?

51. The number of motor vehicles produced in the United States increased steadily by a total of 5,310,000 between the years 1990 and 2008. Compute the rate representing the increase in the number of vehicles produced per year during this time period. (*Source:* American Automobile Manufacturers Association)

52. The total number of prisoners in the United States increased steadily by a total of 344,000 in an 8-year period. Compute the rate representing the increase in the number of prisoners per year.

53. a. The population of Mexico increased steadily by 22 million people in a 10-year period. Compute the rate representing the increase in the population per year.

   b. The population of Brazil increased steadily by 10.2 million in a 5-year period. Compute the rate representing the increase in the population per year.

   c. Which country has a greater increase in population per year?

54. a. The price per share of Microsoft stock rose $18.24 in a 24-month period. Compute the rate representing the increase in the price per month.

   b. The price per share of IBM stock rose $22.80 in a 12-month period. Compute the rate representing the increase in the price per month.

   c. Which stock had a greater rate of increase per month?

55. A cheetah can run 120 m in 4.1 sec. An antelope can run 50 m in 2.1 sec. Compare their unit speeds to determine which animal is faster. Round to the nearest whole unit.

Calculator Connections

Topic: Applications of unit rates

Calculator Exercises

Don Shula coached football for 33 years. He had 328 wins and 156 losses. Tom Landry coached football for 29 years. He had 250 wins and 162 losses. Use this information to answer Exercises 56–57.

56. a. Compute a unit rate representing the average number of wins per year for Don Shula. Round to one decimal place.

   b. Compute a unit rate representing the average number of wins per year for Tom Landry. Round to one decimal place.

   c. Which coach had a better rate of wins per year?

57. a. Compute a unit rate representing the number of wins to the number of losses for Don Shula. Round to one decimal place.

   b. Compute a unit rate representing the number of wins to the number of losses for Tom Landry. Round to one decimal place.

   c. Which coach had a better win/loss rate?

58. Compare three brands of soap. Find the price per ounce and determine the best buy. (Round to two decimal places.)

   a. Dove: $5.99 for a 6-bar pack of 4.5-oz bars

   b. Dial: $2.89 for a 3-bar pack of 4.5-oz bars

   c. Irish Spring: $6.99 for an 8-bar pack of 4.5-oz bars

59. Mayonnaise comes in 32-, 16-, and 8-oz jars. They are priced at $3.61, $2.19, and $1.19, respectively. Find the unit cost of each size jar, to find the best buy. (Round to three decimal places.)

60. Albacore tuna comes in different-size cans. Find the unit cost of each package to find the best buy. (Round to three decimal places.)

   a. $3.59 for a 12-oz can

   b. $4.99 for a 4-pack of 6-oz cans

   c. $2.99 for a 3-pack of 3-oz cans

61. Coca-Cola is sold in a variety of different packages. Find the unit cost of each package, to find the better buy. (Round to three decimal places.)

   a. $4.99 for a case of 24 twelve-oz cans

   b. $5.00 for a 12-pack of 8-oz cans

Section 5.3 Proportions

Objectives

1. **Definition of a Proportion**
2. **Determining Whether Two Ratios Form a Proportion**
3. **Solving Proportions**

1. Definition of a Proportion

A statement indicating that two quantities are equal is called an **equation**. In this section, we are interested in a special type of equation called a proportion. A **proportion** states that two ratios or rates are equal. For example:

$$\frac{1}{4} = \frac{10}{40} \text{ is a proportion.}$$

We know that the fractions $\frac{1}{4}$ and $\frac{10}{40}$ are equal because $\frac{10}{40}$ reduces to $\frac{1}{4}$.

We read the proportion $\frac{1}{4} = \frac{10}{40}$ as follows: "1 is to 4 as 10 is to 40."

We also say that the numbers 1 and 4 are *proportional to* the numbers 10 and 40.

Example 1   Writing Proportions

Write a proportion for each statement.

  a. 5 is to 12 as 30 is to 72.

  b. 240 mi is to 4 hr as 300 mi is to 5 hr.

  c. The numbers 3 and 7 are proportional to the numbers 12 and 28.

Solution:

  a. $\dfrac{5}{12} = \dfrac{30}{72}$ 5 is to 12 as 30 is to 72.

  b. $\dfrac{240 \text{ mi}}{4 \text{ hr}} = \dfrac{300 \text{ mi}}{5 \text{ hr}}$ 240 mi is to 4 hr as 300 mi is to 5 hr.

  c. $\dfrac{3}{7} = \dfrac{12}{28}$ 3 and 7 are proportional to 12 and 28.

Skill Practice

Write a proportion for each statement.

  1. 7 is to 28 as 13 is to 52.
  2. \$17 is to 2 hr as \$102 is to 12 hr.
  3. The numbers 5 and 11 are proportional to the numbers 15 and 33.

TIP: A proportion is an equation and must have an equal sign.

2. Determining Whether Two Ratios Form a Proportion

To determine whether two ratios form a proportion, we must determine whether the ratios are equal. Recall from Section 2.3 that two fractions are equal whenever their cross products are equal. That is,

$$\frac{a}{b} = \frac{c}{d} \quad \text{implies} \quad a \cdot d = b \cdot c \quad \text{(and vice versa).}$$

Example 2   Determining Whether Two Ratios Form a Proportion

Determine whether the ratios form a proportion.

  a. $\dfrac{3}{5} \overset{?}{=} \dfrac{9}{15}$ **b.** $\dfrac{1\frac{2}{3}}{4} \overset{?}{=} \dfrac{10}{5\frac{1}{2}}$

Solution:

  a. $\dfrac{3}{5} \overset{?}{\times} \dfrac{9}{15}$

   $(3)(15) \overset{?}{=} (5)(9)$ Cross-multiply to form the cross products.

   $45 = 45$ ✔

The cross products are equal. Therefore, the ratios form a proportion.

  b. $\dfrac{1\frac{2}{3}}{4} \overset{?}{\times} \dfrac{10}{5\frac{1}{2}}$

   $\left(1\dfrac{2}{3}\right)\left(5\dfrac{1}{2}\right) \overset{?}{=} (4)(10)$ Cross-multiply to form the cross products.

   $\dfrac{5}{3} \cdot \dfrac{11}{2} \overset{?}{=} 40$ Write the mixed numbers as improper fractions.

   $\dfrac{55}{6} \neq 40$ Multiply fractions.

The cross products are not equal. The ratios do not form a proportion.

Skill Practice

Determine whether the ratios form a proportion.

  4. $\dfrac{4}{9} \overset{?}{=} \dfrac{12}{27}$

  5. $\dfrac{3\frac{1}{4}}{5} \overset{?}{=} \dfrac{8}{12\frac{1}{2}}$

Avoiding Mistakes

Cross multiplication can be performed only when there are two fractions separated by an equal sign.

Answers

  1. $\dfrac{7}{28} = \dfrac{13}{52}$ **2.** $\dfrac{\$17}{2 \text{ hr}} = \dfrac{\$102}{12 \text{ hr}}$

  3. $\dfrac{5}{11} = \dfrac{15}{33}$ **4.** Yes   **5.** No

Example 3 **Determining Whether Pairs of Numbers Are Proportional**

Determine whether the numbers 2.7 and 5.3 are proportional to the numbers 8.1 and 15.9.

Solution:

Two pairs of numbers are proportional if their ratios are equal.

$$\frac{2.7}{5.3} \overset{?}{\times} \frac{8.1}{15.9}$$

$(2.7)(15.9) \overset{?}{=} (5.3)(8.1)$ Cross-multiply to form the cross products.

$42.93 = 42.93$ ✔   Multiply decimals.

The cross products are equal. The ratios form a proportion.

3. Solving Proportions

A proportion is made up of four values. If three of the four values are known, we can solve for the fourth.

Consider the proportion $\frac{x}{20} = \frac{3}{4}$. We let the variable x represent the unknown value in the proportion. To solve for x, we can equate the cross products to form an equivalent equation.

$$\frac{x}{20} \overset{\times}{} \frac{3}{4}$$

$4 \cdot x = 3 \cdot 20$ Cross-multiply to form the cross products.

$4 \cdot x = 60$ To determine the correct value of x, we ask, What number multiplied by 4 equals 60? The answer is 15 and can be found mentally by computing $60 \div 4 = 15$.

$\dfrac{4 \cdot x}{4} = \dfrac{60}{4}$ To show this step in the equation, divide both sides of the equation by 4.

$\dfrac{\overset{1}{4} \cdot x}{\underset{1}{4}} = \dfrac{\overset{15}{60}}{\underset{1}{4}}$ Simplify common factors within each fraction.

$1x = 15$

$x = 15$

We can check the value of x in the original proportion.

Check: $\dfrac{x}{20} = \dfrac{3}{4}$ substitute 15 for x $\longrightarrow$ $\dfrac{15}{20} \overset{?}{=} \dfrac{3}{4}$

$(15)(4) \overset{?}{=} (3)(20)$

$60 = 60$ ✔   The solution 15 checks.

This process uses the following important fact. *We may divide both sides of an equation by the same nonzero number.* By so doing, we produce an equivalent equation that has the same solution. In general, to solve an equation, the goal is to isolate the variable on one side of the equation. In the equation $4 \cdot x = 60$, we must eliminate the factor of 4 next to x. By dividing by 4 we form a ratio $\frac{4 \cdot x}{4}$. This reduces to $\frac{1 \cdot x}{1}$ which in turn simplifies to x.

The steps to solve a proportion are summarized next.

PROCEDURE   Solving a Proportion
Step 1 Set the cross products equal to each other.
Step 2 Divide both sides of the equation by the number being multiplied by the variable.
Step 3 Check the solution in the original proportion.

Example 4 Solving a Proportion

Solve the proportion. $\dfrac{x}{13} = \dfrac{6}{39}$

Solution:

$$\frac{x}{13} = \frac{6}{39}$$

$39 \cdot x = (6)(13)$ Set the cross products equal.

$39 \cdot x = 78$ Simplify.

$$\frac{39 \cdot x}{39} = \frac{78}{39}$$ Divide both sides by the number being multiplied by x (in this case, 39).

$$\frac{\overset{1}{\cancel{39}} \cdot x}{\underset{1}{\cancel{39}}} = \frac{\overset{2}{\cancel{78}}}{\underset{1}{\cancel{39}}}$$ Simplify common factors within each fraction.

$x = 2$ Check: $\dfrac{x}{13} = \dfrac{6}{39} \xrightarrow{\text{substitute } x=2} \dfrac{2}{13} \overset{?}{=} \dfrac{6}{39}$

$(2)(39) \overset{?}{=} (6)(13)$

$78 = 78$ ✔

The solution 2 checks in the original proportion.

Skill Practice

Solve the proportion. Be sure to check your answer.

10. $\dfrac{2}{9} = \dfrac{n}{81}$

Example 5 Solving a Proportion

Solve the proportion. $\dfrac{4}{15} = \dfrac{9}{n}$

Solution:

$$\frac{4}{15} = \frac{9}{n}$$ The variable can be represented by any letter.

$4 \cdot n = (9)(15)$ Set the cross products equal.

$4 \cdot n = 135$ Simplify.

$$\frac{4 \cdot n}{4} = \frac{135}{4}$$ Divide both sides by the number being multiplied by n (in this case, 4).

$$\frac{\overset{1}{\cancel{4}} \cdot n}{\underset{1}{\cancel{4}}} = \frac{135}{4}$$

$n = \dfrac{135}{4}$ The fraction $\frac{135}{4}$ is in lowest terms.

Skill Practice

Solve the proportion. Be sure to check your answer.

11. $\dfrac{3}{w} = \dfrac{21}{77}$

Avoiding Mistakes

When solving an equation, don't try to "cancel" across the equal sign.

Answers

10. $n = 18$.   **11.** $w = 11$

The solution may be written as $n = \frac{135}{4}$ or $n = 33\frac{3}{4}$ or $n = 33.75$.

To check the solution in the original proportion, we may use any of the three forms of the answer. We will use the decimal form.

Check: $\dfrac{4}{15} = \dfrac{9}{n}$ $\xrightarrow{\text{substitute } n = 33.75}$ $\dfrac{4}{15} \overset{?}{=} \dfrac{9}{33.75}$

$$(4)(33.75) \overset{?}{=} (9)(15)$$

$$135 = 135 \checkmark \quad \text{The solution checks.}$$

Skill Practice

Solve the proportion. Be sure to check your answer.

12. $\dfrac{0.6}{x} = \dfrac{1.5}{2}$

Example 6 **Solving a Proportion**

Solve the proportion. $\quad \dfrac{0.8}{3.1} = \dfrac{4}{p}$

Solution:

$$\dfrac{0.8}{3.1} = \dfrac{4}{p}$$

$(0.8) \cdot p = (4)(3.1)$ Set the cross products equal.

$0.8p = 12.4$ Notice that in this case we dropped the $\cdot$ symbol for multiplication between 0.8 and p. If a variable and a constant are written adjacent to each other without an operator $(+, -, \times, \text{or} \div)$ between them, the operation is understood to be multiplication. That is, $0.8p = 0.8 \times p$.

$$\dfrac{0.8p}{0.8} = \dfrac{12.4}{0.8}$$ Divide both sides by the number being multiplied by p (in this case, 0.8).

$$\dfrac{\overset{1}{\cancel{0.8}}p}{\underset{1}{\cancel{0.8}}} = \dfrac{12.4}{0.8}$$

$$p = 15.5$$ Divide $12.4 \div 0.8 = 15.5$.

The check is left to the reader.

We chose to give the solution to Example 6 in decimal form because the values in the original proportion are decimal numbers. However, it would be correct to give the solution as a mixed number or fraction. The solution $p = 15.5$ is also equivalent to $p = 15\frac{1}{2}$ or $p = \frac{31}{2}$.

Example 7 **Solving a Proportion**

Solve the proportion. $\quad \dfrac{12}{8} = \dfrac{x}{\frac{2}{3}}$

Answer

12. $x = 0.8$

Solution:

$$\frac{12}{8} = \frac{x}{\frac{2}{3}}$$

$$(12)\left(\frac{2}{3}\right) = 8 \cdot x \qquad \text{Set the cross products equal.}$$

$$\left(\frac{12}{1}\right)\left(\frac{2}{3}\right) = 8x \qquad \text{Write the whole number as an improper fraction.}$$

$$\left(\frac{\overset{4}{\cancel{12}}}{1}\right)\left(\frac{2}{\underset{1}{\cancel{3}}}\right) = 8x \qquad \text{Simplify common factors.}$$

$$\frac{8}{1} = 8x \qquad \text{Multiply fractions.}$$

$$8 = 8x \qquad \text{Simplify.}$$

$$\frac{8}{8} = \frac{8x}{8} \qquad \text{Divide both sides by the number being multiplied by } x \text{ (in this case, 8).}$$

$$\frac{\overset{1}{\cancel{8}}}{\underset{1}{\cancel{8}}} = \frac{\overset{1}{\cancel{8}}x}{\underset{1}{\cancel{8}}} \qquad \text{Simplify common factors.}$$

$$1 = x$$

The solution 1 checks in the original proportion.

Skill Practice

Solve the proportion. Be sure to check your answer.

13. $\dfrac{\frac{1}{2}}{3.5} = \dfrac{x}{14}$

Answer

13. $x = 2$

Section 5.3 Practice Exercises

Boost your GRADE at ALEKS.com!

ALEKS version 3.0

- Practice Problems
- Self-Tests
- NetTutor
- e-Professors
- Videos

Study Skills Exercises

1. You should not try to cram for tests. Instead, math is a subject that should be studied every day. This text gives you opportunities to review and practice as you work through the book. Find the page number for the Cumulative Review exercises for this chapter.

2. Define the key terms.

    a. Equation   **b. Proportion**

Review Exercises

For Exercises 3–8, write as a reduced ratio or rate.

3. 3 ft to 45 ft

4. 3 teachers for 45 students

5. 6 apples for 2 pies

6. 6 days to 2 days

7. 264 mi per 36 gal

8. $264 to $36

Objective 1: Definition of a Proportion

For Exercises 9–20, write a proportion for each statement. **(See Example 1.)**

9. 4 is to 16 as 5 is to 20.

10. 3 is to 18 as 4 is to 24.

11. 25 is to 15 as 10 is to 6.

12. 35 is to 14 as 20 is to 8.

13. The numbers 2 and 3 are proportional to the numbers 4 and 6.

14. The numbers 2 and 1 are proportional to the numbers 26 and 13.

15. The numbers 30 and 25 are proportional to the numbers 12 and 10.

16. The numbers 24 and 18 are proportional to the numbers 8 and 6.

17. $6.25 per hour is proportional to $187.50 per 30 hr.

18. $115 per week is proportional to $460 per 4 weeks.

19. 1 in. is to 7 mi as 5 in. is to 35 mi.

20. 16 flowers is to 5 plants as 32 flowers is to 10 plants.

Objective 2: Determining Whether Two Ratios Form a Proportion

For Exercises 21–28, determine whether the ratios form a proportion. **(See Example 2.)**

21. $\dfrac{5}{18} \stackrel{?}{=} \dfrac{4}{16}$

22. $\dfrac{9}{10} \stackrel{?}{=} \dfrac{8}{9}$

23. $\dfrac{16}{24} \stackrel{?}{=} \dfrac{2}{3}$

24. $\dfrac{4}{5} \stackrel{?}{=} \dfrac{24}{30}$

25. $\dfrac{2\frac{1}{2}}{3\frac{2}{3}} \stackrel{?}{=} \dfrac{15}{22}$

26. $\dfrac{1\frac{3}{4}}{3} \stackrel{?}{=} \dfrac{7}{12}$

27. $\dfrac{2}{3.2} \stackrel{?}{=} \dfrac{10}{16}$

28. $\dfrac{4.7}{7} \stackrel{?}{=} \dfrac{23.5}{35}$

For Exercises 29–34, determine whether the pairs of numbers are proportional. **(See Example 3.)**

29. Are the numbers 48 and 18 proportional to the numbers 24 and 9?

30. Are the numbers 35 and 14 proportional to the numbers 5 and 2?

31. Are the numbers $2\frac{3}{8}$ and $1\frac{1}{2}$ proportional to the numbers $9\frac{1}{2}$ and 6?

32. Are the numbers $1\frac{2}{3}$ and $\frac{5}{6}$ proportional to the numbers 5 and $2\frac{1}{2}$?

33. Are the numbers 6.3 and 9 proportional to the numbers 12.6 and 16?

34. Are the numbers 7.1 and 2.4 proportional to the numbers 35.5 and 10?

Objective 3: Solving Proportions

For Exercises 35–42, what number would you divide by on each side of the equation to solve for the variable?

35. $2x = 8$

36. $3x = 27$

37. $5p = 30$

38. $7w = 49$

39. $32 = 8m$

40. $50 = 25y$

41. $0.15 = 0.6x$

42. $1.4 = 0.4z$

For Exercises 43–46, determine whether the given value is a solution to the proportion.

43. $\dfrac{x}{40} = \dfrac{1}{8}$; $x = 5$

44. $\dfrac{14}{x} = \dfrac{12}{18}$; $x = 21$

45. $\dfrac{12.4}{31} = \dfrac{8.2}{y}$; $y = 20$

46. $\dfrac{4.2}{9.8} = \dfrac{z}{36.4}$; $z = 15.2$

For Exercises 47–70, solve the proportion. Be sure to check your answers. **(See Examples 4–7.)**

47. $\dfrac{12}{16} = \dfrac{3}{x}$

48. $\dfrac{20}{28} = \dfrac{5}{x}$

49. $\dfrac{9}{21} = \dfrac{x}{7}$

50. $\dfrac{15}{10} = \dfrac{3}{x}$

51. $\dfrac{p}{12} = \dfrac{25}{4}$

52. $\dfrac{p}{8} = \dfrac{30}{24}$

53. $\dfrac{6}{n} = \dfrac{4}{8}$

54. $\dfrac{49}{n} = \dfrac{14}{18}$

55. $\dfrac{2}{3} = \dfrac{t}{18}$

56. $\dfrac{34}{51} = \dfrac{2}{t}$

57. $\dfrac{25}{100} = \dfrac{9}{y}$

58. $\dfrac{65}{15} = \dfrac{26}{y}$

59. $\dfrac{17}{12} = \dfrac{4\frac{1}{4}}{x}$

60. $\dfrac{26}{30} = \dfrac{5\frac{1}{5}}{x}$

61. $\dfrac{m}{12} = \dfrac{5}{8}$

62. $\dfrac{16}{12} = \dfrac{21}{a}$

63. $\dfrac{3.125}{5} = \dfrac{18.75}{k}$

64. $\dfrac{4.75}{8} = \dfrac{9.5}{k}$

65. $\dfrac{0.5}{h} = \dfrac{1.8}{9}$

66. $\dfrac{2.6}{h} = \dfrac{1.3}{0.5}$

67. $\dfrac{\frac{3}{8}}{6.75} = \dfrac{x}{72}$

68. $\dfrac{12}{\frac{1}{4}} = \dfrac{120}{y}$

69. $\dfrac{4}{\frac{1}{10}} = \dfrac{\frac{1}{2}}{z}$

70. $\dfrac{6}{\frac{1}{3}} = \dfrac{\frac{1}{2}}{t}$

Applications of Proportions and Similar Figures | Section 5.4

1. Applications of Proportions

Proportions are used in a variety of applications. In Examples 1 through 4, we take information from the wording of a problem and form a proportion.

Objectives

1. Applications of Proportions
2. Similar Figures

Example 1 Using a Proportion in a Consumer Application

Linda drove her Honda Accord 145 mi on 5 gal of gas. At this rate, how far can she drive on 12 gal?

Skill Practice

1. Jacques bought 3 lb of tomatoes for $4.50. At this rate, how much would 7 lb cost?

Solution:

Let x represent the distance Linda can go on 12 gal.

This problem involves two rates. We can translate this to a proportion. Equate the two rates.

distance ⟶ $\dfrac{145 \text{ mi}}{5 \text{ gal}} = \dfrac{x \text{ mi}}{12 \text{ gal}}$ ⟵ distance
number of gallons ⟶ ⟵ number of gallons

Solve the proportion.

$(145)(12) = (5) \cdot x$ Cross-multiply to form the cross products.

$1740 = 5x$

$\dfrac{1740}{5} = \dfrac{\overset{1}{\cancel{5}}x}{\cancel{5}}$ Divide both sides by 5.

$348 = x$ Divide $1740 \div 5 = 348$.

Linda can drive 348 mi on 12 gal of gas.

TIP: Notice that the two rates have the same units in the numerator (miles) and the same units in the denominator (gallons).

Answer

1. The price for 7 lb would be $10.50.

Skill Practice

8. The first- and second-place plaques for a softball tournament have the same shape but are different sizes. Find the length of side x.

10 in.
x
8 in.
5 in.

Example 7 Using Similar Polygons in a Photography Application

A negative for a photograph is 3.5 cm by 2.5 cm. If the width of the resulting picture is 4 in., what is the length of the picture? See Figure 5-4.

2.5 cm
3.5 cm
4 in.
x

Figure 5-4

Solution:

Let x represent the length of the photo.

The photo and its negative are similar polygons.

$$\frac{3.5\text{ cm}}{x\text{ in.}} = \frac{2.5\text{ cm}}{4\text{ in.}} \quad \text{Translate to a proportion.}$$

$$(3.5)(4) = (2.5)\cdot x$$

$$14 = 2.5x$$

$$\frac{14}{2.5} = \frac{\overset{1}{\cancel{2.5}}x}{\underset{1}{\cancel{2.5}}} \quad \text{Divide both sides by 2.5.}$$

$$5.6 = x \quad \text{Divide } 14 \div 2.5 = 5.6.$$

The picture is 5.6 in. long.

Answer

8. Side x is 6.25 in.

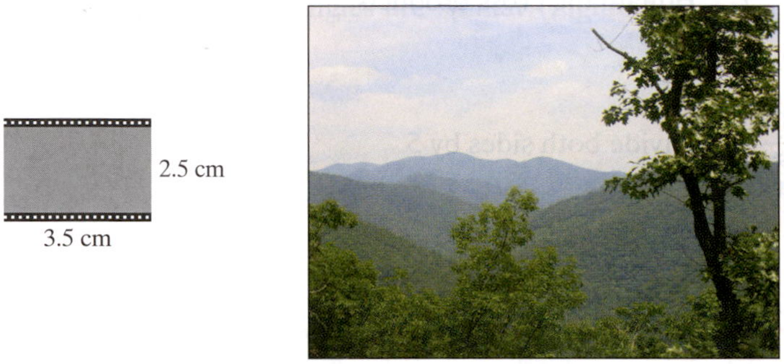

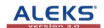

Section 5.4 Practice Exercises

Boost *your* GRADE at ALEKS.com!

ALEKS version 3.0

- Practice Problems
- Self-Tests
- NetTutor
- e-Professors
- Videos

Study Skills Exercise

1. Define the key terms.

a. Similar triangles **b. Similar polygons**

Review Exercises

For Exercises 2–5, use = or ≠ to make a true statement.

2. $\frac{4}{7} \square \frac{12}{21}$ **3.** $\frac{3}{13} \square \frac{15}{65}$ **4.** $\frac{2}{5} \square \frac{21}{55}$ **5.** $\frac{12}{7} \square \frac{35}{19}$

For Exercises 6–12, solve the proportion.

6. $\dfrac{2}{7} = \dfrac{3}{x}$

7. $\dfrac{4}{3} = \dfrac{n}{5}$

8. $\dfrac{p}{9} = \dfrac{1}{6}$

9. $\dfrac{3\frac{1}{2}}{k} = \dfrac{2\frac{1}{3}}{4}$

10. $\dfrac{2.4}{3} = \dfrac{m}{5}$

11. $\dfrac{3}{2.1} = \dfrac{7}{y}$

12. $\dfrac{1.2}{4} = \dfrac{3}{a}$

Objective 1: Applications of Proportions

13. Pam drives her Toyota Prius 244 mi in city driving on 4 gal of gas. At this rate how many miles can she drive on 10 gal of gas? **(See Example 1.)**

14. Didi takes her pulse for 10 sec and counts 13 beats. How many beats per minute is this?

15. To cement a garden path, it takes crushed rock and cement in a ratio of 3.25 kg of rock to 1 kg of cement. If a 24 kg-bag of cement is purchased, how much crushed rock will be needed? **(See Example 2.)**

16. Suppose two adults produce 63.4 lb of garbage in one week. At this rate, how many pounds will 50 adults produce in one week?

17. On a map, the distance from Sacramento, California, to San Francisco, California, is 8 cm. The legend gives the actual distance at 91 mi. On the same map, Faythe measured 7 cm from Sacramento to Modesto, California. What is the actual distance? (Round to the nearest mile.) **(See Example 3.)**

18. On a map, the distance from Nashville, Tennessee, to Atlanta, Georgia, is 3.5 in., and the actual distance is 210 mi. If the map distance between Dallas, Texas, and Little Rock, Arkansas, is 4.75 in., what is the actual distance?

19. At Central Community College, the ratio of female students to male students is 31 to 19. If there are 6200 female students, how many male students are there?

20. Evelyn won an election by a ratio of 6 to 5. If she received 7230 votes, how many votes did her opponent receive?

21. If you flip a coin many times, the coin should come up heads about 1 time out of every 2 times it is flipped. If a coin is flipped 630 times, about how many heads do you expect to come up?

22. A die is a small cube used in games of chance. It has six sides, and each side has 1, 2, 3, 4, 5, or 6 dots painted on it. If you roll a die, the number 4 should come up about 1 time out of every 6 times the die is rolled. If you roll a die 366 times, about how many times do you expect the number 4 to come up?

23. A pitcher gave up 42 earned runs in 126 innings. Approximately how many earned runs will he give up in one game (9 innings)? This value is called the earned run average.

24. In one game Peyton Manning completed 34 passes for 357 yd. At this rate how many yards would be gained for 22 passes?

25. Pierre bought 37.0€ (Euros) with $50 American. At this rate, how many Euros can he buy with $900 American?

26. Erik bought $111 Canadian with $100 American. At this rate, how many Canadian dollars can he buy with $235 American?

27. According to data collected by the U.S. Consumer Product Safety Commission, for persons age 75 and older, nearly 3 out of 4 hospital emergency room visits are due to falls. If a hospital emergency room sees an average of 60 people 75 years of age or older in a month, how many of these visits are expected to be a result of falls?

28. Approximately 24 out of 100 Americans over the age of 12 smoke. How many smokers would you expect in a group of 850 Americans over the age of 12?

29. Park officials stocked a man-made lake with bass last year. To approximate the number of bass this year, a sample of 75 fish is taken out of the lake and tagged. Then later a different sample is taken, and it is found that 21 of 100 fish are tagged. Approximately how many fish are in the lake? Round to the nearest whole unit. **(See Example 4.)**

30. Laws have been instituted in Florida to help save the manatee. To establish the number in Florida, a sample of 150 manatees was marked and let free. A new sample was taken and found that there were 3 marked manatees out of 40 captured. What is the approximate population of manatees in Florida?

31. Yellowstone National Park in Wyoming has the largest population of free-roaming bison. To approximate the number of bison, 200 are captured and tagged and then let free to roam. Later, a sample of 120 bison is observed and 6 have tags. Approximate the population of bison in the park.

32. In Cass County, Michigan, there are about 20 white-tailed deer per square mile. If the county covers 492 mi², about how many white-tailed deer are in the county?

Objective 2: Similar Figures

For Exercises 33–36, the pairs of triangles are similar. Solve for x and y. **(See Example 5.)**

33.
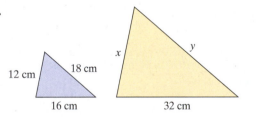

34.
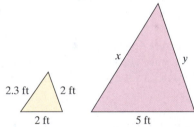

35.
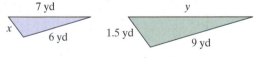

36.
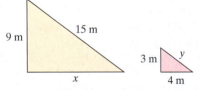

37. The height of a flagpole can be determined by comparing the shadow of the flagpole and the shadow of a yardstick. From the figure, determine the height of the flagpole. **(See Example 6.)**

38. A 15-ft flagpole casts a 4-ft shadow. How long will the shadow be for a 90-ft building?

39. A person 1.6 m tall stands next to a lifeguard observation platform. If the person casts a shadow of 1 m and the lifeguard platform casts a shadow of 1.5 m, how high is the platform?

Figure for Exercise 37

40. A 32-ft tree casts a shadow of 18 ft. How long will the shadow be for a 22-ft tree?

For Exercises 41–44, the pairs of polygons are similar. Solve for the indicated variables. **(See Example 7.)**

41.

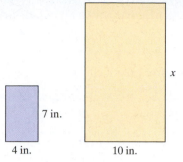

7 in.

4 in.   10 in.

x

42.

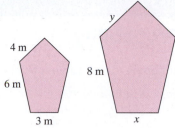

4 m

6 m

3 m

8 m

y

x

43.   2 ft

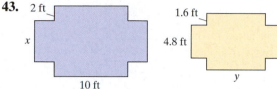

x

10 ft

1.6 ft

4.8 ft

y

44.   30 cm

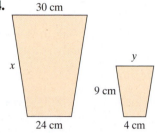

x

24 cm

9 cm

4 cm

y

45. A carpenter makes a schematic drawing of a porch he plans to build. On the drawing, 2 in. represents 7 ft. Find the dimensions of the porch.

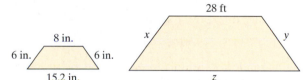

8 in.

6 in.   6 in.

15.2 in.

28 ft

x   y

z

46. The Great Cookie Company has a sign on the front of its store as shown in the figure. The company would like to put the same shape sign in the back, but with dimensions $\frac{1}{3}$ as large. Find the lengths denoted by x and y.

3 yd

The Great Cookie Company

4 yd

x

y

Calculator Connections

Topic: Solving proportions

Calculator Exercises

47. In a recent year the annual crime rate for Oklahoma was 4743 crimes per 100,000 people. If Oklahoma had approximately 3,500,000 people at that time, approximately how many crimes were committed? (*Source:* Oklahoma Department of Corrections)

48. To measure the height of the Washington Monument, a student 5.5 ft tall measures his shadow to be 3.25 ft. At the same time of day, he measured the shadow of the Washington Monument to be 328 ft long. Estimate the height of the monument to the nearest foot.

49. In a recent year the rate of breast cancer in women was 110 cases per 100,000 women. At that time the state of California had approximately 14,000,000 women. How many women in California would be expected to have breast cancer? (*Source:* Centers for Disease Control)

50. In a recent year the rate of prostate disease in U.S. men was 118 cases per 1000 men. At that time the state of Massachusetts had approximately 2,500,000 men. How many men in Massachusetts would be expected to have prostate disease? (*Source:* National Center for Health Statistics)

Group Activity

Investigating Probability

Materials: Paper bags containing 10 white poker chips, 6 red poker chips, and 4 blue poker chips.

Estimated time: 15 minutes

Group Size: 3

1. Each group will receive a bag of poker chips, with 10 white, 6 red, and 4 blue chips.

2. **a.** Write the ratio of red chips in the bag to the total number of chips in the bag. _____ This value represents the *probability* of randomly selecting a red chip from the bag.

   b. Write this fraction in decimal form. _____

   c. Write the decimal from step (b) as a percent. _____
   A probability value indicates the likeliness of an event to occur. For example, to interpret this probability, one might say that there is a 30% chance of selecting a red chip at random from the bag.

3. Determine the probability of selecting a white chip from the bag. Interpret your answer.

4. Determine the probability of selecting a blue chip from the bag. Interpret your answer.

5. Next, have one group member select a chip from the bag at random (without looking), and record the color of the chip. Then replace the chip to the bag. Repeat this step for a total of 20 times (be sure that each student in the group has a chance to pick). Record the total number of red, white, and blue chips selected. Then write the ratio of the number selected out of 20.

	Number of Times Selected	**Ratio of Number Selected Out of 20**
Red		
White		
Blue		

6. How well do your experimental results match the theoretical probabilities found in steps 2–4?

7. The instructor will now pool the data from the whole class. Write the total number of times that red, white, and blue chips were selected, respectively, by the whole class. Then write the ratio of these values to the number of total selections made.

8. How well do the experimental results from the whole class match the theoretical probabilities of selecting a red, white, or blue chip?

Chapter 5   Summary

Key Concepts

A **ratio** is a comparison of two quantities.

The ratio of a to b can be written as follows, provided $b \neq 0$.

  1. a to b 2. $a:b$ 3. $\dfrac{a}{b}$

When we write a ratio in fraction form, we generally simplify it to lowest terms.

Ratios that contain mixed numbers, fractions, or decimals can be simplified to lowest terms with whole numbers in the numerator and denominator.

Examples

Example 1

Three forms of a ratio:

4 to 6   4 : 6 $\dfrac{4}{6}$

Example 2

A hockey team won 4 games out of 6. Write a ratio of games won to total games played and simplify to lowest terms.

$$\dfrac{4 \text{ games won}}{6 \text{ games played}} = \dfrac{2}{3}$$

Example 3

$$\dfrac{2\frac{1}{6}}{\frac{2}{3}} = 2\frac{1}{6} \div \frac{2}{3} = \dfrac{13}{\overset{}{\underset{2}{6}}} \cdot \dfrac{\overset{1}{3}}{2} = \dfrac{13}{4}$$

Example 4

$$\dfrac{2.1}{2.8} = \dfrac{2.1}{2.8} \cdot \dfrac{10}{10} = \dfrac{21}{28} = \dfrac{\overset{3}{21}}{\underset{4}{28}} = \dfrac{3}{4}$$

Section 5.2 Rates

Key Concepts

A **rate** compares two different quantities.

A rate having a denominator of 1 unit is called a **unit rate**. To find a unit rate, divide the numerator by the denominator.

A **unit cost** or unit price is the cost per 1 unit, for example, $1.21/lb or 43¢/oz. Comparing unit prices can help determine the best buy.

Examples

Example 1

New Jersey has 8,470,000 people living in 21 counties. Write a reduced ratio of people per county.

$$\frac{8{,}470{,}000 \text{ people}}{21 \text{ counties}} = \frac{1{,}210{,}000 \text{ people}}{3 \text{ counties}}$$

Example 2

If a race car traveled 1250 mi in 8 hr during a race, what is its speed in miles per hour?

$$\frac{1250 \text{ mi}}{8 \text{ hr}} = 156.25 \text{ mi/hr}$$

Example 3

Tide laundry detergent is offered in two sizes: $10.99 for 150 oz and $7.63 for 100 oz. Find the unit prices to find the best buy.

$$\frac{\$10.99}{150 \text{ oz}} \approx \$0.0733/\text{oz}$$

$$\frac{\$7.63}{100 \text{ oz}} = \$0.0763/\text{oz}$$

The 150-oz package is the better buy because the unit cost is less.

Section 5.3   Proportions

Key Concepts

A statement indicating that two quantities are equal is called an **equation**.

A **proportion** states that two ratios or rates are equal.

$\dfrac{14}{21} = \dfrac{2}{3}$ is a proportion.

To determine if two ratios form a proportion, check to see if the cross products are equal, that is,

$\dfrac{a}{b} = \dfrac{c}{d}$ implies $a \cdot d = b \cdot c$ (and vice versa)

To solve a proportion, solve the equation formed by the cross products.

Examples

Example 1

Write as a proportion.

56 mi is to 2 gal as 84 mi is to 3 gal.

$$\frac{56 \text{ mi}}{2 \text{ gal}} = \frac{84 \text{ mi}}{3 \text{ gal}}$$

Example 2

$$\frac{3}{8} \overset{?}{\times} \frac{2\frac{1}{2}}{6\frac{2}{3}} \qquad 3 \cdot 6\frac{2}{3} \overset{?}{=} 8 \cdot 2\frac{1}{2}$$

$$\frac{3}{1} \cdot \frac{20}{3} \overset{?}{=} \frac{8}{1} \cdot \frac{5}{2}$$

$$20 = 20 \ \checkmark$$

The ratios form a proportion.

Example 3

$$\frac{5}{4} = \frac{18}{x} \qquad \Rightarrow \qquad 5 \cdot x = 4 \cdot 18$$

$$5x = 72$$

$$\frac{\overset{1}{\cancel{5}}x}{\cancel{5}_{1}} = \frac{72}{5}$$

$$x = \frac{72}{5} \text{ or } 14\frac{2}{5} \text{ or } 14.4$$

Section 5.4 Applications of Proportions and Similar Figures

Key Concepts

Example 1 demonstrates an application involving a proportion. Example 2 demonstrates the use of proportions involving similar triangles.

Example 1

According to the National Highway Traffic Safety Administration, 2 out of 5 traffic fatalities involve the use of alcohol. If there were 43,200 traffic fatalities in a recent year, how many involved the use of alcohol?

Let n represent the number of traffic fatalities involving alcohol.

Set up a proportion:

$$\frac{2 \text{ traffic fatalities w/alcohol}}{5 \text{ traffic fatalities}} = \frac{n}{43,200}$$

Solve the proportion:

$$2(43,200) = 5n$$
$$86,400 = 5n$$
$$\frac{86,400}{5} = \frac{\overset{1}{\cancel{5}}n}{\cancel{5}}$$
$$17,280 = n$$

17,280 traffic fatalities involved alcohol.

Examples

Example 2

On a sunny day, a tree 6-ft tall casts a 4-ft shadow. At the same time a telephone pole casts a 10-ft shadow. What is the height of the telephone pole?

A picture illustrates the situation. Let h represent the height of the telephone pole.

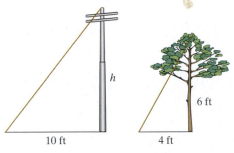

Set up a proportion:

$$\frac{10}{4} = \frac{h}{6}$$

Solve the proportion:

$$(10)(6) = 4h$$
$$60 = 4h$$
$$\frac{60}{4} = \frac{\overset{1}{\cancel{4}}h}{\cancel{4}}$$
$$15 = h \qquad \text{The pole is 15 ft high.}$$

Chapter 5 Review Exercises

Section 5.1

For Exercises 1–3, write the ratios in two other ways.

1. $5:4$ **2.** 3 to 1   **3.** $\dfrac{8}{7}$

For Exercises 4–6, write the ratios in fraction form.

4. Saul had 3 daughters and 2 sons.

    a. Write a ratio of sons to daughters.

    b. Write a ratio of daughters to sons.

    c. Write a ratio of daughters to the total number of children.

5. In his refrigerator, Jonathan has 4 bottles of soda and 5 bottles of juice.

    a. Write a ratio of bottles of soda to bottles of juice.

    b. Write a ratio of bottles of juice to bottles of soda.

    c. Write a ratio of bottles of juice to the total number of bottles.

6. There are 12 face cards in a regular deck of 52 cards.

    a. Write a ratio of face cards to total cards.

    b. Write a ratio of face cards to cards that are not face cards.

For Exercises 7–10, write the ratio in lowest terms.

7. 52 cards to 13 cards   **8.** $21 to $15

9. 80 ft to 200 ft   **10.** 7 days to 28 days

For Exercises 11–14, write the ratio in lowest terms with whole numbers in the numerator and denominator.

11. $1\frac{1}{2}$ hr to $\frac{1}{3}$ hr   **12.** $\frac{2}{3}$ yd to $2\frac{1}{6}$ yd

13. $2.56 to $1.92   **14.** 42.5 mi to 3.25 mi

15. This year a high school had an increase of 320 students. The enrollment last year was 1200 students.

    a. How many students will be attending this year?

    b. Write a ratio of the increase in students to the total enrollment of students this year. Simplify to lowest terms.

16. A living room has dimensions of 3.8 m by 2.4 m. Find the ratio of length to width and reduce to lowest terms.

For Exercises 17–18, refer to the table that shows the number of personnel who smoke in a particular workplace.

	Smokers	Nonsmokers	Totals
Office personnel	12	20	32
Shop personnel	60	55	115

17. Find the ratio of office personnel who smoke to shop personnel who smoke.

18. Find the ratio of the total number of personnel who smoke to the total number of personnel.

Section 5.2

For Exercises 19–22, write each rate in lowest terms.

19. A concession stand sold 20 hot dogs in 45 min.

20. Mike can skate 4 mi in 34 min.

21. The CN Tower in Toronto, Canada, weighs 130,000 tons for a height of approximately 1800 ft.

22. Interpol reports that Denmark has a crime rate of 9460 per 100,000 people.

23. What is the difference between rates in lowest terms and unit rates?

For Exercises 24–27, write each rate as a unit rate.

24. A pheasant can fly 44 mi in $1\frac{1}{3}$ hr.

25. The temperature dropped 14° in 3.5 hr.

26. A hummingbird can flap its wings 2700 times in 30 sec.

27. It takes Perry's lawn company 66 min to cut 6 lawns.

For Exercises 28–29, find the unit costs. Round the answers to three decimal places when necessary.

28. Body lotion costs $5.99 for 10 oz.

29. Three towels cost $10.00.

For Exercises 30–31, compute the unit cost (round to three decimal places). Then determine the best buy.

30. **a.** 48 oz of detergent for $5.99

    b. 60 oz of detergent for $7.19

    c. Which is the best buy?

31. **a.** 32 oz of spaghetti sauce for $2.49

    b. 48 oz of spaghetti sauce for $3.59

    c. Which is the best buy?

32. Suntan lotion costs $5.99 for 8 oz. If Sherri has a coupon for $2.00 off, what will be the unit cost of the lotion after the coupon has been applied?

33. A 24-roll pack of bathroom tissue costs $6.99 without a discount card. The package is advertised at 17¢ per roll if the buyer uses the discount card. What is the difference in price per roll when the buyer uses the discount card? Round to the nearest cent.

34. In Wilmington, North Carolina, Hurricane Floyd dropped 15.06 in. of rain during a 24-hr period. What was the average rainfall per hour? (*Source: National Weather Service*)

35. For a recent year, Toyota steadily increased the number of hybrid vehicles for sale in the United States from 130,000 to 250,000.

   a. What was the increase in the number of hybrid vehicles?

   b. How many additional hybrid vehicles will be available each month?

36. In 1990, Americans ate on average 386 lb of vegetables per year. By 2008, this value increased to 449 lb.

   a. What was the increase in the number of pounds of vegetables?

   b. How many additional pounds of vegetables did Americans add to their diet per year?

Section 5.3

For Exercises 37–42, write a proportion for each statement.

37. 16 is to 14 as 12 is to $10\frac{1}{2}$.

38. 8 is to 20 as 6 is to 15.

39. The numbers 5 and 3 are proportional to the numbers 10 and 6.

40. The numbers 4 and 3 are proportional to the numbers 20 and 15.

41. $11 is to 1 hr as $88 is to 8 hr.

42. 2 in. is to 5 mi as 6 in. is to 15 mi.

For Exercises 43–46, determine whether the ratios form a proportion.

43. $\dfrac{64}{81} \stackrel{?}{=} \dfrac{8}{9}$

44. $\dfrac{3\frac{1}{2}}{7} \stackrel{?}{=} \dfrac{7}{14}$

45. $\dfrac{5.2}{3} \stackrel{?}{=} \dfrac{15.6}{9}$

46. $\dfrac{6}{10} \stackrel{?}{=} \dfrac{6.3}{10.3}$

For Exercises 47–50, determine whether the pairs of numbers are proportional.

47. Are the numbers $2\frac{1}{8}$ and $4\frac{3}{4}$ proportional to the numbers $3\frac{2}{5}$ and $7\frac{3}{5}$?

48. Are the numbers $5\frac{1}{2}$ and 6 proportional to the numbers $6\frac{1}{2}$ and 7?

49. Are the numbers 4.25 and 8 proportional to the numbers 5.25 and 10?

50. Are the numbers 12.4 and 9.2 proportional to the numbers 3.1 and 2.3?

For Exercises 51–56, solve the proportion.

51. $\dfrac{100}{16} = \dfrac{25}{x}$

52. $\dfrac{y}{6} = \dfrac{45}{10}$

53. $\dfrac{1\frac{6}{7}}{b} = \dfrac{13}{21}$

54. $\dfrac{p}{6\frac{1}{3}} = \dfrac{3}{9\frac{1}{2}}$

55. $\dfrac{2.5}{6.8} = \dfrac{5}{h}$

56. $\dfrac{0.3}{1.2} = \dfrac{k}{3.6}$

Section 5.4

57. One year of a dog's life is about the same as 7 years of a human life. If a dog is 12 years old in dog years, how does that equate to human years?

58. Lavu bought 12,000 Japanese yen with $100 American. At this rate, how many yen can he buy with $450 American?

59. The number of births in Alabama in 2007 was approximately 59,800. If the birthrate was about 13 per 1000, what was the approximate population of Alabama? (Round to the nearest person.)

60. If the tax on a $25.00 item is $1.20, what would be the tax on an item costing $145.00?

61. The triangles shown in the figure are similar. Solve for x and y.

62. The height of a building can be approximated by comparing the shadows of the building and of a meter stick. From the figure, find the height of the building.

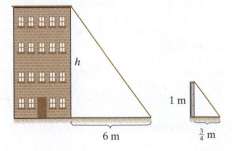

63. The polygons shown in the figure are similar. Solve for *x* and *y*.

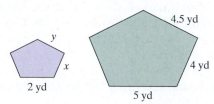

64. The figure shows two picture frames that are similar. Use this information to solve for *x* and *y*.

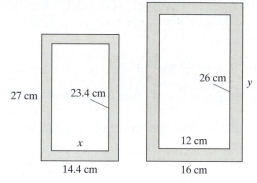

Chapter 5 Test

1. An elementary school has 25 teachers and 521 students. Write a ratio of teachers to students in three different ways.

2. In a marina, there were 17 sailboats out of a total of 23 boats.

   a. Write a ratio of sailboats to total boats.

   b. Write a ratio of sailboats to boats that are not sailboats.

For Exercises 3–4, write as a reduced ratio in fraction form.

3. For the 2006 WNBA season, the Connecticut Sun had a win-loss ratio of 26 to 8.

4. For the 2006 WNBA season, the Houston Comets had a win-loss ratio of 18 to 16.

5. Find the ratio of the shortest side to the longest side. Write the ratio in lowest terms.

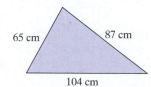

6. a. In a recent year, the number of people in New Mexico whose income was below poverty level was 168 out of every 1000. Write this as a simplified ratio.

   b. The poverty level in Iowa was 72 people to 1000 people. Write this as a simplified ratio.

   c. Compare the ratios and comment.

7. Write as a simplified ratio in two ways: 30 sec to $1\frac{1}{2}$ min

   a. By converting 30 sec to minutes.

   b. By converting $1\frac{1}{2}$ min to seconds.

For Exercises 8–10, write as a rate, simplified to lowest terms.

8. 255 mi per 6 hr

9. 20 lb in 6 weeks

10. 4 g of fat in 8 cookies

For Exercises 11–12, write as a unit rate. Round to the nearest hundredth.

11. The element platinum had density of 2145 g per 100 cm^3.

12. There are approximately 104.8 oz of iron in 45.8 lb of rocks brought back from the moon.

13. What is the unit cost for Raid Ant and Roach spray valued at $6.72 for 30 oz? Round to the nearest cent.

14. A package containing 3 toe rings is on sale for 2 packs for $3.00. What is the cost of 1 toe ring?

15. A generic pain reliever is on sale for 2 bottles for $3. Each bottle contains 30 tablets. Aleve pain reliever is sold in 24-capsule bottles for $1.92. Find the unit cost of each to determine the better buy.

16. What does it mean for two pairs of numbers to be proportional?

For Exercises 17–20, write a proportion for each statement.

17. 42 is to 15 as 28 is to 10.

18. 20 pages is to 12 min as 30 pages is to 18 min.

19. $15 an hour is proportional to $75 for 5 hr.

20. Are the numbers 105 and 55 proportional to the numbers 21 and 10?

For Exercises 21–24, solve the proportion.

21. $\dfrac{25}{p} = \dfrac{45}{63}$ **22.** $\dfrac{32}{20} = \dfrac{20}{x}$

23. $\dfrac{n}{9} = \dfrac{3\frac{1}{3}}{6}$ **24.** $\dfrac{y}{14} = \dfrac{7.2}{16.8}$

25. A computer on dial-up can download 1.6 megabytes (MB) in 2.5 min. How long will it take to download a 4.8-MB file?

26. Cherise is an excellent student and studies 7.5 hr outside of class each week for a 3-credit-hour math class. At this rate, how many hours outside of class does she spend on homework if she is taking 12 credit-hours at school?

27. Ms. Ehrlich wants to approximate the number of goldfish in her backyard pond. She scooped out 8 and marked them. Later she scooped out 10 and found that 3 were marked. Estimate the number of goldfish in her pond. Round to the nearest whole unit.

28. Given that the two triangles are similar, solve for x and y.

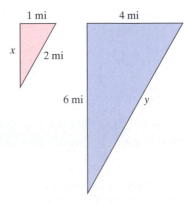

29. Maggie takes a brochure that measures 10 cm by 15 cm and enlarges it on a copy machine. If she wants the height to be 24 cm, how wide will the new image be?

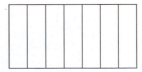

Chapters 1–5   Cumulative Review Exercises

1. Write the number 503,042 in words.

2. Estimate the sum by first rounding the numbers to the nearest hundred.

$$251 + 492 + 631$$

3. Multiply. $226 \times 100{,}000$

4. Divide and write the answer with a remainder.
$355 \div 16$

5. Divide and write the quotient in decimal form.
$355 \div 16$

6. Simplify. $2^2 \times (32 - 11) \div 14$

7. Shade the rectangle to represent the fraction $\frac{3}{7}$.

8. Simplify. $\dfrac{245}{175}$

9. Multiply. $\dfrac{13}{2} \cdot \dfrac{3}{7}$

10. Simplify. $\left(\dfrac{3}{5}\right)^2$

11. Bruce decides to share his 6-in. sub sandwich with his friend. If he gives $\frac{1}{4}$ of it to Dennis, how many inches of the sub does he have left?

12. Simplify. $\dfrac{7}{8} \div \dfrac{3}{4} + \dfrac{5}{6}$

13. Add. $\dfrac{8}{9} + 3$

14. Subtract. $\dfrac{9}{13} - \dfrac{0}{3}$

15. Emil wants to put a wallpaper border in his bathroom. The border will be on three walls (not in the shower) and not over the door. From the figure, determine how much wallpaper border will be needed.

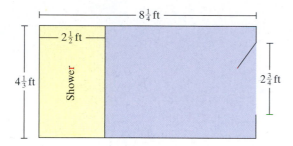

16. Tomoka Consolidated bought $82\frac{1}{4}$ acres of land. If it sold $\frac{3}{4}$ of the land, how much was sold? How much was left?

17. How many ninths are in $6\dfrac{5}{9}$?

18. Write the number 1004.701 in words.

19. Add and subtract. $23.88 + 11.3 - 7.123$

20. Write 4.36 as an improper fraction in lowest terms.

21. Multiply. 43.923×100

22. Divide. $237.9 \div 100$

23. Find the perimeter of the figure. Write the answer in decimal form.

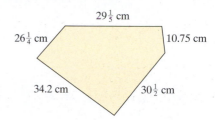

24. Americans buy 61 million newspapers each day and throw out 44 million. Write the ratio 61 to 44 in two other ways.

25. For a recent year at Southeastern Community College in North Carolina, there were 1950 students and 150 faculty. Write the student-to-faculty ratio in lowest terms.

26. In a recent study of 6000 deaths, 840 were due to cancer. Compute the ratio of deaths due to cancer to the total number of deaths studied. Simplify the ratio and interpret the answer in words.

27. Oregon has a land area of approximately 9600 mi^2. The population of Oregon is approximately 1,200,000. Compute the population density (recall that population density is a unit rate given by the number of people per square mile).

28. Determine whether the ratios form a proportion.

  a. $\dfrac{7.5}{10} \stackrel{?}{=} \dfrac{9}{12}$ **b.** $\dfrac{31}{5} \stackrel{?}{=} \dfrac{33}{6}$

29. Solve the proportion. $\dfrac{13}{11.7} = \dfrac{5}{x}$

30. Jim can drive 150 mi on 6 gal of gas. At this rate, how far can he travel on 4 gal?

Percents

6

CHAPTER OUTLINE

Chapter 6

In this chapter we present the concept of percent. Percents are used to measure the number of parts per hundred of some whole amount. As a consumer, it is important to have a working knowledge of percents.

Across

3. The amount by which a retailer increases the cost of an item to make a profit.

5. To change a percent to a fraction or a decimal, we _____ by 100.

Down

1. In a percent proportion, the _____ is the total being considered.

2. To change a fraction or a decimal to a percent, we must _____ by 100%.

4. Percent means per one _____.

Section 6.1	Percents and Their Fraction and Decimal Forms

Objectives

1. **Definition of Percent**
2. **Converting Percents to Fractions**
3. **Converting Percents to Decimals**
4. **Common Percents and Their Fraction and Decimal Forms**
5. **Applications of Percents**

1. Definition of Percent

In this chapter we study the concept of percent. Literally, the word **percent** means *per one hundred.* To indicate percent, we use the percent symbol %. For example, 45% (read as "45 percent") of the land area in South America is rainforest (shaded in green). This means that if South America were divided into 100 squares of equal size, 45 of the 100 squares would cover rainforest. See Figures 6-1 and 6-2.

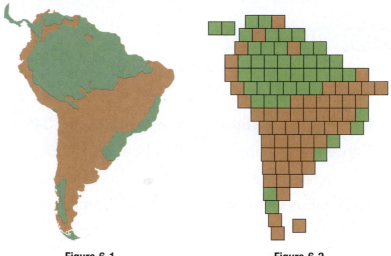

Figure 6-1	Figure 6-2

Consider another example. For a recent year, the population of Virginia could be described as follows.

21%	African American	21 out of 100 Virginians are African American
72%	Caucasian (non-Hispanic)	72 out of 100 Virginians are Caucasian (non-Hispanic)
3%	Asian American	3 out of 100 Virginians are Asian American
3%	Hispanic	3 out of 100 Virginians are Hispanic
1%	Other	1 out of 100 Virginians have other backgrounds

Figure 6-3 represents a sample of 100 residents of Virginia.

AA African American
C Caucasian (non-Hispanic)
A Asian American
H Hispanic
O Other

Figure 6-3

Concept Connections

1. Shade the portion of the figure represented by 18%.

Answer

1.
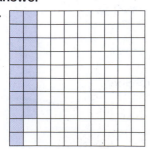

2. Converting Percents to Fractions

By definition, a percent represents a ratio of parts per 100. Therefore, we can write percents as fractions.

Percent		Fraction	Example/Interpretation
7%	=	$\dfrac{7}{100}$	A sales tax of 7% means that 7 cents in tax is charged for every 100 cents spent.
35%	=	$\dfrac{35}{100}$	To say that 35% of households own a cat means that 35 per every 100 households own a cat.

Notice that $35\% = \dfrac{35}{100} = 35 \times \dfrac{1}{100} = 35 \div 100$.

From this discussion we have the following rule for converting percents to fractions.

> **PROCEDURE Converting Percents to Fractions**
>
> **Step 1** Replace the symbol % by $\times \frac{1}{100}$ (or by $\div 100$).
> **Step 2** Simplify the fraction to lowest terms, if possible.

Example 1   Converting Percents to Fractions

Convert each percent to a fraction.

a. 56% **b.** 60% **c.** 125% **d.** 0.4%

Solution:

a. $56\% = 56 \times \dfrac{1}{100}$ Replace the % symbol by $\times \dfrac{1}{100}$.

$= \dfrac{56 \div 4}{100 \div 4}$ Multiply.

$= \dfrac{14}{25}$ Simplify to lowest terms.

b. $60\% = 60 \times \dfrac{1}{100}$ Replace the % symbol by $\times \dfrac{1}{100}$.

$= \dfrac{60 \div 20}{100 \div 20}$ Multiply.

$= \dfrac{3}{5}$ Simplify to lowest terms.

c. $125\% = 125 \times \dfrac{1}{100}$ Replace the % symbol by $\times \dfrac{1}{100}$.

$= \dfrac{125}{100}$ Multiply.

$= \dfrac{5}{4}$ or $1\dfrac{1}{4}$ Simplify to lowest terms.

Skill Practice

Convert each percent to a fraction.
2. 32% **3.** 90%
4. 175% **5.** 0.06%

Answers
2. $\dfrac{8}{25}$ **3.** $\dfrac{9}{10}$
4. $\dfrac{7}{4}$ **5.** $\dfrac{3}{5000}$

d. $0.4\% = 0.4 \times \dfrac{1}{100}$ Replace the % symbol by $\times \dfrac{1}{100}$.

$= \dfrac{4}{10} \times \dfrac{1}{100}$ Write 0.4 in fraction form.

$= \dfrac{4}{1000}$ Multiply.

$= \dfrac{1}{250}$ Simplify to lowest terms.

Concept Connections

Determine whether the percent represents a quantity greater than or less than 1 whole.

6. 1.92%
7. 19.2%
8. 192%

Note that $100\% = 100 \times \frac{1}{100} = 1$. That is, 100% represents 1 whole unit. In Example 1(c), $125\% = \frac{5}{4}$ or $1\frac{1}{4}$. This illustrates that any percent greater than 100% represents a quantity greater than 1 whole. Therefore, its fractional form may be expressed as an improper fraction or as a mixed number.

Note that $1\% = 1 \times \frac{1}{100} = \frac{1}{100}$. In Example 1(d), the value 0.4% represents a quantity less than 1%. Its fractional form is less than one-hundredth.

In Example 2 we convert some common percents to fraction form.

Skill Practice

Convert the percents to fractions.

9. 75%
10. 50%
11. $66\dfrac{2}{3}\%$

Example 2   Converting Percents to Fractions

Convert the percents to fractions.

a. 25%   **b.** 10%   **c.** $33\dfrac{1}{3}\%$

Solution:

a. $25\% = 25 \times \dfrac{1}{100} = \dfrac{25}{100} = \dfrac{1}{4}$ Thus, 25% represents one-quarter of a whole.

b. $10\% = 10 \times \dfrac{1}{100} = \dfrac{10}{100} = \dfrac{1}{10}$ Thus, 10% represents one-tenth of a whole.

c. $33\dfrac{1}{3}\% = 33\dfrac{1}{3} \times \dfrac{1}{100}$ Replace the % symbol by $\times \dfrac{1}{100}$.

$= \dfrac{100}{3} \times \dfrac{1}{100}$ Convert the mixed number to an improper fraction.

$= \dfrac{\overset{1}{\cancel{100}}}{3} \times \dfrac{1}{\underset{1}{\cancel{100}}}$ Simplify common factors.

$= \dfrac{1}{3}$ Thus, $33\dfrac{1}{3}\%$ represents one-third of a whole.

---

3. Converting Percents to Decimals

To express part of a whole unit, we can use a percent, a fraction, or a decimal. We would like to be able to convert from one form to another. The procedure for converting a percent to a decimal is the same as that for converting a

Answers

6. Less than   **7.** Less than
8. Greater than   **9.** $\dfrac{3}{4}$
10. $\dfrac{1}{2}$ **11.** $\dfrac{2}{3}$

percent to a fraction. We replace the % symbol by $\times \frac{1}{100}$. However, when converting to a decimal, it is usually more convenient to use the form $\times 0.01$.

PROCEDURE Converting Percents to Decimals

Replace the % symbol by $\times 0.01$. (This is equivalent to $\times \frac{1}{100}$ and $\div 100$.)

Note: Multiplying a decimal by 0.01 (or dividing by 100) is the same as moving the decimal point 2 places to the left.

Example 3 Converting Percents to Decimals

Convert each percent to its decimal form.

a. 31% **b.** 6.5% **c.** 428% **d.** $1\frac{3}{5}\%$ **e.** 0.05%

Solution:

a. $31\% = 31 \times 0.01$ Replace the % symbol by $\times 0.01$.

$= 0.31$ Move the decimal point 2 places to the left.

b. $6.5\% = 6.5 \times 0.01$ Replace the % symbol by $\times 0.01$.

$= 0.065$ Move the decimal point 2 places to the left.

c. $428\% = 428 \times 0.01$ Because 428% is greater than 100% we expect the decimal form to be a number greater than 1.

$= 4.28$

d. $1\frac{3}{5}\% = 1.6 \times 0.01$ Convert the mixed number to decimal form.

$= 0.016$ Because the percent is just over 1%, we expect the decimal form to be just slightly greater than one-hundredth.

e. $0.05\% = 0.05 \times 0.01$ The value 0.05% is less than 1%. We expect the decimal form to be less than one-hundredth.

$= 0.0005$

Skill Practice

Convert each percent to its decimal form.

12. 67%   **13.** 8.6%

14. 321%   **15.** $6\frac{1}{4}\%$

16. 0.7%

Notice from Examples 1, 2, and 3 that we perform the same procedure to convert a percent to either a decimal or a fraction. In each case we multiply by $\frac{1}{100}$. When converting to a decimal, it is usually easier to use the form $\times 0.01$. When converting to a fraction, it is usually easier to use the form $\times \frac{1}{100}$. In both cases, this operation is also equivalent to dividing by 100.

Answers
12. 0.67   **13.** 0.086   **14.** 3.21
15. 0.0625   **16.** 0.007

4. Common Percents and Their Fraction and Decimal Forms

Table 6-1 shows some common percents and their equivalent fraction and decimal forms.

Table 6-1

Percent	Fraction	Decimal	Example/Interpretation
100%	1	1.00	Of people who give birth, 100% are female.
50%	$\frac{1}{2}$	0.50	Of the population, 50% is male. That is, one-half of the population is male.
25%	$\frac{1}{4}$	0.25	Approximately 25% of the U.S. population smokes. That is, one-quarter of the population smokes.
75%	$\frac{3}{4}$	0.75	Approximately 75% of homes have computers. That is, three-quarters of homes have computers.
10%	$\frac{1}{10}$	0.10	Of the population, 10% is left-handed. That is, one-tenth of the population is left-handed.
1%	$\frac{1}{100}$	0.01	Approximately 1% of babies are born underweight. That is, about 1 in 100 babies is born underweight.
$33\frac{1}{3}\%$	$\frac{1}{3}$	$0.\overline{3}$	A basketball player made $33\frac{1}{3}\%$ of her shots. That is, she made about 1 basket for every 3 shots attempted.
$66\frac{2}{3}\%$	$\frac{2}{3}$	$0.\overline{6}$	Of the population, $66\frac{2}{3}\%$ prefers chocolate ice cream to other flavors. That is, 2 out of 3 people prefer chocolate ice cream.

5. Applications of Percents

The next time you pick up a newspaper or magazine, notice that percents are used in abundance to convey information.

Skill Practice

Find the decimal notation for the percent given in the sentence.

17. The U.S. unemployment rate in 2007 was 4.4%.

18. Satellite Internet subscribers increased by 220% from 2003 to 2007.

Example 4 Converting Percent Notation to Decimal Notation

Find the decimal notation for the percent given in the sentence.

a. Forty-eight percent of applicants to U.S. medical schools in 2006 were female. (*Source:* Association of American Medical Colleges)

b. The price per gallon for regular unleaded gasoline in 2007 was 200.5% of what it was in 1984.

Solution:

a. $48\% = 48 \times 0.01 = 0.48$ Just under one-half of the applicants were female.

b. $200.5\% = 200.5 \times 0.01 = 2.005$ The cost of gas in 2007 was about 2 times as great as in 1984.

Answers

17. 0.044   **18.** 2.2

Example 5   **Converting Percent Notation to Fraction Notation**

Find the fraction form for the percent given in the sentence.

a. Forty-five percent of Americans use the Internet as a resource when planning vacations. (*Source: USA TODAY*)

b. 7.2% of adults suffer from asthma. (*Source:* National Center for Health Statistics)

Solution:

a. $45\% = 45 \times \dfrac{1}{100} = \dfrac{45}{100} = \dfrac{9}{20}$ Just under one-half of Americans planning a vacation use the Internet as a resource.

b. $7.2\% = 7.2 \times \dfrac{1}{100} = \dfrac{72}{10} \times \dfrac{1}{100}$ Write 7.2 in fraction form: $\frac{72}{10}$.

$\qquad\quad = \dfrac{72}{1000} = \dfrac{9}{125}$ Almost one-tenth of adults have asthma.

Skill Practice

Find the fraction form for the percent given in the sentence.

19. In Pennsylvania, 15% of the residents are 65 or older.

20. One study found that teenage substance abuse rises by 40% during the summer months.

Answers

19. $\dfrac{3}{20}$ **20.** $\dfrac{2}{5}$

Section 6.1   Practice Exercises

Boost *your* GRADE at ALEKS.com!

ALEKS version 3.0

- Practice Problems
- Self-Tests
- NetTutor
- e-Professors
- Videos

Study Skills Exercises

1. A test is a *grading* tool for your instructor. How can you turn it into a *learning* tool for you?

2. Define the key term **percent**.

Objective 1: Definition of Percent

For Exercises 3–8, use a percent to express the shaded portion of each drawing.

3. 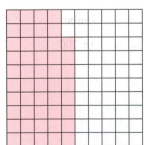   **4.** 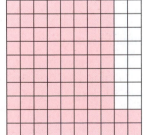   **5.**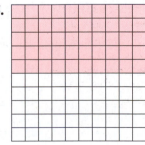

6. 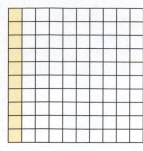   7. 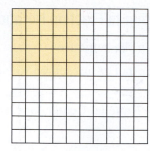   8.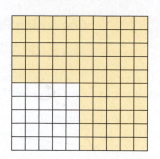

For Exercises 9–12, write a percent for each statement.

9. A bank pays $2 in interest for every $100 deposited.

10. In South Dakota, 5 out of every 100 people work in construction.

11. Out of 100 acres, 70 acres were planted with corn.

12. On TV, 26 out of every 100 minutes are filled with commercials.

Objective 2: Converting Percents to Fractions

13. Explain the procedure to change a percent to a fraction.

14. What fraction represents 50%.

For Exercises 15–34, change the percent to a simplified fraction or mixed number. **(See Examples 1–2.)**

15. 3%   16. 7%   17. 84%   18. 32%

19. 25%   20. 20%   21. 3.4%   22. 5.2%

23. 115%   24. 150%   25. 175%   26. 120%

27. 0.5%   28. 0.2%   29. 0.25%   30. 0.75%

31. $66\frac{2}{3}\%$ 32. $5\frac{1}{6}\%$ 33. $24\frac{1}{2}\%$ 34. $6\frac{1}{4}\%$

Objective 3: Converting Percents to Decimals

35. Explain the procedure to change a percent to a decimal.

For Exercises 36–51, change the percent to a decimal. **(See Example 3.)**

36. 58%   37. 72%   38. 15%   39. 66%

40. 8.5%   41. 12.9%   42. 72.31%   43. 41.05%

44. 142%   45. 201%   46. 0.55%   47. 0.75%

48. $26\frac{2}{5}\%$ 49. $16\frac{1}{4}\%$ 50. $55\frac{1}{20}\%$ 51. $62\frac{1}{5}\%$

Objective 4: Common Percents and Their Fraction and Decimal Forms

For Exercises 52–57, use a percent to express the shaded portion of each drawing.

52.

53.

54.

55.

56.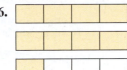

57.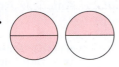

Mixed Exercises

For Exercises 58–63, match the percent with its fraction form.

58. $66\frac{2}{3}\%$ **a.** $\frac{3}{2}$

59. 10% **b.** $\frac{3}{4}$

60. 90% **c.** $\frac{2}{3}$

61. 75% **d.** $\frac{1}{10}$

62. 25% **e.** $\frac{9}{10}$

63. 150% **f.** $\frac{1}{4}$

For Exercises 64–69, match the percent with its decimal form.

64. 30% **a.** 0.01

65. $33\frac{1}{3}\%$ **b.** 0.50

66. 125% **c.** 0.80

67. 50% **d.** $0.\overline{3}$

68. 1% **e.** 0.30

69. 80% **f.** 1.25

70. In which direction do you move the decimal point when you convert a percent to a decimal? By how many places?

Objective 5: Applications of Percents

For Exercises 71–78, find the decimal and fraction equivalent of the percent given in the sentence. (See Examples 4–5.)

71. Between 1990 and 2006 the population in California grew by 22.5%.

72. Las Vegas is considered the fastest-growing city in the United States. Between 1990 and 2005 its population increased by 110%.

73. For a recent year, the unemployment rate in Kansas was 4.3%.

74. For a recent year, the unemployment rate in the United States was 5.8%.

75. From 2003 to 2006, the electricity generated by nuclear power in Illinois decreased by 0.6%. (*Source:* Energy Information Administration)

76. For a recent year the average U.S. income tax rate was 18.2%.

77. Thirty-five percent of Americans say they entertain at home once or twice a year. (*Source: USA TODAY*)

78. Twenty-nine percent of Americans say they entertain at home once a month.

 **79.** The graph represents the percent of dog owners who participate in certain activities to treat their dogs. Write the decimal and fraction forms of the percents given in the graph. (*Source:* American Animal Hospital Association)

Percent of Dog Owners Who Treat Their Dogs

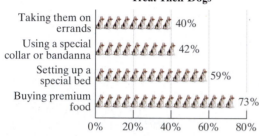

Taking them on errands	40%
Using a special collar or bandanna	42%
Setting up a special bed	59%
Buying premium food	73%

0% 20% 40% 60% 80%

80. The graph represents the percent of people with at least a bachelor's degree for selected large cities. Write the decimal and fraction forms of the percents given in the graph. (*Source:* U.S. Census Bureau)

Large Cities with the Highest Percent of Residents with at Least a Bachelor's Degree

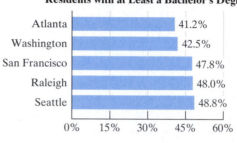

Atlanta	41.2%
Washington	42.5%
San Francisco	47.8%
Raleigh	48.0%
Seattle	48.8%

0% 15% 30% 45% 60%

Section 6.2 Fractions and Decimals and Their Percent Forms

Objectives

1. Converting Fractions and Decimals to Percents
2. Approximating Percents
3. Fractions, Decimals, Percents: A Summary

1. Converting Fractions and Decimals to Percents

In Section 6.1, we converted percents to their equivalent fraction and decimal forms. This is done by replacing the % symbol by $\times \frac{1}{100}$. In this section, we reverse the process. We convert fractions and decimals to percents by multiplying by 100 and applying the % symbol.

> **PROCEDURE** Converting Fractions and Decimals to Percent Form
>
> Multiply the fraction or decimal by 100%.
>
> *Note:* Multiplying a decimal by 100 moves the decimal point 2 places to the right.

Example 1 **Converting Decimals to Percents**

Convert each decimal to its equivalent percent form.

 a. 0.62 **b.** 1.75 **c.** 1 **d.** 0.004 **e.** 8.9

Solution:

a. $0.62 = 0.62 \times 100\%$ Multiply by 100%.

  $= 62\%$ Multiplying by 100 moves the decimal point 2 places to the right.

b. $1.75 = 1.75 \times 100\%$ Multiply by 100%.

  $= 175\%$ The decimal number 1.75 is greater than 1. Therefore, we expect a percent greater than 100%.

c. $1 = 1 \times 100\%$ Multiply by 100%.

  $= 100\%$ Recall that 1 whole is equal to 100%.

d. $0.004 = 0.004 \times 100\%$ Multiply by 100%.

  $= 0.4\%$ Move the decimal point to the right 2 places.

e. $8.9 = 8.90 \times 100\%$ Multiply by 100%.

  $= 890\%$

Skill Practice

Convert each decimal to its percent form.
1. 0.46
2. 3.25
3. 2
4. 0.0006
5. 2.5

TIP: Multiplying a number by 100% is equivalent to multiplying the number by 1. Thus, the value of the number is not changed.

Example 2 **Converting a Fraction to Percent Notation**

Convert the fraction to percent notation.

$$\frac{3}{5}$$

Solution:

$\dfrac{3}{5} = \dfrac{3}{5} \times 100\%$ Multiply by 100%.

$= \dfrac{3}{5} \times \dfrac{100}{1}\%$ Convert the whole number to an improper fraction.

$= \dfrac{3}{\overset{}{\underset{1}{5}}} \times \dfrac{\overset{20}{100}}{1}\%$ Multiply fractions and simplify to lowest terms.

$= 60\%$

TIP: We could also have converted $\frac{3}{5}$ to decimal form first (by dividing the numerator by the denominator) and then converted the decimal to a percent.

   convert to decimal   convert to percent

   $\frac{3}{5} = 0.60$ $=$ $0.60 \times 100\% = 60\%$

Skill Practice

Convert the fraction to percent notation.
6. $\dfrac{7}{10}$

Answers
1. 46% 2. 325% 3. 200%
4. 0.06% 5. 250% 6. 70%

Example 3 **Converting a Fraction to Percent Notation**

Convert the fraction to percent notation.

$$\dfrac{2}{3}$$

Solution:

$$\dfrac{2}{3} = \dfrac{2}{3} \times 100\% \qquad \text{Multiply by 100\%.}$$

$$= \dfrac{2}{3} \times \dfrac{100}{1}\% \qquad \text{Convert the whole number to an improper fraction.}$$

$$= \dfrac{200}{3}\%$$

The number $\dfrac{200}{3}\%$ can be written as $66\frac{2}{3}\%$ or as $66.\overline{6}\%$.

> **TIP:** First converting $\frac{2}{3}$ to a decimal before converting to percent notation is an alternative approach.
>
> $$\underset{\text{convert to decimal}}{\dfrac{2}{3} = 0.\overline{6}} \qquad = \qquad \underset{\text{convert to percent}}{0.666\ldots \times 100\% = 66.\overline{6}\%}$$

In Example 4, we convert an improper fraction and a mixed number to percent form.

Example 4 **Converting Improper Fractions and Mixed Numbers to Percents**

Convert to percent notation.

a. $2\dfrac{1}{4}$ **b.** $\dfrac{13}{10}$

Solution:

a. $2\dfrac{1}{4} = 2\dfrac{1}{4} \times 100\% \qquad \text{Multiply by 100\%.}$

$$= \dfrac{9}{4} \times \dfrac{100}{1}\% \qquad \text{Convert to improper fractions.}$$

$$= \dfrac{9}{\overset{}{\underset{1}{4}}} \times \dfrac{\overset{25}{\cancel{100}}}{1}\% \qquad \text{Multiply and simplify to lowest terms.}$$

$$= 225\%$$

b. $\dfrac{13}{10} = \dfrac{13}{10} \times 100\% \qquad \text{Multiply by 100\%.}$

$$= \dfrac{13}{10} \times \dfrac{100}{1}\% \qquad \text{Convert the whole number to an improper fraction.}$$

$$= \dfrac{13}{\underset{1}{\cancel{10}}} \times \dfrac{\overset{10}{\cancel{100}}}{1}\% \qquad \text{Multiply and simplify to lowest terms.}$$

$$= 130\%$$

Notice that both answers in Example 4 are greater than 100%. This is reasonable because any number greater than 1 whole unit represents a percent greater than 100%.

2. Approximating Percents

In Example 5 we approximate a percent from its fraction form.

Example 5	Approximating a Percent

Convert the fraction $\frac{5}{13}$ to percent notation rounded to the nearest tenth of a percent.

Solution:

$$\frac{5}{13} = \frac{5}{13} \times 100\% \qquad \text{Multiply by 100\%.}$$

$$= \frac{5}{13} \times \frac{100}{1}\% \qquad \text{Write the whole number as an improper fraction.}$$

$$= \frac{500}{13}\%$$

To round to the nearest tenth of a percent, we must divide. We will obtain the hundredths-place digit in the quotient on which to base the decision on rounding.

$$38.4\overset{1}{6} \approx 38.5$$

Thus, $\frac{5}{13} \approx 38.5\%$.

```
         38.46
   13)500.00
      -39
       110
      -104
        60
       -52
        80
       -78
         2
```

3. Fractions, Decimals, Percents: A Summary

The diagram in Figure 6-4 illustrates the methods for converting fractions, decimals, and percents.

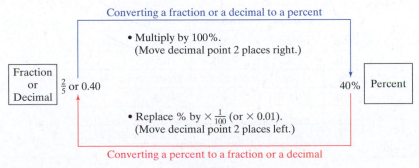

Converting a fraction or a decimal to a percent

- Multiply by 100%.
  (Move decimal point 2 places right.)

Fraction or Decimal $\frac{2}{5}$ or 0.40   40%   Percent

- Replace % by $\times \frac{1}{100}$ (or $\times 0.01$).
  (Move decimal point 2 places left.)

Converting a percent to a fraction or a decimal

Figure 6-4

Example 6   **Converting Fractions, Decimals, and Percents**

Complete the table.

	Fraction	Decimal	Percent
a.		0.55	
b.	$\frac{1}{200}$		
c.			160%
d.		2.4	
e.			$66\frac{2}{3}\%$
f.	$\frac{2}{9}$		

Solution:

a. 0.55 to fraction: $0.55 = \dfrac{55}{100} = \dfrac{11}{20}$

0.55 to percent: $0.55 \times 100\% = 55\%$

b. $\dfrac{1}{200}$ to decimal: $1 \div 200 = 0.005$

$\dfrac{1}{200}$ to percent: $\dfrac{1}{200} \times 100\% = \dfrac{100}{200}\% = 0.5\%$

c. 160% to fraction: $160 \times \dfrac{1}{100} = \dfrac{160}{100} = \dfrac{8}{5}$ or $1\dfrac{3}{5}$

160% to decimal: $160 \times 0.01 = 1.6$

d. 2.4 to fraction: $\dfrac{24}{10} = \dfrac{12}{5}$ or $2\dfrac{2}{5}$

2.4 to percent: $2.4 \times 100\% = 240\%$

e. $66\dfrac{2}{3}\%$ to fraction: $66\dfrac{2}{3} \times \dfrac{1}{100} = \dfrac{\overset{2}{\cancel{200}}}{3} \times \dfrac{1}{\underset{1}{\cancel{100}}} = \dfrac{2}{3}$

$66\dfrac{2}{3}\%$ to decimal: $66\dfrac{2}{3} \times 0.01 = 66.\overline{6} \times 0.01 = 0.\overline{6}$

f. $\dfrac{2}{9}$ to decimal: $2 \div 9 = 0.\overline{2}$

$\dfrac{2}{9}$ to percent: $\dfrac{2}{9} \times 100\% = \dfrac{2}{9} \times \dfrac{100}{1}\% = \dfrac{200}{9}\% = 22\dfrac{2}{9}\%$ or $22.\overline{2}\%$

The completed table is as follows.

Answers

	Fraction	Decimal	Percent
13.	$\frac{141}{100}$ or $1\frac{41}{100}$	1.41	141%
14.	$\frac{1}{50}$	0.02	2%
15.	$\frac{9}{50}$	0.18	18%
16.	$\frac{29}{50}$	0.58	58%
17.	$\frac{1}{3}$	$0.\overline{3}$	$33\frac{1}{3}\%$
18.	$\frac{7}{9}$	$0.\overline{7}$	$77.\overline{7}\%$

	Fraction	Decimal	Percent
a.	$\frac{11}{20}$	0.55	55%
b.	$\frac{1}{200}$	0.005	0.5%
c.	$\frac{8}{5}$ or $1\frac{3}{5}$	1.6	160%
d.	$\frac{12}{5}$ or $2\frac{2}{5}$	2.4	240%
e.	$\frac{2}{3}$	$0.\overline{6}$	$66\frac{2}{3}\%$
f.	$\frac{2}{9}$	$0.\overline{2}$	$22\frac{2}{9}\%$ or $22.\overline{2}\%$

Section 6.2 Practice Exercises

Study Skills Exercise

1. Do you remember your instructor's name, office hours, office location, and office phone? Write them here:

Instructor's name: Instructor's office hours:

Instructor's office location: Instructor's office phone:

Review Exercises

For Exercises 2–5, convert the percent to a fraction or mixed number.

2. 60%   **3.** 130%   **4.** $16\frac{1}{2}\%$ **5.** 0.5%

For Exercises 6–9, convert the percent to a decimal.

6. 80%   **7.** $6\frac{1}{3}\%$ **8.** 143%   **9.** 0.3%

Objective 1: Converting Fractions and Decimals to Percents

For Exercises 10–13, multiply.

10. $0.68 \times 100\%$ **11.** $1.62 \times 100\%$ **12.** $0.005 \times 100\%$ **13.** $0.26 \times 100\%$

14. Write the rule for multiplying a decimal by 100.

For Exercises 15–18, multiply.

15. $\dfrac{5}{4} \times 100\%$ **16.** $\dfrac{2}{5} \times 100\%$ **17.** $\dfrac{77}{100} \times 100\%$ **18.** $\dfrac{113}{100} \times 100\%$

For Exercises 19–30, convert the decimal to a percent. **(See Example 1.)**

19. 0.27   **20.** 0.51   **21.** 0.19   **22.** 0.33

23. 1.75   **24.** 2.8   **25.** 0.124   **26.** 0.277

27. 0.006   **28.** 0.0008   **29.** 1.014   **30.** 2.203

For Exercises 31–42, convert the fraction to a percent. **(See Examples 2–3.)**

31. $\dfrac{71}{100}$ **32.** $\dfrac{89}{100}$ **33.** $\dfrac{19}{20}$ **34.** $\dfrac{7}{20}$

35. $\dfrac{7}{8}$ **36.** $\dfrac{5}{8}$ **37.** $\dfrac{13}{16}$ **38.** $\dfrac{11}{16}$

39. $\dfrac{5}{6}$ **40.** $\dfrac{5}{12}$ **41.** $\dfrac{4}{9}$ **42.** $\dfrac{1}{9}$

For Exercises 43–48, write the fraction as a percent.

43. One-quarter of Americans say they entertain at home 2 or more times a month. (*Source: USA TODAY*)

44. According to the Centers for Disease Control (CDC), $\frac{37}{100}$ of U.S. teenage boys say they rarely or never wear their seatbelts.

45. According to the Centers for Disease Control, $\frac{1}{10}$ of teenage girls in the United States say they rarely or never wear their seatbelts.

46. In Italy, $\frac{3}{50}$ of the country's budget comes from tourism.

47. In a recent year, $\frac{2}{3}$ of the beds in U.S. hospitals were occupied.

48. In Georgia in 2005, almost $\frac{1}{5}$ of the residents were not covered by health insurance. (*Source:* U.S. Bureau of the Census)

For Exercises 49–56, convert to percent notation. **(See Example 4.)**

49. $1\frac{3}{4}$ **50.** $\frac{7}{2}$ **51.** $\frac{27}{20}$ **52.** $2\frac{1}{8}$

53. $\frac{11}{9}$ **54.** $1\frac{5}{9}$ **55.** $1\frac{2}{3}$ **56.** $\frac{7}{6}$

Objective 2: Approximating Percents

For Exercises 57–64, write the fraction in percent notation to the nearest tenth of a percent. **(See Example 5.)**

57. $\frac{3}{7}$ **58.** $\frac{6}{7}$ **59.** $\frac{1}{13}$ **60.** $\frac{3}{13}$

61. $\frac{5}{11}$ **62.** $\frac{8}{11}$ **63.** $\frac{13}{15}$ **64.** $\frac{1}{15}$

Objective 3: Fractions, Decimals, Percents: A Summary

65. Explain the difference between $\frac{1}{2}$ and $\frac{1}{2}\%$.

66. Explain the difference between $\frac{3}{4}$ and $\frac{3}{4}\%$.

67. Explain the difference between 25% and 0.25%.

68. Explain the difference between 10% and 0.10%.

69. Which of the numbers represent 125%?

    a. 1.25   **b.** 0.125   **c.** $\frac{5}{4}$ **d.** $\frac{5}{4}\%$

70. Which of the numbers represent 60%?

    a. 6.0   **b.** 0.60%   **c.** 0.6   **d.** $\frac{3}{5}$

71. Which of the numbers represent 30%?

    a. $\frac{3}{10}$ **b.** $\frac{1}{3}$ **c.** 0.3   **d.** 0.03%

72. Which of the numbers represent 180%?

    a. 18   **b.** 1.8   **c.** $\dfrac{9}{5}$ **d.** $\dfrac{9}{5}\%$

Mixed Exercises

For Exercises 73–76, complete the table. **(See Example 6.)**

73.

	Fraction	Decimal	Percent
a.	$\frac{1}{4}$		
b.		0.92	
c.			15%
d.		1.6	
e.	$\frac{1}{100}$		
f.			0.5%

74.

	Fraction	Decimal	Percent
a.			0.6%
b.	$\frac{2}{5}$		
c.		2	
d.	$\frac{1}{2}$		
e.		0.12	
f.			45%

75.

	Fraction	Decimal	Percent
a.			14%
b.		0.87	
c.		1	
d.	$\frac{1}{3}$		
e.			0.2%
f.	$\frac{19}{20}$		

76.

	Fraction	Decimal	Percent
a.		1.3	
b.			22%
c.	$\frac{3}{4}$		
d.		0.73	
e.			$22.\overline{2}\%$
f.	$\frac{1}{20}$		

Expanding Your Skills

77. Is the number 1.4 less than or greater than 100%?

78. Is the number 0.0087 less than or greater than 1%?

79. Is the number 0.052 less than or greater than 50%?

80. Is the number 25 less than or greater than 25%?

Percent Proportions and Applications

1. Introduction to Percent Proportions

Recall that a percent is a ratio in parts per 100. For example, $50\% = \frac{50}{100}$. However, a percent can be represented by infinitely many equivalent fractions. Thus,

$$50\% = \frac{50}{100} = \frac{1}{2} = \frac{2}{4} = \frac{3}{6} \quad \text{and infinitely many more.}$$

Equating a percent to an equivalent ratio forms a proportion that we call a **percent proportion**. A percent proportion is a proportion in which one ratio is written with a denominator of 100. For example:

$$\frac{50}{100} = \frac{3}{6} \quad \text{is a percent proportion.}$$

Objectives

1. Introduction to Percent Proportions
2. Identifying the Parts of a Percent Proportion
3. Solving Percent Proportions
4. Applications of Percent Proportions

2. Identifying the Parts of a Percent Proportion

We will be using percent proportions to solve a variety of application problems. But first we need to identify and label the parts of a percent proportion.

A percent proportion can be written in the form:

$$\frac{\text{Amount}}{\text{Base}} = p\% \qquad \text{or} \qquad \frac{\text{Amount}}{\text{Base}} = \frac{p}{100}$$

For example:

4 L out of 8 L is 50%
   | | |
amount   base   p

$$\frac{4}{8} = 50\% \qquad \text{or} \qquad \frac{4}{8} = \frac{50}{100}$$

In this example, 8 L is some total (or base) quantity and 4 L is some part (or amount) of that whole. The ratio $\frac{4}{8}$ represents a fraction of the whole equal to 50%. In general, we offer the following guidelines for identifying the parts of a percent proportion.

PROCEDURE Identifying the Parts of a Percent Proportion

A percent proportion can be written as

$$\frac{\text{Amount}}{\text{Base}} = p\% \qquad \text{or} \qquad \frac{\text{Amount}}{\text{Base}} = \frac{p}{100}.$$

- The **base** is the total or whole amount being considered. It often appears after the word *of* within a word problem.
- The **amount** is the part being compared to the base. It sometimes appears with the word *is* within a word problem.

Example 1 Identifying Amount, Base, and *p* for a Percent Proportion

Identify the amount, base, and *p* value, and then set up a percent proportion.

a. 25% of 60 students is 15 students.   **b.** $32 is 50% of $64.

c. 5 of 1000 employees is 0.5%.

Solution:

For each problem, we recommend that you identify *p* first. It is the number in front of the symbol %. Then identify the base. In most cases it follows the word *of*. Then, by the process of elimination, find the amount.

a. 25% of 60 students is 15 students.
      | | |
      p   base   amount
   (before % (after the
    symbol)   word *of*)

$$\text{amount} \rightarrow \frac{15}{60} = \frac{25}{100} \begin{matrix} \leftarrow p \\ \leftarrow 100 \end{matrix}$$
$$\text{base} \rightarrow$$

b. $32 is 50% of $64.
      | | |
   amount   p   base

$$\text{amount} \rightarrow \frac{32}{64} = \frac{50}{100} \begin{matrix} \leftarrow p \\ \leftarrow 100 \end{matrix}$$
$$\text{base} \rightarrow$$

c. 5 of 1000 employees is 0.5%.
     | | |
  amount base   p

$$\text{amount} \rightarrow \frac{5}{1000} = \frac{0.5}{100} \begin{matrix} \leftarrow p \\ \leftarrow 100 \end{matrix}$$
$$\text{base} \rightarrow$$

3. Solving Percent Proportions

In Example 1, we practiced identifying the parts of a percent proportion. Now we consider percent proportions in which one of these numbers is unknown. Furthermore, we will see that the examples come in three types:

- Amount is unknown.
- Base is unknown.
- Value p is unknown.

However, the process for solving in each case is the same.

Example 2	Solving Percent Proportions—Amount Unknown

a. What is 30% of 180?   **b.** 70% of 500 people is how many people?

Solution:

a. What is 30% of 180?   The base and value for p are known.

amount (x) p base

Let x represent the unknown amount.

$$\frac{x}{180} = \frac{30}{100}$$ Set up a percent proportion.

$100 \cdot x = (30)(180)$ Equate the cross products.

$100x = 5400$

$$\frac{\overset{1}{\cancel{100}}x}{\cancel{100}} = \frac{54\cancel{00}}{\cancel{100}}$$ Divide both sides of the equation by 100.

$x = 54$ Simplify to lowest terms.

Therefore, 54 is 30% of 180.

> **TIP:** We can check the answer to Example 2(a) as follows. Ten percent of a number is $\frac{1}{10}$ of the number. Furthermore, $\frac{1}{10}$ of 180 is 18. Thirty percent of 180 must be 3 times this amount.
>
> $3 \times 18 = 54$ ✔

b. 70% of 500 people is how many people?

p base   amount (x)

The base and value for p are known.

Let x represent the unknown amount.

$$\frac{x}{500} = \frac{70}{100}$$ Set up a percent proportion.

$100 \cdot x = (70)(500)$ Equate the cross products.

$100x = 35{,}000$

$$\frac{\overset{1}{\cancel{100}}x}{\cancel{100}} = \frac{35{,}0\cancel{00}}{\cancel{100}}$$ Divide both sides by 100.

$x = 350$ Simplify to lowest terms.

Therefore, 70% of 500 is 350.

> **TIP:** To check the solution to Example 2(b), we can compute 10% of 500, which is 50. To find 70%, multiply this by 7. We have $7(50) = 350$. ✔

Example 3 **Solving Percent Proportions—Base Unknown**

a. 40% of what number is 25?   **b.** $13.50 is 150% of how many dollars?

Solution:

a. 40% of what number is 25?   The amount and value of p are known.

p base (x) amount   Let x represent the unknown base.

$$\frac{25}{x} = \frac{40}{100}$$ Set up a percent proportion.

$(25)(100) = 40 \cdot x$ Equate the cross products.

$2500 = 40x$

$$\frac{2500}{40} = \frac{\overset{1}{\cancel{40}}x}{\cancel{40}}$$ Divide both sides by 40.

$62.5 = x$ Therefore, 40% of 62.5 is 25.

b. $13.50 is 150% of how many dollars?   The amount and value of p are known.

amount p base (x) Let x represent the unknown base.

$$\frac{13.50}{x} = \frac{150}{100}$$ Set up a percent proportion.

$150 \cdot x = (13.50)(100)$ Equate the cross products.

$150x = 1350$

$$\frac{\overset{1}{\cancel{150}}x}{\cancel{150}} = \frac{1350}{150}$$ Divide both sides by 150.

$x = 9$ Therefore, $13.50 is 150% of $9.

Example 4 **Solving Percent Proportions—p Unknown**

a. What percent of 80 mi is 12.4 mi?

b. 48 is what percent of 42? Round to the nearest percent.

Solution:

a. What percent of 80 mi is 12.4 mi?   The amount and base are known.

p base   amount   The value of p is unknown.

$$\frac{12.4}{80} = \frac{p}{100}$$ Set up a percent equation.

$(12.4)(100) = 80 \cdot p$ Equate the cross products.

$1240 = 80p$

$$\frac{1240}{80} = \frac{\overset{1}{\cancel{80}}p}{\cancel{80}}$$ Divide both sides by 80.

$15.5 = p$ Therefore, 15.5% of 80 mi is 12.4 mi.

b. 48 is what percent of 42? Round to the nearest percent.

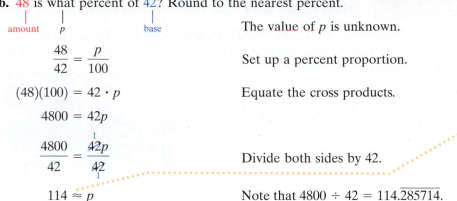

amount p base

$\dfrac{48}{42} = \dfrac{p}{100}$	Set up a percent proportion.
$(48)(100) = 42 \cdot p$	Equate the cross products.
$4800 = 42p$	
$\dfrac{4800}{42} = \dfrac{\overset{1}{\cancel{42}}p}{\underset{1}{\cancel{42}}}$	Divide both sides by 42.
$114 \approx p$	Note that $4800 \div 42 = 114.\overline{285714}$. Rounded to the nearest whole number, $p \approx 114$.

Therefore, 48 is approximately 114% of 42.

4. Applications of Percent Proportions

We now use percent proportions to solve application problems involving percents.

Example 5	Using Percents in Meteorology

Buffalo, New York, receives an average of 94 in. of snow each year. This year it had 120% of the normal annual snowfall. How much snow did Buffalo get this year?

Solution:

This situation can be translated as:

"The amount of snow Buffalo received is 120% of 94 in."

amount (x) p base

$\dfrac{x}{94} = \dfrac{120}{100}$	Set up a percent proportion.
$100 \cdot x = (120)(94)$	Equate the cross products.
$100x = 11{,}280$	
$\dfrac{\overset{1}{\cancel{100}}x}{\underset{1}{\cancel{100}}} = \dfrac{11{,}280}{100}$	Divide both sides by 100.
$x = 112.8$	

This year, Buffalo had 112.8 in. of snow.

Skill Practice

10. In a recent year it was estimated that 24.7% of U.S. adults smoked tobacco products regularly. In a group of 2000 adults, how many would be expected to be smokers?

TIP: In a word problem, it is always helpful to check the reasonableness of your answer. In Example 5, we are looking for 120% of 94 in. But 120% must be *more* than the base amount of 94 in. Therefore, we suspect that our solution is reasonable.

Answer

10. 494 people

Example 6 Using Percents in Statistics

Harvard University's freshman class for 2006 had 18% Asian American students. If this represented 380 students, how many students were admitted to the freshman class? Round to the nearest student.

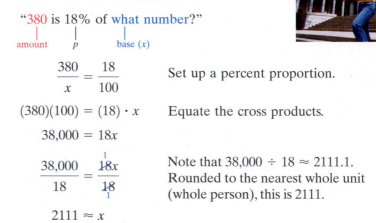

Solution:

This situation can be translated as:

"380 is 18% of what number?"

amount p base (x)

$$\frac{380}{x} = \frac{18}{100}$$ Set up a percent proportion.

$$(380)(100) = (18) \cdot x$$ Equate the cross products.

$$38{,}000 = 18x$$

$$\frac{38{,}000}{18} = \frac{18x}{18}$$ Note that $38{,}000 \div 18 \approx 2111.1$. Rounded to the nearest whole unit (whole person), this is 2111.

$$2111 \approx x$$

The freshman class at Harvard in 2006 had approximately 2111 students.

TIP: We can check the answer to Example 6 by substituting $x = 2111$ back into the original proportion. The cross products will not be exactly the same because we had to round the value of x. However, the cross products should be *close*.

$$\frac{380}{2111} \stackrel{?}{=} \frac{18}{100}$$ Substitute $x = 2111$ into the proportion.

$$(380)(100) \stackrel{?}{=} (18)(2111)$$

$$38{,}000 \approx 37{,}998 \quad ✔ \quad \text{The values are very close.}$$

Example 7 Using Percents in Business

Suppose a tennis pro who is ranked 90th in the world on the men's professional tour earns $280,000 per year in tournament winnings and endorsements. He pays his coach $100,000 per year. What percent of his income goes toward his coach? Round to the nearest tenth of a percent.

Answers
11. 42 students are in the class.
12. Of the donations, approximately 14.1% are in the $100–$199 range.

Solution:

This can be translated as:

"What *percent* of $280,000 is $100,000?"

The value of p is unknown.   base (x) amount

$$\frac{100{,}000}{280{,}000} = \frac{p}{100}$$ Set up a percent proportion.

$$\frac{100{,}000}{280{,}000} = \frac{p}{100}$$ The ratio on the left side of the equation can be simplified by a factor of 10,000. "Strike through" four zeros in the numerator and denominator.

$$\frac{10}{28} = \frac{p}{100}$$

$$(10)(100) = (28) \cdot p$$ Equate the cross products.

$$1000 = 28p$$

$$\frac{1000}{28} = \frac{\overset{1}{28}p}{\underset{1}{28}}$$ Divide both sides by 28.

$$\frac{1000}{28} = p$$

$$35.7 \approx p$$ Dividing $1000 \div 28$, we get approximately 35.7.

The tennis pro spends about 35.7% of his income on his coach.

$$
\begin{array}{r}
35.71 \\
28\overline{)1000.00} \\
-84 \\
\hline
160 \\
-140 \\
\hline
200 \\
-196 \\
\hline
40 \\
-28 \\
\hline
12
\end{array}
$$

Section 6.3   Practice Exercises

Boost your GRADE at ALEKS.com!

- Practice Problems
- Self-Tests
- NetTutor
- e-Professors
- Videos

Study Skills Exercises

1. **a.** Do you believe that you have math anxiety? If yes, why do you think so?

   b. Of the list below, circle the activities that you think can help someone with math anxiety.

   Deep breathing   Reading a book about math anxiety

   Scheduling extra study time   Keeping a positive attitude

2. Define the key terms.

   a. Percent proportion   **b. Base**   **c. Amount**

Review Exercises

For Exercises 3–5, convert the decimal to a percent.

3. 0.55   **4.** 1.30   **5.** 0.0006

For Exercises 6–8, convert the fraction to a percent.

6. $\dfrac{3}{8}$ **7.** $\dfrac{5}{2}$ **8.** $\dfrac{1}{100}$

For Exercises 9–11, convert the percent to a fraction.

9. $62\frac{1}{2}\%$ **10.** 2%   **11.** 77%

For Exercises 12–14, convert the percent to a decimal.

12. 82%   **13.** 0.3%   **14.** 100%

Objective 1: Introduction to Percent Proportions

For Example 15–19, determine if the proportion is a percent proportion.

15. $\dfrac{7}{100} = \dfrac{14}{200}$ **16.** $\dfrac{150}{300} = \dfrac{50}{100}$ **17.** $\dfrac{2}{3} = \dfrac{6}{9}$

18. $\dfrac{1\frac{1}{2}}{100} = \dfrac{3}{200}$ **19.** $\dfrac{\frac{3}{4}}{100} = \dfrac{3}{400}$

For Exercises 20–24, shade the figure to estimate the amount. The first exercise is given as an example.

Example: Find 60% of 80. (Answer: 48)

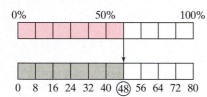

20. Find 40% of 60.

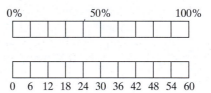

21. Find 75% of 60.

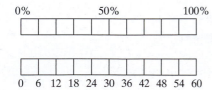

22. Find 15% of 240.

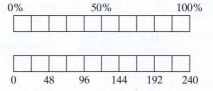

23. Find 80% of 40.

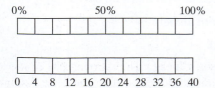

24. Find 10% of 30.

Objective 2: Identifying the Parts of a Percent Proportion

For Exercises 25–30, identify the amount, base, and *p* value. **(See Example 1.)**

25. 12 balloons is 60% of 20 balloons.

26. 25% of 400 cars is 100 cars.

27. $99 of $200 is 49.5%.

28. 45 of 50 children is 90%.

29. 50 hr is 125% of 40 hr.

30. 175% of 2 in. of rainfall is 3.5 in.

For Exercises 31–36, write the percent proportion.

31. 10% of 120 trees is 12 trees.

32. 15% of 20 pictures is 3 pictures.

33. 72 children is 80% of 90 children.

34. 21 dogs is 20% of 105 dogs.

35. 21,684 college students is 104% of 20,850 college students.

36. 103% of $40,000 is $41,200.

Objective 3: Solving Percent Proportions

For Exercises 37–46, solve the percent problems with an unknown amount. **(See Example 2.)**

37. Compute 54% of 200 employees.

38. Find 35% of 412.

39. What is $\frac{1}{2}$% of 40?

40. What is 1.8% of 900 grams?

41. Find 112% of 500.

42. Compute 106% of 1050.

43. Pedro pays 28% of his salary in income tax. If he makes $72,000 in taxable income, how much income tax does he pay?

44. A car dealer sets the sticker price of a car by taking 115% of the wholesale price. If a car sells wholesale at $17,000, what is the sticker price?

45. Jesse Ventura became the 38th governor of Minnesota by receiving 37% of the votes. If approximately 2,060,000 votes were cast, how many did Mr. Ventura get?

46. In a psychology class, 61.9% of the class consists of freshmen. If there are 42 students, how many are freshmen? Round to the nearest whole unit.

For Exercises 47–56, solve the percent problems with an unknown base. **(See Example 3.)**

47. 18 is 50% of what number?

48. 22% of what length is 44 ft?

49. 30% of what weight is 69 lb?

50. 70% of what number is 28?

51. 9 is $\frac{2}{3}$% of what number?

52. 9.5 is 200% of what number?

53. Albert saves $120 per month. If this is 7.5% of his monthly income, how much does he make per month?

54. Janie and Don left their house in South Bend, Indiana, to visit friends in Chicago. They drove 80% of the distance before stopping for lunch. If they had driven 56 mi before lunch, what would be the total distance from their house to their friends' house in Chicago?

55. Amiee read 14 e-mails, which was only 40% of her total e-mails. What is her total number of e-mails?

56. A recent survey found that 5% of the population of the United States is unemployed. If Charlotte, North Carolina, has 32,000 unemployed, what is the population of Charlotte?

For Exercises 57–64, solve the percent problems with *p* unknown. **(See Example 4.)**

57. What percent of $120 is $42?

58. 112 is what percent of 400?

59. 84 is what percent of 70?

60. What percent of 12 letters is 4 letters?

61. What percent of 320 mi is 280 mi?

62. 54¢ is what percent of 48¢?

63. A student answered 29 problems correctly on a final exam of 40 problems. What percent of the questions did she answer correctly?

64. During his college basketball season, Jeff made 520 baskets out of 1280 attempts. What was his shooting percentage? Round to the nearest whole percent.

For Exercises 65–68, use the table given. The data represent 600 police officers broken down by gender and by the number of officers promoted.

	Promoted	Not Promoted	Total
Male	140	340	480
Female	20	100	120
Total	160	440	600

65. What percent of the officers are female?

66. What percent of the officers are male?

67. What percent of the officers were promoted? Round to the nearest tenth of a percent.

68. What percent of the officers were not promoted? Round to the nearest tenth of a percent.

Objective 4: Applications of Percent Proportions (Mixed Exercises)

69. The rainfall at Birmingham Airport in the United Kingdom averages 56 mm per month. In August the amount of rain that fell was 125% of the average monthly rainfall. How much rain fell in August? **(See Example 5.)**

70. In a recent survey 38% of people in the United States say that gas prices have affected the type of vehicle they will buy. In a sample of 500 people who are in the market for a new vehicle, how many would you expect to be influenced by gas prices?

71. Harvard University reported that 209 African American students were admitted to the freshman class in a recent year. If this represents 11% of the total freshman class, how many freshmen were admitted? **(See Example 6.)**

72. Yellowstone National Park has 3366 mi² of undeveloped land. If this represents 99% of the total area, find the total area of the park.

73. During the 2006–2007 basketball season, Dirk Nowitzki of the Dallas Mavericks made 72 three-point shots out of 173 attempts. To the nearest tenth of a percent, find the percent of shots made. **(See Example 7.)**

74. As of the 2006–2007 football season, Peyton Manning had completed 3131 passes out of 4890 attempts. Find his completion percentage to the nearest tenth of a percent.

75. The graph shows the percent of households that own dogs according to the number of people residing in the household. (*Source:* American Veterinary Medical Association)

    a. If 200 five-person households are surveyed, how many would you expect to own dogs?

    b. If 50 three-person households are surveyed, how many would you expect to own dogs?

76. A computer has 74.4 GB (gigabytes) of memory available. If 7.56 GB is used, what percent of the memory is used? Round to the nearest percent.

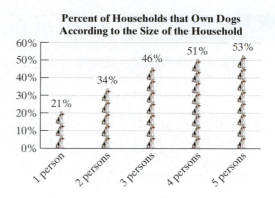

Percent of Households that Own Dogs According to the Size of the Household

A used car dealership sells several makes of vehicles. For Exercises 77–80, refer to the graph. Round the answers to the nearest whole unit.

77. If the dealership sold 215 vehicles in one month, how many were Chevys?

78. If the dealership sold 182 vehicles in one month, how many were Fords?

79. If the dealership sold 27 Hondas in one month, how many total vehicles were sold?

80. If the dealership sold 10 cars in the "Other" category, how many total vehicles were sold?

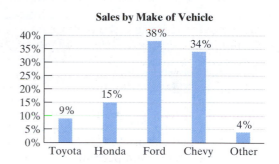

Sales by Make of Vehicle

Expanding Your Skills

81. Carson had $600 and spent 44% of it on clothes. Then he spent 20% of the remaining money on dinner. How much did he spend altogether?

82. Melissa took $52 to the mall and spent 24% on makeup. Then she spent one-half of the remaining money on lunch. How much did she spend altogether?

It is customary to leave a 15–20% tip for the server in a restaurant. However, when you are at a restaurant in a social setting, you probably do not want to take out a pencil and piece of paper to figure out the tip. It is more effective to compute the tip mentally. Try this method.

Step 1: First, if the bill is not a whole dollar amount, simplify the calculations by rounding the bill to the next-higher whole dollar.

Step 2: Take 10% of the bill. This is the same as taking one-tenth of the bill. Move the decimal point to the left 1 place.

Step 3: If you want to leave a 20% tip, double the value found in step 2.

Step 4: If you want to leave a 15% tip, first note that 15% is 5% + 10%. Therefore, add one-half of the value found in step 2 to the number in step 2.

83. Estimate a 20% tip on a bill of $57.65. (*Hint:* Round up to $58 first.)

84. Estimate a 20% tip on a bill of $18.79.

85. Estimate a 15% tip on a dinner bill of $42.00.

86. Estimate a 15% tip on a luncheon bill of $12.00.

Section 6.4   Percent Equations and Applications

Objectives

1. Solving Percent Equations—Amount Unknown
2. Solving Percent Equations—Base Unknown
3. Solving Percent Equations—Percent Unknown
4. Applications of Percent Equations

1. Solving Percent Equations—Amount Unknown

In this section, we investigate an alternative method to solve applications involving percents. We use percent equations. A **percent equation** represents a percent proportion in an alternative form. For example, recall that we can write a percent proportion as follows:

$$\frac{\text{amount}}{\text{base}} = p\% \qquad \text{percent proportion}$$

This is equivalent to writing $\text{Amount} = (p\%) \cdot (\text{base})$ percent equation

To set up a percent equation, it is necessary to translate an English sentence into a mathematical equation. As you read through the examples in this section, you will notice several key words. In the phrase *percent of*, the word *of* implies multiplication. The verb *to be* (am, is, are, was, were, been) often implies = .

Skill Practice

Solve, using a percent equation.
1. What is 40% of 90?

Example 1   Solving a Percent Equation—Amount Unknown

What is 30% of 60?

Solution:

We translate the words to mathematical symbols.

What is 30% of 60?

$x = (30\%) \cdot (60)$

In this context, the word *of* means to multiply.

Let x represent the unknown amount.

To find x, we must multiply 30% by 60. However, 30% means $\frac{30}{100}$ or 0.30. For the purpose of calculation, we *must* convert 30% to its equivalent decimal or fraction form. The equation becomes

$x = (0.30)(60)$

$= 18$

TIP: The solution to Example 1 can be checked by noting that 10% of 60 is 6. Therefore, 30% is equal to (3)(6) = 18.

The value 18 is 30% of 60.

Skill Practice

Solve, using a percent equation.
2. 235% of 60 amounts to what number?

Example 2   Equations—Amount Unknown

142% of 75 amounts to what number?

Solution:

142% of 75 amounts to what number?

$(142\%) \cdot (75) = x$

$(1.42)(75) = x$

$106.5 = x$

Let x represent the unknown amount.
The word *of* implies multiplication.
The phrase *amounts to* implies = .
Convert 142% to its decimal form (1.42).
Multiply.

Therefore, 142% of 75 amounts to 106.5.

Answers
1. 36   2. 141

Examples 1 and 2 illustrate that the percent equation gives us a quick way to find an unknown amount. For example, because $(p\%) \cdot (\text{base}) = \text{amount}$, we have

$$50\% \text{ of } 80 \ = 0.50(80) \ = 40$$

$$25\% \text{ of } 20 \ = 0.25(20) \ = 5$$

$$87\% \text{ of } 600 = 0.87(600) = 522$$

$$250\% \text{ of } 90 = 2.50(90) \ = 225$$

2. Solving Percent Equations—Base Unknown

Examples 3 and 4 illustrate the case in which the base is unknown.

Example 3 Solving a Percent Equation—Base Unknown

225 is 40% of what number?

Solution:

225 is 40% of what number?	Let x represent the base number.
$225 = (0.40) \cdot \quad x$	Notice that we immediately converted 40% to its decimal form 0.40 so that we would not forget.
$225 = 0.40x$	
$\dfrac{225}{0.40} = \dfrac{\overset{1}{\cancel{0.40}}x}{\underset{1}{\cancel{0.40}}}$	Divide both sides of the equation by the number multiplied by the variable x. In this case, divide by 0.40.
$562.5 = x$	Divide: $225 \div 0.40 = 562.5$.

The value 225 is 40% of 562.5.

Skill Practice

Use a percent equation to solve.

3. 94 is 80% of what number?

Example 4 Solving a Percent Equation—Base Unknown

0.19 is 0.2% of what number?

Solution:

0.19 is 0.2% of what number?	Let x represent the base number.
$0.19 = (0.002) \cdot x$	Convert 0.2% to its decimal form 0.002.
$0.19 = 0.002x$	
$\dfrac{0.19}{0.002} = \dfrac{\overset{1}{\cancel{0.002}}x}{\underset{1}{\cancel{0.002}}}$	Divide both sides by the number multiplied by x. In this case, divide by 0.002.
$95 = x$	Divide: $0.19 \div 0.002 = 95$.

Therefore, 0.19 is 0.2% of 95.

Skill Practice

Use a percent equation to solve.

4. 5.6 is 0.8% of what number?

3. Solving Percent Equations—Percent Unknown

Examples 5 and 6 demonstrate the process to find an unknown percent.

Answers

3. 117.5 **4.** 700

Example 5 Solving a Percent Equation—Percent Unknown

75 is what percent of 250?

Solution:

75 is what percent of 250?

$$75 = x \cdot (250)$$ Let x represent the unknown percent.

$$75 = 250x$$

$$\frac{75}{250} = \frac{250x}{250}$$ Divide both sides by the number multiplied by the variable. In this case, divide by 250.

$$0.3 = x$$ Divide: $75 \div 250 = 0.3$.

At this point, we have $x = 0.3$. To write the value of x in percent form, multiply by 100%.

$$x = 0.3$$

$$= 0.3 \times 100\%$$

$$= 30\%$$

Avoiding Mistakes

When solving for an unknown percent using a percent equation, it is necessary to convert x to its percent form.

Thus, 75 is 30% of 250.

Example 6 Solving a Percent Equation—Percent Unknown

What percent of $60 is $92? Round to the nearest tenth of a percent.

Solution:

What percent of $60 is $92?

$$x \cdot (60) = 92$$ Let x represent the unknown percent.

$$60x = 92$$

$$\frac{60x}{60} = \frac{92}{60}$$ Divide both sides by the number multiplied by the variable. In this case, divide by 60.

$$x = 1.5\overline{3}$$ Divide: $92 \div 60 = 1.5\overline{3}$.

At this point, we have $x = 1.5\overline{3}$. To convert x to its percent form, multiply by 100%.

$$x = 1.5\overline{3}$$

$$= 1.5\overline{3} \times 100\%$$ Convert from decimal form to percent form.

$$= (1.53333\ldots) \times 100\%$$

$$= 153.333\ldots\%$$

The hundredths-place digit is less than 5. Discard it and the digits to its right.

Round to the nearest tenth of a percent.

$$\approx 153.3\%$$

Avoiding Mistakes

Notice that in Example 6 we converted the final answer to percent form first *before* rounding. With the number written in percent form, we are sure to round to the nearest tenth of a percent.

Therefore, $92 is approximately 153.3% of $60. (Notice that $92 is just over $1\frac{1}{2}$ times $60, so our answer seems reasonable.)

4. Applications of Percent Equations

In Examples 7, 8, and 9, we use percent equations in application problems. An important part of this process is to extract the base, amount, and percent from the wording of the problem.

Example 7 Using a Percent Equation in Ecology

Forty-six panthers are thought to live in Florida's Big Cypress National Preserve. This represents 53% of the panthers living in Florida. How many panthers are there in Florida? Round to the nearest whole unit. (*Source:* U.S. Fish and Wildlife Services)

Solution:

This problem translates to

"46 is 53% of the number of panthers living in Florida."

$$46 = (0.53) \cdot x \qquad \text{Let } x \text{ represent the total number of panthers.}$$

$$46 = 0.53x$$

$$\frac{46}{0.53} = \frac{\overset{1}{\cancel{0.53}}x}{\underset{1}{\cancel{0.53}}} \qquad \begin{array}{l}\text{Divide both sides by the number multiplied by the variable. In this case, divide by 0.53.}\end{array}$$

$$87 \approx x \qquad \begin{array}{l}\text{Divide: } 46 \div 0.53 \approx 87 \\ \text{(rounded to the nearest whole number).}\end{array}$$

There are approximately 87 panthers in Florida.

Skill Practice

7. Brianna read 143 pages in a book. If this represents 22% of the book, how many pages are in the book?

Example 8 Using a Percent Equation in Sports Statistics

At one time, Steve Young of the San Francisco 49ers was ranked as the NFL's best passer (based on quarterback rating points). For one particular game he completed 23 of 30 passes. What percent of passes did he complete? Round to the nearest tenth of a percent.

Solution:

This problem translates to

"23 is what percent of 30?"

$$23 = \qquad x \qquad \cdot 30 \qquad \text{Let } x \text{ represent the unknown.}$$

$$23 = 30x$$

$$\frac{23}{30} = \frac{30x}{30} \qquad \text{Divide both sides by 30.}$$

$$0.767 \approx x \qquad \text{Divide: } 23 \div 30 \approx 0.767.$$

Skill Practice

8. Brandon had $60 in his wallet to take himself and a date to dinner and a movie. If he spent $28 on dinner and $19 on the movie and popcorn, what percent of his money did he spend? Round to the nearest tenth of a percent.

Answers

7. The book is 650 pages long.
8. Brandon spent about 78.3% of his money.

The decimal value 0.767 has been rounded to 3 decimal places. We did this because the next step is to convert the decimal to a percent. Move the decimal point to the right 2 places and attach the % symbol. We have 76.7% which is rounded to the nearest tenth of a percent.

$$x \approx 0.767$$
$$= 0.767 \times 100\%$$
$$= 76.7\%$$

Steve Young completed approximately 76.7% of his passes.

Skill Practice

9. In a science class, 85% of the students passed the class. If there were 40 people in the class, how many passed?

Example 9 Using a Percent Equation in Voting Statistics

Arnold Schwarzenegger received 50.5% of the votes in the 2003 California recall election for governor. If approximately 7.4 million votes were cast, how many votes did Schwarzenegger receive?

Solution:

This problem translates to

"What number is 50.5% of 7.4?"

$$x = 0.505 \cdot 7.4 \quad \text{Write 50.5\% in decimal form.}$$

$$x = (0.505)(7.4) \quad \text{Let } x \text{ represent the number of votes for Schwarzenegger.}$$

$$x = 3.737 \quad \text{Multiply.}$$

Arnold Schwarzenegger received approximately 3.737 million votes.

Answer

9. 34 students passed the class.

Section 6.4 Practice Exercises

Study Skill Exercises

1. There's a saying, "Leave no stone unturned." In math, this means "leave no homework problem undone." Did you do all the assigned homework in Section 6.3? Do you understand the concepts well enough to move on to the homework in this section?

2. Define the key term **percent equation**.

Review Exercises

3. Explain how to solve the equation $26x = 65$.

4. Explain how to solve the equation $54 = 6x$.

For Exercises 5–10, solve the equation for the variable.

5. $3x = 27$

6. $12x = 48$

7. $0.15x = 45$

8. $0.32x = 60$

9. $1.02x = 841.5$

10. $1.06x = 90.1$

Objective 1: Solving Percent Equations—Amount Unknown

For Exercises 11–16, write the percent equation. Then solve for the unknown amount. **(See Examples 1–2.)**

11. What is 35% of 700?

12. Find 12% of 625.

13. 0.55% of 900 is what number?

14. What is 0.4% of 75?

15. Find 133% of 600.

16. 120% of 40.4 is what number?

17. What is a quick way to find 50% of a number?

18. What is a quick way to find 10% of a number?

19. Compute 200% of 14 mentally.

20. Compute 75% of 80 mentally.

21. Compute 50% of 40 mentally.

22. Compute 10% of 32 mentally.

23. Household bleach is 6% sodium hypochlorite (active ingredient). In a 64-oz bottle, how much is active ingredient?

24. One antifreeze solution is 40% alcohol. How much alcohol is in a 12.5-L mixture?

25. In football, Dan Marino completed 60% of his passes. If he attempted 8358 passes, how many did he complete? Round to the nearest whole unit.

26. To pass an exit exam, a student must pass a 60-question test with a score of 80% or better. What is the minimum number of questions she must answer correctly?

Objective 2: Solving Percent Equations—Base Unknown

For Exercises 27–32, write the percent equation. Then solve for the unknown base. **(See Examples 3–4.)**

27. 18 is 40% of what number?

28. 72 is 30% of what number?

29. 92% of what number is 41.4?

30. 84% of what number is 100.8?

31. 3.09 is 103% of what number?

32. 189 is 105% of what number?

33. In tests of a new anti-inflammatory drug, it was found that 47 subjects experienced nausea. If this represents 4% of the sample, how many subjects were tested?

34. Ted typed 80% of his research paper before taking a break.

   a. If he typed 8 pages, how many total pages are in the paper?

   b. How many pages does he have left to type?

35. In a recent report, approximately 61.6 million Americans had some form of heart and blood vessel disease. If this represents 22% of the population, approximate the total population of the United States.

36. A city has a population of 245,300 which is 110% of the population from the previous year. What was the population the previous year?

Objective 3: Solving Percent Equations—Percent Unknown

For Exercises 37–44, convert the decimal to a percent.

37. 0.13   **38.** 0.4   **39.** 1.08   **40.** 2.2

41. 0.005   **42.** 0.007   **43.** 0.17   **44.** 0.9

For Exercises 45–50 write the percent equation. Then solve for the unknown percent. Round to the nearest tenth of a percent if necessary. **(See Examples 5–6.)**

45. What percent of 480 is 120?   **46.** 180 is what percent of 2000?   **47.** 666 is what percent of 740?

48. What percent of 60 is 2.88?   **49.** What percent of 300 is 400?   **50.** 28 is what percent of 24?

51. At a softball game, the concession stand had 120 hot dogs and sold 84 of them. What percent was sold?

52. The YMCA wants to raise $2500 for its summer program for disadvantaged children. If the YMCA has already raised $900, what percent of its goal has been achieved?

For Exercises 53–54, refer to the table that shows the 1-year absentee record for a business.

53. a. Determine the total number of employees.

   b. What percent missed exactly 3 days of work?

   c. What percent missed between 1 and 5 days, inclusive?

54. a. What percent missed at least 4 days?

   b. What percent did not miss any days?

Number of Days Missed	Number of Employees
0	4
1	2
2	14
3	10
4	16
5	18
6	10
7	6

Objective 4: Applications of Percent Equations (Mixed Exercises)

55. In a recent year, children and adolescents comprised 6.3 million hospital stays. If this represents 18% of all hospital stays, what was the total number of hospital stays? **(See Example 7.)**

56. One fruit drink advertised that it contained "10% real fruit juice." In one bottle, this was found to be 4.8 oz of real juice.

   a. How many ounces of drink does the bottle contain?

   b. How many ounces is something other than fruit juice?

57. Of the 87 panthers living in the wild in Florida, 11 are thought to live in Everglades National Park. To the nearest tenth of a percent, what percent is this? (*Source:* U.S. Fish and Wildlife Services) **(See Example 8.)**

58. Forty-four percent of Americans use online travel sites to book hotel or airline reservations. If 400 people need to make airline or hotel reservations, how many would be expected to use online travel sites?

59. Fifty-two percent of American parents have started to put money away for their children's college educations. In a survey of 800 parents, how many would be expected to have started saving for their children's education? (*Source: USA TODAY*) **(See Example 9.)**

60. The Earth is covered by approximately 360 million km² of water. If the total surface area is 510 km², what percent is water? (Round to the nearest tenth of a percent.)

61. Brian has been saving money to buy a 61-in. Samsung Projection HDTV. He has saved $1440 so far, but this is only 60% of the total cost of the television. What is the total cost?

62. Approximately 3.6 million elementary and secondary school teachers participated in classroom instruction in the fall of 2006. This represents a 119% increase since 1996. (*Source:* National Center for Educational Statistics) How many elementary and secondary teachers were in the classroom in 1996? Round to the nearest million.

63. A television station plays commercials for 26% of its air time. In 60 min, how many minutes of commercials would be expected?

64. Sixty-five percent of the human body is water. For a 150-lb person, how much is water?

For Exercises 65–68, use the graph.

65. If there were 10,000,000 people in the workforce in the 25–34 age group, how many made over $10 per hour?

66. If there were 6,600,000 people in the workforce in the 55–64 age group, how many made over $10 per hour?

67. If 4,000,000 people in the 16–24 age group made $10 or more per hour, how many total workers in this age group are there?

68. If 9,000,000 people in the 45–54 age group made $10 or more per hour, how many total workers in this age group are there?

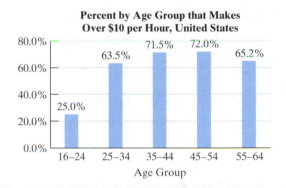

Percent by Age Group that Makes Over $10 per Hour, United States

Expanding Your Skills

The maximum recommended heart rate (in beats per minute) is given by 220 minus a person's age. For aerobic activity, it is recommended that individuals exercise at 60%–85% of their maximum recommended heart rate. This is called the aerobic range. Use this information for Exercises 69–70.

69. a. Find the maximum recommended heart rate for a 20-year-old.

   b. Find the aerobic range for a 20-year-old.

70. a. Find the maximum recommended heart rate for a 42-year-old.

   b. Find the aerobic range for a 42-year-old.

Problem Recognition Exercises

Percents

For Exercises 1–4, determine the percent represented by the shaded portion of the figure.

1.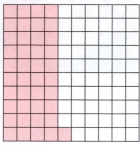

2.

3.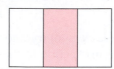

4.

5. Is 104% of 80 less than or greater than 80?

6. Is 8% of 50 less than or greater than 5?

7. Is 11% of 90 less than or greater than 9?

8. Is 52% of 200 less than or greater than 100?

For Exercises 9–34, solve the problem by using a percent proportion or a percent equation.

9. 6 is 0.2% of what number?

10. What percent of 500 is 120?

11. 12% of 40 is what number?

12. 27 is what percent of 180?

13. 150% of what number is 105?

14. What number is 30% of 120?

15. What is 7% of 90?

16. 100 is 40% of what number?

17. 180 is what percent of 60?

18. 0.5% of 140 is what number?

19. 75 is 0.1% of what number?

20. 27 is what percent of 72?

21. What number is 50% of 50?

22. What number is 15% of 900?

23. 50 is 50% of what number?

24. 900 is 15% of what number?

25. What percent of 250 is 2?

26. 75 is what percent of 60?

27. 26 is 10% of what number?

28. 11 is 55% of what number?

29. What number is 10% of 26?

30. What number is 55% of 11?

31. 186 is what percent of 248?

32. 5 is what percent of 20?

33. 248 is what percent of 186?

34. 20 is what percent of 5?

For Exercises 35–40, perform the calculations mentally.

35. What is 10% of 82?

36. What is 5% of 82?

37. What is 20% of 82?

38. What is 50% of 82?

39. What is 200% of 82?

40. What is 15% of 82?

Applications Involving Sales Tax, Commission, Discount, and Markup

Percents are used in an abundance of applications in day-to-day life. In this section, we investigate four common applications of percents:

- Sales tax
- Commission
- Discount
- Markup

Objectives

1. **Applications Involving Sales Tax**
2. **Applications Involving Commission**
3. **Applications Involving Discount and Markup**

1. Applications Involving Sales Tax

The first application involves computing sales tax. **Sales tax** is a tax based on a percent of the cost of merchandise.

> **FORMULA Sales Tax Formula**
>
> $$\left(\begin{array}{c}\text{Amount of}\\\text{sales tax}\end{array}\right) = \left(\begin{array}{c}\text{Tax}\\\text{rate}\end{array}\right) \cdot \left(\begin{array}{c}\text{Cost of}\\\text{merchandise}\end{array}\right)$$

In this formula the tax rate is usually given by a percent. Also note that there are three parts to the formula, just as there are in the general percent equation (see Section 6.4). The sales tax formula is a special case of a percent equation.

Example 1   Computing Sales Tax

Suppose a Toyota Camry sells for $20,000.

a. Compute the sales tax for a tax rate of 5.5%.

b. What is the total price of the car?

Skill Practice

1. A graphing calculator costs $110. The sales tax rate is 4.5%.
   a. Compute the amount of sales tax.
   b. Compute the total cost.

Solution:

a. Let x represent the amount of sales tax.   Label the unknown.

Tax rate $= 5.5\%$ Identify the parts of the formula.

Cost of merchandise $= \$20,000$

$$\left(\begin{array}{c}\text{Amount of}\\\text{sales tax}\end{array}\right) = \left(\begin{array}{c}\text{Tax}\\\text{rate}\end{array}\right) \cdot \left(\begin{array}{c}\text{Cost of}\\\text{merchandise}\end{array}\right)$$

$$x = (5.5\%) \cdot (\$20,000)$$ Substitute values into the sales tax formula.

$$x = (0.055)(\$20,000)$$ Convert the percent to its decimal form.

$$x = \$1100$$

The sales tax on the vehicle is $1100.

b. The total price is $20,000 + $1100 = $21,100.

Avoiding Mistakes

Notice that we must use the decimal form of the sales tax rate in the calculation.

Answers

1. **a.** The tax is $4.95.
   b. The total cost is $114.95.

Example 2 Computing a Sales Tax Rate

Lindsay has just moved and must buy a new refrigerator for her home. The refrigerator costs $1200 and the sales tax is $48. Because she is new to the area, she does not know the sales tax rate. Use these figures to compute the tax rate.

Solution:

Let x represent the sales tax rate.   Label the unknown.

Cost of merchandise = $1200   Identify the parts of the formula.

Amount of sales tax = $48

$$\left(\begin{matrix}\text{Amount of} \\ \text{sales tax}\end{matrix}\right) = \left(\begin{matrix}\text{Tax} \\ \text{rate}\end{matrix}\right) \cdot \left(\begin{matrix}\text{Cost of} \\ \text{merchandise}\end{matrix}\right)$$

$$48 \quad = \quad x \quad \cdot \quad (1200)$$ Substitute values into the sales tax formula.

$$48 = 1200x$$

$$\frac{48}{1200} = \frac{\overset{1}{\cancel{1200}}x}{\underset{1}{\cancel{1200}}}$$ Divide both sides by 1200.

$$0.04 = x$$

The question asks for the tax rate which is given in percent form.

$$x = 0.04$$

$$= 0.04 \times 100\%$$

$$= 4\%$$ The sales tax rate is 4%.

Example 3 Finding Cost of Merchandise

The tax on a new CD comes to $1.05. If the tax rate is 6%, find the cost of the CD before tax.

Solution:

Let x represent the cost of the CD.   Label the unknown.

Tax rate = 6%   Identify the parts of the formula.

Amount of tax = $1.05

$$\left(\begin{matrix}\text{Amount of} \\ \text{sales tax}\end{matrix}\right) = \left(\begin{matrix}\text{Tax} \\ \text{rate}\end{matrix}\right) \cdot \left(\begin{matrix}\text{Cost of} \\ \text{merchandise}\end{matrix}\right)$$

$$1.05 \quad = \quad (0.06) \cdot \quad x$$ Notice that we immediately converted 6% to its percent form.

$$1.05 = 0.06x$$

$$\frac{1.05}{0.06} = \frac{\overset{1}{\cancel{0.06}}x}{\underset{1}{\cancel{0.06}}}$$ Divide both sides by 0.06.

$$17.5 = x$$ The CD costs $17.50.

Answers

2. The tax rate is 6%.
3. The mower costs $430 before tax.

2. Applications Involving Commission

Salespeople often receive all or part of their salary in commission. **Commission** is a form of income based on a percent of sales.

> **FORMULA** Commission Formula
>
> $$\begin{pmatrix} \text{Amount of} \\ \text{commission} \end{pmatrix} = \begin{pmatrix} \text{Commission} \\ \text{rate} \end{pmatrix} \cdot \begin{pmatrix} \text{Total} \\ \text{sales} \end{pmatrix}$$

For example, if a realtor gets a 6% commission on the sale of a $200,000 home, then

$$\text{Commission} = (0.06)(\$200,000)$$
$$= \$12,000$$

Example 4 Finding Commission Rate

Alexis works in real estate sales.

a. If she sells a $150,000 house and earns a commission of $10,500, what is her commission rate?

b. At this rate, how much will she earn by selling a $200,000 house?

Solution:

a. Let x represent the commission rate.   Label the unknown.

Total sales = $150,000   Identify the parts of the formula.

Amount of commission = $10,500

$$\begin{pmatrix} \text{Amount of} \\ \text{commission} \end{pmatrix} = \begin{pmatrix} \text{Commission} \\ \text{rate} \end{pmatrix} \cdot \begin{pmatrix} \text{Total} \\ \text{sales} \end{pmatrix}$$

$$10,500 \quad = \quad x \quad \cdot (150,000) \qquad \text{Substitute values into the commission formula.}$$

$$10,500 = 150,000x$$

$$\frac{10,500}{150,000} = \frac{\overset{1}{\cancel{150,000}}x}{\underset{1}{\cancel{150,000}}} \qquad \text{Divide both sides by } 150,000.$$

$$0.07 = x$$

$$x = 0.07 \times 100\% \qquad \text{Convert to percent form.}$$

$$= 7\%$$

The commission rate is 7%.

b. The commission on a $200,000 house is given by

Amount of commission = $(0.07)(\$200,000) = \$14,000$

Alexis will earn $14,000 by selling a $200,000 house.

Skill Practice

4. Trevor sold a home for $160,000 and earned a $6400 commission. What is his commission rate?

Answer

4. His commission rate is 4%.

Skill Practice

5. A sales rep for a pharmaceutical firm makes $50,000 as his base salary. In addition, he makes 6% commission on sales. If his salary for the year amounts to $98,000, what were his total sales?

Example 5 — Finding Sales Base

Tonya is a real estate agent. She makes $10,000 as her annual base salary for the work she does in the office. In addition, she makes 8% commission on her total sales. If her salary for the year amounts to $106,000, what was her total in sales?

Solution:

First note that her commission is her total salary minus the $10,000 for working in the office. Thus,

Amount of commission = $106,000 − $10,000 = $96,000

Let x represent Tonya's total sales. — Label the unknown.

Amount of commission = $96,000 — Identify the parts of the formula.
Commission rate = 8%

$$\left(\begin{array}{c}\text{Amount of}\\ \text{commission}\end{array}\right) = \left(\begin{array}{c}\text{Commission}\\ \text{rate}\end{array}\right) \cdot \left(\begin{array}{c}\text{Total}\\ \text{sales}\end{array}\right)$$

$$96{,}000 = (0.08) \cdot x$$ — Substitute values into the commission formula.

$$96{,}000 = 0.08x$$

$$\frac{96{,}000}{0.08} = \frac{0.08x}{0.08}$$ — Divide both sides by 0.08.

$$1{,}200{,}000 = x$$

Tonya's sales totaled $1,200,000 ($1.2 million).

3. Applications Involving Discount and Markup

When we go to the store, we often find items discounted or on sale. For example, a printer might be discounted 20%, or a blouse might be on sale for 30% off. We compute the amount of the **discount** (the savings) as follows.

FORMULA Discount Formulas

$$\left(\begin{array}{c}\text{Amount of}\\ \text{discount}\end{array}\right) = \left(\begin{array}{c}\text{Discount}\\ \text{rate}\end{array}\right) \cdot \left(\begin{array}{c}\text{Original}\\ \text{price}\end{array}\right)$$

Sale price = Original price − Amount of discount

Answer
5. He made $800,000 in sales.

Example 6 **Computing Discount Rate**

A gold chain originally priced $500 is marked down to $375. What is the discount rate?

Solution:

First note that the amount of the discount is given by

Discount = Original price − Sale price

$$= \$500 - \$375$$

$$= \$125$$

Let x represent the discount rate.   Label the unknown.

Original price = $500   Identify the parts of the formula.

Amount of discount = $125

$$\begin{pmatrix} \text{Amount of} \\ \text{discount} \end{pmatrix} = \begin{pmatrix} \text{Discount} \\ \text{rate} \end{pmatrix} \cdot \begin{pmatrix} \text{Original} \\ \text{price} \end{pmatrix}$$

$$125 \quad = \quad x \quad \cdot \quad 500$$ Substitute values into discount formula.

$$125 = 500x$$

$$\frac{125}{500} = \frac{\overset{1}{\cancel{500}}x}{\underset{1}{\cancel{500}}}$$ Divide both sides by 500.

$$0.25 = x$$

Converting $x = 0.25$ to percent form, we have $x = 25\%$. The chain has been discounted 25%.

Retailers often buy goods from manufacturers or wholesalers. To make a profit, the retailer must increase the cost of the merchandise before reselling it. This is called **markup**.

FORMULA Markup Formulas

$$\begin{pmatrix} \text{Amount of} \\ \text{markup} \end{pmatrix} = \begin{pmatrix} \text{Markup} \\ \text{rate} \end{pmatrix} \cdot \begin{pmatrix} \text{Original} \\ \text{price} \end{pmatrix}$$

Retail price = Original price + Amount of markup

Example 7 **Computing Markup**

A college bookstore marks up the price of books 40%.

a. What is the markup for a math text that has a manufacturer price of $66?

b. What is the retail price of the book?

c. If there is a 6% sales tax, how much will the book cost to take home?

Solution:

a. Let x represent the amount of markup.   Label the unknown.

   Markup rate = 40%   Identify parts of the formula.

   Original price = $66

$$\left(\begin{array}{c}\text{Amount of}\\\text{markup}\end{array}\right) = \left(\begin{array}{c}\text{Markup}\\\text{rate}\end{array}\right) \cdot \left(\begin{array}{c}\text{Original}\\\text{price}\end{array}\right)$$

$$x \quad = \quad (0.40) \quad \cdot \quad (\$66) \qquad \text{Use the decimal form of 40\%.}$$

$$x = (0.40)(\$66)$$

$$= \$26.40$$

The amount of markup is $26.40.

b. Retail price = Original price + Markup

$$= \$66 + \$26.40$$

$$= \$92.40$$

The retail price is $92.40.

c. Next we must find the amount of the sales tax. This value is added to the cost of the book. The sales tax rate is 6%.

$$\left(\begin{array}{c}\text{Amount of}\\\text{sales tax}\end{array}\right) = \left(\begin{array}{c}\text{Tax}\\\text{rate}\end{array}\right) \cdot \left(\begin{array}{c}\text{Cost of}\\\text{merchandise}\end{array}\right)$$

$$\text{Tax} \quad = (0.06) \cdot \quad (\$92.40)$$

$$\approx \$5.54 \qquad \text{Round the tax to the nearest cent.}$$

The total cost of the book is $92.40 + $5.54 = $97.94.

It is important to note the similarities in the formulas presented in this section. To find the amount of sales tax, commission, discount, or markup, we multiply a rate (percent) by some original amount.

FORMULA   Summary Formulas for Sales Tax, Commission, Discount, and Markup

$$\text{Amount} \quad = \quad \underline{\text{Rate}} \times \underline{\text{Original amount}}$$

Sales tax:
$$\left(\begin{array}{c}\text{Amount of}\\\text{sales tax}\end{array}\right) = \left(\begin{array}{c}\text{Tax}\\\text{rate}\end{array}\right) \cdot \left(\begin{array}{c}\text{Cost of}\\\text{merchandise}\end{array}\right)$$

Commission:
$$\left(\begin{array}{c}\text{Amount of}\\\text{commission}\end{array}\right) = \left(\begin{array}{c}\text{Commission}\\\text{rate}\end{array}\right) \cdot \left(\begin{array}{c}\text{Total}\\\text{sales}\end{array}\right)$$

Discount:
$$\left(\begin{array}{c}\text{Amount of}\\\text{discount}\end{array}\right) = \left(\begin{array}{c}\text{Discount}\\\text{rate}\end{array}\right) \cdot \left(\begin{array}{c}\text{Original}\\\text{price}\end{array}\right)$$

Markup:
$$\left(\begin{array}{c}\text{Amount of}\\\text{markup}\end{array}\right) = \left(\begin{array}{c}\text{Markup}\\\text{rate}\end{array}\right) \cdot \left(\begin{array}{c}\text{Original}\\\text{price}\end{array}\right)$$

Section 6.5 Practice Exercises

Study Skills Exercises

1. Which of the following strategies can help you study for a test? Check all that apply.

   ☐ Read the Chapter Summary. The Chapter Summary for this chapter is on page ___.

   ☐ Do the Review Exercises at the end of the chapter. The Review Exercises for this chapter are on page ___.

   ☐ Do the Chapter Test at the end of the chapter. The Chapter Test for this chapter is on page ___.

2. Define the key terms.

   a. Sales tax   **b. Commission**   **c. Discount**   **d. Markup**

Review Exercises

For Exercises 3–6, find the answer mentally.

3. What is 15% of 80?

4. 20 is what percent of 60?

5. 14 is 50% of what number?

6. What percent of 6 is 12?

For Exercises 7–12, solve the percent problem by using either method from Sections 6.3 and 6.4.

7. 52 is 0.2% of what number?

8. What is 225% of 36?

9. 6 is what percent of 25?

10. 18 is 75% of what number?

11. What is 1.6% of 550?

12. 32.2 is what percent of 28?

Objective 1: Applications Involving Sales Tax

For Exercises 13–20, complete the table.

	Cost of Merchandise	Sales Tax Rate	Amount of Tax	Total Cost
13.	$ 20.00	5%		
14.	$ 56.00	6%		
15.	$ 12.50		$ 0.50	
16.	$212.00		$14.84	
17.		2.5%	$ 2.75	
18.		3%	$18.00	
19.	$ 55.00			$ 58.30
20.	$214.00			$220.42

21. A new coat costs $68.25. If the sales tax rate is 5%, what is the total bill? **(See Example 1.)**

22. Sales tax for a county in Wisconsin is 4.5%. Compute the amount of tax on a new personal MP3 player that sells for $64.

23. The sales tax on a set of luggage is $16.80. If the luggage cost before tax is $240.00, what is the sales tax rate? **(See Example 2.)**

24. A new shirt is labeled at $42.00. Jon purchased the shirt and paid $44.10.

   a. How much was the sales tax?

   b. What is the sales tax rate?

25. The 6% sales tax on a fruit basket came to $2.67. What is the price of the fruit basket? **(See Example 3.)**

26. The sales tax on a bag of groceries came to $1.50. If the sales tax rate is 6%, what was the price of the groceries before tax?

Objective 2: Applications Involving Commission

For Exercises 27–32, complete the table.

	Total Sales	Commission Rate	Amount of Commission
27.	$ 20,000.00	5%	
28.	$ 540.00	16%	
29.	$125,000.00		$10,000.00
30.	$ 800.00		$ 24.00
31.		10%	$ 540.00
32.		15%	$ 159.00

33. Zach works in an insurance office. He receives a commission of 7% on new policies. How much did he make last month in commission if he sold $48,000 in new policies?

34. Marisa makes a commission of 15% on sales over $400. One day she sells $750 worth of merchandise.

   a. How much over $400 did Marisa sell?

   b. How much did she make in commission that day?

35. In one week, Rodney sold $2000.00 worth of sports equipment. He received $300.00 in commission. What is his commission rate? **(See Example 4.)**

36. A realtor sold a townhouse for $95,000. If he received a commission of $7600, what is his commission rate?

37. A realtor makes an annual salary of $25,000 plus a 3% commission on sales. If a realtor's salary is $67,000, what was the amount of her sales? **(See Example 5.)**

38. A salesperson receives a weekly salary of $100, plus a 5.5% commission on sales. Her salary last week was $1090. What were her sales that week?

39. Jeff works as a pharmaceutical representative. He receives a 6% monthly commission on sales up to $60,000. He receives an 8.5% commission on all sales above $60,000. If he sold $86,000 worth of his product one month, how much did he receive in commission?

Objective 3: Applications Involving Discount and Markup

For Exercises 40–47, complete the table.

	Original Price	Discount Rate	Amount of Discount	Sale Price
40.	$ 56.00	20%		
41.	$175.00	15%		
42.	$900.00			$600.00
43.	$900.00			$630.00
44.			$ 8.50	$ 76.50
45.			$ 33.00	$ 77.00
46.		50%	$ 38.00	
47.		40%	$ 23.36	

48. Hospital employees get a 15% discount at the hospital cafeteria. If the lunch bill originally comes to $5.60, what is the price after the discount?

49. A health club membership costs $550 for one year. If a member pays up front in a lump sum, the member will receive a 10% discount.

   a. How much money is discounted?

   b. How much will the yearly membership cost with the discount?

50. A bathing suit is on sale for $45. If the regular price is $60, what is the discount rate?

51. A printer that sells for $229 is on sale for $183.20. What is the discount rate?
(See Example 6.)

52. Find the discount and the sale price of the tent in the given advertisement.

53. A set of dishes had an original price of $112. Then it was discounted 50%. A week later, the new sale price was discounted another 50%. At that time, was the set of dishes free? Explain why or why not.

Explorer 4-person tent
On Sale 30% OFF

Was $269

54. Find the discount and the sale price of the bike in the given advertisement.

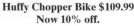
Huffy Chopper Bike $109.99
Now 10% off.

55. Find the discount and the discount rate of the chair from the given advertisement.

Accent Chair was $235.00
and is now $188.00

56. Find the discount and the discount rate of the watch from the given advertisement. Round the discount rate to the nearest tenth of a percent.

For Exercises 57–64, complete the table.

	Original Price	Markup Rate	Amount of Markup	Retail Price
57.	$ 92.00	5%		
58.	$ 25.00	10%		
59.	$110.00			$118.80
60.	$ 50.00			$ 57.50
61.			$ 97.50	$422.50
62.			$175.00	$875.00
63.		20%	$ 9.00	
64.		18%	$ 31.50	

65. A business suit has a wholesale price of $150.00. A department store's markup rate is 18%. **(See Example 7.)**

   a. What is the markup for this suit?

   b. What is the retail price?

   c. If Antonio buys this suit including a 7% sales tax, how much will he pay?

66. An import/export business marks up imported merchandise by 110%. If a wicker chair imported from Singapore originally costs $84 from the manufacturer, what is the retail price?

67. A table is purchased from the manufacturer for $300 and is sold retail at $375. What is the markup rate?

68. A $60 hairdryer is sold for $69. What is the markup rate?

69. A campus bookstore adds $43.20 to the cost of a science text. If the final cost is $123.20, what is the markup rate?

70. The retail price of a golf club is $420.00. If the golf store has marked up the price by $70, what is the markup rate?

Percent Increase and Decrease

1. Definition of Percent Increase and Decrease

Two important applications of percents are finding percent increase and percent decrease. For example:

- The price of gas increased 40% in 4 years.
- After taking a new drug for 3 months, a patient's cholesterol decreased by 35%.

When we compute **percent increase** or **percent decrease**, we are comparing the *change* between two given amounts to the *original amount*. The change (amount of increase or decrease) is found by subtraction. To compute the percent increase or decrease, we use the following formulas. ·

> **FORMULA   Percent Increase or Percent Decrease**
>
> $$\left(\begin{array}{c}\text{Percent}\\\text{increase}\end{array}\right) = \left(\frac{\text{Amount of increase}}{\text{Original amount}}\right) \times 100\%$$
>
> $$\left(\begin{array}{c}\text{Percent}\\\text{decrease}\end{array}\right) = \left(\frac{\text{Amount of decrease}}{\text{Original amount}}\right) \times 100\%$$

The formulas for percent increase and percent decrease are derived from the standard percent equation.

$$(\text{Amount of increase}) = (\text{Rate of increase})(\text{Original amount})$$

We can divide both sides of the equation by original amount:

$$\frac{\text{Amount of increase}}{\text{Original amount}} = \text{Rate of increase}$$

The rate of increase is a rate given in decimal form. To convert it to a percent, we multiply both sides by 100%.

$$\left(\frac{\text{Amount of increase}}{\text{Original amount}}\right) \times 100\% = \text{Percent increase}$$

2. Computing Percent Increase

In Example 1, we apply the percent increase formula.

Example 1   Computing Percent Increase

The price of heating oil climbed from an average of $1.25 per gallon to $1.55 per gallon in a 3-year period. Compute the percent increase.

Solution:

The original price was $1.25 per gallon.   Identify the parts of the
                                                   formula.

The final price after the increase is $1.55.

The amount of increase is given by subtraction.

Skill Practice

1. After a raise, Denisha's salary increased from $42,500 to $44,200. What is the percent increase?

Answer

1. Denisha's salary increased by 4%.

$$\text{Amount of increase} = \$1.55 - \$1.25$$

$$= \$0.30$$

There was a $0.30 increase in price.

$$\left(\begin{array}{c}\text{Percent}\\\text{increase}\end{array}\right) = \left(\frac{\text{Amount of increase}}{\text{Original amount}}\right) \times 100\%$$

$$= \frac{0.30}{1.25} \times 100\%$$

Apply the percent increase formula.

$$= 0.24 \times 100\%$$

$$= 24\%$$

There was a 24% increase in the price of heating oil.

Example 1 could also have been solved by using a percent proportion.

$$\frac{p}{100} = \frac{\text{amount of increase}}{\text{original amount}}$$

$$\frac{p}{100} = \frac{0.30}{1.25}$$

Substitute values into the proportion.

$$1.25p = (0.30)(100)$$

Equate the cross products.

$$1.25p = 30$$

$$\frac{\overset{1}{\cancel{1.25}}p}{\underset{1}{\cancel{1.25}}} = \frac{30}{1.25}$$

Divide both sides by 1.25

$$p = 24$$

There was a 24% increase in heating oil.

3. Computing Percent Decrease

Skill Practice

2. Refer to the graph in Example 2. Compute the percent decrease between day 2 and day 3.

Example 2   Finding Percent Decrease in an Application

The graph in Figure 6-5 represents the closing price of Time Warner stock for a 5-day period. Compute the percent decrease between the first day and the fifth day.

Stock Price

Figure 6-5

Solution:

The amount of decrease is given by: $21.50 − $20.64 = $0.86

The original amount is the closing price on Day 1: $21.50

$$\left(\begin{array}{c}\text{Percent}\\\text{decrease}\end{array}\right) = \left(\frac{\text{Amount of decrease}}{\text{Original amount}}\right) \times 100\%$$

$$= \left(\frac{\$0.86}{\$21.50}\right) \times 100\%$$

$$= (0.04) \times 100\%$$

$$= 4\%$$

The stock fell by 4%.

It is very important to note that percent increase and percent decrease are based on the original amount or starting point. Suppose that a 125 lb college student gains 25 pounds in her first year of college. We would say that she had a percent increase of $\frac{25}{125} = 20\%$. If she then loses 25 lb during her second year, her percent decrease would not be 20% but rather $\frac{25}{150} = 16.\overline{6}\%$. The percent decrease must be calculated from her new starting weight of 150 lb.

Section 6.6   Practice Exercises

Boost your GRADE at ALEKS.com!

• Practice Problems
• Self-Tests
• NetTutor
• e-Professors
• Videos

Study Skills Exercises

1. When you are taking a test, which of the following should you do? Check all that apply. Can you add any other suggestions to this list?

☐ Answer each question.
☐ Write neatly.
☐ Check your work if time allows.

☐ Show all work.
☐ Include your scratch paper with your test.
☐ Correct your test after you get it back so that you do not make the same errors again.

2. Define the key terms.

   a. Percent increase   b. Percent decrease

Review Exercises

3. The price of a sports coat is $65.

   a. If Chris buys the sports coat today, what will the final price be after he adds a 5% sales tax?

   b. Tomorrow the coat goes on sale for 20% off. How much would he spend tomorrow (including the 5% tax)?

   c. How much will Chris save by waiting until tomorrow to buy the coat?

4. The price of a jogging suit is $32.50.

   a. If Kira buys the jogging suit today, what will the final price be after a 6% sales tax is added?

   b. Tomorrow the suit goes on sale for 25% off. How much would she spend tomorrow (including the 6% tax)?

   c. How much will Kira save by waiting until tomorrow to buy the suit?

5. Katie earns a commission of 14% on all merchandise that she sells over $200. What is her commission if she sells a total of $425 of merchandise?

6. Sean earns a salary of $12 per hour and a commission of 3% on all sales. If Sean works a 30-hr week and sells $1290.00 of merchandise, what is his salary for that week?

7. Explain how to change a decimal into a percent.

For Exercises 8–11, change the decimal to a percent.

   8. 0.23   **9.** 0.05   **10.** 0.88   **11.** 0.12

Objective 1: Definition of Percent Increase and Decrease

12. a. To find the amount of increase in the price of a product, should you subtract the original price from the increased price or the increased price from the original price?

   b. To find the amount of decrease in the price of a product, should you subtract the original price from the decreased price or the decreased price from the original price?

For Exercises 13–20, (a) identify if there is an increase or decrease and (b) find the amount of increase or decrease.

   13. 48 to 59   **14.** 78 to 123   **15.** 145 to 135   **16.** 190 to 109

   17. 654 to 645   **18.** 24 to 42   **19.** 67 to 79   **20.** 205 to 105

Objective 2: Computing Percent Increase

21. Select the correct percent increase for a price that is double the original amount. For example, a book that originally cost $30 now costs $60.

   a. 200%   **b.** 2%

   c. 100%   **d.** 150%

22. Select the correct percent increase for a price that is greater by $\frac{1}{2}$ of the original amount. For example, an employee made $20 per hour and now makes $30 per hour.

   a. 150%   **b.** 50%

   c. $\frac{1}{2}$% **d.** 200%

23. The U.S. government classified 8 million documents as secret in 2001. By 2003 (2 years after the attacks on 9-11), this number had increased to 14 million. What is the percent increase? (*Source: Time,* April 12, 2004)
   (See Example 1.)

24. One of the top-selling cars is the Honda Civic. In 1 year, the sales increased from 300,000 to 309,000. Compute the percent increase.

25. The number of accidents from all-terrain vehicles that required emergency room visits for children under 16 increased from 21,000 to 42,000 in an 10-year period. What was the percent increase?

26. The number of deaths from alcohol-induced causes rose from approximately 20,200 to approximately 20,700 in a 10-year period. (*Source:* Centers for Disease Control) What is the percent increase? Round to the nearest tenth of a percent.

27. Robin's health-care premium increased from $5000 per year to $5500 per year. What is the percent increase?

28. The yearly deductible for Diane's health care plan rose from $800 to $1000. What is the percent increase?

29. Lynn is an accountant and charges $165 per hour. If she raises her hourly rate to $170 per hour, what is the percent increase? Round to the nearest percent.

30. Joel's yearly salary went from $42,000 to $45,000. What is the percent increase? Round to the nearest percent.

Objective 3: Computing Percent Decrease

31. Select the correct percent decrease for a price that is one-half of the original amount. For example, a bathing suit that originally cost $62 now costs $31.

   a. 50%   **b.** $\frac{1}{2}$%   **c.** 100%   **d.** 5%

32. Select the correct percent decrease for a price that is one-quarter of the original amount. For example, a T-shirt that originally cost $40 now costs $10.

   a. $\frac{1}{4}$%   **b.** 25%   **c.** $\frac{3}{4}$%   **d.** 75%

33. A stock closed at $12.60 per share on Monday. By Friday, the closing price was $11.97 per share. What was the percent decrease? **(See Example 2.)**

34. During a 5-year period, the number of participants collecting food stamps went from 27 million to 17 million. What is the percent decrease? Round to the nearest whole percent. (*Source:* U.S. Department of Agriculture)

35. Julie bought a new water efficient toilet for her house. Her old toilet used 5 gal of water per flush. The new toilet uses only 1.6 gal of water per flush. What is the percent decrease in water per flush?

36. Nancy put new insulation in her attic and discovered that her heating bill for December decreased from $160 to $140. What is the percent decrease?

37. A paper shredder was marked down from $79 to $59. What is the percent decrease in price? Round to the nearest tenth of a percent.

38. A 19-in. computer monitor is marked down from $279 to $249. What is the percent decrease in price? Round to the nearest tenth of a percent.

39. Gus, the cat, originally weighed 12 lb. He was diagnosed with a thyroid disorder, and Dr. Smith the veterinarian found that his weight had decreased to 10.2 lb. What percent of his body weight did Gus lose?

40. To lose weight, Kelly reduced her Calories from 3000 per day to 1800 per day. What is the percent decrease in Calories?

Calculator Connections

Topic: Using a calculator to compute percent increase and percent decrease

Calculator Exercises

For Exercises 41–48, round the percent increase or decrease to the nearest tenth of a percent.

	Country	Population in 2000 (Millions)	Population in 2005 (Millions)	Change (Millions)	Percent Increase or Decrease
41.	Mexico	100.3	110.8		
42.	France	60.0	61.4		
43.	Bulgaria	8.15	8.11		
44.	Trinidad	1.09	1.075		

	Item	Value in 2000	Value in 2005	Change	Percent Increase or Decrease
45.	Number of unemployed people in the U.S.	5.6 million	7.8 million		
46.	Total pounds of fish caught commercially in the U.S.	9.1 billion	9.8 billion		
47.	U.S. federal debt	$5.7 trillion	$8.8 trillion		
48.	Number of motor vehicle deaths in the U.S.	42,500	42,100		

Section 6.7   Simple and Compound Interest

Objectives

1. Simple Interest
2. Compound Interest
3. Using the Compound Interest Formula

1. Simple Interest

In this section, we use percents to compute simple and compound interest on an investment or a loan.

Banks hold large quantities of money for their customers. They keep some cash for day-to-day transactions, but invest the remaining portion of the money. As a result, banks often pay customers interest.

When making an investment, **simple interest** is the money that is earned on principal (**principal** is the original amount of money invested). When people take out a loan, the amount borrowed is the principal. The interest is a percent of the amount borrowed that you must pay back in addition to the principal.

The following formulas can be used to compute simple interest for an investment or a loan and to compute the total amount in the account.

FORMULA Simple Interest Formulas

Simple interest = Principal × Rate × Time $I = Prt$

Total amount = Principal + Interest $A = P + I$

where

I = amount of interest
P = amount of principal
r = annual interest rate (in decimal form)
t = time (in years)
A = total amount in an account

Concept Connections

1. Consider this statement: "$8000 invested at 4% for 3 years yields $960." Identify the principal, interest, interest rate, and time.

The time, t, is expressed in years because the rate, r, is an *annual* interest rate. If we were given a monthly interest rate, then the time, t, should be expressed in months.

Example 1 Computing Simple Interest

Suppose $2000 is invested in an account that earns 7% simple interest.

a. How much interest is earned after 3 years?

b. What is the total value of the account after 3 years?

Solution:

a. Principal: $P = \$2000$ Identify the parts of the formula.

Annual interest rate: $r = 7\%$

Time (in years): $t = 3$

Let I represent the amount of interest.   Label the unknown.

$I = Prt$

$\quad = (\$2000)(7\%)(3)$ Substitute values into the formula.

$\quad = (2000)(0.07)(3)$ Convert 7% to decimal form.

$\quad = \quad 140(3)$ Apply the order of operations.

$\quad = 420$ Multiply from left to right.

The amount of interest earned is $420.

b. The total amount in the account is given by

$A = P + I$

$\quad = \$2000 + \$420$

$\quad = \$2420$

The total amount in the account is $2420.

Avoiding Mistakes

It is important to use the decimal form of the interest rate when calculating interest.

Skill Practice

2. Suppose $1500 is invested in an account that earns 6% simple interest.
   a. How much interest is earned in 5 years?
   b. What is the total value of the account after 5 years?

Concept Connections

3. To find the interest earned on $6000 at 5.5% for 6 years, what number should be substituted for r: 5.5 or 0.055?

When applying the simple interest formula, it is important that time be expressed in years. This is demonstrated in Example 2.

Answers

1. Principal = $8000; interest = $960; rate = 4% (or 0.04 in decimal form); time = 3 years
2. **a.** $450 in interest is earned.
   b. The total account value is $1950.
3. 0.055

Skill Practice

4. Morris takes out a loan for $10,000. He pays simple interest at 7% for 66 months.
   a. Write 66 months in terms of years.
   b. How much money does he pay in interest?
   c. How much total money must he repay to pay off the loan?

Example 2 Computing Simple Interest

Clyde takes out a loan for $3500. He pays simple interest at a rate of 6% for 4 years 3 months.

a. How much money does he pay in interest?

b. How much total money must he pay to pay off the loan?

Solution:

a. $P = \$3500$ Identify parts of the formula.

$r = 6\%$

$t = 4\frac{1}{4}$ years or 4.25 years   3 months $= \frac{3}{12}$ year $= \frac{1}{4}$ year

$I = Prt$

$= (\$3500)(0.06)(4.25)$ Substitute values into the interest formula. Convert 6% to decimal form 0.06.

$= \$892.50$ Multiply.

The interest paid is $892.50.

b. To find the total amount that must be paid, we have

$A = P + I$

$= \$3500 + \$892.50$

$= \$4392.50$ The total amount that must be paid is $4392.50.

2. Compound Interest

Simple interest is based only on a percent of the original principal. However, many day-to-day applications involve compound interest. **Compound interest** is based on both the original principal and the interest earned.

To compare the difference between simple and compound interest, consider this scenario in Example 3.

Skill Practice

5. Suppose $2000 is invested at 5% interest for 3 years.
   a. Compute the total amount after 3 years, using simple interest.
   b. Compute the total amount after 3 years of compounding interest annually.

Example 3 Comparing Simple Interest and Compound Interest

Suppose $1000 is invested at 8% interest for 3 years.

a. Compute the total amount in the account after 3 years, using simple interest.

b. Compute the total amount in the account after 3 years of compounding interest annually.

Solution:

a. $P = \$1000$

$r = 8\%$

$t = 3$ years

$I = Prt$

$= (\$1000)(0.08)(3)$

$= \$240$

The amount of simple interest earned is $240. The total amount in the account is $1000 + $240 = $1240.

Answers

4. a. 66 months = 5.5 years
   b. Morris pays $3850 in interest.
   c. Morris must pay a total of $13,850 to pay off the loan.
5. a. $2300   b. $2315.25

b. To compute interest compounded annually over a period of 3 years, compute the interest earned in the first year. Then add the principal plus the interest earned in the first year. This value then becomes the principal on which to base the interest earned in the second year. We repeat this process, finding the interest for the second and third years based on the principal and interest earned in the preceding years. This process is outlined in a table.

Year	Interest Earned $I = Prt$	Total Amount in Account
First year	$I = (\$1000)(0.08)(1) = \$80$	$\$1000 + \$80 = \$1080$
Second year	$I = (\$1080)(0.08)(1) = \$86.40$	$\$1080 + \$86.40 = \$1166.40$
Third year	$I = (\$1166.40)(0.08)(1) = \$93.31$	$\$1166.40 + 93.31 = \mathbf{\$1259.71}$

The total amount in the account by compounding interest annually is $1259.71.

Notice that in Example 3 the final amount in the account is greater for the situation where interest is compounded. The difference is $1259.71 − $1240 = $19.71. By compounding interest we earn more money.

Interest may be compounded more than once per year.

Annually   1 time per year
Semiannually   2 times per year
Quarterly   4 times per year
Monthly   12 times per year
Daily   365 times per year

To compute compound interest, the calculations become very tedious. Banks use computers to perform the calculations quickly. You may want to use a calculator if calculators are allowed in your class.

3. Using the Compound Interest Formula

As you can see from Example 3, computing compound interest by hand is a cumbersome process. Can you imagine computing daily compound interest (365 times a year) by hand!

We now use a formula to compute compound interest. This formula requires the use of a scientific or graphing calculator. In particular, the calculator must have an exponent key $\boxed{y^x}$, $\boxed{x^y}$, or $\boxed{\wedge}$.

Let A = total amount in an account

P = principal

r = annual interest rate

t = time in years

n = number of compounding periods per year

Then $A = P \cdot \left(1 + \dfrac{r}{n}\right)^{n \cdot t}$ computes the total amount in an account.

To use this formula, note the following guidelines:

- Rate r must be expressed in decimal form.
- Time t must be the total time of the investment in *years*.
- Number n is the number of compounding periods per year.

Annual	$n = 1$
Semiannual	$n = 2$
Quarterly	$n = 4$
Monthly	$n = 12$
Daily	$n = 365$

Skill Practice

6. Suppose $2000 is invested at 5% interest compounded annually for 3 years. Use the formula.

$$A = P \cdot \left(1 + \frac{r}{n}\right)^{n \cdot t}$$

to find the total amount after 3 years. Compare the answer to margin exercise 5(b).

Example 4   Computing Compound Interest by Using the Compound Interest Formula

Suppose $1000 is invested at 8% interest compounded annually for 3 years. Use the compound interest formula to find the total amount in the account after 3 years. Compare the result to the answer from Example 3(b).

Solution:

$P = \$1000$ Identify the parts of the formula.

$r = 8\%$ (0.08 in decimal form)

$t = 3$ years

$n = 1$ (annual compound interest is compounded 1 time per year)

$$A = P \cdot \left(1 + \frac{r}{n}\right)^{n \cdot t}$$

$$= 1000 \cdot \left(1 + \frac{0.08}{1}\right)^{1 \cdot 3}$$ Substitute values into the formula.

$$= 1000(1 + 0.08)^3$$ Apply the order of operations. Divide within parentheses. Simplify the exponent.

$$= 1000(1.08)^3$$ Add within parentheses.

$$= 1000(1.259712)$$ Evaluate $(1.08)^3 = (1.08)(1.08)(1.08) = 1.259712$.

$$= 1259.712$$ Multiply.

$$\approx 1259.71$$ Round to the nearest cent.

Answer

6. $2315.25; this is the same as the result in margin exercise 5(b).

The total amount in the account after 3 years is $1259.71. This is the same value obtained in Example 3(b).

Calculator Connections

Topic: Using a Calculator to Compute Compound Interest

To enter the expression from Example 4 into a calculator, follow these keystrokes.

Expression	Keystrokes	Result
$1000 \cdot (1 + 0.08)^3$	1000 [×] [(] 1 [+] 0.08 [)] [y^x] 3 [=]	1259.712
or	1000 [×] [(] 1 [+] 0.08 [)] [∧] 3 [ENTER]	1259.712

Example 5 **Computing Compound Interest by Using the Compound Interest Formula**

Suppose $8000 is invested in an account that earns 5% interest compounded quarterly for $1\frac{1}{2}$ years. Use the compound interest formula to compute the total amount in the account after $1\frac{1}{2}$ years.

Solution:

$P = \$8000$ Identify the parts of the formula.

$r = 5\%$ (0.05 in decimal form)

$t = 1.5$ years

$n = 4$ (quarterly interest is compounded 4 times per year)

$$A = P \cdot \left(1 + \frac{r}{n}\right)^{n \cdot t}$$

$$= 8000 \cdot \left(1 + \frac{0.05}{4}\right)^{(4)(1.5)}$$ Substitute values into the formula.

$$= 8000(1 + 0.0125)^6$$ Apply the order of operations. Divide within parentheses. Simplify the exponent.

$$= 8000(1.0125)^6$$ Add within parentheses.

$$\approx 8000(1.077383181)$$ Evaluate $(1.0125)^6$. If your teacher allows the use of a calculator, consider using the exponent key.

$$\approx 8619.07$$ Multiply and round to the nearest cent.

The total amount in the account after $1\frac{1}{2}$ years is $8619.07.

Skill Practice

7. Suppose $5000 is invested at 9% interest compounded monthly for 30 years. Use the formula for compound interest to find the total amount in the account after 30 years.

Calculator Connections

Topic: Using a Calculator to Compute Compound Interest

To enter the expression from Example 5 into a calculator, follow these keystrokes.

Expression	Keystrokes		Result

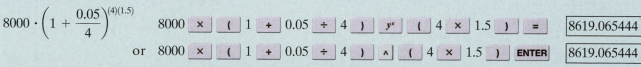

Expression

$8000 \cdot \left(1 + \dfrac{0.05}{4}\right)^{(4)(1.5)}$

8000 ✕ (1 + 0.05 ÷ 4) yˣ (4 ✕ 1.5) = | 8619.065444 |

or 8000 ✕ (1 + 0.05 ÷ 4) ^ (4 ✕ 1.5) ENTER | 8619.065444 |

Note: It is mandatory to insert parentheses ( ) around the product in the exponent.

Answer

7. The account is worth $73,652.88 after 30 years.

Section 6.7   Practice Exercises

Study Skills Exercises

1. Instructors vary in what they emphasize on tests. For example, test material may come from the textbook, notes, handouts, homework, etc. What do you find that your instructor emphasizes?

2. Define the key terms.

    a. **Simple interest**   b. **Principal**   c. **Compound interest**

Review Exercises

For Exercises 3–6, find the percent increase or the percent decrease. Round to the nearest whole percent.

	U.S. National Parks	Visitors in 2000 (Thousands)	Visitors in 2004 (Thousands)	Change	Percent Increase or Decrease
3.	Bryce Canyon, UT	1099	987		
4.	Petrified Forest, AZ	605	580		
5.	Denali, AK	364	404		
6.	Glacier, MT	1729	2034		

Objective 1: Simple Interest

For Exercises 7–14, find the simple interest and the total amount including interest. **(See Examples 1 and 2.)**

	Principal	Annual Interest Rate	Time, Years	Interest	Total Amount
7.	$6,000	5%	3	_____	_____
8.	$4,000	3%	2	_____	_____
9.	$5,050	6%	4	_____	_____
10.	$4,800	4%	3	_____	_____
11.	$12,000	4%	$4\frac{1}{2}$	_____	_____
12.	$6,230	7%	$6\frac{1}{3}$	_____	_____
13.	$10,500	4.5%	4	_____	_____
14.	$9,220	8%	4	_____	_____

15. Dale deposited $2500 in an account that pays $3\frac{1}{2}\%$ simple interest for 4 years.

   a. How much interest will he earn in 4 years?

   b. What will be the total value of the account after 4 years?

16. Charlene invested $3400 at 4% simple interest for 5 years.

   a. How much interest will she earn in 5 years?

   b. What will be the total value of the account after 5 years?

17. Gloria borrowed $400 for 18 months at 8% simple interest.

   a. How much interest will Gloria have to pay?

   b. What will be the total amount that she has to pay back?

18. Floyd borrowed $1000 for 2 years 3 months at 8% simple interest.

   a. How much interest will Floyd have to pay?

   b. What will be the total amount that he has to pay back?

19. Jozef deposited $10,300 into an account paying 4% simple interest 5 years ago. If he withdraws the entire amount of money, how much will he have?

20. Heather invested $20,000 in an account that pays 6% simple interest. If she invests the money for 10 years, how much will she have?

21. Anne borrowed $4500 from a bank that charges 10% simple interest. If she repays the loan in $2\frac{1}{2}$ years, how much will she have to pay back?

22. Dan borrowed $750 from his brother who is charging 8% simple interest. If Dan pays his brother back in 6 months, how much does he have to pay back?

Objective 2: Compound Interest

23. If a bank compounds interest semiannually for 3 years, how many total compounding periods are there?

24. If a bank compounds interest quarterly for 2 years, how many total compounding periods are there?

25. If a bank compounds interest monthly for 2 years, how many total compounding periods are there?

26. If a bank compounds interest monthly for $1\frac{1}{2}$ years, how many total compounding periods are there?

27. Mary Ellen deposited $500 in a bank. **(See Example 3.)**

   a. If the bank offers 4% simple interest, compute the amount in the account after 3 years.

   b. Now suppose the bank offers 4% interest compounded annually. Complete the table to determine the amount in the account after 3 years.

Year	Interest Earned	Total Amount in Account
1		
2		
3		

28. Fatima deposited $12,000 in an account.

   a. If the bank offers 5% simple interest, compute the amount in the account after 3 years.

   b. Now suppose the bank offers 5% interest compounded annually, Complete the table to determine the amount in the account after 3 years.

Year	Interest Earned	Total Amount in Account
1		
2		
3		

29. The amount of $8000 is invested at 4% for 3 years.

   a. Compute the ending balance if the bank calculates simple interest.

   b. Compute the ending balance if the bank calculates interest compounded annually.

Year	Interest Earned	Total Amount in Account
1		
2		
3		

   c. How much more interest is earned in the account with compound interest?

30. The amount of $12,000 is invested at 8% for 1 year.

   a. Compute the ending balance if the bank calculates simple interest.

   b. Compute the ending balance if the bank calculates interest compounded quarterly.

Year	Interest Earned	Total Amount in Account
1st		
2nd		
3rd		
4th		

   c. How much more interest is earned in the account with compound interest?

Objective 3: Using the Compound Interest Formula

31. For the formula $A = P \cdot \left(1 + \dfrac{r}{n}\right)^{n \cdot t}$, identify what each variable means.

32. If $1000 is deposited in an account paying 8% interest compounded monthly for 3 years, label the following variables: P, r, n, and t.

Calculator Connections

Topic: Computing Compound Interest

Calculator Exercises

For Exercises 33–40, find the total amount for the investment, using compound interest. **(See Examples 4–5.)**

	Principal	Annual Interest Rate	Time, Years	Compounded	Total Amount
33.	$5,000	4.5%	5	Annually	_____
34.	$12,000	5.25%	4	Annually	_____
35.	$6,000	5%	2	Semiannually	_____
36.	$4,000	3%	3	Semiannually	_____
37.	$10,000	6%	$1\frac{1}{2}$	Quarterly	_____
38.	$9,000	4%	$2\frac{1}{2}$	Quarterly	_____
39.	$14,000	4.5%	3	Monthly	_____
40.	$9,000	8%	2	Monthly	_____

Group Activity

Tracking Stocks

Materials: Computer with online access or the financial page of a newspaper.

Estimated time: 15 minutes

Group Size: 4

1. Each member of the group will choose a stock that is listed on the New York Stock Exchange. You can find prices in the newspaper or online at http://finance.yahoo.com/.

Here are some possible stocks to track.

Apple Computer	Home Depot
Best Buy	Microsoft
Walgreens	Nike
General Motors	Hershey
Exxon Mobil	Coca-Cola

2. Beginning on a Monday, record the closing price of the stock. [The closing price is the price at the end of the day at 4:00 P.M. Eastern Standard Time (EST).] Then, each day for 2 weeks, record the closing price of the stock along with the difference in price from the previous day. Record an increase in price in green with an up arrow and a decrease in price in red with a down arrow as shown below. For example:

Date	Price ($)	Increase or Decrease ($)	Percent Increase/Decrease
April 5	34.67		
April 6	34.82	0.15 ⇑	0.4%
April 7	33.98	0.84 ⇓	2.4%

Date	Price ($)	Increase or Decrease ($)	Percent Increase/Decrease

3. For each stock, calculate the amount of increase or decrease in price from the previous day.

4. For each stock, calculate the *percent* increase or decrease in price from the previous day.

5. Which stock was the best investment during this 2-week period? (This is generally considered to be the stock with the greatest percent increase.)

Chapter 6   Summary

Section 6.1   Percents and Their Fraction and Decimal Forms

Key Concepts

The word **percent** means *per one hundred.*

Converting Percents to Fractions

1. Replace the % symbol by $\times \frac{1}{100}$ (or by $\div 100$).
2. Simplify the fraction to lowest terms, if possible.

Converting Percents to Decimals

Replace the % symbol by $\times 0.01$. (This is equivalent to $\times \frac{1}{100}$ and $\div 100$).

Note: Multiplying a decimal by 0.01 is the same as moving the decimal point 2 places to the left.

Examples

Example 1

40% means 40 per 100 or $\frac{40}{100}$.

Example 2

$$84\% = 84 \times \frac{1}{100} = \frac{84}{100} = \frac{21}{25}$$

Example 3

$$24.5\% = 24.5 \times 0.01 = 0.245$$

Example 4

$$0.07\% = 0.07 \times 0.01 = 0.0007$$

(Move the decimal point 2 places to the left.)

Section 6.2   Fractions and Decimals and Their Percent Forms

Key Concepts

Converting Fractions and Decimals to Percent Form

Multiply the fraction or decimal by 100%. (100% = 1.)

Examples

Example 1

$$\frac{1}{5} = \frac{1}{5} \times 100\% = \frac{100}{5}\% = 20\%$$

Example 2

$$1.14 = 1.14 \times 100\% = 114\%$$

Example 3

$$\frac{2}{3} = 0.\overline{6} \times 100\% = 66.\overline{6}\% \text{ or } 66\tfrac{2}{3}\%$$

Section 6.3   Percent Proportions and Applications

Key Concepts

A **percent proportion** is a proportion that equates a percent to an equivalent ratio.

A percent proportion can be written in the form

$$\frac{\text{Amount}}{\text{Base}} = p\% \quad \text{or} \quad \frac{\text{Amount}}{\text{Base}} = \frac{p}{100}$$

The **base** is the total or whole amount being considered. The **amount** is the part being compared to the base.

To solve a percent proportion, equate the cross products and divide by the factor with the variable. The variable can represent the amount, base, or p. Examples 3–5 demonstrate each type of percent problem.

Example 4

Of a sample of 400 people, 85% found relief using a particular pain reliever. How many people found relief?

Solve the proportion: $\dfrac{x}{400} = \dfrac{85}{100}$

$$85 \cdot 400 = 100x$$

$$\frac{34{,}000}{100} = \frac{100x}{100}$$

$$340 = x$$

340 people found relief.

Examples

Example 1

$\dfrac{36}{100} = \dfrac{9}{25}$ is a percent proportion.

Example 2

For the percent proportion $\dfrac{12}{200} = \dfrac{6}{100}$,

12 is the amount, 200 is the base, and $p = 6$.

Example 3

44% of what number is 275?

Solve the proportion: $\dfrac{275}{x} = \dfrac{44}{100}$

$$44x = 275 \cdot 100$$

$$\frac{\overset{1}{\cancel{44}}x}{\underset{1}{\cancel{44}}} = \frac{27{,}500}{44}$$

$$x = 625$$

Example 5

There are approximately 750,000 career employees in the U.S. Postal Service. If 60,000 are mail handlers, what percent does this represent?

Solve the proportion: $\dfrac{60{,}000}{750{,}000} = \dfrac{p}{100}$

$$750{,}000p = 60{,}000 \cdot 100$$

$$\frac{\overset{1}{\cancel{750{,}000}}p}{\underset{1}{\cancel{750{,}000}}} = \frac{6{,}000{,}000}{750{,}000}$$

$$p = 8$$

Of postal employees, 8% are mail handlers.

Section 6.4 — Percent Equations and Applications

Key Concepts

A **percent equation** represents a percent proportion in an alternative form:

Amount $= (p\%) \cdot$ (base)

Examples 1–3 demonstrate three types of percent problems.

Example 2

Of the car repairs performed on a certain day, 21 were repairs on transmissions. If 60 cars were repaired, what percent involved transmissions?

Solve the equation: $21 = 60x$

$$\frac{21}{60} = \frac{\overset{1}{\cancel{60}}x}{\underset{1}{\cancel{60}}}$$

$$0.35 = x$$

Because the problem asks for a percent, we have
$x = 0.35$

$= 0.35 \times 100\%$

$= 35\%$

Therefore, 35% of cars repaired involved transmissions.

Examples

Example 1

Of all breast cancer cases, 99% occur in women. Out of 2700 cases of breast cancer reported, how many are expected to occur in women?

Solve the equation: $x = (99\%)(2700)$

$$x = (0.99)(2700)$$

$$x = 2673$$

About 2673 cases are expected to occur in women.

Example 3

There are 599 endangered plants in the U.S. This represents 60.7% of the total number of endangered species. Find the total number of endangered species. Round to the nearest whole number.

Solve the equation: $599 = 0.607x$

$$\frac{599}{0.607} = \frac{\overset{1}{\cancel{0.607}}x}{\underset{1}{\cancel{0.607}}}$$

$$987 \approx x$$

There is a total of approximately 987 endangered species in the U.S.

| **Section 6.5** | **Applications Involving Sales Tax, Commission, Discount, and Markup** |

Key Concepts

To find **sales tax**, use the formula

$$\binom{\text{Amount of}}{\text{sales tax}} = \binom{\text{Tax}}{\text{rate}} \cdot \binom{\text{Cost of}}{\text{merchandise}}$$

To find a **commission**, use the formula

$$\binom{\text{Amount of}}{\text{commission}} = \binom{\text{Commission}}{\text{rate}} \cdot \binom{\text{Total}}{\text{sales}}$$

To find **discount** and sale price, use the formulas

$$\binom{\text{Amount of}}{\text{discount}} = \binom{\text{Discount}}{\text{rate}} \cdot \binom{\text{Original}}{\text{price}}$$

Sale price = Original price − Amount of discount

To find **markup** and retail price, use the formulas

$$\binom{\text{Amount of}}{\text{markup}} = \binom{\text{Markup}}{\text{rate}} \cdot \binom{\text{Original}}{\text{price}}$$

Retail price = Original price + Amount of markup

Examples

Example 1

A DVD is priced at $16.50, and the total amount paid is $17.82. To find the sales tax rate, first find the amount of tax.

$17.82 − $16.50 = $1.32

To compute the sales tax rate, solve:

$$1.32 = x \cdot 16.50$$

$$\frac{1.32}{16.50} = \frac{x \cdot \overset{1}{\cancel{16.50}}}{\underset{1}{\cancel{16.50}}}$$

$$0.08 = x$$

The sales tax rate is 8%.

Example 2

Fletcher makes 13% commission on the sale of all merchandise. If he sells $11,290 worth of merchandise, find how much Fletcher will earn.

$$x = (0.13)(11{,}290)$$

$$= 1467.7$$

Fletcher will earn $1467.70 in commission.

Example 3

Margaret found a ring that was originally $425 but is on sale for 30% off. To find the sale price, first find the amount of discount.

$$a = (0.30) \cdot (425)$$

$$= 127.5$$

The sale price is $425 − $127.50 = $297.50.

Example 4

A wholesale coat company marks up its coats 20% before selling them retail. To find the retail price of a $340 coat, first find the amount of markup.

$$a = (0.20) \cdot (340)$$

$$= 68$$

The retail price is $340 + $68 = $408.

Section 6.6   Percent Increase and Decrease

Key Concepts

Percent increase or **percent decrease** compares the *change* between two given amounts to the *original amount*.

Computing Percent Increase or Decrease

$$\begin{pmatrix} \text{Percent} \\ \text{increase} \end{pmatrix} = \begin{pmatrix} \dfrac{\text{Amount of increase}}{\text{Original amount}} \end{pmatrix} \times 100\%$$

$$\begin{pmatrix} \text{Percent} \\ \text{decrease} \end{pmatrix} = \begin{pmatrix} \dfrac{\text{Amount of decrease}}{\text{Original amount}} \end{pmatrix} \times 100\%$$

Examples

Example 1

In one year a child grows from 35 in. to 42 in. The increase is $42 - 35 = 7$ in. The percent increase is

$$\frac{7}{35} \times 100\% = 0.20 \times 100\%$$

$$= 20\%$$

Section 6.7   Simple and Compound Interest

Key Concepts

To find the **simple interest** made on a certain **principal**, use the formula $I = Prt$

where I = amount of interest

P = amount of principal

r = annual interest rate

t = time, years

The formula for the total amount in an account is $A = P + I$, where A = total amount in an account.

Many day-to-day applications involve compound interest. **Compound interest** is based on both the original principal and the interest earned.

The formula $A = P \cdot \left(1 + \dfrac{r}{n}\right)^{n \cdot t}$ computes the total amount in an account that uses compound interest

where A = total amount in an account

P = principal

r = annual interest rate

t = time, years

n = number of compounding periods per year

Examples

Example 1

Betsey deposited $2200 in her account which pays 5.5% simple interest. To find the simple interest she will earn after 4 years, use the formula $I = Prt$ and solve for I.

$$I = (2200)(0.055)(4)$$

$$= 484 \quad \text{She will earn \$484 interest.}$$

To find the balance or total amount of her account, apply the formula $A = P + I$.

$$A = 2200 + 484$$

$$= 2684 \quad \text{Betsey's balance will be \$2684.}$$

Example 2

Gene borrows $1000 at 6% interest compounded semi-annually. If he pays off the loan in 3 years, how much will he have to pay?

We are given $P = 1000$, $r = 0.06$, $n = 2$ (semiannually means twice a year), and $t = 3$.

$$A = 1000\left(1 + \frac{0.06}{2}\right)^{2 \cdot 3}$$

$$= 1000(1.03)^6$$

$$\approx 1194.05$$

Gene will have to pay $1194.05 to pay off the loan with interest.

Chapter 6   Review Exercises

Section 6.1

For Exercises 1–4, use a percent to express the shaded portion of each drawing.

1.

2.

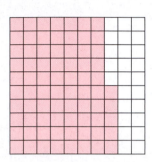

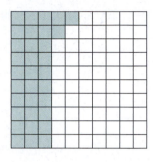

3.

4.

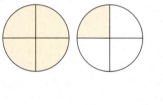

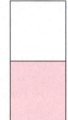

5. 68% can be expressed as which of the following forms? Identify all that apply.

   a. $\dfrac{68}{1000}$ **b.** $\dfrac{68}{100}$

   c. 0.68   **d.** 0.068

6. 0.4% can be expressed as which of the following forms? Identify all that apply.

   a. $\dfrac{4}{100}$ **b.** 0.04

   c. $\dfrac{0.4}{100}$ **d.** 0.004

For Exercises 7–12, match the percent with its fraction form.

7. 30%   **a.** $\dfrac{33}{100}$

8. $33\frac{1}{3}$%   **b.** $\dfrac{1}{2}$

9. 33%   **c.** $\dfrac{2}{3}$

10. 50%   **d.** $\dfrac{1}{3}$

11. $66\frac{2}{3}$%   **e.** $\dfrac{3}{5}$

12. 60%   **f.** $\dfrac{3}{10}$

For Exercises 13–18, match the percent with its decimal form.

13. 7.5%   **a.** 1

14. 75%   **b.** 0.25

15. 50%   **c.** 0.75

16. 100%   **d.** 0.0025

17. 0.25%   **e.** 0.075

18. 25%   **f.** 0.5

For Exercises 19–20, write the percent as a fraction and as a decimal.

19. Out of all the phone calls received per week, 42% were from solicitors.

20. In an hour-long TV show, 20% of the time is devoted to commercials.

The graph represents the average annual population growth rate for four of the fastest-growing cities. Use this graph for Exercises 21–24.

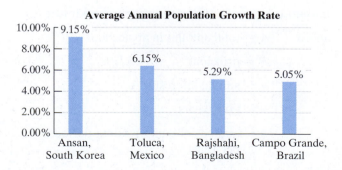

21. Write the rate for Toluca, Mexico, in decimal form.

22. Write the rate for Rajshahi, Bangladesh, in decimal form.

23. Write the rate for Ansan, South Korea, in fraction form.

24. Write the rate for Campo Grande, Brazil, in fraction form.

Section 6.2

For Exercises 25–28, convert the fraction to a percent.

25. $\dfrac{17}{100}$ **26.** $\dfrac{22}{50}$

27. $\dfrac{4}{5}$ **28.** $\dfrac{7}{4}$

For Exercises 29–32, convert the decimal to a percent.

29. 0.12   **30.** 1.1

31. 0.005   **32.** 0.4

For Exercises 33–36, write the fraction as a percent.

33. In a classroom the ratio of girls to boys is $\frac{14}{16}$.

34. In 2008, $\frac{19}{25}$ of the U.S. population were Internet users.

35. It is estimated that in 2010, $\frac{3}{5}$ of the population will be between the ages of 18 and 64 years.

36. In 2007, over $\frac{1}{10}$ of births were from teenage mothers.

For Exercises 37–42, complete the table.

	Fraction	Decimal	Percent
37.	$\frac{9}{20}$		
38.		1	
39.			6%
40.		1.2	
41.	$\frac{9}{1000}$		
42.			75%

Section 6.3

For Exercises 43–46, identify the amount, base, and p values for the percent proportion $\dfrac{\text{Amount}}{\text{Base}} = \dfrac{p}{100}$.

43. 45% of $150 is $67.50.

44. 360 births is 12% of 3000 births.

45. 30.24 m² of 144 m² is 21%.

46. 106% of 30 gal is 31.8 gal.

For Exercises 47–50, write the percent proportion.

47. 6 books of 8 books is 75%

48. 15% of 180 lb is 27 lb.

49. 200% of $420 is $840.

50. 6 pine trees out of 2000 pine trees is 0.3%.

For Exercises 51–56, solve the percent problems, using proportions.

51. What is 12% of 50?

52. $5\frac{3}{4}$% of 64 is what number?

53. 11 is what percent of 88?

54. 8 is what percent of 2500?

55. 13 is $33\frac{1}{3}$% of what number?

56. 24 is 120% of what number?

57. Based on recent statistics, one airline expects that 4.2% of its customers will be "no-shows." If the airline sold 260 seats, how many people would the airline expect as no-shows? Round to the nearest whole unit.

58. In a survey of college students, 58% said that they wore their seatbelts regularly. If this represents 493 people, how many people were surveyed?

59. Victoria spends $720 per month on rent. If her monthly take-home pay is $1800, what percent does she pay in rent? Round to the nearest whole percent.

60. Of the rental cars at the U-Rent-It company, 40% are compact cars. If this represents 26 cars, how many cars are on the lot?

Section 6.4

For Exercises 61–66, write as a percent equation and solve.

61. 18% of 900 is what number?

62. What number is 29% of 404?

63. 18.90 is what percent of 63?

64. What percent of 250 is 86?

65. 30 is 25% of what number?

66. 26 is 130% of what number?

67. A student buys a used book for $54.40. This is 80% of the original price. What was the original price? Round to the nearest cent.

68. Veronica has read 330 pages of a 600-page novel. What percent of the novel has she read?

69. Elaine tries to keep her fat intake to no more than 30% of her total calories. If she has a 2400-calorie diet, how many fat calories can she consume to stay within her goal?

70. It is predicted that by 2010, 13% of Americans will be over the age of 65. By 2050 that number could rise to 20%. Suppose that the U.S. population is 300,000,000 in 2010 and 404,000,000 in 2050.

    a. Find the number of Americans over 65 in the year 2010.

    b. Find the number of Americans over 65 in the year 2050.

Section 6.5

For Exercises 71–74, solve the problem involving sales tax.

71. A Plasma TV costs $1279. Find the sales tax if the rate is 6%.

72. The sales tax on a sofa is $47.95. If the sofa costs $685.00 before tax, what is the sales tax rate?

73. To get a roll of film developed, it costs $6.75. The total bill came to $7.29.

    a. How much is the sales tax?

    b. What is the sales tax rate?

74. A resort hotel charges an 11% resort tax along with the 6% sales tax. If the hotel's one-night accommodation is $125.00, what will a tourist pay for 4 nights, including tax?

For Exercises 75–78, solve the problems involving commission.

75. At a recent auction, *Boy with a Pipe,* an early work by Pablo Picasso, sold for $104 million. The commission for the sale of the work was $11 million. What was the rate of commission? Round to the nearest tenth of a percent. (*Source: The New York Times*)

76. Andre earns a commission of 12% on sales of restaurant supplies. If he sells $4075 in one week, how much commission will he earn?

77. Sela sells sportswear at a department store. She earns an hourly wage of $8, and she gets a 5% commission on all merchandise that she sells over $200. If Sela works an 8-hr day and sells $420 of merchandise, how much will she earn that day?

78. A house is sold for $160,000, and the real estate agent earned $9600 in commission. What is the commission rate?

For Exercises 79–82, solve the problems involving discount and markup.

79. Find the discount and the sale price of the movie if the regular price is $28.95.

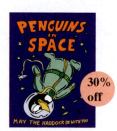

80. This notebook computer was originally priced at $1747. How much is the discount? After the $50 rebate, how much will a person pay for this computer?

81. A rug manufacturer sells a rug to a retail store for $160. The store then marks up the rug to $208. What is the markup rate?

82. Peg sold some homemade baskets to a store for $50 each. The store marks up all merchandise by 18%. What will be the retail price of the baskets after the markup?

Section 6.6

For Exercises 83–84, (**a**) identify if there is an increase or decrease and (**b**) find the percent of increase or decrease.

83. 86 to 107.5   **84.** 410 to 82

85. The number of species of animals on the endangered species list went from 263 in 1990 to 410 in 2006. Find the percent increase. Round to the nearest tenth of a percent. (*Source:* U.S. Fish and Wildlife Services)

86. The number of corded phones went from 30 million in 2000 to 22 million in 2003. Find the percent decrease. Round to the nearest whole percent. (*Source:* Telecommunications Association)

The number of subscribers to cellular telephones has increased since 1985. Use the information in the table to answer Exercises 87–88. In each case, round to the nearest whole percent.

Year	Number of Subscribers (1000s)
1985	300
1989	3,500
1993	16,000
1997	55,000
2001	128,000
2006	224,000

87. What was the percent increase in cellular phone subscribers between 2001 and 2006?

88. What was the percent increase in cellular phone subscribers between 1985 and 1989?

Section 6.7

For Exercises 89–90, find the simple interest and the total amount including interest.

	Principal	Annual Interest Rate	Time, Years	Interest	Total Amount
89.	$10,200	3%	4	_____	_____
90.	$7000	4%	5	_____	_____

91. Jean-Luc borrowed $2500 at 5% simple interest. What is the total amount that he will pay back at the end of 18 months (1.5 years)?

92. Kyle loaned his brother $800 for 2 years. He is charging 2.5% simple interest. How much will his brother owe Kyle in 2 years?

93. Sydney deposited $6000 in a certificate of deposit that pays 4% interest compounded annually. Complete the table to determine her balance after 3 years.

Year	Interest	Total
1		
2		
3		

94. Nell deposited $10,000 in a money market account that pays 3% interest compounded semiannually. Complete the table to find her balance after 2 years.

Compound Periods	Interest	Total
Period 1 (end of first 6 months)		
Period 2 (end of year 1)		
Period 3 (end of 18 months)		
Period 4 (end of year 2)		

For Exercises 95–98, find the total amount for the investment, using compound interest. Use the formula

$$A = P \cdot \left(1 + \frac{r}{n}\right)^{n \cdot t}.$$

	Principal	Annual Interest Rate	Time in Years	Compounded	Total Amount
95.	$850	8%	2	Quarterly	_____
96.	$2050	5%	5	Semiannually	_____
97.	$11,000	7.5%	6	Annually	_____
98.	$8200	4.5%	4	Monthly	_____

Chapter 6   Test

1. Write a percent to express the shaded portion of the figure.

2. Shade the figure so that it represents 85%.

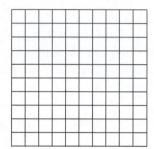

3. Write the percent in decimal form and in fraction form.

   a. For a recent year, the unemployment rate of Illinois was 5.4%.

   b. The incidence of breast cancer increased by 0.15% between 2003 and 2005.

   c. For a certain city, gas prices increased by 170% in 10 years.

4. Write the following percents in fraction form.

   a. 1%   **b.** 25%

   c. $33\frac{1}{3}$%   **d.** 50%

   e. $66\frac{2}{3}$%   **f.** 75%

   g. 100%   **h.** 150%

For Exercises 5–6, write the percent as a decimal and as a fraction.

5. The incidence of colon cancer decreased by 1.5% between 2003 and 2005. (*Source:* National Cancer Institute)

6. In 1950, 9.9% of the U.S. population was made up of African Americans.

7. Explain the process to write a fraction as a percent.

For Exercises 8–11, write the fraction as a percent. Round to the nearest tenth of a percent if necessary.

8. $\dfrac{3}{5}$ **9.** $\dfrac{1}{250}$

10. $\dfrac{7}{4}$ **11.** $\dfrac{5}{7}$

12. Explain the process to write a decimal as a percent.

For Exercises 13–16, write the decimal as a percent.

13. 0.32   **14.** 0.052

15. 1.3   **16.** 0.006

For Exercises 17–22, solve the percent problems.

17. What is 24% of 150?

18. What is 120% of 16?

19. 21 is 6% of what number?

20. 40% of what number is 80?

21. What percent of 220 is 198?

22. 75 is what percent of 150?

23. At McDonald's, a side salad without dressing has 10 mg of sodium. With a serving of Newman's Own Low-Fat Balsamic Dressing, the sodium content of the salad is 740 mg. (*Source:* www.mcdonalds.com)

   a. How much sodium is in the dressing itself?

   b. What percent of the sodium content in a side salad with dressing is from the dressing? Round to the nearest tenth of a percent.

The composition of the lower level of the Earth's atmosphere is given in the figure (other gases are present in minute quantities). For Exercises 24–25, use the information in the graph.

Composition of the Earth's Atmosphere

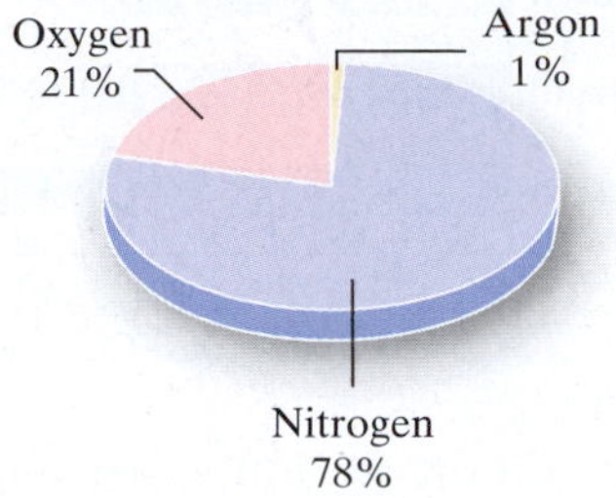
Oxygen
21%
Argon
1%
Nitrogen
78%

24. How much nitrogen would be expected in 500 m^3 of atmosphere?

25. How much oxygen would be expected in 2000 m^3 of atmosphere?

26. Brad bought a pair of blue jeans that cost \$30.00. He wrote his check for \$32.10.

a. What is the amount of sales tax that he paid?

b. What is the sales tax rate?

27. Charles earns a salary of \$400 per week and gets a bonus of 6% commission on all merchandise that he sells. If Charles sells \$3500 worth of merchandise, how much will he earn in that week?

28. Find the discount rate of the product in the advertisement.

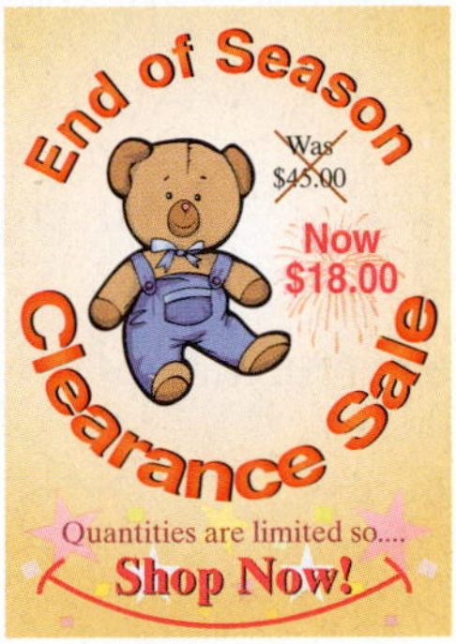
End of Season
Was
$45.00
Now
$18.00
Clearance Sale
Quantities are limited so....
Shop Now!

29. The price of hogs in 2002 was \$0.32 per pound. In 2004, the price rose to \$0.44 per pound. What was the percent increase?

30. Maury borrowed \$5000 at 8% simple interest. He plans to pay back the loan in 3 years.

a. How much interest will he have to pay?

b. What is the total amount that he has to pay back?

31. Use the formula $A = P \cdot \left(1 + \frac{r}{n}\right)^{n \cdot t}$ to calculate the total amount in an account that began with \$25,000 invested at 4.5% compounded quarterly for 5 years.

Chapters 1–6 Cumulative Review Exercises

1. What is the name of the place value for the digit 6 in the number 26,009,235?

2. Fill in the table with either the word name for the number or the number in standard form.

	Country	Area (mi^2) Standard Form	Area (mi^2) Words
a.	United States		Three million, five hundred thirty-nine thousand, two hundred forty-five
b.	Saudi Arabia	830,000	
c.	Falkland Islands	4,700	
d.	Colombia		Four hundred one thousand, forty-four

3. Multiply: $\begin{array}{r} 34{,}882 \\ \times\ 100 \\ \hline \end{array}$

4. Divide: $9\overline{)783}$

5. Add: $234 + 44 + 6 + 2901$

6. Simplify: $\sqrt{16} - 6 \div 3 + 3^2$

7. Identify whether the fraction is proper or improper.

a. $\frac{6}{6}$ **b.** $\frac{10}{7}$

c. $\frac{7}{10}$ **d.** $\frac{1}{100}$

For Exercises 8–11, multiply or divide as indicated. Simplify the fraction to lowest terms.

8. $\dfrac{3}{8} \times \dfrac{32}{9}$

9. $\dfrac{42}{25} \div \dfrac{7}{100}$

10. $\dfrac{21}{2} \div 7$

11. $16 \times \dfrac{1}{24}$

12. Find the area of the figure.

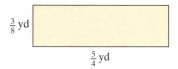

$\frac{3}{8}$ yd

$\frac{5}{4}$ yd

13. Find the perimeter of the figure.

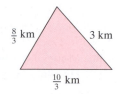

$\frac{8}{3}$ km   3 km

$\frac{10}{3}$ km

14. Add: $\dfrac{3}{10} + \dfrac{17}{100} + \dfrac{3}{1000}$

15. The value of the Daxor stock rose $\$1\frac{2}{5}$ from $\$14\frac{7}{10}$. What is the current price?

16. A sheet of paper has the dimensions $13\frac{1}{2}$ in. by 17 in. What is the area of the paper?

17. a. List four multiples of 18.

 b. List all factors of 18.

 c. Write the prime factorization of 18.

18. Write a fraction that represents the shaded portion of each figure.

 a.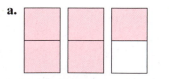

 b.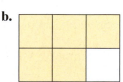

For Exercises 19–22, write the fraction as a decimal.

19. $\dfrac{3}{8}$

20. $\dfrac{4}{3}$

21. $\dfrac{7}{9}$

22. $\dfrac{3}{4}$

For Exercises 23–24, refer to the chart.

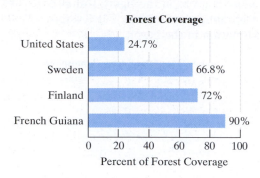

Forest Coverage

United States — 24.7%
Sweden — 66.8%
Finland — 72%
French Guiana — 90%

0 20 40 60 80 100
Percent of Forest Coverage

23. What is the difference in the percent of forest coverage in French Guiana and the United States?

24. What is the difference in the percent of forest coverage in Sweden and the United States?

For Exercises 25–28, multiply and divide by powers of 10.

25. 85×0.001

26. 85×100

27. $85 \div 10$

28. $85 \div 0.0001$

For Exercises 29–32, solve the proportions.

29. $\dfrac{3}{4} = \dfrac{15}{p}$

30. $\dfrac{2.5}{6} = \dfrac{p}{9}$

31. $\dfrac{4\frac{1}{3}}{p} = \dfrac{12}{18}$

32. $\dfrac{p}{100} = \dfrac{21.87}{81}$

33. If it takes $\frac{1}{2}$ hr to read 1 chapter of a book, how long will it take to read 5 chapters?

34. A 9.2-oz bag of candy costs $\$2.30$. Find the unit cost, that is, the price per ounce.

35. A computer can download 1.6 megabytes (MB) in 2.5 min. How long will it take to download a 4.6-MB file? Round to the nearest tenth of a minute.

36. A DC-10 aircraft flew 1799 mi in 3.5 hr. Find the unit rate in miles per hour.

37. By 2008 it is projected that online grocery sales will reach $17.4 billion. In 2003, sales were only $3.7 billion. Find the percent increase. Round to the nearest whole percent. (*Source: USA TODAY*)

38. New York City has the greatest population of any U.S. city. In 1900 it had 3.4 million people. By 2000, it had 8.0 million.

   a. Compute the difference in population between the year 2000 and the year 1900.

   b. Assuming a steady increase, compute the rate representing the increase in population per year. (*Source:* U.S. Bureau of the Census)

39. Kevin deposited $13,000 into a certificate of deposit that pays 3.2% simple interest. How much will Kevin have if he keeps the certificate for 5 years?

40. A mortgage charges 8% interest that is compounded monthly. Use the formula $A = P \cdot \left(1 + \dfrac{r}{n}\right)^{n \cdot t}$ to find the total amount of interest paid for a 10-year mortgage on $75,000.

3. Converting U.S. Customary Units of Length by Using Unit Ratios

Example 1 demonstrates how substitution can be used to convert between two units of measure. Another method is to multiply by a **conversion factor**. A conversion factor is a ratio of equivalent measures.

For example, note that 1 yd = 3 ft. Therefore, $\dfrac{1 \text{ yd}}{3 \text{ ft}} = 1$ and $\dfrac{3 \text{ ft}}{1 \text{ yd}} = 1$.

These ratios are called **unit ratios** (or **unit fractions**). In a unit ratio, the quotient is 1 because we are dividing measurements of equal length. To convert from one unit of measure to another, we can multiply by a unit ratio. We offer these guidelines to determine the proper unit ratio to use.

> **PROCEDURE   Choosing a Unit Ratio as a Conversion Factor**
>
> In a unit ratio,
> - The unit of measure in the numerator should be the new unit you want to convert *to*.
> - The unit of measure in the denominator should be the original unit you want to convert *from*.

Example 2   Converting Units of Length by Using Unit Ratios

a. 1500 ft = _____ yd   **b.** 9240 yd = _____ mi   **c.** 8.2 mi = _____ ft

Solution:

a. From Table 7-1, we have 1 yd = 3 ft.

$$1500 \text{ ft} = 1500 \text{ ft} \cdot \frac{1 \text{ yd}}{3 \text{ ft}}$$

← new unit to convert to
← unit to convert from

$$= \frac{1500 \text{ ft}}{1} \cdot \frac{1 \text{ yd}}{3 \text{ ft}}$$

Notice that the original units of ft reduce or "cancel" in much the same way as simplifying fractions. The unit yd remains in the final answer.

$$= \frac{1500}{3} \text{ yd}$$

$$= 500 \text{ yd}$$

b. From Table 7-1, we have 1 mi = 1760 yd.

$$9240 \text{ yd} = 9240 \text{ yd} \cdot \frac{1 \text{ mi}}{1760 \text{ yd}}$$

← new unit to convert to
← unit to convert from

$$= \frac{9240 \text{ yd}}{1} \cdot \frac{1 \text{ mi}}{1760 \text{ yd}}$$

The units of yd reduce, leaving the answer in miles.

$$= \frac{9240}{1760} \text{ mi}$$

Multiply fractions.

$$= 5.25 \text{ mi}$$

Simplify.

Concept Connections

Complete the unit ratio.

3. $\dfrac{1 \text{ ft}}{\text{in.}}$ **4.** $\dfrac{\text{yd}}{3 \text{ ft}}$

Skill Practice

Convert, using unit ratios.

5. 720 in. = _____ ft
6. 4224 ft = _____ mi
7. 8 mi = _____ yd

TIP: It is important to write the units associated with the numbers. The units can help you select the correct unit ratio.

Answers

3. $\dfrac{1 \text{ ft}}{12 \text{ in.}}$ **4.** $\dfrac{1 \text{ yd}}{3 \text{ ft}}$ **5.** 60 ft
6. 0.8 mi   **7.** 14,080 yd

Measurement

7

CHAPTER OUTLINE

Chapter 7

In this chapter we present units of measurement in both the U.S. Customary System of measure and the metric system. It is useful to have a mental concept of these measurements for estimating in many real world situations. We will also learn how to convert between different units of measurement.

Locate the following terms from this chapter in the word search puzzle.

ton   pound   ounce   quart   gallon   cup
pint   mile   foot   yard   inch

```
t n i p a v m q r
o i t o h h a u s
o b o u n c e g t
f i m n g l n y r
j h m d e o w i y
d n m i l e o k n
c r i l o c l t o
s d a r q u a r t
n g c y t p i m o
```

Section 7.1 Converting U.S. Customary Units of Length

Objectives

1. **U.S. Customary Units of Length**
2. **Converting U.S. Customary Units of Length by Using Substitution**
3. **Converting U.S. Customary Units of Length by Using Unit Ratios**
4. **Adding and Subtracting Mixed Units**
5. **Applications**

1. U.S. Customary Units of Length

In many applications in day-to-day life, we need to measure things. To measure an object means to assign it a number and a **unit of measure**. We first present common units of measure for length which include inches (in.), feet (ft), yards (yd), and miles (mi). These units are part of the **U.S. Customary System of measurement** (sometimes called the English system of measurement).

The U.S. Customary units of length and some common equivalents are given in Table 7-1.

Table 7-1   U.S. Customary Units of Length and Their Equivalents

1 ft = 12 in.	1 in. = $\frac{1}{12}$ ft
1 yd = 3 ft	1 ft = $\frac{1}{3}$ yd
1 mi = 5280 ft	1 ft = $\frac{1}{5280}$ mi
1 mi = 1760 yd	1 yd = $\frac{1}{1760}$ mi

Note that sometimes units of feet are denoted with the ′ symbol. That is, 3 ft = 3′. Similarly, sometimes units of inches are denoted with the ″ symbol. That is, 4 in. = 4″.

2. Converting U.S. Customary Units of Length by Using Substitution

Example 1 demonstrates how we can use substitution to convert between two units of length.

Skill Practice

Use substitution to convert units.
1. 6 ft = _____ in.
2. 12.6 ft = _____ yd

Example 1   Converting Units of Length

a. 4 ft = _____ in.   b. 66 in. = _____ ft

Solution:

a. First note that 4 ft = 4 × 1 ft. Then we can substitute 1 ft = 12 in.

$4 \text{ ft} = 4 \times 1 \text{ ft}$

$= 4 \times 12 \text{ in.}$ Substitute 1 ft = 12 in.

$= 48 \text{ in.}$

b. 66 in. = 66 × 1 in.

$= 66 \times \frac{1}{12} \text{ ft}$ Substitute 1 in. = $\frac{1}{12}$ ft.

$= \frac{66}{12} \text{ ft}$ Multiply.

$= \frac{11}{2} \text{ ft or } 5\frac{1}{2} \text{ ft or 5.5 ft}$

TIP: In Example 1(a) we converted feet to inches by multiplying by 12. In Example 1(b) we reversed this process to convert inches to feet. In this case we multiplied by $\frac{1}{12}$ which is the same as *dividing* by 12.

Answers
1. 72 in.   2. 4.2 yd

---

c. From Table 7-1 we have 1 mi = 5280 ft.

$8.2 \text{ mi} = 8.2 \text{ mi} \cdot \dfrac{5280 \text{ ft}}{1 \text{ mi}}$ ← new unit to convert to
    ← unit to convert from

$= \dfrac{8.2 \text{ mi}}{1} \cdot \dfrac{5280 \text{ ft}}{1 \text{ mi}}$ The units of mi reduce, leaving the answer in feet.

$= 43,296 \text{ ft}$

Skill Practice

Convert, using unit ratios.
8. 6.2 yd = _____ in.
9. 6336 in. = _____ mi

Example 3   Making Multiple Conversions of Length

a. 0.25 mi = _____ in.   b. 22 in. = _____ yd

Solution:

a. To convert miles to inches, we use two conversion factors. The first unit ratio converts miles to feet. The second unit ratio converts feet to inches.

converts mi to ft   converts ft to in.

$0.25 \text{ mi} = 0.25 \text{ mi} \cdot \dfrac{5280 \text{ ft}}{1 \text{ mi}} \cdot \dfrac{12 \text{ in.}}{1 \text{ ft}}$

$= \dfrac{0.25 \text{ mi}}{1} \cdot \dfrac{5280 \text{ ft}}{1 \text{ mi}} \cdot \dfrac{12 \text{ in.}}{1 \text{ ft}}$ The units mi and ft reduce, leaving the answer in inches.

$= 15,840 \text{ in.}$

converts in. to ft   converts ft to yd

b. $22 \text{ in.} = 22 \text{ in.} \cdot \dfrac{1 \text{ ft}}{12 \text{ in.}} \cdot \dfrac{1 \text{ yd}}{3 \text{ ft}}$ Multiply by two conversion factors. The first converts inches to feet. The second converts feet to yards.

$= \dfrac{22 \text{ in.}}{1} \cdot \dfrac{1 \text{ ft}}{12 \text{ in.}} \cdot \dfrac{1 \text{ yd}}{3 \text{ ft}}$ The units in. and ft reduce, leaving the answer in yards.

$= \dfrac{22}{36} \text{ yd}$ Multiply fractions.

$= \dfrac{11}{18} \text{ yd} \quad \text{or} \quad 0.6\overline{1} \text{ yd}$ Simplify.

4. Adding and Subtracting Mixed Units

To add and subtract measurements, we must have like units. For example:

$$3 \text{ ft} + 8 \text{ ft} = 11 \text{ ft}$$

Sometimes, however, measurements have mixed units. For example, a drainpipe might be 4 ft 6 in. long or symbolically 4′6″. Measurements and calculations with mixed units can be handled in much the same way as mixed numbers.

Answers
8. 223.2 in.   9. 0.1 mi

Example 4 | **Adding and Subtracting Mixed Units of Measurement**

a. Add 4′6″ + 2′9″.

b. Subtract 8′2″ − 3′6″.

Solution:

a. 4′6″ + 2′9″ =   4 ft +   6 in.
   +2 ft +   9 in.
   _____
   6 ft + 15 in.   Add like units.

= 6 ft + 1 ft + 3 in.   Because 15 in. is more than 1 ft, we can write 15 in. = 1 ft + 3 in.

= 7 ft 3 in.   or   7′3″

b. 8′2″ − 3′6″ =   8 ft + 2 in. = $\overset{7}{8}$ ft + $\overset{12\,in.}{2}$ in.   Borrow 1 ft = 12 in.
   −(3 ft + 6 in.)   −(3 ft + 6 in.)

=   7 ft + 14 in.
   −(3 ft +   6 in.)
   _____
   4 ft +   8 in.   or   4′8″

5. Applications

In Example 5 we convert to a common unit of measurement to find the perimeter of an object.

Example 5 | **Finding Perimeter**

Find the perimeter in feet.

2.2 mi

1056 ft

Solution:

The perimeter is found by adding the lengths of the sides. Before we begin, the length and width must have the same units. We will convert 2.2 mi to feet.

First note that 2.2 mi = 2.2 mi · $\frac{5280 \text{ ft}}{1 \text{ mi}}$

2.2 mi = 11,616 ft

1056 ft

= 11,616 ft

Perimeter = 11,616 ft + 1056 ft + 11,616 ft + 1056 ft

= 25,344 ft

Answers
10. 13′2″   11. 1′6″
12. 42 ft

Skill Practice

13. Shemika cut 10 pieces of rope that are each 2 ft 3 in. How much total rope was cut?

Example 6 Applying U.S. Customary Units of Length

A plumber needs 5 pipes that are 8 ft 4 in. each. What is the total length of piping?

Solution:

First, note that 8 ft 4 in. = 8 ft + 4 in. Second, recall that multiplication represents repeated addition. Therefore, the total length is 5(8 ft + 4 in.). See Figure 7-1. We can apply the distributive property introduced in Section 1.5. That is, $a(b + c) = ab + ac$.

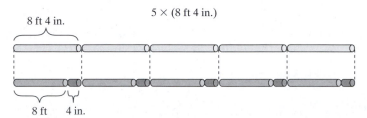

Figure 7-1

$$5(8 \text{ ft} + 4 \text{ in.}) = 5(8 \text{ ft}) + 5(4 \text{ in.})$$
$$= 40 \text{ ft} + 20 \text{ in.}$$
$$= 40 \text{ ft} + 1 \text{ ft} + 8 \text{ in.} \qquad \text{Note that } 20 \text{ in.} = 1 \text{ ft} + 8 \text{ in.}$$
$$= 41 \text{ ft} + 8 \text{ in.}$$

The total length of piping is 41 ft 8 in.

TIP: Example 6 could have been solved by first converting 8 ft 4 in. to either inches or feet and then multiplying by 5. Converting to inches, we have

$$8 \text{ ft } 4 \text{ in.} = 96 \text{ in.} + 4 \text{ in.} = 100 \text{ in.}$$

Now multiply by 5. $5(100 \text{ in.}) = 500 \text{ in.}$

$$= 500 \text{ in.} \cdot \left(\frac{1 \text{ ft}}{12 \text{ in.}} \right) \longleftarrow \text{Convert back to feet.}$$

$$= \frac{500}{12} \text{ ft}$$

$$= 41 \text{ ft. } 8 \text{ in.} \qquad \text{Convert to mixed units by dividing.}$$

$$\begin{array}{r} 41 \\ 12\overline{)500} \\ \underline{48} \\ 20 \\ \underline{12} \\ 8 \end{array}$$

Answer

13. 22 ft 6 in.

Example 7   **Applying U.S. Customary Units of Length in Construction**

A carpenter has a board 6′10″ in length. He must cut the board into two pieces of equal length. How long is each piece?

Solution:

The distance 6′10″ is equal to 6 ft 10 in. or 6 ft + 10 in. We can divide this total distance in half by dividing each individual unit by 2. That is,

$$(6 \text{ ft} + 10 \text{ in.}) \div 2 = \frac{6 \text{ ft}}{2} + \frac{10 \text{ in.}}{2} = 3 \text{ ft} + 5 \text{ in.} \quad \text{or} \quad 3'5''$$

Skill Practice

14. A 12′6″ length of ribbon is cut into 3 pieces of equal length. Find the length of each piece.

Answer

14. 4 ft 2 in.

Section 7.1   Practice Exercises

Boost your GRADE at ALEKS.com!

- Practice Problems
- Self-Tests
- NetTutor
- e-Professors
- Videos

Study Skills Exercises

1. Careless mistakes are usually caused by losing focus on what you are doing. What are some of the distractions that you encounter when doing homework?

List some ways you can avoid distractions while doing your homework.

2. Define the key terms.

    a. Unit of measurement

    b. U.S. Customary System of measurement

    c. Conversion factor

    d. Unit ratio (or unit fraction)

Objective 1: U.S. Customary Units of Length

For Exercises 3–8, fill in the blanks with the correct units. Refer to Table 7-1.

3. 5280 ft = 1 ____

4. $\frac{1}{12}$ ft = 1 ____

5. 1 yd = 3 ____

6. 1 ft = 12 ____

7. 1 ft = $\frac{1}{3}$ ____

8. 1 ft = $\frac{1}{5280}$ ____

Objective 2: Converting U.S. Customary Units of Length by Using Substitution

For Exercises 9–20, convert the units of length by using substitution. **(See Example 1.)**

9. 2 yd = ____ ft

10. 2 mi = ____ yd

11. 6 ft = ____ in.

12. 1.25 mi = ____ ft

13. 2 mi = ____ ft

14. 5 ft = ____ in.

15. 24 ft = ____ yd

16. 36 in. = ____ ft

17. 9 in. = ____ ft

18. 10 ft = ____ yd

19. 1760 ft = ____ mi

20. 880 yd = ____ mi

Objective 3: Converting U.S. Customary Units of Length by Using Unit Ratios

21. Identify an appropriate ratio to convert feet to inches by using multiplication.

   a. $\dfrac{12 \text{ ft}}{1 \text{ in.}}$ **b.** $\dfrac{12 \text{ in.}}{1 \text{ ft}}$ **c.** $\dfrac{1 \text{ ft}}{12 \text{ in.}}$ **d.** $\dfrac{1 \text{ in.}}{12 \text{ ft}}$

22. Identify an appropriate ratio to convert yards to miles by using multiplication.

   a. $\dfrac{1 \text{ mi}}{5280 \text{ ft}}$ **b.** $\dfrac{1760 \text{ mi}}{1 \text{ yd}}$ **c.** $\dfrac{1 \text{ mi}}{1760 \text{ yd}}$ **d.** $\dfrac{1760 \text{ yd}}{1 \text{ mi}}$

23. Identify an appropriate ratio to convert yards to feet by using multiplication.

   a. $\dfrac{3 \text{ ft}}{1 \text{ yd}}$ **b.** $\dfrac{1 \text{ yd}}{3 \text{ ft}}$ **c.** $\dfrac{3 \text{ yd}}{1 \text{ ft}}$ **d.** $\dfrac{1 \text{ ft}}{3 \text{ yd}}$

24. Identify an appropriate ratio to convert feet to miles by using multiplication.

   a. $\dfrac{5280 \text{ ft}}{1 \text{ mi}}$ **b.** $\dfrac{1 \text{ ft}}{5280 \text{ mi}}$ **c.** $\dfrac{5280 \text{ mi}}{1 \text{ ft}}$ **d.** $\dfrac{1 \text{ mi}}{5280 \text{ ft}}$

For Exercises 25–36, convert the units of length by using unit ratios. **(See Example 2.)**

25. 9 ft = _____ yd

26. $2\dfrac{1}{3}$ yd = _____ ft

27. 3.5 ft = _____ in.

28. $4\dfrac{1}{2}$ in. = _____ ft

29. 11,880 ft = _____ mi

30. 0.75 mi = _____ ft

31. 6 yd = _____ ft

32. 5280 yd = _____ mi

33. 14 ft = _____ yd

34. 75 in. = _____ ft

35. 320 mi = _____ yd

36. $3\dfrac{1}{4}$ ft = _____ in.

For Exercises 37–45, convert the units of length, involving multiple conversions. **(See Example 3.)**

37. 171 in. = _____ yd

38. 0.3 mi = _____ in.

39. 2 yd = _____ in.

40. 12,672 in. = _____ mi

41. 0.8 mi = _____ in.

42. 900 in. = _____ yd

43. 12,672 in. = _____ mi

44. 6 yd = _____ in.

45. 1.6 mi = _____ in.

Objective 4: Adding and Subtracting Mixed Units

46. a. Convert 6′4″ to inches.

   b. Convert 6′4″ to feet.

47. a. Convert 10 ft 8 in. to inches.

   b. Convert 10 ft 8 in. to feet.

48. a. Convert 2 yd 2 ft to feet.

   b. Convert 2 yd 2 ft to yards.

49. a. Convert 3′6″ to feet.

   b. Convert 3′6″ to inches.

For Exercises 50–58, add or subtract as indicated. **(See Example 4.)**

50. 1′3″ + 6′4″

51. 6′2″ + 4′6″

52. 2 ft 8 in. + 3 ft 4 in.

53. 5 ft 2 in. + 6 ft 10 in.

54. 4′10″ + 6′4″

55. 4′9″ + 3′9″

56. 8 ft 8 in. − 5 ft 4 in.

57. 3 ft 2 in. − 1 ft 5 in.

58. 9′2″ − 4′10″

For Exercises 59–66, multiply or divide as indicated.

59. 2(4 ft 5 in.)

60. 4(5 ft 1 in.)

61. 6(4 ft 8 in.)

62. 8(2 ft 5 in.)

63. (6′4″) ÷ 2

64. (16′8″) ÷ 8

65. $\dfrac{18 \text{ ft } 3 \text{ in.}}{3}$

66. $\dfrac{10 \text{ ft } 10 \text{ in.}}{5}$

Objective 5: Applications (Mixed Exercises)

67. Find the perimeter in feet. **(See Example 5.)**

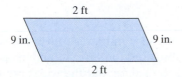

68. Find the perimeter in yards.

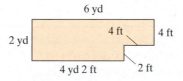

69. The garden pictured needs a decorative border. The border comes in pieces that are 1.5 ft long. How many pieces of border are needed?

70. Monte fences all sides of a field with panels of fencing that are 2 yd wide. How many panels of fencing does he need?

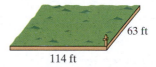

71. A carpenter needs eight pieces of molding that are 6 ft 4 in. each. What is the total length? **(See Example 6.)**

72. To gift wrap a package, 3′2″ of ribbon is needed. How much ribbon is needed to wrap 10 packages?

73. A plumber used two pieces of pipe for a job. One piece was 4′6″ and the other was 2′8″. How much pipe was used?

74. A carpenter needs to put wood molding around three sides of a room. Two sides are 6′8″ long, and the third side is 10′ long. How much molding should the carpenter purchase?

75. If you have 4 yd of rope and you use 5 ft, how much is left over? Express the answer in feet.

76. In 2002, the Blaisdell Arena football field, home of the Hawaiian Islanders football team, was discovered to be smaller than the official dimensions. The width was measured to be 82 ft 10 in. Regulation width is 85 ft. What is the difference between the widths of a regulation field and the field at Blaisdell Arena?

77. A cable 6 ft 9 in. is cut into three pieces of equal length. How long is each piece? **(See Example 7.)**

78. A piece of rope 8′6″ in length is cut in half. How long is each piece?

79. A picnic table requires 5 boards that are 6′ long, 4 boards that are 3′3″ long, and 2 boards that are 18″ long. Find the total length of lumber required.

80. A roll of ribbon is 60 yd. If you wrap 12 packages that each use 2.5 ft of ribbon, how much ribbon is left over?

81. Jessica wants to put adhesive outdoor tread on the 14 steps leading to her front door. Her plan is to center two strips of tread, 4 in. from the edges of each step, as shown in the figure. If the tread is packaged in 5-yd rolls, how many rolls should she buy?

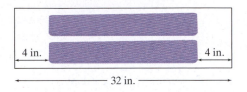

4 in.   4 in.

32 in.

82. Each year the Moon's orbit moves 1.5 in. farther away from Earth. At this rate, how far will the Moon's orbit move in the next 100 years? Express your answer in feet.

Expanding Your Skills

In Section 1.5 we learned that area is measured in square units such as in.^{2}, ft^2, yd^2, and mi^2. Converting square units involves a different set of conversion factors. For example, 1 yd = 3 ft, but 1 yd^2 = 9 ft^2. To understand why, recall that the formula for the area of a rectangle is $A = l \times w$. In a square, the length and the width are the same distance s. Therefore, the area of a square is given by the formula $A = s \times s$, where s is the length of a side. Thus,

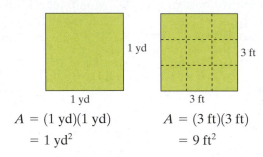

1 yd

1 yd

$A = (1\text{ yd})(1\text{ yd})$

$= 1\text{ yd}^2$

3 ft

3 ft

$A = (3\text{ ft})(3\text{ ft})$

$= 9\text{ ft}^2$

Instead of learning a new set of conversion factors, we can use multiples of the conversion factors that we mastered in Section 7.1.

Example: Converting Area

Convert 4 yd^2 to square feet.

Solution: We will use the unit ratio of $\dfrac{3\text{ ft}}{1\text{ yd}}$ twice.

$$\frac{4\text{ yd}^2}{1} \cdot \underbrace{\frac{3\text{ ft}}{1\text{ yd}} \cdot \frac{3\text{ ft}}{1\text{ yd}}}_{\text{Multiply first.}} = \frac{4\text{ yd}^2}{1} \cdot \frac{9\text{ ft}^2}{1\text{ yd}^2} = 36\text{ ft}^2$$

TIP: We use the unit ratio $\frac{3\text{ ft}}{1\text{ yd}}$ twice because there are two dimensions. Length and width must both be converted to feet.

For Exercises 83–90, convert the units of area by using multiple factors of the given unit ratio.

83. 54 ft^2 = ____ yd^2 (Use two factors of the ratio $\dfrac{1\text{ yd}}{3\text{ ft}}$.)

84. 108 ft^2 = ____ yd^2

85. 432 in.^{2} = ____ ft^2 (Use two factors of the ratio $\dfrac{1\text{ ft}}{12\text{ in.}}$.)

86. 720 in^2 = ____ ft^2

87. 5 ft^2 = ____ in.^{2} (Use two factors of $\dfrac{12\text{ in.}}{1\text{ ft}}$.)

88. 7 ft^2 = ____ in.^{2}

89. 3 yd^2 = ____ ft^2 (Use two factors of $\dfrac{3\text{ ft}}{1\text{ yd}}$.)

90. 10 yd^2 = ____ ft^2

Converting U.S. Customary Units of Time, Weight, and Capacity

1. U.S. Customary Units of Time, Weight, and Capacity

In this section we convert U.S. Customary units of time, weight, and capacity. Table 7-2 summarizes several common units and their equivalent measures. These relationships should be memorized as common knowledge.

Table 7-2 **Summary of U.S. Customary Units of Length, Time, Weight, and Capacity**

Length	Time
1 foot (ft) = 12 inches (in.)	1 year (yr) = 365 days
1 yard (yd) = 3 feet (ft)	1 week (wk) = 7 days
1 mile (mi) = 5280 feet (ft)	1 day = 24 hours (hr)
1 mile (mi) = 1760 yards (yd)	1 hour (hr) = 60 minutes (min)
	1 minute (min) = 60 seconds (sec)
Capacity	**Weight**
3 teaspoons (tsp) = 1 tablespoon (T)	1 pound (lb) = 16 ounces (oz)
1 cup (c) = 8 fluid ounces (fl oz)	1 ton = 2000 pounds (lb)
1 pint (pt) = 2 cups (c)	
1 quart (qt) = 2 pints (pt)	
1 quart (qt) = 4 cups (c)	
1 gallon (gal) = 4 quarts (qt)	

2. Converting Units of Time

In Example 1, we convert from one unit of time to another.

Example 1 Converting Units of Time

a. 32 hr = ___ days   **b.** 36 hr = ___ sec

Solution:

a. $32 \text{ hr} = \dfrac{32 \text{ hr}}{1} \cdot \dfrac{1 \text{ day}}{24 \text{ hr}}$ ← new unit to convert to
← unit to convert from

Recall that
1 day = 24 hr.

$= \dfrac{32}{24} \text{ days}$ Multiply fractions.

$= \dfrac{4}{3} \text{ days or } 1\dfrac{1}{3} \text{ days}$ Simplify.

converts   converts
hr to min   min to sec

b. $36 \text{ hr} = \dfrac{36 \text{ hr}}{1} \cdot \dfrac{60 \text{ min}}{1 \text{ hr}} \cdot \dfrac{60 \text{ sec}}{1 \text{ min}}$ Multiply by two conversion factors.

$= 129{,}600 \text{ sec}$ Simplify.

Example 2   **Converting Units of Time**

After running a marathon, Dave crossed the finish line and noticed that the race clock read 2:20:30. Convert this time to minutes.

Solution:

The notation 2:20:30 means 2 hr 20 min 30 sec. We must convert 2 hr to minutes and 30 sec to minutes. Then we add the total number of minutes.

$$2 \text{ hr} = \frac{2 \text{ hr}}{1} \cdot \frac{60 \text{ min}}{1 \text{ hr}} = 120 \text{ min}$$

$$30 \text{ sec} = \frac{30 \text{ sec}}{1} \cdot \frac{1 \text{ min}}{60 \text{ sec}} = \frac{30}{60} \text{ min} = \frac{1}{2} \text{ min} \quad \text{or} \quad 0.5 \text{ min}$$

The total number of minutes is 120 min + 20 min + 0.5 min = 140.5 min Dave finished the race in 140.5 min.

3. Converting U.S. Customary Units of Weight

Measurements of weight record the force of an object subject to gravity. In Example 3 we convert from one unit of weight to another.

Example 3   **Converting Units of Weight**

a. The average weight of an adult male African elephant is 12,400 lb. Convert this value to tons.

b. Convert the weight of a 7-lb 3-oz baby to ounces.

Solution:

a. Recall that 1 ton = 2000 lb.

$$12,400 \text{ lb} = \frac{12,400 \text{ lb}}{1} \cdot \frac{1 \text{ ton}}{2000 \text{ lb}}$$

$$= \frac{12,400}{2000} \text{ tons} \qquad \text{Multiply fractions.}$$

$$= \frac{31}{5} \text{ tons} \quad \text{or} \quad 6.2 \text{ tons}$$

An adult male African elephant weighs 6.2 tons.

b. To convert 7 lb 3 oz to ounces, we must convert 7 lb to ounces.

$$7 \text{ lb} = \frac{7 \text{ lb}}{1} \cdot \frac{16 \text{ oz}}{1 \text{ lb}} \qquad \text{Recall that 1 lb = 16 oz.}$$

$$= 112 \text{ oz}$$

The baby's total weight is 112 oz + 3 oz = 115 oz.

Example 4   **Applying U.S. Customary Units of Weight**

Jessica lifts four boxes of books. The boxes have the following weights: 16 lb 4 oz, 18 lb 8 oz, 12 lb 5 oz, and 22 lb 9 oz. How much weight did she lift altogether?

Solution:

$$\begin{array}{r} 16\ \text{lb}\quad 4\ \text{oz} \\ 18\ \text{lb}\quad 8\ \text{oz} \\ 12\ \text{lb}\quad 5\ \text{oz} \\ +\ 22\ \text{lb}\quad 9\ \text{oz} \\ \hline 68\ \text{lb}\quad 26\ \text{oz} \end{array}$$

Add like units in columns.

$$68\ \text{lb}\ 26\ \text{oz} = 68\ \text{lb} + 26\ \text{oz}$$
$$= 68\ \text{lb} + 1\ \text{lb} + 10\ \text{oz} \qquad \text{Recall that } 1\ \text{lb} = 16\ \text{oz}.$$
$$= 69\ \text{lb}\ 10\ \text{oz}$$

Jessica lifted 69 lb 10 oz of books.

Skill Practice

6. A set of triplets weighed 4 lb 3 oz, 3 lb 9 oz, and 4 lb 5 oz. What is the total weight of all three babies?

4. Converting U.S. Customary Units of Capacity

A typical can of soda contains 12 fl oz. This is a measure of capacity. Capacity is the volume or amount that a container can hold. The U.S. Customary units of capacity are fluid ounces (fl oz), cup (c), pint (pt), quart (qt), and gallon (gal).

   One fluid ounce is approximately the amount of liquid that two large spoonfuls will hold. One cup is the amount in an average-size cup of tea. While Table 7-2 summarizes the relationships among units of capacity, we also offer an illustration (Figure 7-2).

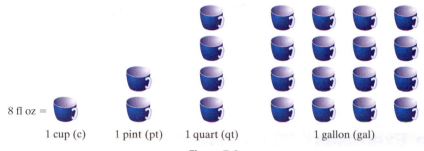

8 fl oz = | 1 cup (c) | 1 pint (pt) | 1 quart (qt) | 1 gallon (gal)

Figure 7-2

Concept Connections

7. From Figure 7-2, determine how many cups are in 1 gal.
8. From Figure 7-2, determine how many pints are in 1 gal.

Example 5 Converting Units of Capacity

a. 1.25 pt = _____ qt   **b.** 2 gal = _____ c   **c.** 48 fl oz = _____ gal

Solution:

a. $1.25\ \text{pt} = \dfrac{1.25\ \text{pt}}{1} \cdot \dfrac{1\ \text{qt}}{2\ \text{pt}}$ Recall that 1 qt = 2 pt.

$\qquad = \dfrac{1.25}{2}\ \text{qt}$ Multiply fractions.

$\qquad = 0.625\ \text{qt}$ Simplify.

b. $2\ \text{gal} = 2\ \text{gal} \cdot \dfrac{4\ \text{qt}}{1\ \text{gal}} \cdot \dfrac{4\ \text{c}}{1\ \text{qt}}$ Use two conversion factors. The first unit ratio converts gallons to quarts. The second converts quarts to cups.

$\qquad = \dfrac{2\ \text{gal}}{1} \cdot \dfrac{4\ \text{qt}}{1\ \text{gal}} \cdot \dfrac{4\ \text{c}}{1\ \text{qt}}$

$\qquad = 32\ \text{c}$ Multiply.

Skill Practice

Convert.

9. 8.5 gal = _____ qt
10. 2.25 qt = _____ c
11. 40 fl oz = _____ qt

Answers

6. 12 lb 1 oz   7. 16 c in 1 gal
8. 8 pt in 1 gal   9. 34 qt
10. 9 c   11. 1.25 qt

Objective 3: Converting Metric Units of Length

For Exercises 25–30, complete the unit ratios.

25. $\dfrac{1\ \text{km}}{\underline{\quad}\ \text{m}}$

26. $\dfrac{1\ \text{hm}}{\underline{\quad}\ \text{m}}$

27. $\dfrac{1\ \text{m}}{\underline{\quad}\ \text{cm}}$

28. $\dfrac{1\ \text{m}}{\underline{\quad}\ \text{mm}}$

29. $\dfrac{1\ \text{m}}{\underline{\quad}\ \text{dm}}$

30. $\dfrac{1\ \text{dam}}{\underline{\quad}\ \text{m}}$

For Exercises 31–50, convert metric units of length by using unit ratios or the prefix line. **(See Examples 2–3.)**

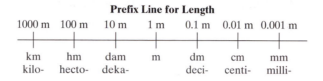

Prefix Line for Length

1000 m	100 m	10 m	1 m	0.1 m	0.01 m	0.001 m
km	hm	dam	m	dm	cm	mm
kilo-	hecto-	deka-		deci-	centi-	milli-

31. 2430 m = _____ km

32. 52 hm = _____ m

33. 103 dm = _____ m

34. 1251 mm = _____ m

35. 50 m = _____ mm

36. 1.3 m = _____ mm

37. 4 km = _____ m

38. 5 m = _____ cm

39. 4.31 cm = _____ mm

40. 18 cm = _____ mm

41. 3328 dm = _____ km

42. 128 hm = _____ km

43. 345 mm = _____ m

44. 450 mm = _____ dm

45. 0.25 km = _____ m

46. 3 hm = _____ m

47. 4003 cm = _____ dm

48. 6.8 m = _____ cm

49. 0.07 mm = _____ cm

50. 8 m = _____ cm

51. One of the tallest sand sculptures was 20.91 m tall. How many centimeters is this? **(See Example 4.)**

52. The smallest steam engine is 16.24 mm long. How many centimeters is this? (*Source: Guinness Book of World Records*)

53. The lowest point in Antarctica is 2538 m below sea level. How many kilometers is this?

54. The lowest point in South America is 40 m below sea level. How many kilometers is this?

55. The Renaissance Tower in Dallas is 270 m tall. How many kilometers is this?

56. The Trump Building in New York is 283 m tall. How many kilometers is this?

Objective 4: Applications

57. Veronique has a piece of molding 1 m long. Does she have enough to cut four pieces to frame the picture shown in the figure? **(See Example 5.)**

58. Rosanna has material 1.5 m long for a window curtain. If the window is 90 cm and she needs 10 cm for a hem at the bottom and 12 cm for finishing the top, does Rosanna have enough material?

12 cm

40 cm

59. A square tile is 110 mm in length. If they are placed side by side, how many tiles will it take to cover a length of wall 1.43 m long?

60. Two Olympic speed skating races for women are 500 m and 5 km. What is the difference (in meters) between the lengths of these races?

61. A parking area in an apartment complex is 0.108 km long. If parking spaces are 4.5 m wide, how many can fit along the length of the lot?

4.5 m

0.108 km

62. Find the missing length.

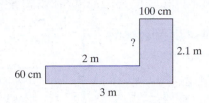

100 cm

?

2.1 m

2 m

60 cm

3 m

Expanding Your Skills

In the Expanding Your Skills of Section 7.1, we converted U.S. Customary units of area. We use the same procedure to convert metric units of area. This procedure involves multiplying by two unit ratios of length.

Example: Converting area

Convert 1000 mm² to square centimeters.

Solution: $\dfrac{1000\ \text{mm}^2}{1} \cdot \dfrac{1\ \text{cm}}{10\ \text{mm}} \cdot \dfrac{1\ \text{cm}}{10\ \text{mm}} = \dfrac{1000\ \text{mm}^2}{1} \cdot \dfrac{1\ \text{cm}^2}{100\ \text{mm}^2} = \dfrac{1000\ \text{cm}^2}{100} = 10\ \text{cm}^2$

Multiply first.

For Exercises 63–66, convert the units of area, using two factors of the given unit ratio.

63. 30,000 mm² = _____ cm² $\left(\text{Use } \dfrac{1\ \text{cm}}{10\ \text{mm}}.\right)$

64. 65,000,000 m² = _____ km² $\left(\text{Use } \dfrac{1\ \text{km}}{1000\ \text{m}}.\right)$

65. 4.1 m² = _____ cm² $\left(\text{Use } \dfrac{100\ \text{cm}}{1\ \text{m}}.\right)$

66. 5600 cm² = _____ m² $\left(\text{Use } \dfrac{1\ \text{m}}{100\ \text{cm}}.\right)$

Metric Units of Mass, Capacity, and Medical Applications

Section 7.4

1. Converting Metric Units of Mass

In Section 7.2 we learned that the pound and ton are two measures of weight in the U.S. Customary System. Measurements of weight give the force of an object under the influence of gravity. The mass of an object is related to its weight, however, mass is not affected by gravity. Thus, the weight of an object will be different on Earth than on the Moon because the effect of gravity is different. However, the mass of the object will stay the same.

The fundamental unit of mass in the metric system is the **gram** (g). A penny is approximately 2.5 g (Figure 7-6). A paper clip is approximately 1 g (Figure 7-7).

≈ 2.5 g

Figure 7-6

≈ 1 g

Figure 7-7

Objectives

1. **Converting Metric Units of Mass**
2. **Converting Metric Units of Capacity**
3. **Summary of Metric Conversions**
4. **Medical Applications**

Concept Connections

1. Which object could have a mass of 2 g?
   a. Rubber band
   b. Can of tuna fish
   c. Cell phone

Answer

1. a

Other common metric units of mass are given in Table 7-4. Once again, notice that the metric units of mass are related to the gram by powers of 10.

Concept Connections

Fill in the blank with < or >.

2. 1 g ☐ 1 kg

3. 1 g ☐ 1 cg

Table 7-4   Metric Units of Mass and Their Equivalents

1 kilogram (kg) = 1000 g	
1 hectogram (hg) = 100 g	
1 dekagram (dag) = 10 g	
1 gram (g) = 1 g	
1 decigram (dg) = 0.1 g	$\left(\frac{1}{10}\ g\right)$
1 centigram (cg) = 0.01 g	$\left(\frac{1}{100}\ g\right)$
1 milligram (mg) = 0.001 g	$\left(\frac{1}{1000}\ g\right)$

TIP: In addition to the key facts presented in Table 7-4, the following equivalences are useful.

$$100\ cg = 1\ g$$
$$1000\ mg = 1\ g$$

On the surface of Earth, 1 kg of mass is equivalent to approximately 2.2 lb of weight. Therefore, a 180-lb man has approximately 81.8 kg of mass.

$$180\ lb \approx 81.8\ kg$$

The metric prefix line for mass is shown in Figure 7-8. This can be used to convert from one unit of mass to another.

Prefix Line

1000 g	100 g	10 g	1 g	0.1 g	0.01 g	0.001 g
kg	hg	dag	g	dg	cg	mg
kilo-	hecto-	deka-		deci-	centi-	milli-

Figure 7-8

Skill Practice

Convert.

4. 80 kg = _____ g

5. 49 cg = _____ g

Example 1   Converting Metric Units of Mass

a. 1.6 kg = _____ g   **b.** 1400 mg = _____ g

Solution:

a. $1.6\ kg = \dfrac{1.6\ \cancel{kg}}{1} \cdot \dfrac{1000\ g}{1\ \cancel{kg}}$ ← new unit to convert to ← unit to convert from

$= 1600\ g$

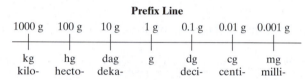

Start here.

$$1.6\ kg = 1.600\ kg = 1600\ g$$

b. $1400\ mg = \dfrac{1400\ \cancel{mg}}{1} \cdot \dfrac{1\ g}{1000\ \cancel{mg}}$ ← new unit to convert to ← unit to convert from

$= \dfrac{1400}{1000}\ g$

$= 1.4\ g$

Start here.

$$1400\ mg = 1400\ mg = 1.4\ g$$

Answers

2. <   **3.** >

4. 80,000   **5.** 0.49 g

2. Converting Metric Units of Capacity

The basic unit of capacity in the metric system is the **liter** (L). One liter is slightly more than 1 qt. Other common units of capacity are given in Table 7-5.

Table 7-5 **Metric Units of Capacity and Their Equivalents**

1 kiloliter (kL) = 1000 L	
1 hectoliter (hL) = 100 L	
1 dekaliter (daL) = 10 L	
1 liter (L) = 1 L	
1 deciliter (dL) = 0.1 L	$\left(\frac{1}{10} L\right)$
1 centiliter (cL) = 0.01 L	$\left(\frac{1}{100} L\right)$
1 milliliter (mL) = 0.001 L	$\left(\frac{1}{1000} L\right)$

TIP: In addition to the key facts presented in Table 7-5, the following equivalences are useful.

$$100 \text{ cL} = 1 \text{ L}$$
$$1000 \text{ mL} = 1 \text{ L}$$

1 mL is also equivalent to a **cubic centimeter** (**cc** or **cm³**). The unit cc is often used to measure dosages of medicine. For example, after having an allergic reaction to a bee sting, a patient might be given 1 cc of adrenalin.

The metric prefix line for capacity is similar to that of length and mass (Figure 7-9). It can be used to convert between metric units of capacity.

1 cc = 1 ml

Prefix Line

1000 L   100 L   10 L   1 L   0.1 L   0.01 L   0.001 L

kL	hL	daL	L	dL	cL	mL
kilo-	hecto-	deka-		deci-	centi-	milli-

Figure 7-9

Example 2   **Converting Metric Units of Capacity**

a. 5.5 L = _____ mL

b. 150 cL = _____ L

Solution:

a. $5.5 \text{ L} = \dfrac{5.5 \text{ } \cancel{L}}{1} \cdot \dfrac{1000 \text{ mL}}{1 \text{ } \cancel{L}}$ ← new unit to convert to
← unit to convert from

= 5500 mL

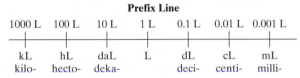

| kL | hL | daL | L | dL | cL | mL |

Start here.

5.5 L = 5.500 L = 5500 mL

b. $150 \text{ cL} = \dfrac{150 \text{ } \cancel{cL}}{1} \cdot \dfrac{1 \text{ L}}{100 \text{ } \cancel{cL}}$ ← new unit to convert to
← unit to convert from

$= \dfrac{150}{100} \text{ L}$

= 1.5 L

| kL | hL | daL | L | dL | cL | mL |

Start here.

150 cL = 1.50 L = 1.5 L

Skill Practice

Convert.

11. 0.5 cc =_____ mL

12. 0.04 L =_____ cc

Example 3 Converting Metric Units of Capacity

a. 15 cc = _____ mL   **b.** 0.8 cL = _____ cc

Solution:

a. Recall that 1 cc = 1 mL. Therefore 15 cc = 15 mL.

b. We must convert from centiliters to milliliters, and then from milliliters to cubic centimeters.

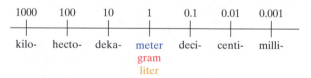

$$0.8 \text{ cL} = \frac{0.8 \text{ cL}}{1} \cdot \frac{10 \text{ mL}}{1 \text{ cL}}$$

$$= 8 \text{ mL}$$

$$= 8 \text{ cc} \qquad \text{Recall that } 1 \text{ mL} = 1 \text{ cc.}$$

0.8 cL = 8 mL = 8 cc

3. Summary of Metric Conversions

The prefix line in Figure 7-10 summarizes the relationships learned thus far.

1000	100	10	1	0.1	0.01	0.001
kilo-	hecto-	deka-	meter gram liter	deci-	centi-	milli-

Figure 7-10

Skill Practice

13. The distance between Savannah and Hinesville is 64 km. How many meters is this?

14. A bottle of water holds 1420 mL. How many liters is this?

15. The mass of a box of cereal is 680 g. Convert this to kilograms.

16. A cat receives 1 mL of an antibiotic solution. Convert this to cc.

Example 4 Converting Metric Units

a. The distance between San Jose and Santa Clara is 26 km. Convert this to meters.

b. A bottle of canola oil holds 946 mL. Convert this to liters.

c. The mass of a bag of rice is 90,700 cg. Convert this to grams.

d. A dose of an antiviral medicine is 0.5 cc. Convert this to milliliters.

Solution:

a. 26 km = 26,000 m

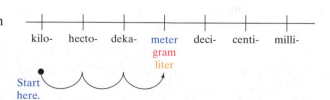

b. 946 mL = 0.946 L

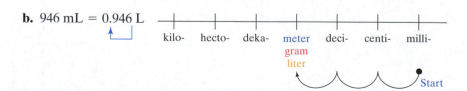

Answers

11. 0.5 mL   **12.** 40 cc

13. 64,000 m   **14.** 1.42 L

15. 0.68 kg   **16.** 1 cc

c. 90,700 cg = 907.00 g = 907 g

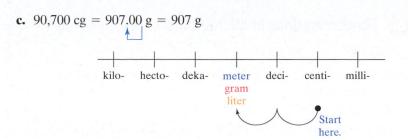

d. Recall that 1 cc = 1 mL. Therefore, 0.5 cc = 0.5 mL.

4. Medical Applications

Example 5   Applying Metric Units of Measure to Medicine

A doctor orders the antibiotic oxacillin for a child. The dosage is 12.5 mg of the drug per kilogram of the child's body mass. This dosage is given 4 times a day.

a. How much of the drug should a 24-kg child get in one dose?

b. How much of the drug would the child get if she were on a 10-day course of the antibiotic?

Solution:

a. We need to multiply the unit rate of 12.5 mg per kilogram times the child's body mass.

$$\text{Single dose} = (12.5 \text{ mg/kg})(24 \text{ kg})$$
$$= 300 \text{ mg}$$

b. For a 10-day course, we need to multiply 300 g by the number of doses per day (4), and the total number of days (10).

$$\text{Total amount of drug} = (300 \text{ mg})(4)(10)$$
$$= 12,000 \text{ mg} \quad \text{or equivalently 12 g.}$$

Sometimes doctors prescribe medicines in very small amounts. In these cases, it is sometimes more convenient to use units of **micrograms**. The abbreviation for microgram is mcg or sometimes μg. Furthermore,

1000 mcg = 1 mg   It takes 1 thousand micrograms to equal 1 milligram.

1,000,000 mcg = 1 g   It takes 1 million micrograms to equal 1 gram.

Skill Practice

17. A child is to receive 0.5 mg of a drug per kilogram of the child's body mass. If the child is 27 kg, how much of the drug should the child receive?

Answer

17. 13.5 mg

Skill Practice

Convert.

18. 0.04 mg = _____ mcg

19. 95,000 mcg = _____ mg

Example 6 **Converting Units of Micrograms**

a. Convert 0.85 mg = _____ mcg

b. A doctor gives a heart patient an initial dose of 200 mcg of nitroglycerin. How many milligrams is this?

Solution:

a. $0.85 \text{ mg} = \dfrac{0.85 \text{ mg}}{1} \cdot \dfrac{1000 \text{ mcg}}{1 \text{ mg}}$ ⟵ new unit to convert to
⟵ unit to convert from

$= 850 \text{ mcg}$

b. $200 \text{ mcg} = \dfrac{200 \text{ mcg}}{1} \cdot \dfrac{1 \text{ mg}}{1000 \text{ mcg}}$ ⟵ new unit to convert to
⟵ unit to convert from

$= 0.2 \text{ mg}$

Answers

18. 40 mcg **19.** 95 mg

Section 7.4 Practice Exercises

Boost your GRADE at ALEKS.com!

ALEKS version 3.0

- Practice Problems
- Self-Tests
- NetTutor
- e-Professors
- Videos

Study Skills Exercises

1. To help you stay motivated for this class, list three reasons why you are taking the class.

2. Define the key terms.

    a. Gram   **b. Liter**   **c. Cubic centimeter**   **d. Microgram**

Review Exercises

For Exercises 3–8, complete the table.

	Object	mm	cm	m	km
3.	Distance between Orlando and Miami				670
4.	Length of the Mississippi River				3766
5.	Length of a screw		2.5		
6.	Thickness of a pizza		3.2		
7.	Thickness of a dime	1.35			
8.	Diameter of a quarter	24.3			

Objective 1: Converting Metric Units of Mass

For Exercises 9–18, convert the units of mass. **(See Example 1.)**

9. 539 g = _____ kg

10. 328 mg = _____ g

11. 2.5 kg = _____ g

12. 2011 g = _____ kg

 **13.** 0.0334 g = _____ mg

14. 0.38 dag = _____ dg

15. 90 hg = _____ kg

16. 0.003 kg = _____ g

17. 45 dg = _____ g

18. 409 cg = _____ g

For Exercises 19–26, complete the table.

	Object	mg	cg	g	kg
19.	Bag of cat food				1.58
20.	Bag of flour				2.26
21.	Can of tuna			170	
22.	Bag of rice			907	
23.	Box of raisins		42,500		
24.	Hockey puck		17,000		
25.	Dose of acetaminophen	325			
26.	Olive	12			

Objective 2: Converting Metric Units of Capacity

For Exercises 27–32, fill in the blank with >, <, or =.

27. 1 cL _____ 1 L

28. 1 L _____ 1 mL

29. 1 mL _____ 1 cc

30. 1 L _____ 1 cc

31. 1 cL _____ 1 kL

32. 1 mL _____ 1 cL

33. What does the abbreviation *cc* represent?

34. Which of the following are measures of capacity? Circle all that apply.

 a. cm   **b.** cc   **c.** cL   **d.** cg

For Exercises 35–44, convert the units of capacity. **(See Examples 2–3.)**

35. 3200 mL = _____ L

36. 280 L = _____ kL

37. 7 L = _____ cL

38. 0.52 L = _____ mL

39. 42 mL = _____ dL

40. 0.88 L = _____ hL

41. 64 cc = _____ mL

42. 125 mL = _____ cc

43. 0.04 L = _____ cc

44. 38 cc = _____ L

For Exercises 45–52, complete the table.

	Object	mL	cL	L	kL
45.	1 Tablespoon	15			
46.	Bottle of vanilla extract	59			
47.	Bottle of vinegar		35.5		
48.	Bottle of soy sauce		29.6		
49.	Bottle of soda pop			2	
50.	Bottle of water			1	
51.	Capacity of a cooler				0.0377
52.	Capacity of a gasoline tank				0.0757

Objective 3: Summary of Metric Conversions (Mixed Exercises)

53. Identify the unit that applies to length.

    a. L   **b.** g   **c.** m

54. Identify the unit that applies to capacity.

    a. L   **b.** g   **c.** m

55. Identify the unit that applies to mass.

    a. L   **b.** g   **c.** m

56. Identify the units that apply to length.

    a. mL   **b.** mm   **c.** hg   **d.** cc   **e.** kg   **f.** hm   **g.** cL

57. Identify the units that apply to capacity.

    a. kg   **b.** km   **c.** cL   **d.** cc   **e.** hm   **f.** dag   **g.** mm

58. Identify the units that apply to mass.

    a. dg   **b.** hm   **c.** kL   **d.** cc   **e.** dm   **f.** kg   **g.** cL

For Exercises 59–64, convert the metric units as indicated. **(See Example 4.)**

59. The height of the tallest living tree is 112.014 m. Convert this to dekameters.

60. The Congo River is 4669 km long. Convert this to meters.

61. There is 600 mg of calcium in a multivitamin. Convert this to grams.

62. A can of soup contains 305 g. Convert this to milligrams.

63. A gasoline can has a capacity of 19 L. Convert this to kiloliters.

64. The capacity of a coffee cup is 0.25 L. Convert this to milliliters.

65. In one day, Stacy gets 600 mg of calcium in her daily vitamin, 500 mg in her calcium supplement, and 250 mg in the dairy products she ingests. How many grams of calcium will she ingest in one week?

66. Cliff drives his children to their sports activities outside of school. When he drives his son to baseball practice, it is a 6-km round trip. When he drives his daughter to basketball practice, it is a 1800-m round trip. If basketball practice is 3 times a week and baseball practice is twice a week, how many kilometers does Cliff drive?

67. A gas tank holds 45 L. If it costs $74.25 to fill up the tank, what is the price per liter?

68. A can of paint holds 120 L. How many kiloliters are contained in 8 cans?

69. A bottle of water holds 710 mL. How many liters are in a 6-pack?

70. A bottle of olive oil has 33 servings of 15 mL each. How many centiliters of oil does the bottle contain?

71. A quart of milk has 130 mg of sodium per cup. How much sodium is in the whole bottle?

72. A ½-c serving of cereal has 180 mg of potassium. This is 5% of the recommended daily allowance of potassium. How many cups of cereal are needed to get 100% of the recommended daily allowance of potassium?

Objective 4: Medical Applications

73. The drug amoxicillin is an antibiotic used to treat bacterial infections. A doctor orders 250 mg every 8 hr. How many grams of the drug would be given in 1 wk?

74. The drug acetaminophen is used as a pain reliever. A patient takes 325 mg every 6 hr. How many grams of the drug would be taken in a 3-day period?

75. Dr. Boyd gives a patient 2 cc of Zantac. How many milliliters is this?

76. If a nurse mixed 11.5 mL of sterile water with 1.5 mL of oxacillin, how many cubic centimeters will this produce?

77. A tetanus vaccine was purchased by a group of family practice doctors. They purchased 1 L of the vaccine. How many patients can be vaccinated if the normal dose is 2 cc?

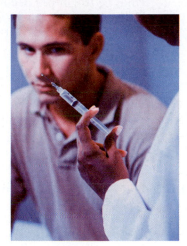

78. A pharmacist has a 1-L bottle of cough syrup. How many 20 cL bottles can she make?

79. A doctor orders 0.2 mg of a drug per kilogram of a patient's body mass. How much of the drug should be given to a patient who is 48 kg?

80. The dosage for a painkiller is 0.05 mg per kilogram of a patient's body mass. How much of the drug should be administered to a patient who is 90 kg?

81. The drug Zovirax is sometimes used to treat chicken pox in children. One doctor recommended 20 mg of the drug per kilogram of the child's body mass, 4 times daily. **(See Example 5.)**

   a. How much of the drug should a 20-kg child receive for one dose?

   b. How much of the drug would be given over a 5-day period?

82. The drug Amoxil is sometimes used to treat children with bacterial infections. One doctor prescribed 40 mg of the drug per kilogram of the child's body mass, 3 times daily.

   a. How much of the drug should a 15-kg child receive for one dose?

   b. How much of the drug would be given to the child over a 10-day period?

83. Convert 0.01 mg to micrograms. **(See Example 6.)**

84. Convert 0.0004 cg to micrograms.

85. A doctor orders 0.2 mg of the drug atropine given by injection. How many micrograms is this?

86. The drug Synthroid is used to treat thyroid disease. A patient is sometimes started on a dose of 0.05 mg/day. How many micrograms is this?

87. The drug cyanocobalamin is prescribed by one doctor in the amount of 1000 mcg. How many milligrams is this?

88. An injection of naloxone is given in the amount of 800 mcg. How many milligrams is this?

89. A nurse must administer 45 mg of a drug. The drug is available in a liquid form with a concentration of 15 mg per milliliter of the solution. How many milliliters of the solution should the nurse give?

90. A patient must receive 500 mg of medication in a solution that has a strength of 250 mg per 5 milliliter of solution. How many milliliters of solution should be given?

Expanding Your Skills

91. A normal value of hemoglobin in the blood for an adult male is 18 gm/dL (that is, 18 grams per deciliter). How much hemoglobin would be expected in 20 mL of a males's blood?

92. A normal value of hemoglobin in the blood for an adult female is 15 gm/dL (that is, 15 gm per deciliter). How much hemoglobin would be expected in 40 mL of a female's blood?

In the U.S. Customary System of measurement, 1 ton = 2000 lb. In the metric system, 1 metric ton = 1000 kg. Use this information to answer Exercises 93–96.

93. Convert 3300 kg to metric tons.

94. Convert 5780 kg to metric tons.

95. Convert 10.9 metric tons to kilograms.

96. Convert 8.5 metric tons to kilograms.

Problem Recognition Exercises

U.S. Customary and Metric Conversions

For Exercises 1–30, convert the units as indicated.

1. 36 c = ____ qt

2. 220 cm = ____ m

3. $\frac{3}{4}$ lb = ____ oz

4. 0.3 L = ____ mL

5. 12 ft = ____ yd

6. 6.03 kg = ____ g

7. 45 dm = ____ m

8. 9 in. = ____ ft

9. $\frac{1}{2}$ mi = ____ ft

10. 6000 lb = ____ tons

11. 8 pt = ____ qt

12. 1.5 tsp = ____ T

13. 21 m = ____ km

14. 68 mg = ____ cg

15. 36 mL = ____ cc

16. 64 oz = ____ lb

17. 4322 g = ____ kg

18. 5 m = ____ mm

19. 20 fl oz = ____ c

20. 510 sec = ____ min

21. 4 pt = ____ gal

22. 26 fl oz = ____ c

23. 5.46 kg = ____ g

24. 9.02 L = ____ cL

25. 9.1 mi = ____ yd

26. 48 oz = ____ lb

27. 1.62 tons = ____ lb

28. 4.6 km = ____ m

29. 60 hr = ____ days

30. 8 cc = ____ mL

Converting Between U.S. Customary and Metric Units

1. Summary of U.S. Customary and Metric Unit Equivalents

In this section, we learn how to convert between U.S. Customary and metric units of measure. Suppose, for example, that you take a trip to Europe. A street sign indicates that the distance to Paris is 45 km (Figure 7-11). This distance may be unfamiliar to you until you convert to miles.

| Paris | 45 km |
| Le Havre | 135 km |

Figure 7-11

Objectives

1. Summary of U.S. Customary and Metric Unit Equivalents
2. Converting U.S. Customary and Metric Units
3. Applications
4. Units of Temperature

Example 1   Converting Metric Units to U.S. Customary Units

Use the fact that 1 mi ≈ 1.61 km to convert 45 km to miles. Round to the nearest mile.

Solution:

$$45 \text{ km} \approx \frac{45 \text{ km}}{1} \cdot \frac{1 \text{ mi}}{1.61 \text{ km}}$$ Set up a unit ratio to convert kilometers to miles.

$$= \frac{45}{1.61} \text{ mi}$$ Multiply fractions.

$$\approx 28 \text{ mi}$$ Divide and round to the nearest mile.

The distance of 45 km to Paris is approximately 28 mi.

Skill Practice

1. Use the fact that 1 mi ≈ 1.61 km to convert 184 km to miles. Round to the nearest mile.

Table 7-6 summarizes some common metric and U.S. Customary equivalents.

Table 7-6

Length	Weight/Mass (on Earth)	Capacity
1 in. = 2.54 cm	1 lb ≈ 0.45 kg	1 qt ≈ 0.95 L
1 ft ≈ 0.305 m	1 oz ≈ 28 g	1 fl oz ≈ 30 mL = 30 cc
1 yd ≈ 0.914 m		
1 mi ≈ 1.61 km		

2. Converting U.S. Customary and Metric Units

Using the U.S. Customary and metric equivalents given in Table 7-6, we can create unit ratios to convert between units.

Answer

1. 114 mi

Example 2   **Converting Units of Length**

Fill in the blanks. Round to two decimal places, if necessary.

a. 18 cm = _____ in.   **b.** 15 yd ≈ _____ m   **c.** 82 m ≈ _____ ft

Solution:

a. $18 \text{ cm} = \dfrac{18 \text{ cm}}{1} \cdot \dfrac{1 \text{ in.}}{2.54 \text{ cm}}$ From Table 7-6, we know 1 in. = 2.54 cm.

$= \dfrac{18}{2.54} \text{ in.}$ Multiply fractions.

$\approx 7.09 \text{ in.}$ Divide and round to two decimal places.

b. $15 \text{ yd} \approx \dfrac{15 \text{ yd}}{1} \cdot \dfrac{0.914 \text{ m}}{1 \text{ yd}}$ From Table 7-6, we know 1 yd ≈ 0.914 m.

$= 13.71 \text{ m}$ Multiply.

c. $82 \text{ m} \approx \dfrac{8.2 \text{ m}}{1} \cdot \dfrac{1 \text{ ft}}{0.305 \text{ m}}$ From Table 7-6, we know 1 ft ≈ 0.305 m.

$= \dfrac{8.2}{0.305} \text{ ft}$ Multiply.

$\approx 26.89 \text{ ft}$ Divide and round to two decimal places.

Example 3   **Converting Units of Weight and Mass**

Fill in the blank. Round to one decimal place, if necessary.

a. 180 g ≈ _____ oz   **b.** 5.25 tons ≈ _____ kg

Solution:

a. $180 \text{ g} \approx \dfrac{180 \text{ g}}{1} \cdot \dfrac{1 \text{ oz}}{28 \text{ g}}$ From Table 7-6, we know 1 oz ≈ 28 g.

$= \dfrac{180}{28} \text{ oz}$

$\approx 6.4 \text{ oz}$ Divide and round to one decimal place.

b. We can first convert 5.25 tons to pounds. Then we can use the fact that 1 lb ≈ 0.45 kg.

$5.25 \text{ tons} = \dfrac{5.25 \text{ tons}}{1} \cdot \dfrac{2000 \text{ lb}}{1 \text{ ton}}$ Convert tons to pounds.

$= 10{,}500 \text{ lb}$

$\approx 10{,}500 \text{ lb} \cdot \dfrac{0.45 \text{ kg}}{1 \text{ lb}}$ Convert pounds to kilograms.

$= 4725 \text{ kg}$

Answers

2. 13.1 ft   **3.** 7.6 cm   **4.** 5941 m
5. 17.8 lb   **6.** 3600 kg

Example 4 Converting Units of Capacity

For each of the following conversions, round to two decimal places, if necessary.

a. 75 mL ≈ _____ fl oz   b. 3 qt ≈ _____ L

Solution:

a. 75 mL ≈ $\dfrac{75 \text{ mL}}{1} \cdot \dfrac{1 \text{ fl oz}}{30 \text{ mL}}$ From Table 7-6, we know 1 fl oz ≈ 30 mL.

$= \dfrac{75}{30}$ fl oz   Multiply fractions.

$= 2.5$ fl oz   Divide.

b. 3 qt ≈ $\dfrac{3 \text{ qt}}{1} \cdot \dfrac{0.95 \text{ L}}{1 \text{ qt}}$ From Table 7-6, we know 1 qt ≈ 0.95 L.

$= 2.85$ L   Multiply.

Skill Practice

Convert. Round to one decimal place, if necessary.

7. 120 mL ≈ _____ fl oz
8. 4 qt ≈ _____ L

3. Applications

Example 5 Converting Units in an Application

A 2-L bottle of soda sells for $2.19. A 32-oz bottle of soda sells for $1.59. Compare the price per quart of each bottle to determine the better buy.

Solution:

Note that 1 qt = 2 pt = 4 c = 32 fl oz. So a 32-oz bottle of soda costs $1.59 per quart. Next, if we can convert 2 L to quarts, we can compute the unit cost per quart and compare the results.

$2 \text{ L} \approx \dfrac{2 \text{ L}}{1} \cdot \dfrac{1 \text{ qt}}{0.95 \text{ L}}$ Recall that 1 qt = 0.95 L.

$\approx \dfrac{2}{0.95}$ qt   Multiply fractions.

≈ 2.11 qt   Divide and round to 2 decimal places.

Now find the cost per quart. $\dfrac{\$2.19}{2.11 \text{ qt}} \approx \$1.04$ per quart

The cost for the 2-L bottle is $1.04 per quart, whereas the cost for 32 oz is $1.59 per quart. Therefore, the 2-L bottle is the better buy.

Skill Practice

9. A 720-mL bottle of water sells for $0.79. A 32-oz bottle of water sells for $1.29. Compare the price per ounce to determine the better buy.

Answers

7. 4 fl oz   8. 3.8 L
9. 720 mL is 24 oz. The cost per ounce is $0.033. The unit price for the 32-oz bottle is $0.040 per ounce. The 720-mL bottle is the better buy.

Example 6   **Converting Units in an Application**

In track and field, the 1500-m race is slightly less than 1 mi. How many yards less is it? Round to the nearest yard.

Solution:

We know that 1 mi = 1760 yd. If we can convert 1500 m to yards, then we can subtract the results.

1 mi = 1760 yd

1500 m = ? yd

$$1500 \text{ m} \approx \frac{1500 \text{ m}}{1} \cdot \frac{1 \text{ yd}}{0.914 \text{ m}}$$ Recall that 1 yd = 0.914 m.

$$\approx \frac{1500}{0.914} \text{ yd}$$ Multiply fractions.

$$\approx 1641 \text{ yd}$$ Divide and round to the nearest yard.

Therefore, the difference between 1 mi and 1500 m is:

(1 mi) − (1500 m)

1760 yd − 1641 yd = 119 yd

4. Units of Temperature

In the United States, the **Fahrenheit** scale is used most often to measure temperature. On this scale, water freezes at 32°F and boils at 212°F. The symbol ° stands for "degrees," and °F means "degrees Fahrenheit."

Another scale used to measure temperature is the **Celsius** temperature scale. On this scale, water freezes at 0°C and boils at 100°C. The symbol °C stands for "degrees Celsius."

Figure 7-12 shows the relationship between the Celsius scale and the Fahrenheit scale.

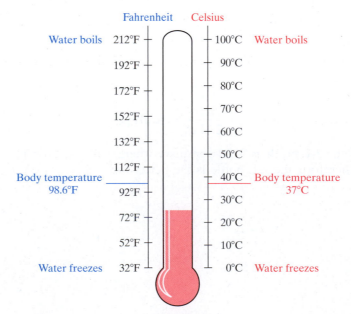

Figure 7-12

To convert back and forth between the Fahrenheit and Celsius scales, we use the following formulas.

> ### FORMULA Conversions for Temperature Scale
>
> To convert from °C to °F: To convert from °F to °C:
>
> $$F = \frac{9}{5}C + 32 \qquad\qquad\qquad C = \frac{5}{9}(F - 32)$$
>
> *Note:* Using decimal notation we can write the formulas as
>
> $$F = 1.8C + 32 \qquad\qquad\qquad C = \frac{F - 32}{1.8}$$

Example 7   Converting Units of Temperature

Convert a body temperature of 98.6°F to degrees Celsius.

Solution:

Because we want to convert degrees Fahrenheit to degrees Celsius, we use the formula $C = \frac{5}{9}(F - 32)$.

$$C = \frac{5}{9}(F - 32)$$

$$= \frac{5}{9}(98.6 - 32) \qquad \text{Substitute } F = 98.6.$$

$$= \frac{5}{9}(66.6) \qquad\qquad \text{Perform the operation inside parentheses first.}$$

$$= \frac{(5)(66.6)}{9}$$

$$= 37$$

Body temperature is 37°C.

Answer

13. 28.9°C

Skill Practice

14. The high temperature on a day in March for Raleigh, North Carolina, was 10°C. Convert this to degrees Fahrenheit.

Example 8 Converting Units of Temperature

Convert the temperature inside a refrigerator, 5°C, to degrees Fahrenheit.

Solution:

Because we want to convert degrees Celsius to degrees Fahrenheit, we use the formula $F = \dfrac{9}{5}C + 32$

$$F = \frac{9}{5}C + 32$$

$$= \frac{9}{5} \cdot 5 + 32 \qquad \text{Substitute } C = 5.$$

$$= \frac{9}{\cancel{5}} \cdot \frac{\cancel{5}}{1} + 32$$

$$= 9 + 32$$

$$= 41$$

The temperature inside the refrigerator is 41°F.

Answer

14. 50°F

Section 7.5 Practice Exercises

Boost your GRADE at ALEKS.com!

- Practice Problems
- Self-Tests
- NetTutor
- e-Professors
- Videos

Study Skills Exercises

1. Make a list of all the section titles in the chapter that you are studying. Write each section title on a separate sheet of paper or index card. Go back and fill in the list of objectives under each section title. When you are studying for the test, try to make up an exercise that corresponds to each objective and then work the exercise. To get started, write a problem for the objective of converting metric units of capacity from Section 7.4.

2. Define the key terms.

    a. Celsius   **b. Fahrenheit**

Review Exercises

For Exercises 3–6, select the equivalent amounts of mass. (*Hint:* There may be more than one answer for each exercise.)

3. 500 g   **a.** 500,000 g

                                   b. 5 g

4. 500 mg   **c.** 500,000,000 mg

                                   d. 0.5 kg

5. 500 cg   **e.** 5000 mg

                                   f. 50,000 cg

6. 500 kg   **g.** 0.5 g

                                   h. 50 cg

For Exercises 7–10, select the equivalent amounts of capacity.

7. 200 L
8. 200 kL
9. 200 mL
10. 200 cL

a. 2000 mL
b. 200 cc
c. 0.2 kL
d. 20,000,000 cL
e. 200,000 L
f. 200,000 mL
g. 0.2 L
h. 2 L

Objective 1: Summary of U.S. Customary and Metric Unit Equivalents

11. Identify an appropriate ratio to convert 5 yards to meters by using multiplication.

a. $\dfrac{1 \text{ yd}}{0.914 \text{ m}}$ b. $\dfrac{0.914 \text{ m}}{1 \text{ yd}}$ c. $\dfrac{0.914 \text{ yd}}{1 \text{ m}}$ d. $\dfrac{1 \text{ m}}{0.914 \text{ yd}}$

12. Identify an appropriate ratio to convert 3 pounds to kilograms by using multiplication.

a. $\dfrac{0.45 \text{ lb}}{1 \text{ kg}}$ b. $\dfrac{1 \text{ kg}}{0.45 \text{ lb}}$ c. $\dfrac{1 \text{ lb}}{0.45 \text{ kg}}$ d. $\dfrac{0.45 \text{ kg}}{1 \text{ lb}}$

13. Identify an appropriate ratio to convert 2 quarts to liters by using multiplication.

a. $\dfrac{0.95 \text{ L}}{1 \text{ qt}}$ b. $\dfrac{1 \text{ qt}}{0.95 \text{ L}}$ c. $\dfrac{0.95 \text{ qt}}{1 \text{ L}}$ d. $\dfrac{1 \text{ L}}{0.95 \text{ qt}}$

14. Identify an appropriate ratio to convert 10 miles to kilometers by using multiplication.

a. $\dfrac{1 \text{ mi}}{1.61 \text{ km}}$ b. $\dfrac{1 \text{ km}}{1.61 \text{ mi}}$ c. $\dfrac{1.61 \text{ km}}{1 \text{ mi}}$ d. $\dfrac{1.61 \text{ mi}}{1 \text{ km}}$

Objective 2: Converting U.S. Customary and Metric Units

For Exercises 15–23, convert the units of length. Round the answer to one decimal place, if necessary. **(See Examples 1–2.)**

15. 2 in. ≈ _____ cm
16. 120 km ≈ _____ mi
17. 8 m ≈ _____ yd

18. 4 ft ≈ _____ m
19. 400 ft ≈ _____ m
20. 0.75 m ≈ _____ yd

21. 45 in ≈ _____ m
22. 150 cm ≈ _____ ft
23. 0.5 ft ≈ _____ cm

For Exercises 24–32, convert the units of weight and mass. Round the answer to one decimal place, if necessary. **(See Example 3.)**

24. 6 oz ≈ _____ g
25. 6 lb ≈ _____ kg
26. 4 kg ≈ _____ lb

27. 10 g ≈ _____ oz
28. 14 g ≈ _____ oz
29. 0.54 kg ≈ _____ lb

30. 0.3 lb ≈ _____ kg
31. 2.2 tons ≈ _____ kg
32. 4500 kg ≈ _____ tons

For Exercises 33–38, convert the units of capacity. Round the answer to one decimal place, if necessary. **(See Example 4.)**

33. 6 qt ≈ _____ L

34. 5 fl oz ≈ _____ mL

35. 120 mL ≈ _____ fl oz

36. 19 L ≈ _____ qt

37. 960 cc ≈ _____ fl oz

38. 0.5 fl oz ≈ _____ cc

Objective 3: Applications

For Exercises 39–54, refer to Table 7-6 on page 443.

39. A 2-lb box of sugar costs $3.19. A box that contains single-serving packets contains 354 g and costs $1.49. Find the unit costs in dollars per ounce to determine the better buy. **(See Example 5.)**

40. At the grocery store, Debbie compares the prices of a 2-L bottle of water and a 6-pack of bottled water. The 2-L bottle is priced at $1.59. The 6-pack costs $3.60 and each bottle in the package contains 24 fl oz. Compare the cost of water per quart to determine which is a better buy.

41. A cross-country skiing race is 30 km long. Is this length more or less than 18 mi? **(See Example 6.)**

42. A can of cat food is 85 g. How many ounces is this? Round to the nearest ounce.

43. Carly Patterson of the U.S. Olympic gymnastic team weighed 97 lb when she was the Olympic All-Around champion. How many kilograms is this?

44. Warren's dad ran the 100-yd dash when he was in high school. Suppose Warren runs the 100-*meter* dash on his track team.

   a. Who runs the longer race?

   b. Find the difference between the lengths in meters.

45. In a recent year, the price of gas in Germany was $1.90 per liter. What is the price per gallon?

46. A jar of spaghetti sauce is 2 lb 8 oz. How many kilograms is this? Round to two decimal places.

47. The thickness of a hockey puck is 2.54 cm. How many inches is this?

48. A bottle of grape juice contains 1.9 L of juice. Is there enough juice to fill 10 glasses that hold 6 oz each? If yes, how many ounces will be left over?

49. Football player Tony Romo weighs 99,790 g. How many pounds is this? Round to the nearest pound.

50. The distance between two lightposts is 6 m. How many feet is this? Round to the nearest foot.

51. A nurse gives a patient 45 cc of saline solution. How many fluid ounces does this represent?

52. Cough syrup comes in a bottle that contains 4 fl oz. How many milliliters is this?

53. Suppose this figure is a drawing of a room where 1 cm represents 2 ft. If you were to install molding along the edge of the floor, how many feet would you need?

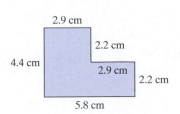

2.9 cm

2.2 cm

4.4 cm

2.9 cm

2.2 cm

5.8 cm

54. Suppose this figure is a drawing of a room where 1 cm represents 3 ft. If you were to install molding along the edge of the floor, how many feet would you need?

4.3 cm

3.2 cm

Objective 4: Units of Temperature

For Exercises 55–60, convert the temperatures by using the appropriate formula: $F = \frac{9}{5}C + 32$ or $C = \frac{5}{9}(F - 32)$. **(See Examples 7–8.)**

55. $25°C = $ _____ °F

56. $113°F = $ _____ °C

57. $68°F = $ _____ °C

58. $15°C = $ _____ °F

59. $30°C = $ _____ °F

60. $104°F = $ _____ °C

61. The boiling point of the element boron is 4000°C. Find the boiling point in degrees Fahrenheit.

62. The melting point of the element copper is 1085°C. Find the melting point in degrees Fahrenheit.

63. If the outdoor temperature is 35°C, is it a hot day or a cold day?

64. The high temperature in London, England, on a typical September day was 18°C, and the low was 13°C. Convert these temperatures to degrees Fahrenheit.

65. Use the fact that water boils at 100°C to show that the boiling point is 212°F.

66. Use the fact that water freezes at 0°C to show that the temperature at which water freezes is 32°F.

Expanding Your Skills

In the U.S. Customary System of measurement, 1 ton = 2000 lb. In the metric system, 1 metric ton = 1000 kg. Use this information to answer Exercises 67–70.

67. A Lincoln Navigator weighs 5700 lb. How many metric tons is this?

68. An elevator has a maximum capacity of 1200 lb. How many metric tons is this?

69. The average mass of a blue whale (the largest mammal in the world) is approximately 108 metric tons. How many pounds is this?

70. The mass of a Mini-Cooper is 1.25 metric tons. How many pounds is this? Round to the nearest pound.

Group Activity

Remodeling the Classroom

Materials: A tape measure for measuring the size of the classroom.
Advertisements from the newspaper for carpet and paint.

Estimated time: 30 minutes

Group Size: 3–4

In this activity, your group will determine the cost for updating your classroom with new paint and new carpet.

1. Measure and record the dimensions of the room and also the height of the walls. You may want to sketch the floor and walls and then label their dimensions on the figure.

2. Calculate and record the area of the floor.

3. Calculate and record the total area of the walls. Subtract any area taken up by doors, windows, and chalkboards.

4. Look through advertisements for carpet that would be suitable for your classroom. You may have to look online for a better choice. Choose a carpet.

5. Calculate how much carpet is needed based on your measurements. To do this, take the area of the floor found in step 2 and add 10% of that figure to allow for waste.

6. Determine the cost to carpet the classroom. Do not forget to include carpet padding and labor to install the carpet. You may have to look online to find prices for padding and installation if they are not included in the price. Also include sales tax for your area.

7. Look through advertisements for paint that would be suitable for your classroom. You may have to look online for a better choice. Choose a paint.

8. Calculate the number of gallons of paint needed for your classroom. You may assume that 1 gal of paint will cover 400 ft^2. Calculate the cost of paint for the classroom. Include sales tax, but do not include labor costs for painting. You will do the painting yourself!

9. Calculate the total cost for carpeting and painting the classroom.

Chapter 7   Summary

Weight, and Capacity

Section 7.1   Converting U.S. Customary Units of Length

Key Concepts

The **U.S. Customary System** for measuring length is commonly used in the United States. Conversion of length can be done by substitution using these equivalents.

$$1 \text{ ft} = 12 \text{ in.} \qquad 1 \text{ in.} = \tfrac{1}{12} \text{ ft}$$

$$1 \text{ yd} = 3 \text{ ft} \qquad 1 \text{ ft} = \tfrac{1}{3} \text{ yd}$$

$$1 \text{ mi} = 5280 \text{ ft} \qquad 1 \text{ ft} = \tfrac{1}{5280} \text{ mi}$$

$$1 \text{ mi} = 1760 \text{ yd} \qquad 1 \text{ yd} = \tfrac{1}{1760} \text{ mi}$$

Conversion of length can also be done by multiplying by an appropriate conversion factor, such as $\tfrac{12 \text{ in.}}{1 \text{ ft}}$ or $\tfrac{1 \text{ mi}}{5280 \text{ ft}}$.

To choose a **conversion factor**, follow these guidelines:

- The unit of measure in the numerator is the new unit you want to convert *to*.
- The unit of measure in the denominator is the original unit you want to convert *from*.

When adding and subtracting measurements, add or subtract like units.

Examples

Example 1

To convert 18 in. to feet, write

$$18 \text{ in.} = 18 \times 1 \text{ in.}$$

$$= \frac{18}{1} \times \frac{1}{12} \text{ ft}$$

$$= \frac{18}{12} \text{ ft} = \frac{3}{2} \text{ ft or } 1\frac{1}{2} \text{ ft}$$

Example 2

To convert 8 yd to feet, multiply.

$$8 \text{ yd} \cdot \frac{3 \text{ ft}}{1 \text{ yd}} = \frac{8 \cancel{\text{ yd}}}{1} \cdot \frac{3 \text{ ft}}{1 \cancel{\text{ yd}}} = 24 \text{ ft}$$

Example 3

To convert 60 in. to yards, multiply.

$$60 \text{ in.} \cdot \frac{1 \text{ ft}}{12 \text{ in.}} \cdot \frac{1 \text{ yd}}{3 \text{ ft}}$$

$$= \frac{60 \cancel{\text{ in.}}}{1} \cdot \frac{1 \cancel{\text{ ft}}}{12 \cancel{\text{ in.}}} \cdot \frac{1 \text{ yd}}{3 \cancel{\text{ ft}}}$$

$$= \frac{60 \text{ yd}}{36}$$

$$= \frac{5}{3} \text{ yd or } 1\frac{2}{3} \text{ yd}$$

Example 4

To add 3 ft 9 in. + 2 ft 10 in., add like terms.

$$\begin{array}{r} 3 \text{ ft} + 9 \text{ in.} \\ + 2 \text{ ft} + 10 \text{ in.} \\ \hline 5 \text{ ft} + 19 \text{ in.} = 5 \text{ ft} + 1 \text{ ft} + 7 \text{ in.} \\ = 6 \text{ ft } 7 \text{ in.} \end{array}$$

| Section 7.4 | **Metric Units of Mass, Capacity, and Medical Applications** |

Key Concepts

The prefix line can be used to convert metric units for mass, capacity, and length.

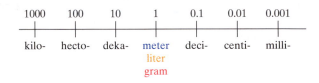

The metric unit conversions for mass are given in **grams**.

1 kilogram (kg) = 1000 g

1 hectogram (hg) = 100 g

1 dekagram (dag) = 10 g

1 gram (g) = 1 g

1 decigram (dg) = 0.1 g $\left(\frac{1}{10} \text{ g}\right)$

1 centigram (cg) = 0.01 g $\left(\frac{1}{100} \text{ g}\right)$

1 milligram (mg) = 0.001 g $\left(\frac{1}{1000} \text{ g}\right)$

The metric unit conversions for capacity are given in **liters**.

1 kiloliter (kL) = 1000 L

1 hectoliter (hL) = 100 L

1 dekaliter (daL) = 10 L

1 liter (L) = 1 L

1 deciliter (dL) = 0.1 L $\left(\frac{1}{10} \text{ L}\right)$

1 centiliter (cL) = 0.01 L $\left(\frac{1}{100} \text{ L}\right)$

1 milliliter (mL) = 0.001 L $\left(\frac{1}{1000} \text{ L}\right)$

Note that 1 mL = 1 cc.

In medical applications the **microgram** is often used.

1000 mcg = 1 mg

Examples

Example 1

To convert 0.962 kg to grams, we can use a unit ratio:

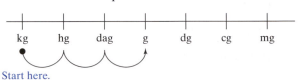

$$= 962 \text{ g}$$

Or we can use the prefix line.

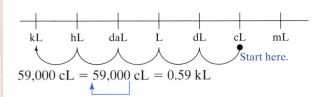

$$0.962 \text{ kg} = 0.962 \text{ kg} = 962 \text{ g}$$

Example 2

To convert 59,000 cL to kL, we can use unit ratios:

$$59,000 \text{ cL} = \frac{59,000 \text{ cL}}{1} \cdot \frac{1 \text{ L}}{100 \text{ cL}} \cdot \frac{1 \text{ kL}}{1000 \text{ L}}$$

$$= 0.59 \text{ kL}$$

Or we can use the prefix line.

59,000 cL = 59,000 cL = 0.59 kL

Section 7.5   Converting Between U.S. Customary and Metric Units

Key Concepts

The common conversions between the U.S. Customary and metric systems are given.

Length

1 in. = 2.54 cm

1 ft ≈ 0.305 m

1 yd ≈ 0.914 m

1 mi ≈ 1.61 km

Weight/Mass (on Earth)

1 lb ≈ 0.45 kg

1 oz ≈ 28 g

Capacity

1 qt ≈ 0.95 L

1 fl oz ≈ 30 mL = 30 cc

To convert U.S. Customary units to metric or metric units to U.S. Customary units, use unit ratios.

The U.S. Customary System uses the **Fahrenheit** scale (°F) to measure temperature. The metric system uses the **Celsius** scale (°C). The conversions are given.

To convert from °C to °F: $F = \dfrac{9}{5}C + 32$

To convert from °F to °C: $C = \dfrac{5}{9}(F - 32)$

Examples

Example 1

Convert 1200 yd to meters by using a unit ratio.

$$1200 \text{ yd} \approx \frac{1200 \text{ yd}}{1} \cdot \frac{0.914 \text{ m}}{1 \text{ yd}} = 1096.8 \text{ m}$$

Example 2

To convert 900 cc to fluid ounces, recall that 1 cc = 1 mL. Therefore, 900 cc = 900 mL. Then use a unit ratio to convert to fluid ounces.

$$900 \text{ mL} \approx \frac{900 \text{ mL}}{1} \cdot \frac{1 \text{ fl oz}}{30 \text{ mL}} = 30 \text{ fl oz}$$

Example 3

The average January temperature in Havana, Cuba, is 21°C. The average January temperature in Johannesburg, South Africa, is 69°F. Which temperature is warmer?

Convert 21°C to degrees Fahrenheit:

$$F = \frac{9}{5}C + 32$$

$$= \frac{9}{5}(21) + 32$$

$$= 37.8 + 32 = 69.8$$

The value 21°C = 69.8°F, which is 0.8 degree warmer than the temperature in Johannesburg.

Chapter 7   Review Exercises

Section 7.1

For Exercises 1–8, convert the units of length.

1. 48 in. = ___ ft   **2.** $3\frac{1}{4}$ ft = ___ in.

3. 2 mi = ___ yd   **4.** 2200 yd = ___ mi

5. 7040 ft = ___ mi   **6.** $\frac{1}{2}$ mi = ___ ft

7. 2 yd = ___ in.   **8.** 6336 in. = ___ mi

For Exercises 9–16, perform the indicated operations.

9. 3 ft 9 in. + 5 ft 6 in.   **10.** 4'11" + 1'5"

11. 5'3" − 2'5"   **12.** 12 ft 7 in. − 8 ft 10 in.

13. 4 × (5'3")   **14.** 2 × (4 ft 8 in.)

15. 6 ft 3 in. ÷ 3   **16.** 6 yd 2 ft ÷ 2

17. Find the perimeter in feet.

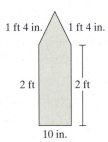

1 ft 4 in.   1 ft 4 in.

2 ft   2 ft

10 in.

18. A roll of wire contains 50 yd of wire. If Ivan uses 48 ft, how much wire is left?

Section 7.2

For Exercises 19–30, convert the units of time, weight, and capacity.

19. 72 hr = ___ days   **20.** 6 min = ___ sec

21. 5 lb = ___ oz   **22.** 1 wk = ___ hr

23. 12 fl oz = ___ c   **24.** 0.25 ton = ___ lb

25. 3500 lb = ___ tons   **26.** 150 min = ___ hr

27. 1800 sec = ___ hr   **28.** 2 gal = ___ pt

29. 12 oz = ___ lb   **30.** 16 qt = ___ gal

31. A runner finished a race with a time of 2:24:30. Convert the time to minutes.

32. Margaret Johansson gave birth to triplets who weighed 3 lb 10 oz, 4 lb 2 oz, and 4 lb 1 oz. What was the total weight of the triplets?

33. One and one-half tons of dirt are to be equally distributed to 8 locations. How many pounds will go to each location?

34. A case of 12-fl-oz cans of soda contains 24 cans. How many gallons of soda are in the case?

Section 7.3

For Exercises 35–38, select the most reasonable measurement.

35. A pencil is _____ long.

    a. 16 mm   **b.** 16 cm

    c. 16 m   **d.** 16 km

36. A mosquito is _____ long.

    a. 12 mm   **b.** 12 cm

    c. 12 m   **d.** 12 km

37. A two-story house is _____ tall.

    a. 9 mm   **b.** 9 cm

    c. 9 m   **d.** 9 km

38. The distance between Houston and Dallas is _____ .

    a. 362 mm   **b.** 362 cm

    c. 362 m   **d.** 362 km

For Exercises 39–48, convert the metric units of length.

39. 52 cm = ___ mm   **40.** 91 mm = ___ cm

41. 2.338 km = ___ m   **42.** 93 m = ___ km

43. 34 dm = ___ m   **44.** 2.1 m = ___ dam

45. 4 cm = ___ m   **46.** 3 m = ___ cm

47. 1.2 m = ___ mm   **48.** 4023 hm = ___ km

49. The highest point in Arizona is 3.851 km above sea level. The highest point in Louisiana is 163 m. What is the difference between these elevations in meters?

50. Determine the perimeter of the triangle in centimeters.

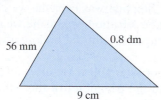

56 mm   0.8 dm

9 cm

Section 7.4

For Exercises 51–56, convert the metric units of mass.

51. 6.1 g = ___ cg   **52.** 420 g = ___ kg

53. 3212 mg = ___ g   **54.** 0.7 hg = ___ g

55. 5 cg = ___ mg   **56.** 0.1 dag = ___ cg

For Exercises 57–62, convert the metric units of capacity.

57. 300 mL = ___ L   **58.** 2.4 hL = ___ L

59. 830 cL = ___ L   **60.** 124 mL = ___ cc

61. 225 cc = ___ cL   **62.** 0.49 kL = ___ L

63. The dimensions of a dining room table are 2 m by 125 cm. Convert the units to meters and find the perimeter and area of the tabletop.

64. A bottle of apple juice contains 1.2 L of juice. If a glass holds 24 cL, how many glasses can be filled from this bottle?

65. An adult has a mass of 68 kg. A baby has a mass of 3200 g. What is the difference in their masses, in kilograms?

66. From a wooden board 2 m long, Jesse needs to cut 3 pieces that are each 75 cm long. Is the 2-m length of board long enough for the 3 pieces?

67. A physician prescribes a drug based on a patient's mass. The dosage is given as 0.04 mg of the drug per kilogram of the patient's mass.

 a. How much would the physician prescribe for an 80-kg patient?

 b. If the dosage was to be given twice a day, how much of the drug would the patient take in a week?

68. Convert 0.45 mg to micrograms (mcg).

69. A standard hypodermic syringe holds 3 cc of fluid. If a nurse uses 1.8 mL of the fluid, how much is left in the syringe?

70. A medication comes in 250-mg capsules. If Clayton took 3 capsules a day for 10 days, how many grams of the medication did he take?

Section 7.5

For Exercises 71–80, refer to page 443. Convert the units of length, capacity, mass, and weight. Round to the nearest hundredth, if necessary.

71. 6.2 in. ≈ _____ cm   **72.** 75 mL ≈ _____ fl oz

73. 140 g ≈ _____ oz   **74.** 5 L ≈ _____ qt

75. 3.4 ft ≈ _____ m   **76.** 100 lb ≈ _____ kg

77. 120 km ≈ _____ mi   **78.** 6 qt ≈ _____ L

79. 1.5 fl oz ≈ _____ cc   **80.** 12.5 tons ≈ _____ kg

81. The height of a computer desk is 30 in. The height of the chair is 38 cm. What is the difference in height between the desk and chair, in centimeters?

82. A bag of snack crackers contains 7.2 oz. If one serving is 30 g, approximately how many servings are in one bag?

83. A prescription for cough syrup indicates that 30 mL should be taken twice a day for 7 days. What is the total amount, in liters, of cough syrup to be taken?

84. The Boston Marathon is 42.195 km long. Convert this distance to miles. Round to the tenths place.

85. Write the formula to convert degrees Fahrenheit to degrees Celsius.

86. When roasting a turkey, the meat thermometer should register between 180°F and 185°F to indicate that the turkey is done. Convert these temperatures to degrees Celsius. Round to the nearest tenth, if necessary.

87. Write the formula to convert degrees Celsius to degrees Fahrenheit.

88. The average October temperature for Toronto, Ontario, Canada, is 8°C. Convert this temperature to degrees Fahrenheit.

Chapter 7   Test

1. Identify the units that apply to measuring length. Circle all that apply.

   a. Pound   **b.** Ounce   **c.** Meter

   d. Mile   **e.** Gram   **f.** Pint

   g. Feet   **h.** Liter   **i.** Fluid ounce

   j. Kilometer

2. Identify the units that apply to measuring capacity. Circle all that apply.

   a. Pound   **b.** Ounce   **c.** Meter

   d. Mile   **e.** Gram   **f.** Pint

   g. Foot   **h.** Liter   **i.** Fluid ounce

   j. Kilometer

3. Identify the units that apply to measuring mass or weight. Circle all that apply.

   a. Pound   **b.** Ounce   **c.** Meter

   d. Mile   **e.** Gram   **f.** Pint

   g. Foot   **h.** Liter   **i.** Fluid ounce

   j. Kilometer

4. A backyard needs 25 ft of fencing. How many yards is this?

5. It's estimated that an adult *Tyrannosaurus Rex* weighed approximately 11,000 lb. How many tons is this?

6. Two exits on the highway are 52,800 ft apart. How many miles is this?

7. A recipe for brownies calls for $\frac{3}{4}$ c of milk and 4 oz of water. What is the total amount of liquid in ounces?

8. A television show has 1200 sec of commercials. How many minutes is this?

9. Find the perimeter of the rectangle in feet.

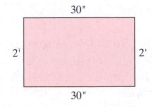

10. A decorator wraps a gift, using two pieces of ribbon. One is 1'10" and the other is 2'4". Find the total length of ribbon used.

11. When Stephen was born, he weighed 8 lb 1 oz. When he left the hospital, he weighed 7 lb 10 oz. How much weight did he lose after he was born?

12. A decorative pillow requires 3 ft 11 in. of fringe around the perimeter of the pillow. If 5 pillows are produced, how much fringe is required?

13. Josh ran a race and finished with the time of 1:15:15. Convert this time to minutes.

14. Approximate the width of the nut in centimeters and millimeters.

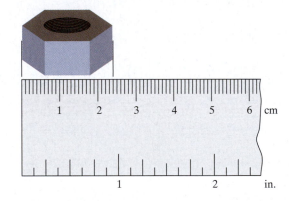

15. Select the most reasonable measurement for the length of a living room.

    a. 5 mm   **b.** 5 cm   **c.** 5 m   **d.** 5 km

16. The span of the Mackinac Bridge in Michigan is 1158 m. What is this length in kilometers?

17. A tablespoon (1 T) contains 0.015 L. How many milliliters is this?

18. **a.** What does the abbreviation *cc* stand for?

    b. Convert 235 mL to cubic centimeters.

    c. Convert 1 L to cubic centimeters.

19. A can of diced tomatoes is 411 g. Convert 411 g to centigrams.

20. A box of crackers is 210 g. If a serving of crackers is 30,000 mg, how many servings are in the box?

For Exercises 21–26, refer to Table 7-6 on page 443.

21. A bottle of Sprite contains 2 L. What is the capacity in quarts? Round to the nearest tenth.

22. Maurice Greene was one of the premier sprinters in U.S. track and field. His best race is the 100-m. How many yards is this? Round to the nearest yard.

23. The distance between two exits on a highway is 4.5 km. How far is this in miles? Round to the nearest tenth.

24. Breckenridge, Colorado, is 9603 ft above sea level. What is this height in meters? Round to the nearest meter.

25. A snowy egret stands about 20 in. tall and has a 38-in. wingspan. Convert both values to centimeters.

26. The mass of a laptop computer is 5000 g. What is the weight in pounds? Round to the nearest pound.

27. The oven temperature needed to bake cookies is 375°F. What is this temperature in degrees Celsius? Round to the nearest tenth.

28. The average January temperature in Albuquerque, New Mexico, is 2°C. Convert this temperature to degrees Fahrenheit.

29. A patient is supposed to get 0.1 mg of a drug for every kilogram of body mass, four times a day. How much of the drug would a 70-kg woman get each day?

30. Gus, the cat, had an overactive thyroid gland. The vet prescribed 0.125 mg of Methimazole every 12 hr. How many micrograms is this per week?

Chapters 1–7   Cumulative Review Exercises

1. Round each number to the indicated place.

   a. 2499; thousands place   **b.** 42,099; tens place

For Exercises 2–3, refer to the figure.

2. Find the perimeter.

3. Find the area.

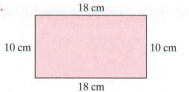

4. Simplify. $144 \div 9 \div (17 - 3 \cdot 5)^2$

5. The table gives the amount of money spent on research and development for five major U.S. companies. (*Source:* The Financial Times Ltd.)

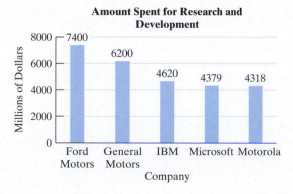

Amount Spent for Research and Development

a. Which company spends the most on research and development? What is that amount?

b. What is the difference between the amount spent at IBM and the amount spent at Motorola?

c. What is the total amount spent for research and development for these five companies?

6. Write an equivalent fraction with the indicated denominator. $\frac{2}{13} = \frac{}{39}$

7. Is 32,542 divisible by 3? Explain why or why not.

8. Write the prime factorization of 108.

9. Find the area of the triangle.

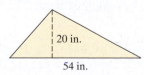

10. Keesha wants to bake oatmeal cookies, but she has only 1 c of oatmeal. The recipe calls for 3 c. Does Keesha have enough oatmeal to make $\frac{1}{4}$ of the recipe?

11. Find the LCD of $\dfrac{1}{2}, \dfrac{6}{5}$, and $\dfrac{3}{10}$.

12. Simplify. $\dfrac{1}{2} + \dfrac{6}{5} - \dfrac{3}{10}$

For Exercises 13–16, perform the indicated operations with mixed numbers.

13. $6\dfrac{2}{3} + 2\dfrac{5}{6}$

14. $6\dfrac{2}{3} \cdot 2\dfrac{5}{6}$

15. $6\dfrac{2}{3} \div 2\dfrac{5}{6}$

16. $6\dfrac{2}{3} - 2\dfrac{5}{6}$

For Exercises 17–22, complete the table.

	Fraction	Decimal
17.	$\dfrac{1}{3}$	
18.		0.45
19.		1.25
20.	$\dfrac{7}{2}$	
21.	$\dfrac{3}{8}$	
22.		0.04

23. A football team won 6 games and lost 5.

   a. Write a ratio of wins to losses.

   b. Write a ratio of wins to total games played.

24. A pharmacist has a 1-L bottle of an antacid solution. How many smaller bottles can he fill if they contain 25 mL each?

25. If a car dealership sells 18 cars in a 5-day period, how many cars can the dealership expect to sell in 25 days?

26. A hospital ward has 40 beds and employs 6 nurses. What is the unit rate of beds per nurse? Round to the nearest tenth.

27. Are 6 and 8 proportional to 2 and 3?

28. Four-fifths of the drinks sold at a movie theater were soda. What percent is this?

29. Assume that these figures are similar. Solve for x.

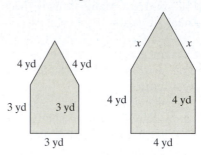

30. Of the trees in a forest, 62% were saved from a fire. If this represents 1420 trees, how many trees were originally in the forest? Round to the nearest whole tree.

31. Of 60 people, 45% had eaten at least one meal at McDonald's this week. How many people ate at McDonald's?

32. The sales tax on a $21 meal is $1.26. What is the sales tax rate?

33. The commission rate that Yvonne earns is 14% of sales. If she earned $2100 in commission, how much did she sell?

34. If $5000 is invested at 3.4% simple interest for 6 years, how much interest is earned?

For Exercises 35–40, convert the units of capacity, length, mass, and weight. Round to one decimal place, if necessary.

35. 5800 g = ___ kg

36. 5.8 kg ≈ ___ lb

37. 72 in. ≈ ___ cm

38. 72 in. = ___ ft

39. $3\dfrac{1}{2}$ qt = ___ pt

40. $3\dfrac{1}{2}$ qt ≈ ___ L

Geometry

8

CHAPTER OUTLINE

Chapter 8

In this chapter we study geometry. We begin by categorizing familiar figures such as squares, rectangles, triangles and circles. Then we study the concepts of perimeter and area. We identify familiar three-dimensional figures and learn how to find their volumes.

Before you begin work in this chapter, try to match as many formulas as possible. Then as you go through the chapter, check to see if you got them all correct.

1. $A = \pi r^2$ _____

2. $V = lwh$ _____

3. $P = 4s$ _____

4. $C = 2\pi r$ _____

5. $A = bh$ _____

6. $V = \pi r^2 h$ _____

7. $P = 2l + 2w$ _____

8. $A = lw$ _____

9. $V = \dfrac{4}{3}\pi r^3$ _____

10. $A = \dfrac{1}{2}(b_1 + b_2)h$ _____

a. Volume:

b. Area:

c. Area:

d. Perimeter:

e. Area:

f. Area:

g. Circumference:

h. Perimeter:

i. Volume:

j. Volume:

Section 8.1   Lines and Angles

Objectives

1. Basic Definitions
2. Naming and Measuring Angles
3. Complementary and Supplementary Angles
4. Parallel and Perpendicular Lines

1. Basic Definitions

In this chapter, we will introduce some basic concepts of geometry.

A **point** is a specific location in space. We often symbolize a point by a dot and label it with a capital letter such as P.

• P

A **line** consists of infinitely many points that follow a straight path. A line extends forever in both directions. This is illustrated by arrowheads at both ends. Figure 8-1 shows a line through points A and B. The line can be represented as $\overleftrightarrow{AB}$ or as $\overleftrightarrow{BA}$.

Figure 8-1

A **line segment** is a part of a line between and including two distinct endpoints. A line segment with endpoints P and Q can be denoted $\overline{PQ}$ or $\overline{QP}$. See Figure 8-2.

Figure 8-2

Concept Connections

1. Are the rays $\overrightarrow{AB}$ and $\overrightarrow{BA}$ the same?
2. Are the lines $\overleftrightarrow{PQ}$ and $\overleftrightarrow{QP}$ the same?

A **ray** is the part of a line that includes an endpoint and all points on one side of the endpoint. In Figure 8-3, ray $\overrightarrow{PQ}$ is named by using the endpoint P and another point Q on the ray. Notice that the rays $\overrightarrow{PQ}$ and $\overrightarrow{QP}$ are different because they extend in different directions.

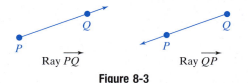

Ray $\overrightarrow{PQ}$ Ray $\overrightarrow{QP}$

Figure 8-3

TIP: A ray has only one endpoint, which is always written first.

Skill Practice

Identify each as a point, line, line segment, or ray.

3. $\overline{RS}$ 4. • Q
5. $\overrightarrow{XY}$ 6. $\overleftrightarrow{TV}$

Example 1   Identifying Points, Lines, Line Segments, and Rays

Identify each as a point, line, line segment, or ray.

a. $\overleftrightarrow{MN}$ b. $\overrightarrow{NM}$ c. $\overline{MN}$ d. • S

Solution:

a. The double arrowheads indicate that $\overleftrightarrow{MN}$ is a line.

b. The single arrowhead indicates that $\overrightarrow{NM}$ is a ray with endpoint N.

c. The bar drawn above the letters $\overline{MN}$ indicates a line segment with endpoints M and N.

d. The dot represents a point.

2. Naming and Measuring Angles

An **angle** is a geometric figure formed by two rays that share a common endpoint. The common endpoint is called the **vertex** of the angle. In Figure 8-4, the rays $\overrightarrow{PR}$ and $\overrightarrow{PQ}$ share the endpoint P. These rays form the sides of the angle denoted by $\angle QPR$

Answers

1. No   2. Yes
3. Line segment
4. Point   5. Ray   6. Line

or $\angle RPQ$. Notice that when we name an angle, the vertex must be the middle letter. Sometimes a small arc ⌒ is drawn to illustrate the location of an angle. In such a case, the angle may be named by using the symbol $\angle$ along with the letter of the vertex.

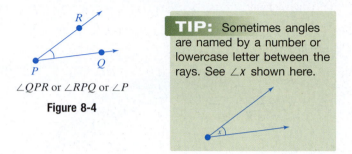

$\angle QPR$ or $\angle RPQ$ or $\angle P$

Figure 8-4

TIP: Sometimes angles are named by a number or lowercase letter between the rays. See $\angle x$ shown here.

Avoiding Mistakes

The ° symbol can be used for temperature or for measuring angles. The context of the problem tells us whether temperature or angle measure is implied by the symbol.

The most common unit to measure an angle is the degree, denoted by °. To become familiar with the measure of angles, consider the following benchmarks. Two rays that form a quarter turn of a circle make a 90° angle. A 90° angle is called a **right angle** and is often depicted with a □ symbol. Two rays that form a half turn of a circle make a 180° angle. A 180° angle is called a **straight angle** because it appears as a straight line. A full circle has 360°. For example, the second hand of a clock sweeps out an angle of 360° in 1 minute. See Figure 8-5.

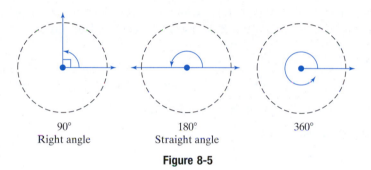

90°
Right angle

180°
Straight angle

360°

Figure 8-5

We can approximate the measure of an angle by using a tool called a *protractor*, shown in Figure 8-6. A protractor uses equally spaced tick marks around a semicircle to measure angles from 0° to 180°.

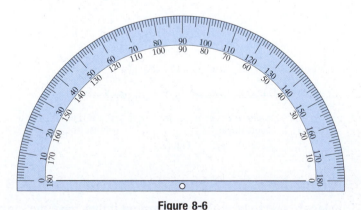

Figure 8-6

Example 2 shows how we can use a protractor to measure several angles. To denote the measure of an angle, we use the symbol m, written in front of the name of the angle. For example, if the measure of angle A is 30°, we write $m(\angle A) = 30°$.

Skill Practice

Read the protractor to determine the measure of each angle.

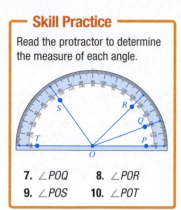

7. ∠POQ   **8.** ∠POR
9. ∠POS   **10.** ∠POT

Example 2 **Measuring Angles**

Read the protractor to determine the measure of each angle.

a. ∠AOB   **b.** ∠AOC   **c.** ∠AOD   **d.** ∠AOE

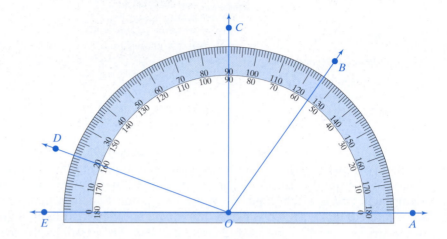

Solution:

We will use the inner scale on the protractor. This is done because we are measuring the angles in a counterclockwise direction, beginning at 0° along ray $\overrightarrow{OA}$.

a. $m(\angle AOB) = 55°$ On the inner scale, ray $\overrightarrow{OA}$ is aligned with 0° and ray $\overrightarrow{OB}$ passes through 55°. Therefore, $m(\angle AOB) = 55°$.

b. $m(\angle AOC) = 90°$ ∠AOC is a right angle.

c. $m(\angle AOD) = 160°$

d. $m(\angle AOE) = 180°$ ∠AOE is a straight angle.

Concept Connections

Answer true or false.

11. An angle whose measure is 102° is obtuse.
12. An angle whose measure is 98° is acute.

An angle is said to be an **acute angle** if its measure is between 0° and 90°. An angle is said to be an **obtuse angle** if its measure is between 90° and 180°. See Figure 8-7.

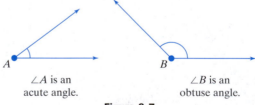

∠A is an acute angle.   ∠B is an obtuse angle.

Figure 8-7

3. Complementary and Supplementary Angles

- Two angles are said to be equal or **congruent** if they have the same measure.
- Two angles are said to be **complementary** if the sum of their measures is 90°. In Figure 8-8, the complement of a 60° angle is a 30° angle, and vice versa.

Answers

7. 20°   **8.** 45°   **9.** 125°
10. 180°   **11.** True   **12.** False

- Two angles are said to be **supplementary** if the sum of their measures is 180°. In Figure 8-9, the supplement of a 60° angle is a 120° angle, and vice versa.

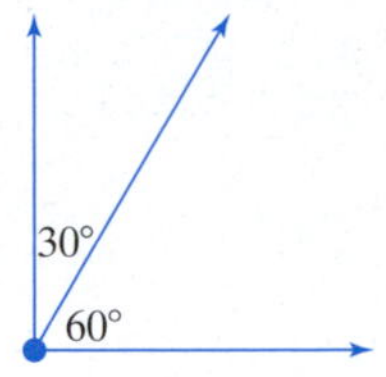

Complementary angles

30°
60°

$60° + 30° = 90°$

Figure 8-8

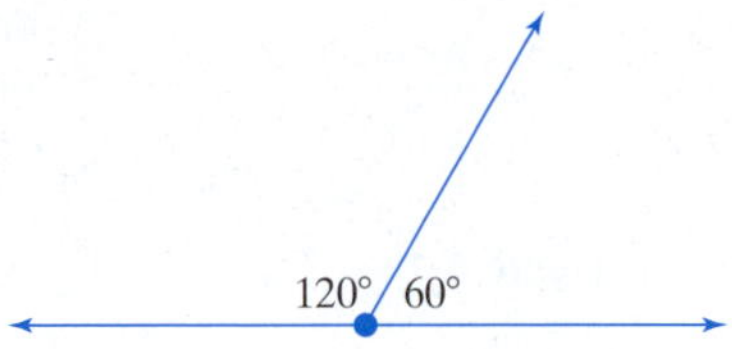

Supplementary angles

120° 60°

$60° + 120° = 180°$

Figure 8-9

| **Example 3** | **Identifying Supplementary and Complementary Angles** |

a. What is the supplement of a 105° angle?

b. What is the complement of a 12° angle?

Skill Practice

13. What is the supplement of a 35° angle?
14. What is the complement of a 52° angle?

Solution:

a. The sum of a 105° angle and its supplement must equal 180°. Writing the related subtraction problem, we have

$$180° - 105° = 75°$$ The supplement is a 75° angle.

b. The sum of a 12° angle and its complement must equal 90°. Writing the related subtraction problem, we have

$$90° - 12° = 78°$$ The complement is a 78° angle.

4. Parallel and Perpendicular Lines

Two lines may intersect (cross) or may be parallel. **Parallel lines** lie on the same flat surface, but never intersect. See Figure 8-10.

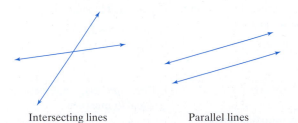

TIP: Sometimes we use the symbol ‖ to denote parallel lines.

Intersecting lines   Parallel lines

Figure 8-10

Notice that two intersecting lines form four angles. In Figure 8-11, $\angle a$ and $\angle c$ are **vertical angles**. They appear on opposite sides of the vertex. Likewise, $\angle b$ and $\angle d$ are vertical angles. Vertical angles are equal in measure. That is $m(\angle a) = m(\angle c)$ and $m(\angle b) = m(\angle d)$.

Angles that share a side are called *adjacent* angles. One pair of *adjacent* angles in Figure 8-11 is $\angle a$ and $\angle b$.

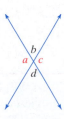

b
a c
d

Figure 8-11

Answers

13. 145°   14. 38°

If two lines intersect at a right angle, they are **perpendicular lines**. See Figure 8-12.

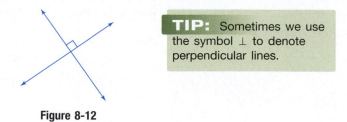

TIP: Sometimes we use the symbol ⊥ to denote perpendicular lines.

Figure 8-12

In Figure 8-13, lines L_1 and L_2 are parallel lines. If a third line m intersects the two parallel lines, eight angles are formed. Suppose we label the eight angles formed by lines L_1, L_2, and m with the numbers 1–8.

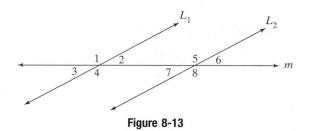

Figure 8-13

These angles have the special properties found in Table 8-1.

Table 8-1

Lines L_1 and L_2 are Parallel; Line m is an Intersecting Line	Name of Angles	Property
	The following pairs of angles are called **alternate interior angles**: ∠2 and ∠7 ∠4 and ∠5	**Alternate interior angles are equal in measure.** $m(\angle 2) = m(\angle 7)$ $m(\angle 4) = m(\angle 5)$
	The following pairs of angles are called **alternate exterior angles**: ∠1 and ∠8 ∠3 and ∠6	**Alternate exterior angles are equal in measure.** $m(\angle 1) = m(\angle 8)$ $m(\angle 3) = m(\angle 6)$
	The following pairs of angles are called **corresponding angles**: ∠1 and ∠5 ∠2 and ∠6 ∠3 and ∠7 ∠4 and ∠8	**Corresponding angles are equal in measure.** $m(\angle 1) = m(\angle 5)$ $m(\angle 2) = m(\angle 6)$ $m(\angle 3) = m(\angle 7)$ $m(\angle 4) = m(\angle 8)$

| Example 4 | Finding the Measure of Angles in a Diagram |

Assume that lines L_1 and L_2 are parallel. Find the measure of each angle, and explain how the angle is related to the given angle of $65°$.

a. $\angle a$

b. $\angle b$

c. $\angle c$

d. $\angle d$

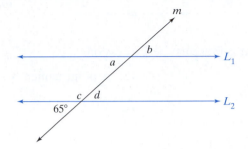

Skill Practice

Assume that lines L_1 and L_2 are parallel. Find the measure of each angle. Explain how the angle is related to the given angle.

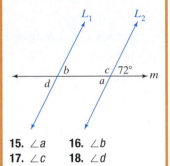

15. $\angle a$ **16.** $\angle b$
17. $\angle c$ **18.** $\angle d$

Solution:

a. $m(\angle a) = 65°$ $\angle a$ is a corresponding angle to the given angle.

b. $m(\angle b) = 65°$ $\angle b$ is an alternate exterior angle to the given angle.

c. $m(\angle c) = 115°$ $\angle c$ is the supplement to the given angle.

d. $m(\angle d) = 65°$ $\angle d$ and the given angle are vertical angles.

Answers

15. $72°$; vertical angles
16. $72°$; corresponding angles
17. $108°$; supplementary angles
18. $72°$; alternate exterior angles

Section 8.1   Practice Exercises

Boost your GRADE at ALEKS.com!

ALEKS® version 3.0

• Practice Problems   • e-Professors
• Self-Tests   • Videos
• NetTutor

Study Skills Exercises

1. For your next test, make a memory sheet: On a 3×5 card (or several 3×5 cards), write all the formulas and rules that you need to know. Memorize all this information. Then when your instructor hands you the test, write down all the information that you can remember before you begin the test. Then you can take the test without worrying that you will forget something important. This process is referred to as a "memory dump." What important definitions and concepts have you learned in this section of the text?

2. Define the key terms.

a. Point	**b.** Line	**c.** Line segment	**d.** Ray
e. Angle	**f.** Vertex	**g.** Right angle	**h.** Straight angle
i. Acute angle	**j.** Obtuse angle	**k.** Congruent angles	**l.** Complementary angles
m. Supplementary angles	**n.** Parallel lines	**o.** Vertical angles	**p.** Perpendicular lines
q. Alternate exterior angles	**r.** Alternate interior angles	**s.** Corresponding angles	

Objective 1: Basic Definitions

3. Explain the difference between a line and a line segment.

4. Explain the difference between a line and a ray.

For Exercises 5–10, identify each figure as a line, line segment, ray, or point. (See Example 1.)

5. K ⟶ H

6. K — H

7. H •

8. H K ⟶

9. ⟷ H K

10. K •

For Exercises 11–14, draw a figure that represents the expression.

11. $\overline{XY}$

12. A point named X

13. $\overleftrightarrow{YX}$

14. $\overrightarrow{XY}$

Objective 2: Naming and Measuring Angles

15. Sketch a right angle.

16. Sketch a straight angle.

17. Sketch an acute angle.

18. Sketch an obtuse angle.

For Exercises 19–24, use the protractor to determine the measure of each angle. (See Example 2.)

19. $m(\angle AOB)$

20. $m(\angle AOC)$

21. $m(\angle AOD)$

22. $m(\angle AOE)$

23. $m(\angle AOF)$

24. $m(\angle AOG)$

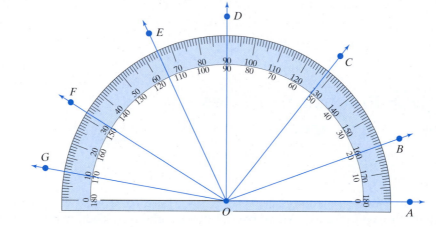

For Exercises 25–32, label each as an obtuse angle, acute angle, right angle, or straight angle.

25. $m(\angle A) = 90°$ **26.** $m(\angle E) = 91°$ **27.** $m(\angle B) = 98°$ **28.** $m(\angle F) = 30°$

29. $m(\angle C) = 2°$ **30.** $m(\angle G) = 130°$ **31.** $m(\angle D) = 180°$ **32.** $m(\angle H) = 45°$

Objective 3: Complementary and Supplementary Angles

For Exercises 33–40, the measure of an angle is given. Find the measure of the complement. (See Example 3.)

33. $80°$ **34.** $5°$ **35.** $27°$ **36.** $64°$

37. $29.5°$ **38.** $13.2°$ **39.** $89°$ **40.** $1°$

For Exercises 41–48, the measure of an angle is given. Find the measure of the supplement. (See Example 3.)

41. $80°$ **42.** $5°$ **43.** $127°$ **44.** $124°$

45. $37.4°$ **46.** $173.9°$ **47.** $179°$ **48.** $1°$

49. Can two supplementary angles both be obtuse? Why or why not?

50. Can two supplementary angles both be acute? Why or why not?

51. Can two complementary angles both be acute? Why or why not?

52. Can two complementary angles both be obtuse? Why or why not?

53. What angle is its own supplement?

54. What angle is its own complement?

Objective 4: Parallel and Perpendicular Lines

55. Sketch two lines that are parallel.

56. Sketch two lines that are *not* parallel.

57. Sketch two lines that are perpendicular.

58. Sketch two lines that are *not* perpendicular.

For Exercises 59–62, find the measure of angles *a*, *b*, *c*, and *d*.

59.

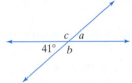

60.

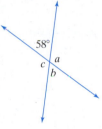

61.

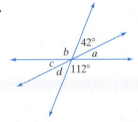

62.

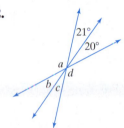

63. If two intersecting lines form vertical angles and each angle measures 90°, what can you say about the lines?

64. Can two adjacent angles formed by two intersecting lines be complementary, supplementary, or neither?

For Exercises 65–66, refer to the figure.

65. Describe the pair of angles, ∠*a* and ∠*c* as complementary, vertical, or supplementary angles.

66. Describe the pair of angles, ∠*b* and ∠*c* as complementary, vertical, or supplementary angles.

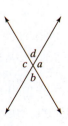

For Exercises 67–70, refer to the figure.

67. Identify a pair of vertical angles.

68. Identify a pair of alternate interior angles.

69. Identify a pair of alternate exterior angles.

70. Identify a pair of corresponding angles.

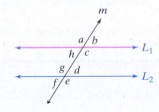

For Exercises 71–72, find the measure of angles *a–g* in the figure. Assume that L_1 and L_2 are parallel and that *m* is an intersecting line. (See Example 4.)

71.

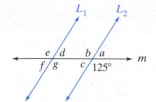

72.

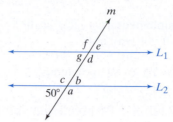

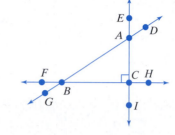

For Exercises 73–82, refer to the figure and answer true or false.

73. $\overleftrightarrow{AC}$ and $\overleftrightarrow{BC}$ are perpendicular lines.

74. $\overleftrightarrow{AB}$ and $\overleftrightarrow{AC}$ are perpendicular lines.

75. ∠*GBF* is an acute angle.

76. ∠*EAD* is an acute angle.

77. ∠*EAD* and ∠*DAC* are complementary angles.

78. ∠*GBF* and ∠*FBA* are complementary angles.

79. ∠*EAD* and ∠*CAB* are vertical angles.

80. ∠*ABC* and ∠*FBG* are vertical angles.

81. The point *B* is on $\overline{GA}$.

82. The point *C* is on $\overline{BH}$.

For Exercises 83–86, find the measure of ∠*XYZ*.

83.

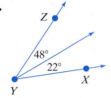

84.

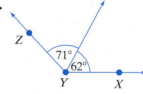

85.

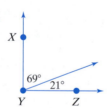

86.

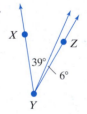

87. Use the figure to find the measure of each angle.

   a. ∠*AOB*

   b. ∠*EOD*

   c. ∠*AOE*

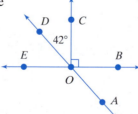

88. Use the figure to find the measure of each angle.

   a. ∠*AOB*

   b. ∠*EOD*

   c. ∠*AOE*

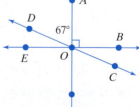

Expanding Your Skills

The second hand on a clock sweeps out a complete circle in 1 min. A circle forms a 360° arc. Use this information for Exercises 89–92.

89. How many degrees does a second hand on a clock move in 30 sec?

90. How many degrees does a second hand on a clock move in 15 sec?

91. How many degrees does a second hand on a clock move in 20 sec?

92. How many degrees does a second hand on a clock move in 45 sec?

Triangles and the Pythagorean Theorem

1. Categorizing Triangles

A triangle is a three-sided polygon. Furthermore, the sum of the measures of the angles within a triangle is 180°. Teachers often demonstrate this fact by tearing a triangular sheet of paper as shown in Figure 8-14. Then they align the **vertices** (points) of the triangle to form a straight angle.

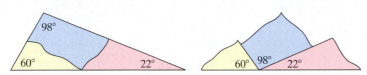

Figure 8-14

> **PROPERTY   Angles of a Triangle**
>
> The sum of the measures of the angles of a triangle equals 180°.

Example 1   **Finding the Measure of Angles Within a Triangle**

Find the measure of angles a and b.

a.

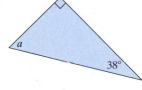

b.

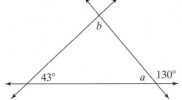

Skill Practice

Find the measures of angles a and b.

1.

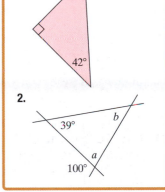

2.

Solution:

a. Recall that the □ symbol represents a 90° angle.

$38° + 90° + m(\angle a) = 180°$ The sum of the angles within a triangle is 180°.

$128° + m(\angle a) = 180°$ Add the measures of the two known angles.

$m(\angle a) = 180° - 128°$ Write the related subtraction problem to find $m(\angle a)$.

$m(\angle a) = 52°$

b. $\angle a$ is the supplement of the 130° angle. Thus $m(\angle a) = 50°$.

$43° + 50° + m(\angle b) = 180°$ The sum of the angles within a triangle is 180°.

$93° + m(\angle b) = 180°$ Add the measures of the two known angles.

$m(\angle b) = 180° - 93°$ Write the related subtraction problem to find $m(\angle b)$.

$m(\angle b) = 87°$

Answers

1. $m(\angle a) = 48°$
2. $m(\angle a) = 80°$
   $m(\angle b) = 61°$

Concept Connections

Match the triangle with the appropriate category, **a**, **b**, or **c**.

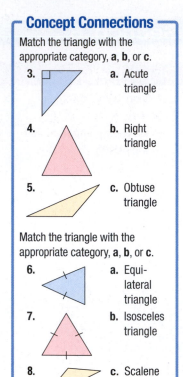

3. ☐⟍   **a.** Acute triangle

4. △   **b.** Right triangle

5. ◢   **c.** Obtuse triangle

Match the triangle with the appropriate category, **a**, **b**, or **c**.

6. ◁   **a.** Equilateral triangle

7. △   **b.** Isosceles triangle

8. ◿   **c.** Scalene triangle

Triangles may be categorized by the measures of their angles and by the number of equal sides or angles (Figures 8-15 and 8-16).

- An **acute triangle** is a triangle in which all three angles are acute.
- A **right triangle** is a triangle in which one angle is a right angle.
- An **obtuse triangle** is a triangle in which one angle is obtuse.

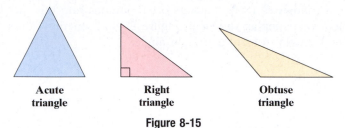

Acute triangle   Right triangle   Obtuse triangle

Figure 8-15

- An **equilateral triangle** is a triangle in which all three sides (and all three angles) are equal in measure.
- An **isosceles triangle** is a triangle in which two sides are equal in length (the angles opposite the equal sides are also equal in measure).
- A **scalene triangle** is a triangle in which no sides (or angles) are equal in measure.

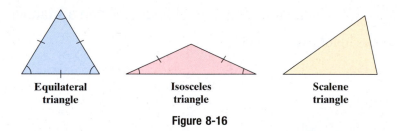

Equilateral triangle   Isosceles triangle   Scalene triangle

Figure 8-16

TIP: Sometimes we use tick marks / to denote segments of equal length. Similarly, we sometimes use a small arc ⟩ to denote angles of equal measure.

2. Square Roots

In this section we present an important theorem called the Pythagorean theorem. To understand this theorem, we first need some background definitions.

Recall from Section 1.7 that to square a number means to find the product of the number and itself. Thus, $b^2 = b \cdot b$. For example:

$$6^2 = 6 \cdot 6 = 36$$

We now want to reverse this process by finding a square root of a number. Recall that this is denoted by the radical sign $\sqrt{}$. For example, $\sqrt{36}$ reads as "the positive square root of 36." Thus,

$$\sqrt{36} = 6 \qquad \text{because } 6^2 = 6 \cdot 6 = 36.$$

Answers

3. b   **4.** a   **5.** c

6. b   **7.** a, b   **8.** c

Example 2 **Evaluating Squares and Square Roots**

Simplify.

a. $\sqrt{64}$ **b.** $\sqrt{100}$ **c.** 100^2 **d.** $\sqrt{1}$

Solution:

a. $\sqrt{64} = 8$ because $8 \cdot 8 = 64$

b. $\sqrt{100} = 10$ because $10 \cdot 10 = 100$

c. $100^2 = 100 \cdot 100$

$= 10{,}000$

d. $\sqrt{1} = 1$ because $1 \cdot 1 = 1$

Table 8-2 gives a list of several whole numbers, their squares, and the corresponding square roots.

Table 8-2

$0^2 = 0 \rightarrow \sqrt{0} = 0$	$8^2 = 64 \rightarrow \sqrt{64} = 8$
$1^2 = 1 \rightarrow \sqrt{1} = 1$	$9^2 = 81 \rightarrow \sqrt{81} = 9$
$2^2 = 4 \rightarrow \sqrt{4} = 2$	$10^2 = 100 \rightarrow \sqrt{100} = 10$
$3^2 = 9 \rightarrow \sqrt{9} = 3$	$11^2 = 121 \rightarrow \sqrt{121} = 11$
$4^2 = 16 \rightarrow \sqrt{16} = 4$	$12^2 = 144 \rightarrow \sqrt{144} = 12$
$5^2 = 25 \rightarrow \sqrt{25} = 5$	$13^2 = 169 \rightarrow \sqrt{169} = 13$
$6^2 = 36 \rightarrow \sqrt{36} = 6$	$14^2 = 196 \rightarrow \sqrt{196} = 14$
$7^2 = 49 \rightarrow \sqrt{49} = 7$	$15^2 = 225 \rightarrow \sqrt{225} = 15$

3. Pythagorean Theorem and Applications

Recall that a right triangle is a triangle with a 90° angle. The two sides forming the right angle are called the **legs**. The side opposite the right angle is called the **hypotenuse**. Note that the hypotenuse is always the longest side. See Figure 8-17. We often use the letters a and b to represent the legs of a right triangle. The letter c is used to label the hypotenuse.

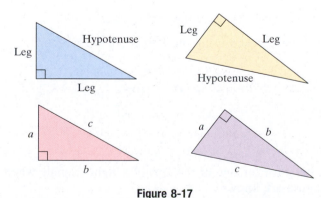

Figure 8-17

For any right triangle, the **Pythagorean theorem** gives us the following important relationship among the lengths of the sides.

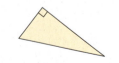
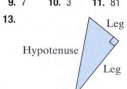

DEFINITION Pythagorean Theorem

For any right triangle,

$$(\text{Leg})^2 + (\text{Leg})^2 = (\text{Hypotenuse})^2$$

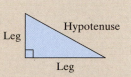

Using the letters a, b, and c to represent the legs and hypotenuse, respectively, we have

$$a^2 + b^2 = c^2$$

Skill Practice

15. Find the length of the hypotenuse.

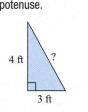

Example 3 **Finding the Length of the Hypotenuse of a Right Triangle**

Find the length of the hypotenuse of the right triangle.

Solution:

The lengths of the legs are given.

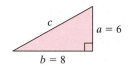

Label the triangle, using a, b, and c. It does not matter which leg is labeled a and which is labeled b.

$a^2 + b^2 = c^2$ Apply the Pythagorean theorem.

$(6)^2 + (8)^2 = c^2$ Substitute $a = 6$ and $b = 8$.

$36 + 64 = c^2$ Simplify.

$100 = c^2$ The solution to this equation is the positive number, c, that when squared equals 100.

$\sqrt{100} = c$

$10 = c$ Simplify the square root of 100.

The solution may be checked using the Pythagorean theorem.

$$a^2 + b^2 = c^2$$
$$(6)^2 + (8)^2 \stackrel{?}{=} (10)^2$$
$$36 + 64 = 100 ✔$$

In Example 4, we solve for one of the legs of a right triangle when the other leg and the hypotenuse are known.

Answer

15. 5 ft

| Example 4 | **Finding the Length of a Leg in a Right Triangle** |

Find the length of the unknown side of the right triangle.

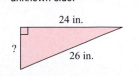

Solution:

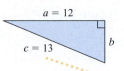

Label the triangle, using a, b, and c. One of the legs is unknown. It doesn't matter whether we call the unknown leg a or b.

$a^2 + b^2 = c^2$ Apply the Pythagorean theorem.

$(12)^2 + b^2 = (13)^2$ Substitute $a = 12$ and $c = 13$.

$144 + b^2 = 169$ Simplify.

Our goal, when solving an equation, is to isolate the variable. For this addition equation, we can write the related subtraction equation.

$b^2 = 169 - 144$ Related subtraction equation.

$b^2 = 25$ The solution to this equation is the positive number b that when squared equals 25.

$b = \sqrt{25}$

$b = 5$ Simplify the square root of 25.

The solution may be checked by using the Pythagorean theorem.

$$a^2 + b^2 = c^2$$
$$(12)^2 + (5)^2 \stackrel{?}{=} (13)^2$$
$$144 + 25 = 169 \checkmark$$

In Example 5 we use the Pythagorean theorem in an application.

| Example 5 | **Using the Pythagorean Theorem in an Application** |

When Barb swam across a river, the current carried her 300 yd downstream from her starting point. If the river is 400 yd wide, how far did Barb swim?

Solution:

We first familiarize ourselves with the problem and draw a diagram (Figure 8-18). The distance Barb actually swims is the hypotenuse of the right triangle. Therefore, we label this distance c.

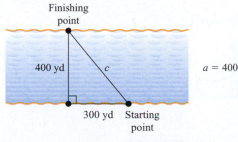

Figure 8-18

Skill Practice

16. Find the length of the unknown side.

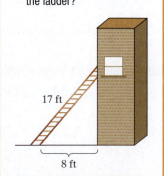

Avoiding Mistakes

Always remember that the hypotenuse (longest side) is given the letter "c" when applying the Pythagorean theorem.

Skill Practice

17. The bottom of a 17-ft ladder is placed 8 ft from the bottom of a building. How far up the building is the top of the ladder?

Answers

16. 10 in.   **17.** 15 ft

$$a^2 + b^2 = c^2 \qquad \text{Apply the Pythagorean theorem.}$$

$$(400)^2 + (300)^2 = c^2 \qquad \text{Substitute } a = 400 \text{ and } b = 300.$$

$$160{,}000 + 90{,}000 = c^2 \qquad \text{Simplify.}$$

$$250{,}000 = c^2 \qquad \text{Add. The solution to this equation is the positive number } c \text{ that when squared equals } 250{,}000.$$

$$\sqrt{250{,}000} = c$$

$$500 = c \qquad \text{Simplify.}$$

The distance that Barb swims is 500 yd.

Section 8.2 Practice Exercises

Boost your GRADE at ALEKS.com!

ALEKS version 3.0

- Practice Problems
- Self-Tests
- NetTutor
- e-Professors
- Videos

Study Skills Exercises

1. When you are solving an application involving geometry, be sure to draw a picture of the situation and label the known quantities with numbers and the unknown quantities with variables. This will help to solve the problem. After reading this section, what geometric figure do you think you will be drawing most often in this section?

2. Define the key terms.

   a. **Vertices**

   b. **Acute triangle**

   c. **Right triangle**

   d. **Obtuse triangle**

   e. **Equilateral triangle**

   f. **Isosceles triangle**

   g. **Scalene triangle**

   h. **Pythagorean theorem**

   i. **Hypotenuse**

   j. **Legs of a right triangle**

Review Exercises

3. Do $\angle ACB$ and $\angle BCA$ represent the same angle?

4. Is line segment $\overline{MN}$ the same as the line segment $\overline{NM}$?

5. Is ray $\overrightarrow{AB}$ the same as ray $\overrightarrow{BA}$?

6. Is the line $\overleftrightarrow{PQ}$ the same as the line $\overleftrightarrow{QP}$?

7. Is a right angle an obtuse angle?

8. Can two acute angles be supplementary?

Objective 1: Categorizing Triangles

For Exercises 9–16, find the measures of angles a and b. **(See Example 1.)**

9.
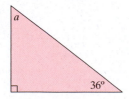

10.
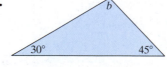

11.
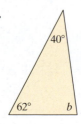

12.

13.

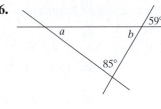

14.

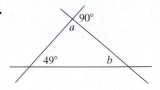

15.

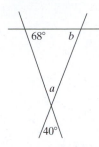

16.

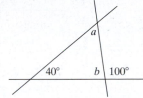

For Exercises 17–22, choose all figures that apply. The tick marks / denote segments of equal length, and small arcs ⟩ denote angles of equal measure.

17. Acute triangle

18. Obtuse triangle

19. Right triangle

20. Scalene triangle

21. Isosceles triangle

22. Equilateral triangle

a.

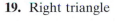

b.

c.

d.

e.

f.

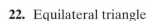

Objective 2: Square Roots

For Exercises 23–38, simplify the squares and the square roots. **(See Example 2.)**

23. $\sqrt{49}$

24. $\sqrt{64}$

25. 7^2

26. 8^2

27. 4^2

28. 5^2

29. $\sqrt{16}$

30. $\sqrt{25}$

31. $\sqrt{36}$

32. $\sqrt{100}$

33. 6^2

34. 10^2

35. 9^2

36. 3^2

37. $\sqrt{81}$

38. $\sqrt{9}$

Objective 3: Pythagorean Theorem and Applications

For Exercises 39–42, find the length of the unknown side. **(See Examples 3–4.)**

39.

40.

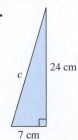

41.

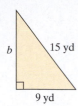

b   15 yd

9 yd

42.

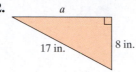

a

17 in.   8 in.

For Exercises 43–46, find the length of the unknown leg or hypotenuse.

43. Leg = 24 ft, hypotenuse = 26 ft

44. Leg = 9 km, hypotenuse = 41 km

45. Leg = 32 in., leg = 24 in.

46. Leg = 16 m, leg = 30 m

47. Find the length of the supporting brace. **(See Example 5.)**

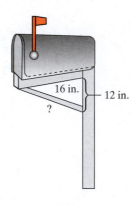

16 in.   12 in.

?

48. Find the length of the ramp.

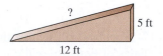

?   5 ft

12 ft

49. Find the height of the airplane above the ground.

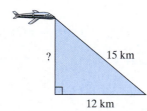

?   15 km

12 km

50. A 25-in. television measures 25 in. across the diagonal. If the width is 20 in., find the height.

25 in.   ?

20 in.

51. A car travels east 24 mi and then south 7 mi. How far is the car from its starting point?

52. A 26-ft-long wire is to be tied from a stake in the ground to the top of a 24-ft pole. How far from the pole should the stake be placed?

For Exercises 53–56, find the perimeter.

53.

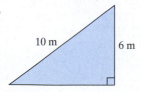

10 m   6 m

54.

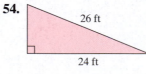

26 ft

24 ft

55.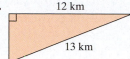

56.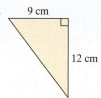

Expanding Your Skills

57. Find the length of side c by dividing this figure into a rectangle and a right triangle. Then find the perimeter of the figure.

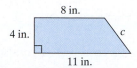

58. Find the perimeter of the figure.

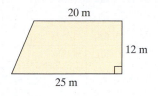

59. Find the perimeter of the figure.

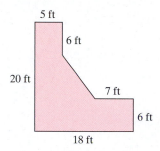

60. Tyler drives 20 mi north, 8 mi east, 4 mi south, and 4 mi east. How far is Tyler from his starting point?

Calculator Connections

Topic: Entering Square Roots on a Calculator

Many square roots cannot be written as a whole number. For example, there is no whole number that when squared equals 26. However, we might speculate that $\sqrt{26}$ is a number slightly greater than 5 because $\sqrt{25} = 5$. A decimal approximation can be made by using a calculator.

$$\sqrt{26} \approx 5.099 \qquad \text{because } 5.099^2 = 25.999801 \approx 26$$

To enter a square root on a calculator, use the $\boxed{\sqrt{x}}$ key. On some calculators, the $\boxed{\sqrt{x}}$ function is associated with the x^2 key. In such a case, it is necessary to press $\boxed{\text{2}^{\text{nd}}}$ or $\boxed{\text{SHIFT}}$ first, followed by the $\boxed{x^2}$ key. Some calculators require the square root key to be entered first, before the number, while with others we enter the number first followed by $\boxed{\sqrt{x}}$.

Expression	Keystrokes	Result
$\sqrt{26}$	26 $\boxed{\sqrt{x}}$ or $\boxed{\sqrt{x}}$ 26 $\boxed{=}$	5.099019514
$\sqrt{9325}$	9325 $\boxed{\sqrt{x}}$ or $\boxed{\sqrt{x}}$ 9325 $\boxed{=}$	96.56603958
$\sqrt{100}$	100 $\boxed{\sqrt{x}}$ or $\boxed{\sqrt{x}}$ 100 $\boxed{=}$	10

For Exercises 61–66, complete the table. For the estimate, find two consecutive whole numbers between which the square root lies. The first row is done for you.

	Square Root	Estimate	Calculator Approximation (Round to 3 Decimal Places)
	$\sqrt{50}$	is between 7 and 8	7.071
61.	$\sqrt{10}$	is between ____ and ____	
62.	$\sqrt{90}$	is between ____ and ____	
63.	$\sqrt{116}$	is between ____ and ____	
64.	$\sqrt{65}$	is between ____ and ____	
65.	$\sqrt{5}$	is between ____ and ____	
66.	$\sqrt{48}$	is between ____ and ____	

For Exercises 67–74, use a calculator to approximate the positive square roots to three decimal places.

67. $\sqrt{427.75}$ **68.** $\sqrt{3184.75}$ **69.** $\sqrt{1,246,000}$ **70.** $\sqrt{50,416,000}$

71. $\sqrt{0.49}$ **72.** $\sqrt{0.25}$ **73.** $\sqrt{0.56}$ **74.** $\sqrt{0.82}$

Topic: Pythagorean Theorem

For Exercises 75–80, find the length of the unknown side. Round to three decimal places if necessary.

75.
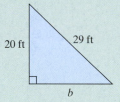

76.
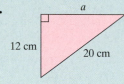

77. Leg = 5 mi, leg = 10 mi

78. Leg = 2 m, leg = 8 m

79. Leg = 12 in., hypotenuse = 22 in.

80. Leg = 15 ft, hypotenuse = 18 ft

81. A square tile is 1 ft on each side. What is the length of the diagonal? Round to the nearest hundredth of a foot.

82. A tennis court is 120 ft long and 60 ft wide. What is the length of the diagonal? Round to the nearest hundredth of a foot.

83. A contractor plans to construct a cement patio for one of the houses that he is building. The patio will be a square, 25 ft by 25 ft. After the contractor builds the frame for the cement, he checks to make sure that it is square by measuring the diagonals. Use the Pythagorean theorem to determine what the length of the diagonals should be if the contractor has constructed the frame correctly. Round to the nearest hundredth of a foot.

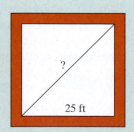

Quadrilaterals, Perimeter, and Area

1. Quadrilaterals

Recall that a **polygon** is a flat figure formed by line segments connected at their ends. A four-sided polygon is called a **quadrilateral**. Some quadrilaterals fall in the following categories.

A **parallelogram** is a quadrilateral with opposite sides parallel. It follows that opposite sides must be equal in length.

Objectives

1. **Quadrilaterals**
2. **Perimeter**
3. **Area**
4. **Applications of Area**

Parallelogram

A **rectangle** is a parallelogram with four right angles.

Rectangle

A **square** is a rectangle with sides of equal length.

Square

A **rhombus** is a parallelogram with sides of equal length. The angles are not necessarily equal.

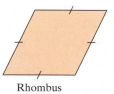

Rhombus

A **trapezoid** is a quadrilateral with one pair of parallel sides.

Trapezoid

Notice that some figures belong to more than one category. For example, a square is also a rectangle and a parallelogram.

2. Perimeter

Recall that the **perimeter** of a polygon is the distance around the figure. For example, we use perimeter to find the amount of fencing needed to enclose a yard. The perimeter of a polygon is found by adding the lengths of the sides. However, with some geometric figures we can shorten the process by using a formula.

FORMULA   Perimeter of a Square

$$P = s + s + s + s$$
$$= 4s$$

FORMULA   Perimeter of a Rectangle

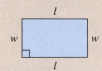

$$P = l + l + w + w$$
$$= 2l + 2w$$

We usually do not give perimeter formulas for other polygons. It is generally easier simply to add the lengths of the sides than to memorize numerous formulas.

Skill Practice

1. Find the perimeter.

4.3 in.

2. Find the perimeter in feet.

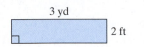
3 yd
2 ft

Example 1   Finding Perimeter

Use an appropriate formula to find the perimeter of each figure.

a.

8.2 yd

b. 9 in.

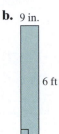
6 ft

Solution:

a. $P = 4s$ The figure is a square. Use $P = 4s$.

  $= 4(8.2 \text{ yd})$ Substitute $s = 8.2$ yd.

  $= 32.8 \text{ yd}$ Simplify.

b. First note that to add the lengths of the sides, we must have like units.

$$9 \text{ in.} = \frac{9 \text{ in.}}{1} \cdot \frac{1 \text{ ft}}{12 \text{ in.}} = \frac{9}{12} \text{ ft} = \frac{3}{4} \text{ ft} \text{ or } 0.75 \text{ ft}$$

  $P = 2l + 2w$ The figure is a rectangle. Use $P = 2l + 2w$.

  $= 2(6 \text{ ft}) + 2(0.75 \text{ ft})$ Substitute $l = 6$ ft and $w = 0.75$ ft.

  $= 12 \text{ ft} + 1.5 \text{ ft}$ Simplify.

  $= 13.5 \text{ ft}$

Concept Connections

3. How many square centimeters are enclosed in the figure?

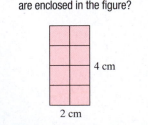
4 cm
2 cm

Answers

1. 17.2 in.   **2.** 22 ft   **3.** 8 cm²

3. Area

The area of a region is the number of square units that can be enclosed within the region. The rectangle shown in Figure 8-19 encloses 6 square inches (in.²). We would compute area, for example, if we wanted to determine how much sod was needed to cover a yard.

We offer the following convenient formulas to find the area enclosed within various figures. We begin with the familiar formulas for the area of a rectangle and square. These formulas were first introduced in Section 1.5.

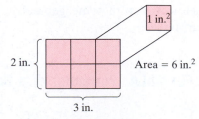
1 in.²
2 in.
Area = 6 in.²
3 in.

Figure 8-19

FORMULA   Area of a Rectangle

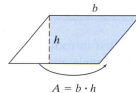

$A = \text{length} \times \text{width}$
$= lw$

FORMULA   Area of a Square

A square is also a rectangle. Therefore, the area of a square is

$A = \text{length} \times \text{width}$
$= s \cdot s$
$= s^2$

To find the area of a parallelogram, consider cutting it into two pieces, as shown in Figure 8-20. Then realign the pieces to form a rectangle.

$A = b \cdot h$ $A = b \cdot h$

Figure 8-20

The area of a parallelogram is the corresponding area of the rectangle: $A = \text{base} \times \text{height}$, or simply $A = bh$. The height h is the distance between the base and its opposite side.

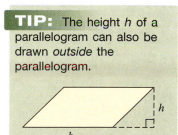

TIP: The height h of a parallelogram can also be drawn *outside* the parallelogram.

FORMULA   Area of a Parallelogram

$A = \text{base} \times \text{height}$
$= bh$

Example 2   **Finding Area**

a. Find the area of the field.

b. Find the area of the matting.

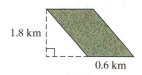

1.8 km

0.6 km

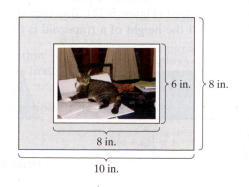

6 in.   8 in.

8 in.

10 in.

Study Skills Exercises

1. It may help to remember formulas if you understand how they were derived. For example, the perimeter of a square has the formula $P = 4s$. It was derived from the fact that perimeter measures the distance around a figure. Observe:

   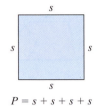

   Explain how the formula for the perimeter of a rectangle ($P = 2l + 2w$) was derived.

   $P = s + s + s + s$
   $= 4s$

2. Define the key terms.

   a. **Polygon**   b. **Quadrilateral**   c. **Parallelogram**   d. **Rectangle**

   e. **Square**   f. **Rhombus**   g. **Trapezoid**   h. **Perimeter**

Review Exercises

For Exercises 3–8, state the characteristics of each triangle.

3. Isosceles triangle

4. Right triangle

5. Acute triangle

6. Equilateral triangle

7. Obtuse triangle

8. Scalene triangle

Objective 1: Quadrilaterals

9. Write the definition of a quadrilateral.

For Exercises 10–14, state the characteristics of each quadrilateral.

10. Square

11. Trapezoid

12. Parallelogram

13. Rectangle

14. Rhombus

Objective 2: Perimeter

For Exercises 15–20, find the perimeter. **(See Example 1.)**

15. Rectangle

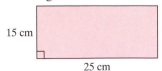

15 cm
25 cm

16. Square

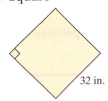

32 in.

17. Square

65 mm

18. Rectangle

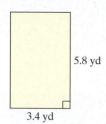

5.8 yd

3.4 yd

19. Trapezoid

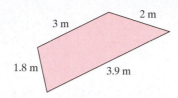

3 m 2 m

1.8 m 3.9 m

20. Trapezoid

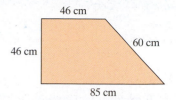

46 cm

46 cm 60 cm

85 cm

21. Find the perimeter of a triangle with sides 3 ft 8 in., 2 ft 10 in., and 4 ft.

22. Find the perimeter of a triangle with sides 4 ft 2 in., 3 ft, and 2 ft 9 in.

23. Find the perimeter of a rectangle with length 2 ft and width 6 in.

24. Find the perimeter of a rectangle with length 4 m and width 85 cm.

25. Find the lengths of the two missing sides labeled x and y. Then find the perimeter of the figure.

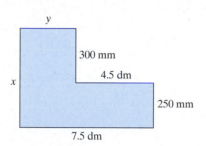

y

300 mm

4.5 dm

x

250 mm

7.5 dm

26. Find the lengths of the two missing sides labeled a and b. Then find the perimeter of the figure.

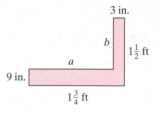

3 in.

b

$1\frac{1}{2}$ ft

a

9 in.

$1\frac{3}{4}$ ft

27. Rain gutters are going to be installed around the perimeter of the house. How many feet of rain gutters are needed?

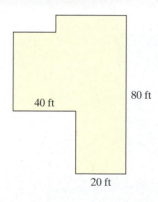

40 ft

80 ft

20 ft

28. Wood molding needs to be installed around the perimeter of a living room floor. With no wood molding needed in the doorway, how much wood molding is needed?

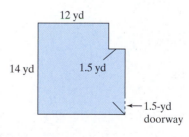

12 yd

14 yd 1.5 yd

←1.5-yd doorway

Objective 3: Area

For Exercises 29–38, find the area. **(See Examples 2–3.)**

29.

24 yd

30.

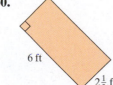

6 ft

$2\frac{1}{3}$ ft

31.

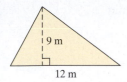

32.

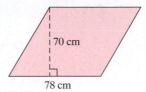

33.

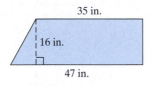

34.

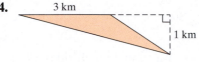

 35.

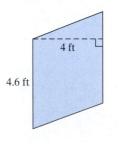

36.

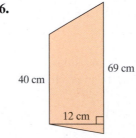

37. Write the answer in square feet.

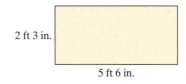

38. Write the answer in square feet.

Objective 4: Applications of Area

For Exercises 39–44, find the area of the shaded region. **(See Example 4.)**

39.

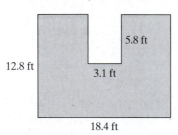

40.

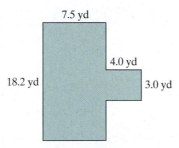

41.

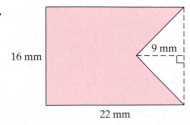

42.

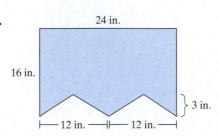

43.

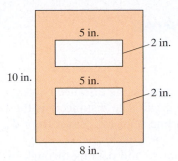

44.

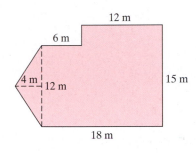

45. A rectangular living room is all to be carpeted except for the tiled portion in front of the fireplace. What is the area to be carpeted? What is the area to be tiled?

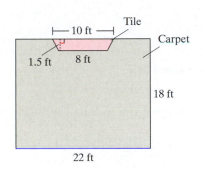

46. A patio area is to be covered with outdoor tile. What is the area of the patio?

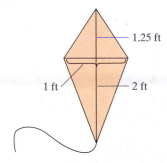

47. Find the area of the sign. **(See Example 5.)**

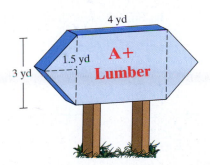

48. Find the area of the kite.

49. The King family plans to paint their garage floor with paint that resists gas, oil, and dirt from tires. The garage is 21 ft wide and 23 ft long. The paint kit they plan to use will cover approximately 250 square feet.

    a. What is the area of the garage floor?

    b. How many kits will be needed to paint the entire garage floor?

Expanding Your Skills

50. If the lengths of the sides of a square are doubled, by how many times is the area increased?

51. If the lengths of the sides of a square are tripled, by how many times is the area increased?

For Exercises 52–55, answer true or false.

52. All rectangles are parallelograms.

53. All trapezoids are parallelograms.

54. All rhombi (plural of rhombus) are squares.

55. All squares are rhombi.

Section 8.4   Circles, Circumference, and Area

Objectives

1. **Basic Definitions**
2. **Circumference**
3. **Area**
4. **Applications**

1. Basic Definitions

A **circle** is a figure consisting of all points located the same distance r from a fixed point C. The point C is called the **center** of the circle. The distance r is the length of a radius of the circle. A **radius** of a circle is a line segment drawn from the center of a circle to a point on the circle.

A line segment connecting two points on a circle and passing through the center is called a **diameter** of the circle. In Figure 8-25, $\overline{AC}$ is a radius of the circle, and $\overline{AB}$ is a diameter.

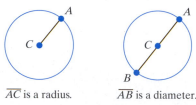

$\overline{AC}$ is a radius. $\overline{AB}$ is a diameter.

Figure 8-25

Notice that the length of a diameter is twice the radius. Therefore, we have

$$d = 2r \quad \text{or equivalently} \quad r = \frac{1}{2}d$$

Skill Practice

1. Find the length of a radius.

2. Find the length of a diameter.

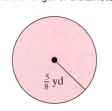

Example 1   Finding Diameter and Radius

a. Find the length of a radius.

b. Find the length of a diameter.

Solution:

a. $r = \dfrac{1}{2}d = \dfrac{1}{2}(4.6 \text{ cm}) = 2.3 \text{ cm}$

b. $d = 2r = \overset{1}{2}\left(\dfrac{3}{\underset{2}{4}} \text{ ft}\right) = \dfrac{3}{2} \text{ ft or } 1.5 \text{ ft}$

---

2. Circumference

The distance around a circle is called the **circumference**. Early mathematicians discovered that if you measure the circumference (C) and diameter (d) of any circle, the ratio of C to d always has the same value. We say that $\frac{C}{d}$ is constant. This is true for any size circle. For example, the circumference of a can of beans is approximately 9.25 in., and its diameter is 3 in. The ratio $\frac{C}{d} = \frac{9.25 \text{ in.}}{3 \text{ in.}} \approx 3.1$. The same ratio is found for a can of paint. See Figure 8-26.

Answers

1. 5.2 m   **2.** $\dfrac{5}{4}$ yd

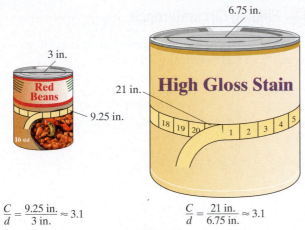

$$\frac{C}{d} = \frac{9.25 \text{ in.}}{3 \text{ in.}} \approx 3.1 \qquad \frac{C}{d} = \frac{21 \text{ in.}}{6.75 \text{ in.}} \approx 3.1$$

Figure 8-26

The constant value given by $\frac{C}{d}$ is called the number π (read "pi"). The number π in decimal form is 3.1415926535..., which goes on forever without a repeating pattern. We approximate π by 3.14 or $\frac{22}{7}$ to make it easier to use in calculations. The relationship between the circumference and diameter of a circle gives us the following formulas for the circumference of a circle.

FORMULA Circumference of a Circle

The circumference C of a circle is given by

$$C = \pi d \qquad \text{or} \qquad C = 2\pi r$$

where π is approximately 3.14 or $\frac{22}{7}$.

Example 2 **Finding Circumference**

Find the circumference.

a. Give the exact answer in terms of π.

b. Approximate the answer using 3.14 for π.

Solution:

a. The diameter is given, $d = 8$ m.

$C = \pi d$

$\quad = \pi(8 \text{ m})$ Substitute $d = 8$ m.

$\quad = 8\pi \text{ m}$ The circumference is exactly 8π meters.

b. $C = 8\pi \text{ m}$

$\quad \approx 8(3.14) \text{ m}$ Approximate π with 3.14.

$\quad = 25.12 \text{ m}$ The circumference is approximately 25.12 meters.

Skill Practice

5. Find the circumference. Find the exact answer, and then approximate the answer using 3.14 for π.

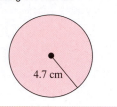

Example 3 Finding Circumference

Find the circumference. Give the exact answer in terms of π, and then approximate the answer using 3.14 for π.

Solution:

The radius is given, $r = 5.1$ ft.

$$C = 2\pi r$$

$= 2\pi(5.1 \text{ ft})$ Substitute $r = 5.1$ ft.

$= 10.2\ \pi$ ft   This is the exact circumference.

$\approx 10.2(3.14 \text{ ft})$ Approximate by using 3.14 for π.

$= 32.028$ ft   The circumference is approximately 32.028 ft.

3. Area

The circumference of a circle is given by $C = 2\pi r$. The length of a **semicircle** (one-half of a circle) is one-half of this amount: $\frac{1}{2}2\pi r = \pi r$. To visualize the formula for the area of a circle, consider the bottom half and top half of a circle cut into pie-shaped wedges. Unfold the figure as shown (Figure 8-27).

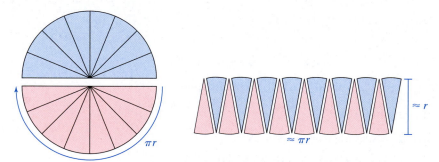

Figure 8-27

The resulting figure is nearly a parallelogram, with base $\approx \pi r$ and height approximately equal to the radius of the circle. The area is (base) $\times$ (height) $\approx (\pi r) \cdot r = \pi r^2$. This is the area formula for a circle.

TIP: To express the formula for the circumference of a circle, we can use either the radius ($C = 2\pi r$) or the diameter ($C = \pi d$).

To find the area of a circle, we will always use the radius ($A = \pi r^2$).

FORMULA Area of a Circle

The area A of a circle is given by

$$A = \pi r^2$$

Answer

5. 9.4π cm ≈ 29.516 cm

| Example 4 | **Finding the Area of a Circle** |

Find the area.

a. Give the exact answer in terms of π.

b. Approximate the answer by using $\frac{22}{7}$ for π.

14 m

Solution:

a. $A = \pi r^2$

$\quad = \pi (14 \text{ m})^2 \qquad$ Substitute $r = 14$ m.

$\quad = \pi (196 \text{ m}^2) \qquad$ Square the radius.

$\quad = 196\pi \text{ m}^2 \qquad$ The exact area is 196π m^2.

b. $A = 196\pi \text{ m}^2$

$\quad \approx 196\left(\dfrac{22}{7}\right) \text{m}^2 \qquad$ Approximate π with $\dfrac{22}{7}$.

$\quad = \dfrac{\overset{28}{196}}{1} \cdot \dfrac{22}{\underset{1}{7}} \text{ m}^2 \qquad$ Multiply the fractions.

$\quad = 616 \text{ m}^2 \qquad$ The area is approximately 616 m^2.

Skill Practice

6. Find the area of the circle.
   a. Give the exact answer in terms of π.
   b. Approximate the answer by using $\frac{22}{7}$ for π.

21 in.

| Example 5 | **Finding the Area of a Circle** |

Find the area of a wading pool that has diameter 4 ft. Give the exact answer and then approximate using 3.14 for π.

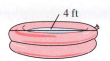
4 ft

Solution:

To compute the area, we first find the radius of the circle.

$$r = \frac{1}{2}d = \frac{1}{2}(4 \text{ ft}) = 2 \text{ ft}$$

$A = \pi r^2$

$\quad = \pi (2 \text{ ft})^2 \qquad$ Substitute $r = 2$ ft.

$\quad = \pi (4 \text{ ft}^2) \qquad$ Square the radius.

$\quad = 4\pi \text{ ft}^2 \qquad$ The exact area is 4π ft^2.

$\quad \approx 4(3.14) \text{ ft}^2 \qquad$ Approximate π with 3.14.

$\quad = 12.56 \text{ ft}^2 \qquad$ The area is approximately 12.56 ft^2.

Skill Practice

7. Find the area of the clock face. Give the exact answer, and then approximate by using 3.14 for π. Round to the nearest whole unit.

$\longmapsto$ 9.4 in. $\longmapsto$

Answers

6. a. 441π in.^{2} **b.** ≈ 1386 in.^{2}

7. 22.09π in.$^2 \approx 69$ in.^{2}

4. Applications

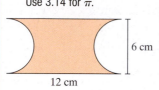
Example 6   Finding Perimeter and Area of a Composite Figure

The region shown is formed by a rectangle and a semicircle. Find the perimeter and area. Use 3.14 for π.

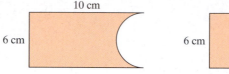   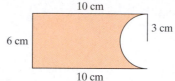

Solution:

The figure consists of a rectangle and a semicircle. We can label the figure further to identify the length of each side and the radius of the semicircle.

The perimeter is the sum of the three sides of the rectangle and the distance around the semicircle.

$$\overbrace{\text{sum of the 3 sides}} \qquad \overbrace{\text{distance around semicircle } (\tfrac{1}{2} \cdot 2\pi r)}$$

$$P \approx 6 \text{ cm} + 10 \text{ cm} + 10 \text{ cm} + \frac{1}{2} \cdot 2(3.14)(3 \text{ cm})$$

$$= 26 \text{ cm} + \frac{1}{2} \cdot 2(3.14)(3 \text{ cm}) \qquad \text{Multiply fractions.}$$

$$= 26 \text{ cm} + 9.42 \text{ cm} \qquad \text{Add.}$$

$$= 35.42 \text{ cm} \qquad \text{The perimeter is approximately 35.42 cm.}$$

The area is the difference of the area of the rectangle and the area of the semicircle.

$$\overbrace{\text{area of rectangle}} \quad \overbrace{\text{area of semicircle } (\tfrac{1}{2}\pi r^2)}$$

$$A \approx (10 \text{ cm})(6 \text{ cm}) - \frac{1}{2}(3.14)(3 \text{ cm})^2$$

$$= 60 \text{ cm}^2 - \frac{1}{2}(3.14)(9 \text{ cm}^2) \qquad \text{Multiply.}$$

$$= 60 \text{ cm}^2 - (1.57)(9 \text{ cm}^2)$$

$$= 60 \text{ cm}^2 - 14.13 \text{ cm}^2 \qquad \text{Subtract like units.}$$

$$= 45.87 \text{ cm}^2 \qquad \text{The area is approximately 45.87 cm}^2.$$

Section 8.4 Practice Exercises

Boost your GRADE at ALEKS.com!

- Practice Problems
- Self-Tests
- NetTutor
- e-Professors
- Videos

Study Skills Exercises

1. To help remember the formulas for circles, list them together and note the similarities and differences in the formulas.

Circumference = _____

Area = _____

Similarities: Differences:

2. Define the key terms.

   a. Circle   **b.** Center   **c.** Radius

   d. Diameter   **e.** Circumference   **f.** Semicircle

Review Exercises

3. Find the area of a rectangle with length 42 cm and width 30 cm.

4. Find the area of a parallelogram with base 42 cm and height 30 cm.

5. Find the area of a triangle with base 42 cm and height 30 cm.

6. How do the areas found in Exercises 3–5 compare to one another?

7. Could the formula $A = bh$ apply to finding the area of a rectangle? Explain.

Objective 1: Basic Definitions

8. How does the length of a radius of a circle compare to the length of a diameter?

For Exercises 9–12, find the length of a diameter. **(See Example 1.)**

9.

6 in.

10.

44 mm

11.

$\frac{3}{2}$ m

12.

$\frac{1}{4}$ yd

For Exercises 13–16, find the length of a radius. **(See Example 1.)**

13.

8 in.

14.

20 cm

15.

16.6 m

16.

52.2 mm

Objective 2: Circumference

17. Circumference is similar to which type of measure? (Circle the correct answer.)

   a. Area   **b.** Capacity   **c.** Perimeter   **d.** Weight

18. Indicate the type of units that could be associated with measuring circumference. Circle all that apply.

   a. ft   **b.** m^2 **c.** Liters   **d.** Meters   **e.** Grams

   f. $in.^2$ **g.** Miles   **h.** Kilometers   **i.** Cubic centimeters

19. Define π in terms of the circumference and diameter of a circle.

20. Which of the following are *not* good approximations for π? Circle all that apply.

   a. 31.4   **b.** 3.14   **c.** $\dfrac{22}{7}$ **d.** $22\frac{1}{7}$

For Exercises 21–28, find the circumference of the circle. (a) Give the exact answer in terms of π and (b) approximate the answer by using 3.14 for π. **(See Examples 2–3.)**

21.

2 m

22.

5 ft

23.

20 cm

24.

12 yd

25.

2.1 cm

26.

6.3 in.

27.

$2\frac{1}{2}$ km

28.

$1\frac{1}{4}$ m

For Exercises 29–34, use 3.14 for π.

29. Find the circumference of the can of soda.

├─ 6 cm ─┤

30. Find the circumference of a can of tuna.

├── 8.5 cm ──┤

31. Find the circumference of a compact disk.

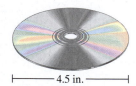

├──── 4.5 in. ────┤

32. Find the outer circumference of a pipe with 3.5-in. diameter.

3.5 in.

33. Find the outer circumference of a washer 2.2 cm in diameter.

34. Find the circumference of a pencil 5 mm in diameter.

Objective 3: Area

For Exercises 35–38, find the area of the circle, (a) give the exact answer in terms of π and (b) approximate the answer by using $\frac{22}{7}$ for π. **(See Examples 4–5.)**

35.
7 m

36.
$\frac{7}{2}$ km

37.
42 in.

38.
21 cm

For Exercises 39–42, find the area, (a) give the exact answer in terms of π and (b) approximate the answer by using 3.14 for π. Round to the nearest whole unit.

39.
25 mm

40.
10 ft

41.
6.2 ft

42.
2.9 m

Objective 4: Applications (Mixed Exercises)

For Exercises 43–51, find the area of the shaded region. Use 3.14 for π. **(See Example 6.)**

43.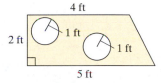
4 ft
2 ft
1 ft
1 ft
5 ft

44.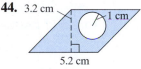
3.2 cm
1 cm
5.2 cm

45.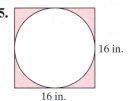
16 in.
16 in.

46.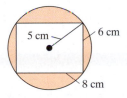
5 cm
6 cm
8 cm

47.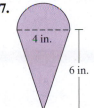
4 in.
6 in.

48.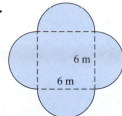
6 m
6 m

49.
10 mm
8 mm

50.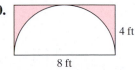
4 ft
8 ft

51.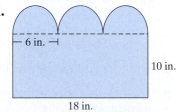
6 in.
10 in.
18 in.

For Exercises 52–57, use 3.14 for π.

52. A roller hockey rink is a rectangle with a semicircle at each end.

   a. How much will it cost to put up a rail around the rink if railing costs $2.59 per *foot*?

   b. How much will it cost to put down the floor if flooring costs $8.00 per square yard?

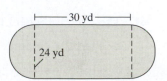
30 yd
24 yd

53. The Large Hadron Collider (LHC) is a particle accelerator and collider built to detect subatomic particles. The accelerator is in a huge circular tunnel that straddles the border between France and Switzerland. The diameter of the tunnel is 5.3 mi. Find the circumference of the tunnel. (*Source:* European Organization for Nuclear Research)

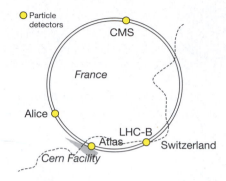

54. A ceiling fan blade rotates in a full circle. If the fan blades are 2 ft long, what is the area covered by the fan blades?

55. An outdoor torch lamp shines light a distance of 30 ft in all directions. What is the total ground area lighted?

56. How many times larger is the area of Circle 2 than Circle 1? **(See figure.)**

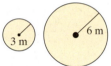

Circle 1   Circle 2

57. Hurricane Katrina's eye was 32 mi wide. The eye of a storm of similar intensity is usually only 10 mi wide. (*Source:* Associated Press 10/8/05 "Mapping Katrina's Storm Surge")

  a. What area was covered by the eye of Katrina? Round to the nearest square mile.

  b. What is the usual area of the eye of a similar storm?

Expanding Your Skills

58. A hula hoop has a 20-in. diameter.

  a. Find the circumference. Use 3.14 for π.

  b. How many times will the hula hoop have to turn to roll down a 40-yd driveway? Round to the nearest whole unit.

59. A bicycle wheel has a 26-in. diameter.

  a. Find the circumference. Use 3.14 for π.

  b. How many times will the wheel have to turn to go a distance of 1000 ft (12,000 in.)? Round to the nearest whole unit.

60. Latasha has a bicycle, and the wheel has a 22-in. diameter. If the wheels of the bike turned 1000 times, how far did she travel? Use 3.14 for π. Give the answer to the nearest inch and to the nearest foot.

61. The exercise wheel for Al's dwarf hamster has a diameter of 6.75 in.

  a. Find the circumference. Use 3.14 for π and round to the nearest inch.

  b. How far does Al's hamster travel if he completes 25 revolutions? Write the answer in feet and round to one decimal place.

Calculator Connections

Topic: Using the π Key

When finding the circumference or the area of a circle, we can use the $\boxed{\pi}$ key on the calculator to lend more accuracy to our calculations. If you press the $\boxed{\pi}$ key on the calculator, the display will show 3.141592654. This number is not the exact value of π (remember that in decimal form π is a nonterminating and nonrepeating decimal). However, using the $\boxed{\pi}$ key provides more accuracy than by using 3.14. For example, suppose we want to find the area of a circle of radius 3 ft. Compare the values by using 3.14 for π versus using the $\boxed{\pi}$ key.

Expression	Keystrokes	Result
$3.14 \cdot 3^2$	3.14 $\boxed{\times}$ 3 $\boxed{x^2}$ $\boxed{=}$	28.26
$\pi \cdot 3^2$	$\boxed{\pi}$ $\boxed{\times}$ 3 $\boxed{x^2}$ $\boxed{=}$	28.27433388

Again, it is important to note that neither of these answers is the exact area. The only way to write the exact value is to express the answer in terms of π. The exact area is 9π ft^2.

For Exercises 62–65, find the area and circumference rounded to four decimals places. Use the $\boxed{\pi}$ key on your calculator.

62.

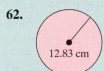

12.83 cm

63.

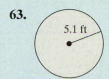

5.1 ft

64.

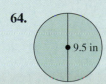

9.5 in

65.

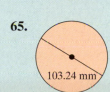

103.24 mm

66. Find the area for each pizza using the π key on your calculator. Then compute the unit cost per square inch. If the pizzas have the same thickness, which pizza is a better buy?

Diameter	Cost	Area	Cost per in.^{2}
8 in.	$ 6.50		
12 in.	12.40		

Problem Recognition Exercises

Area, Perimeter, and Circumference

For Exercises 1–14, find the area and the perimeter or circumference for each figure.

1.
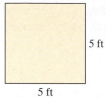
5 ft
5 ft

2.
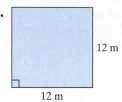
12 m
12 m

3.
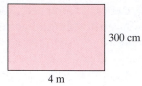
300 cm
4 m

4.

6 in.
2 ft

5.

1 ft
1.5 ft
1 yd

6.
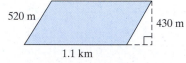
520 m
430 m
1.1 km

7.
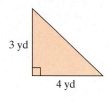
3 yd
4 yd

8.
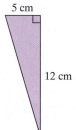
5 cm
12 cm

9.
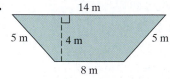
14 m
5 m
4 m
5 m
8 m

10.
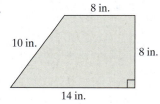
8 in.
10 in.
8 in.
14 in.

11. Use 3.14 for π.

6 yd

12. Use 3.14 for π.

40 cm

13. Use $\frac{22}{7}$ for π.

7 cm

14. Use $\frac{22}{7}$ for π.

14 ft

Volume

1. Introduction to Volume

In this section, we learn how to compute volume. Volume is another word for capacity. We use volume, for example, to determine how much can be held in a moving van.

In addition to the units of capacity learned in Sections 7.2 and 7.4, volume can be measured in cubic units. For example, a cube that is 1 cm on a side has a volume of 1 cubic centimeter (1 cm^3 or cc). A cube that is 1 in. on a side has a volume of 1 cubic inch (1 in.^{3}). See Figure 8-28. Additional units of volume include cubic feet (ft^3), cubic yards (yd^3), cubic meters (m^3), and so on.

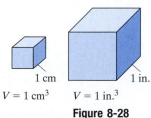

1 cm

$V = 1$ cm^3

1 in.

$V = 1$ in.^{3}

Figure 8-28

> **TIP:** Recall that 1 cubic centimeter can also be denoted as 1 cc. Furthermore, 1 cc = 1 mL.

2. Volume Formulas for Selected Solids

The formulas used to compute the volume of several common solids are given.

FORMULA

Rectangular Solid	**Cube**	**Right Circular Cylinder**

$V = lwh$ $V = s^3$ $V = \pi r^2 h$

Notice that the volume formulas for these three figures are given by the product of the area of the base and the height of the figure:

$V = lw\mathbf{h}$ $V = s \cdot s \cdot \mathbf{s}$ $V = \pi r^2 \mathbf{h}$

↑   ↑   ↑

area of rectangular base   area of square base   area of circular base

---

Example 1   Finding Volume

Find the volume. Round to the nearest whole unit.

Solution:

$V = lwh$ Use the volume formula for a rectangular solid. Identify the length, width, and height.

$= (4 \text{ in.})(3 \text{ in.})(5 \text{ in.})$ $l = 4$ in., $w = 3$ in., and

$= 60 \text{ in.}^3$ $h = 5$ in.

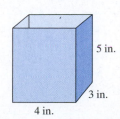

5 in.

3 in.

4 in.

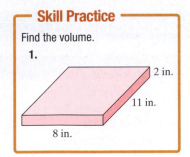
We can visualize the volume by "layering" cubes that are each 1 in. high (Figure 8-29). The number of cubes in each layer is equal to $4 \times 3 = 12$. Each layer has 12 cubes, and there are 5 layers. Thus, the total number of cubes is $12 \times 5 = 60$ for a volume of 60 in.^{3}

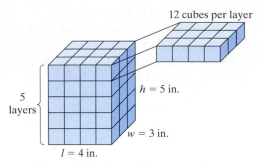

12 cubes per layer

5 layers

$h = 5$ in.

$w = 3$ in.

$l = 4$ in.

Figure 8-29

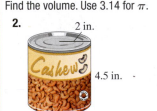
Example 2   **Finding Volume**

Find the volume. Use 3.14 for π. Round to the nearest whole unit.

3.7 cm

11.2 cm

Solution:

$V = \pi r^2 h$ Use the formula for the volume of a right circular cylinder.

$\approx (3.14)(3.7 \text{ cm})^2(11.2 \text{ cm})$ Substitute 3.14 for π, $r = 3.7$ cm, and $h = 11.2$ cm.

$= (3.14)(13.69 \text{ cm}^2)(11.2 \text{ cm})$ Simplify exponents first.

$= 481.44992 \text{ cm}^3$ Multiply from left to right.

$\approx 481 \text{ cm}^3$ Round to the nearest whole unit.

A right circular cone has the shape of a party hat. A sphere has the shape of a ball. To compute the volume of a cone and a sphere, we use the following formulas.

FORMULA

Right Circular Cone   **Sphere**

h

r

r

$V = \frac{1}{3}\pi r^2 h$ $V = \frac{4}{3}\pi r^3$

TIP: Notice that the formula for the volume of a right circular cone is $\frac{1}{3}$ that of a right circular cylinder.

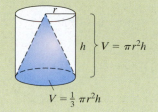

$h \quad \left.\right\} V = \pi r^2 h$

$V = \frac{1}{3}\pi r^2 h$

TIP: A **hemisphere** is one-half of a sphere. Therefore, the volume of a hemisphere is one-half that of a full sphere.

Hemisphere

$V = \frac{1}{2} \cdot \left(\frac{4}{3}\pi r^3\right)$

Example 3   Finding the Volume of a Sphere

Find the volume. Use 3.14 for π. Round to one decimal place.

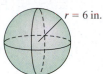

$r = 6$ in.

Solution:

$V = \dfrac{4}{3}\pi r^3$ \qquad Use the formula for the volume of a sphere.

$\approx \dfrac{4}{3}(3.14)(6 \text{ in.})^3$ \qquad Substitute 3.14 for π and $r = 6$ in.

$= \dfrac{4}{3}(3.14)(216 \text{ in.}^3)$ \qquad Simplify exponents first. $(6 \text{ in.})^3 = (6 \text{ in.})(6 \text{ in.})(6 \text{ in.}) = 216 \text{ in.}^3$

$= \dfrac{4}{3}\left(\dfrac{3.14}{1}\right)\left(\dfrac{216 \text{ in.}^3}{1}\right)$ \qquad Multiply fractions.

$= \dfrac{4}{\overset{}{\underset{1}{3}}}\left(\dfrac{3.14}{1}\right)\left(\dfrac{\overset{72}{\cancel{216}} \text{ in.}^3}{1}\right)$ \qquad Simplify to lowest terms.

$= 904.32 \text{ in.}^3$ \qquad Multiply from left to right.

$\approx 904.3 \text{ in.}^3$ \qquad Round to one decimal place.

Skill Practice

Find the volume. Use 3.14 for π.

3. $r = 3$ cm

Example 4   Finding the Volume of a Cone

Find the volume. Use 3.14 for π. Round to one decimal place.

Solution:

$V = \dfrac{1}{3}\pi r^2 h$ \qquad Use the formula for the volume of a right circular cone.

8 in.

5 in.

Skill Practice

Find the volume. Use 3.14 for π.

4.   8 cm

18 cm

Answers

3. 113.04 cm³   **4.** 301.44 cm³

To find the radius we have $r = \dfrac{1}{2}d = \dfrac{1}{2}(5 \text{ in.}) = 2.5 \text{ in.}$

$V \approx \dfrac{1}{3}(3.14)(2.5 \text{ in.})^2(8 \text{ in.})$ Substitute 3.14 for π, $r = 2.5$ in., and $h = 8$ in.

$= \dfrac{1}{3}\left(\dfrac{3.14}{1}\right)\left(\dfrac{6.25 \text{ in.}^2}{1}\right)\left(\dfrac{8 \text{ in.}}{1}\right)$ Simplify exponents first.

$= \dfrac{157}{3} \text{ in.}^3$ Multiply fractions.

$\approx 52.3 \text{ in.}^3$ Round to one decimal place.

3. Volumes of Composite Figures

Skill Practice

5.

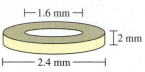

a. Find the radius of the outer cylinder.
b. Find the radius of the inner cylinder.
c. Find the volume of the washer. Use 3.14 for π. Round to the nearest whole unit.

Example 5   Finding the Volume of a Composite Figure

Find the volume of the HEPA filter (Figure 8-30). Use 3.14 for π and round the answer to the nearest whole unit.

Figure 8-30

Solution:

The solid consists of an outer cylinder with a cylindrical core cut out of the center.

To find the volume, we can find the volume of the outer cylinder and subtract the volume of the inner cylinder.

The radius of the outer cylinder is

$r = \dfrac{1}{2}d = \dfrac{1}{2}(14 \text{ in.}) = 7 \text{ in.}$

The radius of the inner cylinder is

$r = \dfrac{1}{2}d = \dfrac{1}{2}(10 \text{ in.}) = 5 \text{ in.}$

The volume of the outer cylinder is

$V = \pi r^2 h$

$\approx (3.14)(7 \text{ in.})^2(10 \text{ in.})$

$= (3.14)(49 \text{ in.}^2)(10 \text{ in.})$

$= 1538.6 \text{ in.}^3$

The volume of the inner cylinder is

$V = \pi r^2 h$

$\approx (3.14)(5 \text{ in.})^2(10 \text{ in.})$

$= (3.14)(25 \text{ in.}^2)(10 \text{ in.})$

$= 785 \text{ in.}^3$

The volume of the HEPA filter is the difference of the outer cylinder and the inner cylinder.

volume of outer cylinder   volume of inner cylinder

Volume of filter $= 1538.6 \text{ in.}^3 - 785 \text{ in.}^3$

$= 753.6 \text{ in.}^3$

$\approx 754 \text{ in.}^3$ Round to the nearest whole unit.

Answers

5 a. 1.2 mm
  b. 0.8 mm
  c. $\approx 5 \text{ mm}^3$

Section 8.5   Practice Exercises

Boost *your* GRADE at ALEKS.com!   • Practice Problems   • e-Professors
• Self-Tests   • Videos
• NetTutor

Study Skills Exercises

1. Apply what you have learned to real life situations. This can help you remember formulas and methods as well as give some meaning to math. Write down one real-life application of geometry.

2. Define the key terms.

   a. Rectangular solid   **b. Cube**   **c. Right circular cylinder**

   d. Right circular cone   **e. Sphere**   **f. Hemisphere**

Review Exercises

For Exercises 3–4, find the circumference, C, and area, A, of the circles. Use 3.14 for π.

3.

4 in.

4.

10 cm

5. Find the area of the shaded region.

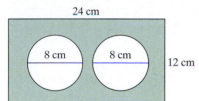

24 cm
8 cm   8 cm   12 cm

6. Find the area.

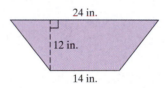

24 in.
12 in.
14 in.

Objective 1: Introduction to Volume

7. Which of the units denote volume? Circle all that apply.

   a. ft^2 **b.** m^3 **c.** in.   **d.** cc   **e.** mi

8. Which of the units denote volume? Circle all that apply.

   a. yd   **b.** yd^2 **c.** yd^3 **d.** km   **e.** km^3

For Exercises 9–12, determine the area of the square and the volume of the cube.

9.

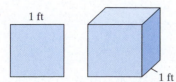

1 ft
1 ft

10.

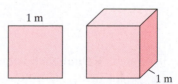

1 m
1 m

11.

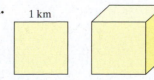

1 km
1 km

12.

1 mi
1 mi

Objective 2: Volume Formulas for Selected Solids

For Exercises 13–24, find the volume. Use 3.14 for π where necessary. **(See Examples 1–4.)**

13.

1.4 cm
1.4 cm
1.4 cm

14.

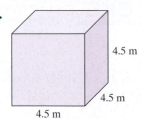

4.5 m
4.5 m
4.5 m

15.

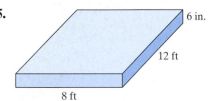

6 in.
12 ft
8 ft

16.

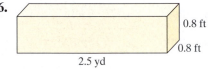

0.8 ft
0.8 ft
2.5 yd

17.

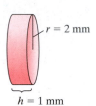

$r = 2$ mm
$h = 1$ mm

18.

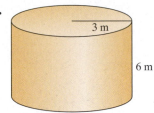

3 m
6 m

19.

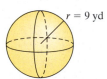

$r = 9$ yd

20.

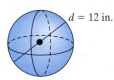

$d = 12$ in.

21.

9 cm
5 cm

22.

12 ft

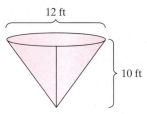

10 ft

23.

12 ft

24.

15 cm

For Exercises 25–32, use 3.14 for π. Round each value to the nearest whole unit.

25. The diameter of a volleyball is 8.2 in. Find the volume.

26. The diameter of a basketball is 9 in. Find the volume.

27. Find the volume of the sand pile.

12 ft

10 ft

28. In decorating cakes, many people use an icing bag that has the shape of a cone. Find the volume of the icing bag.

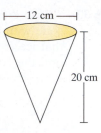

12 cm

20 cm

29. Find the volume of water (in cubic feet) that the pipe can hold.

50 ft

6 in.

30. Find the volume of the wastebasket that has the shape of a cylinder with the height of 2 ft and diameter of 3 ft.

3 ft

2 ft

31. Sam bought an above ground swimming pool with diameter 27 ft and height 54 in.

a. Approximate the volume of the pool in cubic feet using 3.14 for π.

b. How many gallons of water will it take to fill the pool? (*Hint:* 1 gal ≈ 0.1337 ft^3.)

32. Richard needs 3 in. of topsoil for his vegetable garden that is in the shape of a rectangle, 15 ft by 20 ft.

a. Find the amount of topsoil needed in cubic feet.

b. If top soil can be purchased in bags containing 2 ft^3, how many bags must Richard purchase?

Objective 3: Volumes of Composite Figures

For Exercises 33–38, use 3.14 for π. Round each value to the nearest whole unit.

33. A machine part is in the shape of a cylinder with a hole drilled through the center. Find the volume of the machine part.
(See Example 5.)

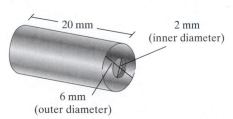

20 mm

2 mm
(inner diameter)

6 mm
(outer diameter)

34. To insulate pipes, a cylinder of Styrofoam has a hole drilled through it to fit around a pipe. What is the volume of this piece of insulation?

4 in.

30 in.

6 in.

35. A gasoline storage tank is in the shape of a cylinder with hemispheres on each end. Find the volume.

2 ft

9 ft

36. A silo is in the shape of a cylinder with a hemisphere on the top. Find the volume.

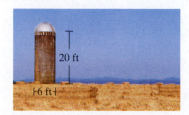

20 ft

6 ft

37. An ice cream cone is in the shape of a cone with a sphere on top. Assuming that ice cream is packed inside the cone, find the volume of the ice cream.

2 in.

5.5 in.

38. A birdbath is made from a hemisphere on a pedestal. Find the volume of water that the birdbath will hold.

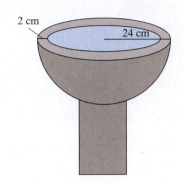

2 cm

24 cm

For Exercises 39–44, find the volume of the shaded region. Use 3.14 for π if necessary.

39.

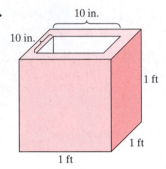

10 in.
10 in.
1 ft
1 ft
1 ft

40.

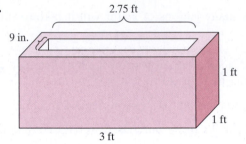

2.75 ft
9 in.
1 ft
1 ft
3 ft

41.

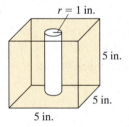

$r = 1$ in.
5 in.
5 in.
5 in.

42.

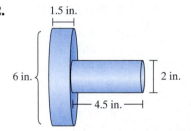

1.5 in.
6 in.
2 in.
4.5 in.

43.

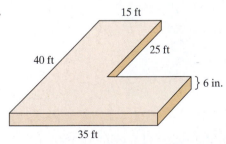

15 ft
25 ft
40 ft
6 in.
35 ft

44.

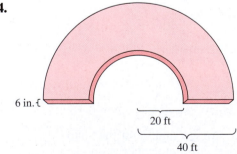

6 in.
20 ft
40 ft

Expanding Your Skills

The volume formulas for right circular cylinders and right circular cones are the same for slanted cylinders and cones. For Exercises 45–48, find the volume. Use 3.14 for π.

45.

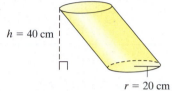

$h = 9$ in.

$r = 3$ in.

46.

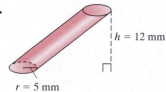

$h = 12$ mm

$r = 5$ mm

47.

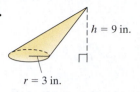

$h = 40$ cm

$r = 20$ cm

48.

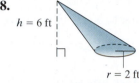

$h = 6$ ft

$r = 2$ ft

Group Activity

Constructing Geometric Figures Using a Compass and a Straightedge

Materials: A compass and straightedge

Estimated time: 15 minutes

Group Size: 2

Over the years, mathematicians have discovered precise methods to construct geometric figures by using only a compass and a straightedge. Here are some for you to try.

1. Constructing a line perpendicular to a given line segment.

Steps	
1. Use the straightedge to draw a line segment. Label two points *A* and *B* on the segment (Figure 8-31).	**Figure 8-31**
2. Set the compass at a radius that is more than halfway from *A* to *B*. Using point *A* as the center, use the compass to construct an arc above the line segment and another below the line segment (Figure 8-32).	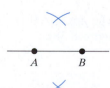 **Figure 8-32**
3. Set the compass at point *B* and draw an arc above the line segment and another below the line segment as in step 2.	

4. Use the straightedge to construct a line through the points where the arcs intersect. This line is perpendicular to segment $\overline{AB}$ (Figure 8-33).

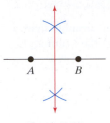

Figure 8-33

2. Constructing an angle congruent (equal in measure) to a given angle.

Steps

1. Construct an angle, $\angle A$, of any size. Use the straightedge to form the two rays (Figure 8-34).

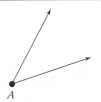

Figure 8-34

2. To construct an angle congruent to $\angle A$, begin by constructing a ray with endpoint D (Figure 8-35).

Figure 8-35

3. Place the compass at point A and draw an arc that intersects both sides of $\angle A$. Label the points B and C. With the compass set at exactly the same measure, place the compass at D and construct an arc that intersects the ray. Label the point of intersection E (Figure 8-36).

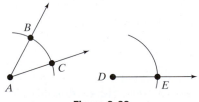

Figure 8-36

4. Use the compass to set the distance from B to C. With this setting, place the compass at E and construct an arc that intersects the arc from step 3. Label the point of intersection F (Figure 8-37).

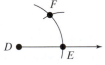

Figure 8-37

5. Construct ray $\overrightarrow{DF}$. The angle $\angle D$ is congruent to $\angle A$ (Figure 8-38).

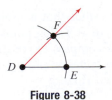

Figure 8-38

3. Constructing a line parallel to a given line.

Steps

1. Construct a line. Choose a point on the line and label it A. Choose a point not on the line and label it B (Figure 8-39).

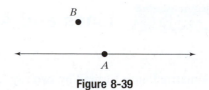

Figure 8-39

2. Set the compass at the distance from A to B and construct two circles, one with center at A and one with center at B (Figure 8-40).

3. Label point C where circle A intersects the line *outside* of circle B (Figure 8-40).

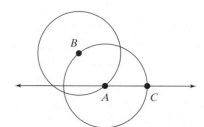

Figure 8-40

4. Construct a circle centered at C with the same radius as circle A and circle B. Label the point where circle C intersects circle A as point D (Figure 8-41).

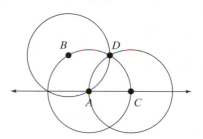

Figure 8-41

5. Construct the line that passes through points B and D. This line is parallel to the original line (Figure 8-42).

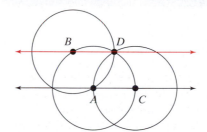

Figure 8-42

Chapter 8 Summary

Section 8.1   Lines and Angles

Key Concepts

An **angle** is a geometric figure formed by two rays that share a common endpoint. The common endpoint is called the **vertex** of the angle.

An angle is **acute** if its measure is between 0° and 90°. An angle is **obtuse** if its measure is between 90° and 180°.

Two angles are said to be **complementary** if the sum of their measures is 90°. Two angles are said to be **supplementary** if the sum of their measures is 180°.

Given two intersecting lines, **vertical angles** are angles that appear on opposite sides of the vertex.

When two parallel lines are crossed by another line eight angles are formed.

Examples

Example 1

$\angle DEF$

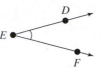

Example 2

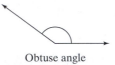

Acute angle   Obtuse angle

Example 3

The complement of a 32° angle is a 58° angle. The supplement of a 32° angle is a 148° angle.

Example 4

Intersecting lines:

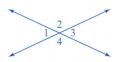

$\angle 1$ and $\angle 3$ are vertical angles and are congruent. Also $\angle 2$ and $\angle 4$ are vertical angles and are congruent.

Example 5

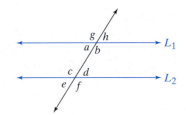

$m(\angle a) = m(\angle d)$ because they are **alternate interior angles**.
$m(\angle e) = m(\angle h)$ because they are **alternate exterior angles**.
$m(\angle c) = m(\angle g)$ because they are **corresponding angles**.

Section 8.2 Triangles and the Pythagorean Theorem

Key Concepts

The sum of the measures of the angles of any triangle is 180°.

Examples

Example 1

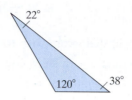

$22° + 120° + 38° = 180°$

An **acute triangle** is a triangle in which all three angles are acute.

A **right triangle** is a triangle in which one angle is a right angle.

An **obtuse triangle** is a triangle in which one angle is obtuse.

Example 2

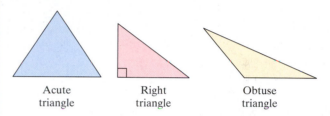

Acute	Right	Obtuse
triangle	triangle	triangle

An **equilateral triangle** is a triangle in which all three sides (and all three angles) are equal in measure.

An **isosceles triangle** is a triangle in which two sides are equal in length (the angles opposite the equal sides are also equal in measure).

A **scalene triangle** is a triangle in which no sides (or angles) are equal in measure.

Example 3

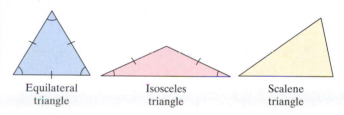

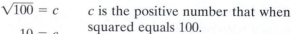

Equilateral	Isosceles	Scalene
triangle	triangle	triangle

Pythagorean Theorem

The sum of the squares of the **legs of a right triangle** equals the square of the **hypotenuse**.

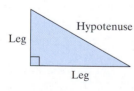 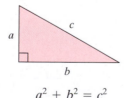

$a^2 + b^2 = c^2$

Example 4

To find the length of the hypotenuse, solve for c.

$6^2 + 8^2 = c^2$

$36 + 64 = c^2$

$100 = c^2$

$\sqrt{100} = c$ c is the positive number that when squared equals 100.

$10 = c$

The length of the hypotenuse is 10 cm.

Section 8.3   Quadrilaterals, Perimeter, and Area

Key Concepts

A four-sided **polygon** is called a **quadrilateral**.

A **parallelogram** is a quadrilateral with opposite sides parallel.

A **rectangle** is a parallelogram with four right angles.

A **square** is a rectangle with sides equal in length.

A **rhombus** is a parallelogram with sides equal in length.

A **trapezoid** is a quadrilateral with one pair of parallel sides.

Perimeter is the distance around a figure.

Perimeter of a square: $P = 4s$

Perimeter of a rectangle: $P = 2l + 2w$

Area is the number of square units that can be enclosed by a figure.

Area of a rectangle: $A = lw$

Area of a square: $A = s^2$

Area of a parallelogram: $A = bh$

Area of a triangle: $A = \frac{1}{2}bh$

Area of a trapezoid: $A = \frac{1}{2}(a + b)h$

Examples

Example 1

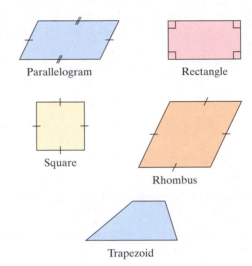

Parallelogram   Rectangle

Square

Rhombus

Trapezoid

Example 2

Given:

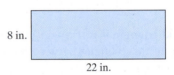

8 in.

22 in.

The perimeter of the rectangle is

$P = 2(22 \text{ in.}) + 2(8 \text{ in.})$

$= 44 \text{ in.} + 16 \text{ in.}$

$= 60 \text{ in.}$

The perimeter is 60 in.

Example 3

Given:

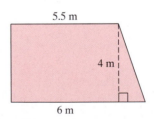

5.5 m

4 m

6 m

The area of the trapezoid is

$A = \frac{1}{2}(6 \text{ m} + 5.5 \text{ m})(4 \text{ m})$

$= \frac{1}{2}(11.5 \text{ m})(4 \text{ m})$

$= 23 \text{ m}^2$

The area is 23 m².

Section 8.4   Circles, Circumference, and Area

Key Concepts

A **circle** is a figure consisting of all points located the same distance r from a fixed point C. The fixed point C is called the **center** of the circle. The line segment from the center to any point on the circle is called a **radius** of the circle. A **diameter** of a circle is a line segment whose endpoints are on the circle and that passes through the center. The number $\pi = \frac{\text{circumference}}{\text{diameter}}$. We often use 3.14 or $\frac{22}{7}$ to approximate π.

The length of a radius is one-half the length of a diameter.

$$r = \frac{1}{2}d \quad \text{or} \quad d = 2r$$

The **circumference** of a circle is the distance around the circle and can be found by using the formula

$$C = 2\pi r \quad \text{or} \quad C = \pi d$$

The area of a circle is found by using the formula

$$A = \pi r^2$$

Examples

Example 1

20 in.

a. Find the radius.

$$r = \frac{1}{2}d = \frac{1}{2}(20 \text{ in.}) = 10 \text{ in.}$$

b. Find the circumference.

$$C = 2\pi r$$
$$= 2\pi \,(10 \text{ in.})$$
$$= 20\pi \text{ in.} \quad \text{(exact value)}$$
$$\approx 20(3.14) \text{ in.}$$
$$= 62.8 \text{ in.} \quad \text{(approximate value)}$$

c. Find the area.

$$A = \pi r^2$$
$$= \pi (10 \text{ in.})^2$$
$$= \pi (100 \text{ in.}^2)$$
$$= 100\pi \text{ in.}^2 \quad \text{(exact value)}$$
$$\approx 100(3.14) \text{ in.}^2$$
$$= 314 \text{ in.}^2 \quad \text{(approximate value)}$$

Section 8.5   Volume

Key Concepts

Volume is another word for capacity.

Formulas for selected solids are given.

Rectangular solid

$V = lwh$

Cube

$V = s^3$

Right circular cylinder

$V = \pi r^2 h$

Right circular cone

$V = \dfrac{1}{3} \pi r^2 h$

Sphere

$V = \dfrac{4}{3} \pi r^3$

Examples

Example 1

Find the volume of a tissue box with dimensions 23.5 cm by 12 cm by 12 cm.

Volume of a rectangular solid: $V = lwh$

$V = (23.5 \text{ cm})(12 \text{ cm})(12 \text{ cm})$

$ = 3384 \text{ cm}^3$

The volume is 3384 cm³.

Example 2

Find the volume of the cone.

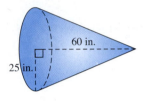

$V = \dfrac{1}{3} \pi r^2 h$

$ \approx \dfrac{1}{3}(3.14)(25 \text{ in.})^2(60 \text{ in.})$

$ = 39{,}250 \text{ in.}^3$

The volume is approximately 39,250 in.³

Chapter 8   Review Exercises

Section 8.1

For Exercises 1–4, match the symbol with a description.

1. $\overleftrightarrow{AB}$ **a.** Ray AB

2. $\overrightarrow{AB}$ **b.** Line segment AB

3. $\overrightarrow{BA}$ **c.** Ray BA

4. $\overline{AB}$ **d.** Line AB

5. Describe the measure of an acute angle.

6. Describe the measure of an obtuse angle.

7. Describe the measure of a straight angle.

8. Describe the measure of a right angle.

9. Let $m(\angle X) = 33°$.

   a. Find the complement of $\angle X$.

   b. Find the supplement of $\angle X$.

10. Let $m(\angle T) = 20°$.

   a. Find the complement of $\angle T$.

   b. Find the supplement of $\angle T$.

For Exercises 11–14, refer to the figure to determine the measure of the indicated angle.

11. $m(\angle ABE)$

12. $m(\angle DBC)$

13. $m(\angle ABG)$

14. $m(\angle ABC)$

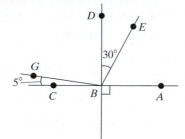

For Exercises 15–17, select the figure or figures that apply.

15. Two lines that are parallel.

16. Two lines that are *not* perpendicular.

17. Two lines that are intersecting.

   a.   **b.**   **c.**

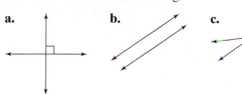

For Exercises 18–24, refer to the figure. Find the measures of the angles.

18. $m(\angle a)$

19. $m(\angle b)$

20. $m(\angle c)$

21. $m(\angle d)$

22. $m(\angle e)$

23. $m(\angle f)$ **24.** $m(\angle g)$

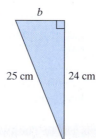

L_1 is parallel to L_2

Section 8.2

For Exercises 25–26, find the measures of the angles x and y.

25.   **26.**

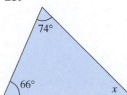

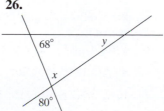

For Exercises 27–32, describe the characteristics of each type of triangle.

27. Obtuse triangle

28. Equilateral triangle

29. Right triangle

30. Acute triangle

31. Isosceles triangle

32. Scalene triangle

For Exercises 33–36, simplify the square roots.

33. $\sqrt{25}$ **34.** $\sqrt{49}$

35. $\sqrt{100}$ **36.** $\sqrt{64}$

37. State the Pythagorean theorem in words.

For Exercises 38–39, find the length of the unknown side.

38.   **39.**

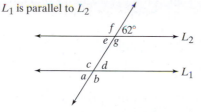

40. Kayla is flying a kite. At one point the kite is 5 m from Kayla horizontally and 12 m above her (see figure). How much string will be extended at this time? (Assume there is no slack.)

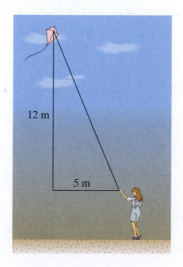

Section 8.3

For Exercises 41–44, indicate the similarities and differences of the quadrilaterals.

41. A rhombus and a square

42. A trapezoid and a parallelogram

43. A rectangle and a square

44. A rectangle and a parallelogram

45. Find the perimeter of the figure.

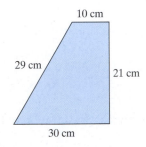

46. Find the perimeter of a triangle with sides of 4.2 m, 6.1 m, and 7.0 m.

47. Find the perimeter of a rectangle with length 16 mi and width 12 mi.

48. How much fencing is required to put up a chain link fence around a 120-yd by 80-yd playground?

49. The perimeter of a rectangle is 120 ft. The two shorter sides add up to 36 ft. What is the length of each of the longer sides?

50. The perimeter of a square is 62 ft. What is the length of each side?

51. Find the area of the triangle.

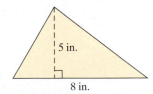

52. Fatima has a Persian rug 8.5 ft by 6 ft. What is the area?

53. A lot is 150 ft by 80 ft. Within the lot, there is a 12-ft easement along all edges. An easement is the portion of the lot on which nothing may be built. What is the area of the portion that may be used for building?

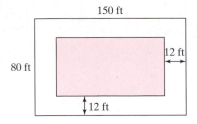

54. Find the area and the perimeter of the shaded triangle.

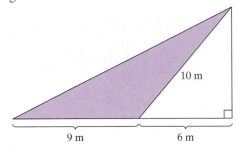

Section 8.4

55. Find the diameter of a circle whose radius is 45 mm.

56. Find the diameter of a circle whose radius is 3.2 ft.

57. Find the radius of a circle whose diameter is 45 mm.

58. Find the radius of a circle whose diameter is 3.2 ft.

For Exercises 59–62, find the circumference C and the area A of the circle.

59. Use 3.14 for π.

60. Use $\dfrac{22}{7}$ for π.

61. Use $\frac{22}{7}$ for π.

140 in.

62. Use 3.14 for π.

40 ft

63. Find the area of the shaded region. Use 3.14 for π.

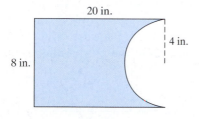

20 in.

4 in.

8 in.

64. The diameter of Emilio's pocket watch is 6 cm. The diameter of his wristwatch is 3 cm.

   a. Find the area of the pocket watch. Use 3.14 for π.

   b. Find the area of the wristwatch.

   c. Is the area of the pocket watch twice the area of the wristwatch?

65. A sign is constructed from a square with a side length of 2 yd. The square has a semicircle on top with diameter the same length as the side of the square. What is the area of the sign?

SUNSHINE
CAFE

Section 8.5

For Exercises 66–69, find the volume. Use 3.14 for π.

66.

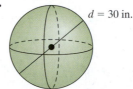

40 cm

25 cm

25 cm

67.

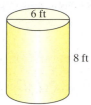

6 ft

8 ft

68.

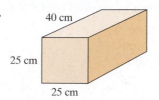

$d = 30$ in.

69.

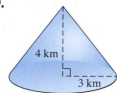

4 km

3 km

70. Find the volume of a can of paint if the can is a cylinder with radius 6.5 in. and height 7.5 in. Round to the nearest whole unit.

71. Find the volume of a ball if the diameter of the ball is approximately 6 in. Round to the nearest whole unit.

For Exercises 72–73, find the volume of the shaded region. Use 3.14 for π if necessary. Round to the nearest whole unit.

72.

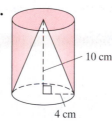

10 cm

4 cm

73.

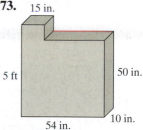

15 in.

5 ft

50 in.

54 in.

10 in.

74. A microwave oven is a rectangular solid with dimensions 1 ft by 1 ft 9 in. by 1 ft 4 in. Find the volume in cubic feet.

Chapter 8   Test

1. Which is a correct representation of the line shown?

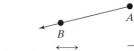

   a. $\overline{PQ}$ **b.** $\overrightarrow{PQ}$ **c.** $\overrightarrow{QP}$ **d.** $\overleftrightarrow{PQ}$

2. Which is a correct representation of the ray pictured?

   a. $\overline{AB}$ **b.** $\overleftrightarrow{AB}$ **c.** $\overrightarrow{AB}$ **d.** $\overrightarrow{BA}$

3. What is the complement of a 16° angle?

4. What is the supplement of a 147° angle?

5. Find the missing angle.

For Exercises 6–7, refer to the figure.

6. Find the diameter.

7. Find the circumference. Use $\dfrac{22}{7}$ for π.

8. A farmer uses a rotating sprinkler to water his crops. If the spray of water extends 150 ft, find the area of one such region. Use 3.14 for π.

9. Find the area of the shaded region.

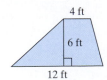

10. Simplify.

   a. $\sqrt{4}$ **b.** 4^2

For Exercises 11–18, identify the angles as acute, obtuse, right, or straight.

11. $m(\angle A) = 100°$ **12.** $m(\angle B) = 89°$

13. $m(\angle C) = 73°$ **14.** $m(\angle D) = 99°$

15. $m(\angle E) = 90°$ **16.** $m(\angle F) = 1°$

17. $m(\angle G) = 180°$ **18.** $m(\angle H) = 45°$

19. Determine the measure of angles x and y. Assume that line 1 is parallel to line 2.

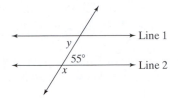

20. Given that the lengths of $\overline{AB}$ and $\overline{BC}$ are equal, what are the measures of $\angle A$ and $\angle C$?

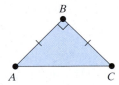

21. From the figure, determine $m(\angle S)$.

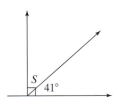

22. What is the sum of all the angles of a triangle?

23. What is the measure of $\angle A$?

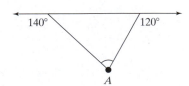

24. A firefighter places a 13-ft ladder against a wall of a burning building. If the bottom of the ladder is 5 ft from the base of the building, how far up the building will the ladder reach?

25. José is a landscaping artist and wants to make a walkway through a rectangular garden, as shown. What is the length of the walkway?

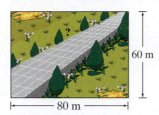

For Exercises 26–31, match the formula with the description.

26. Area of a trapezoid

a. $A = lw$

27. Area of a triangle

b. $P = 4s$

28. Perimeter of a rectangle

c. $A = \frac{1}{2}bh$

29. Perimeter of a square

d. $A = \frac{1}{2}(a + b) \cdot h$

30. Area of a rectangle

e. $A = bh$

31. Area of a parallelogram

f. $P = 2l + 2w$

32. A *regular* octagon has eight sides of equal length. A stop sign is in the shape of a regular octagon. Find the perimeter of the stop sign.

12 in.

33. Jayne wants to put up a wallpaper border for the perimeter of a 12-ft by 15-ft room. The border comes in 6-yd rolls. How many rolls would be needed?

34. Find the area of the ceiling fan blade shown in the figure.

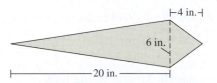

35. Which gives more pizza, a 12-in. by 8-in. rectangular pizza or a 12-in.-diameter round pizza? By how much? Assume that the pizzas have equal thicknesses. Use 3.14 for π.

36. Find the volume of the child's wading pool shown in the figure. Use 3.14 for π and round the answer to the nearest whole unit.

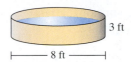

37. Find the volume of the briefcase.

38. Find the volume of the solid. Use $\frac{22}{7}$ for π.

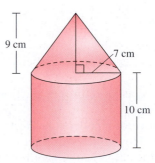

Chapters 1-8   Cumulative Review Exercises

For Exercises 1–3, divide.

1. $80,535 \div 21$ **2.** $0 \div 21$ **3.** $21 \div 0$

For Exercises 4–5, refer to the table.

State	Population
Maine	1,275,000
New Hampshire	1,236,000
Vermont	609,000

4. Find the difference in the populations of Maine and Vermont.

5. Find the sum of the populations of Maine and New Hampshire.

6. Rank the fractions from least to greatest. $\frac{2}{3}, \frac{5}{6}$, and $\frac{3}{5}$

7. There is 14 oz of ketchup in a bottle. If $\frac{1}{4}$ is used, how many ounces are left?

For Exercises 8–10, simplify the expressions.

8. $6 \div \frac{1}{3}$ **9.** $\frac{1}{3} \div 6$ **10.** $\frac{2}{7} \div \frac{3}{7} \cdot \frac{9}{5}$

11. Find the LCM of 6, 4, and 10.

12. Add: $\frac{1}{6} + \frac{1}{4} + \frac{7}{10}$

13. Subtract: $\frac{13}{6} - \frac{3}{4} - \frac{3}{10}$

14. Write $\frac{132}{8}$ as a mixed number.

15. Write $5\frac{1}{9}$ as an improper fraction.

16. The price of a collectible Three Stooges glass is $11.99. How much will a set of four cost?

17. A sale advertises "Buy 2 get 1 free." If Geraldo buys three shirts and spends $26.98, how much money is he saving?

For Exercises 18–20, complete the table.

	Fraction	Decimal
18.	$\frac{3}{8}$	
19.		$0.\overline{2}$
20.		0.02

21. Simplify the ratio: $\dfrac{2\frac{1}{2}}{3\frac{3}{4}}$

22. Solve the proportion: $\dfrac{2}{9} = \dfrac{8.3}{n}$

23. A party consisting of 25 people requires about 7 pizzas. How many pizzas should be purchased if 60 people are expected? (Round to the nearest whole pizza.)

24. Diane drives her Honda Civic 408 mi on one tank of gas. If a tank contains 12 gal, write the unit rate that gives the miles per gallon.

25. The operating cost for a Boeing 727 aircraft is approximately $8590 for 2.5 hr of operation. Find the unit rate in dollars per hour.

26. What is 22% of 240?

27. 65% of what number is 46.8?

28. 65 is what percent of 50?

29. A park bench costs $150 and is marked up to sell for $180. What is the percent markup?

30. A jacket normally sells for $85. If it is on sale for $71.40, what is the percent discount?

31. Find the perimeter in feet.

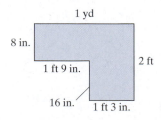

32. A piece of material is 60 in. wide. A sewing pattern requires a width of $4\frac{1}{2}$ ft. Is the material wide enough?

33. A recipe requires $\frac{1}{2}$ cup (c) of milk and 6 fl oz of pineapple juice. How many total cups of liquid are required for this recipe?

34. In Canada, just outside of London, Ontario, the speed limit is posted as 100 kilometers per hour (kph). Convert 100 km to miles to find the equivalent speed in miles per hour (1 mi = 1.61 km). Round to the nearest whole unit.

35. If the temperature in Paris on a winter day is 5°C, what is the temperature in Fahrenheit? ($F = \frac{9}{5}c + 32$)

For Exercises 36–37, find the perimeter.

36.

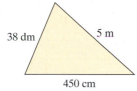

38 dm   5 m   450 cm

37.

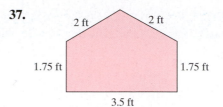

2 ft   2 ft   1.75 ft   1.75 ft   3.5 ft

For Exercises 38–39, find the area. Use 3.14 for π if necessary.

38.

40 cm

39.

3 yd   3 ft

40. Find the volume of the hemisphere. Use 3.14 for π and round to the nearest whole unit.

$r = 6$ in.

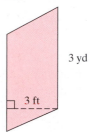

Introduction to Statistics

9

Chapter 9

This chapter introduces the study of statistics. This includes interpreting and constructing a variety of statistical graphs such as bar graphs, line graphs, circle graphs, pictographs, and histograms. We also learn how to compute the mean, median, and mode to measure the "center" of a set of values. We conclude with an introduction to probability, which measures the likelihood of an event to occur.

The crossword puzzle below contains terms from this chapter. Give it a try.

Across

2. The sum of values divided by the number of values in a set of data.
5. The value that occurs most often in a set of data.
6. The branch of mathematics that involves collecting, organizing, and analyzing data.

Down

1. The number of elements in an event divided by the number of elements in the sample space.
3. Any part of a sample space.
4. The middle number in an ordered list of numbers.

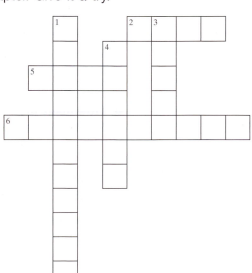

Objective 1: Introduction to Data and Tables

For Exercises 3–6 refer to the table. The table represents the Seven Summits (the highest peaks from each continent). **(See Example 1.)**

Mountain	Continent	Height (ft)
Mt. Kilimanjaro	Africa	19,340
Elbrus	Europe	18,510
Aconcagua	South America	22,834
Denali	North America	20,320
Vinson Massif	Antarctica	16,864
Mt. Kosciusko	Australia	7,310
Mt. Everest	Asia	29,035

3. In which continent does the highest mountain lie?

4. Which mountain among those listed is the lowest? In which continent does it lie?

5. How much higher is Aconcagua than Denali?

6. What is the difference between the heights of the highest mountain in Europe and the highest mountain in Australia?

For Exercises 7–12 refer to the table. The table gives the average ages (in years) for U.S. women and men married for the first time for selected years. (*Source:* U.S. Census Bureau)

	Men	Women
1940	24.3	21.5
1960	22.8	20.3
1980	24.7	22.0
2000	26.8	25.1

7. By how much has the average age for women increased between 1940 and 2000?

8. By how much has the average age for men increased between 1940 and 2000?

9. What is the difference between the men's and women's average age at first marriage in 1940?

10. What is the difference between the men's and women's average age at first marriage in 2000?

11. Which group, men or women, had the consistently higher age at first marriage?

12. Which group, men or women, had a greater increase in age between 1940 and 2000?

13. The following data were taken from a survey of a third-grade class. The survey denotes the gender of a student and whether the student owned a dog, a cat, or neither. Complete the table. Be sure to label the rows and columns. **(See Example 2.)**

Boy–dog	Boy–dog	Boy–cat	Boy–neither
Girl–dog	Girl–neither	Boy–dog	Girl–cat
Girl–neither	Girl–neither	Girl–dog	Girl–cat
Boy–dog	Girl–cat	Boy–neither	Girl–dog
Boy–neither	Girl–neither	Girl–cat	Girl–neither

	Dog	Cat	Neither
Boy			
Girl			

14. In a group of 20 women, 10 were given an experimental drug to lower cholesterol. The other 10 were given a placebo. The letter "D" indicates that the person got the drug, and the letter "P" indicates that the person received the placebo. The values "yes" or "no" indicate whether the person's cholesterol was lowered. Complete the table.

	Yes	No
Drug (D)		
Placebo (P)		

D–yes	D–yes	P–no	D–no	P–yes	D–yes	P–no	P–no	D–yes	D–no
P–yes	P–no	D–yes	P–yes	P–yes	D–no	D–yes	P–no	D–yes	P–no

Objective 2: Bar Graphs

15. The table shows the number of students per computer in U.S. public schools for selected years. (*Source:* National Center for Education Statistics) **(See Example 3.)**

   a. In which year did students have the best access to a computer?

   b. Draw a bar graph with vertical bars to illustrate these data.

Year	Students
1992	16
1995	10
1998	5.7
2001	4.9
2004	4.2
2007	4.0

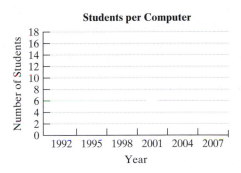

16. The number of new jobs for selected industries are given in the table. (*Source:* Bureau of Labor Statistics)

   a. Which category has the greatest number of new jobs? How many new jobs is this?

   b. Draw a bar graph with vertical bars to illustrate these data.

Industry	Number of New Jobs
Health care	219,400
Temporary help	212,000
Construction	173,000
Food service	167,600
Retail	78,600

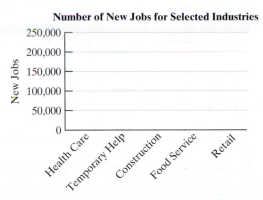

17. The table shows the number of cellular telephone subscriptions by year in the United States. (*Source:* U.S. Bureau of the Census) Construct a bar graph with horizontal bars. The length of each bar represents the number of cellular phone subscriptions for the given year.

Year	Number of Subscriptions (millions)
2002	141
2003	159
2004	182
2005	208
2006	238

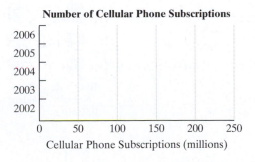

18. The table represents the world's major consumers of primary energy for a recent year. All measurements are in quadrillions of Btu. *Note:* 1 quadrillion = 1,000,000,000,000,000. (*Source:* Energy Information Administration, U.S. Department of Energy) Construct a bar graph using horizontal bars. The length of each bar gives the amount of energy consumed for that country.

Country	Amount of Energy Consumed (Quadrillions of Btu)
Germany	14
Japan	22
Russia	28
China	37
United States	99

Objective 3: Pictographs

19. A local ice cream stand kept track of its ice cream sales for one weekend, as shown in the figure. **(See Example 4.)**

  a. What does each ice cream icon represent?

  b. From the graph, estimate the number of servings of ice cream sold on Saturday.

  c. Which day had approximately 275 servings of ice cream sold?

20. Adults access the Internet to see weather updates and check on current news. The pictograph displays the percent of adult Internet users who access these topics.

  a. What does each computer icon represent?

  b. From the graph, estimate the percent of adult users that access the Internet for weather.

  c. Which type of news is accessed about 45% of the time?

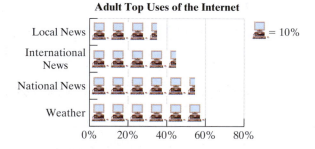

21. The figure displays the annual sales of books for three major companies.

  a. Estimate the sales for the company with the greatest annual sales.

  b. Estimate the total sales for all three companies.

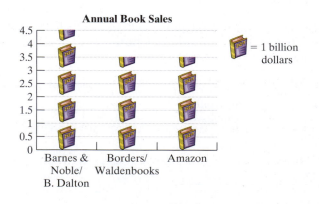

22. Recently the largest populations of senior citizens were in California, Florida, New York, Texas, and Pennsylvania, as shown in the figure. (*Source:* U.S. Bureau of the Census)

    a. Estimate the number of senior citizens living in Texas.

    b. How many more senior citizens are living in California than in Pennsylvania?

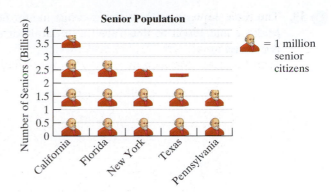

Objective 4: Line Graphs

For Exercises 23–28, use the graph provided. The graph shows the trend depicted by the percent of men and women over 65 years old in the labor force. (*Source:* Bureau of the Census) (See Example 5.)

23. What was the difference in the percent of men and the percent of women over 65 in the labor force in the year 1920?

24. What was the difference in the percent of men and the percent of women over 65 in the labor force in the year 2000?

25. What was the overall trend in the percent of women over 65 in the labor force for the years shown in the graph?

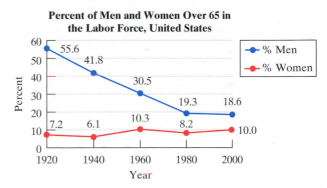

26. What was the overall trend in the percent of men over 65 in the labor force for the years shown in the graph?

27. Use the graph to predict the number of men over 65 in the labor force in the year 2020. Answers will vary.

28. Use the graph to predict the number of women over 65 in the labor force in the year 2020. Answers will vary.

For Exercises 29–34, refer to the graph representing the number of hybrid cars sold in January in the United States for the given years. (*Source:* U.S. Energy Information Administration)

29. In which year were the most hybrid cars sold in January? How many were sold?

30. In which year was the least number of hybrid cars sold in January? How many were sold?

31. What is the difference between the January sales in 2006 and 2005?

32. What is the difference between the January sales in 2007 and 2006?

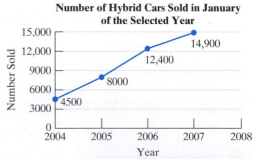

33. Between which two years was the increase in January sales the greatest?

34. Between which two years was the increase in January sales the least?

35. The data shown here give the average height for girls based on age. (*Source:* National Parenting Council) Make a line graph to illustrate these data. For each age value, plot a point for the corresponding height. (See Example 6.)

Age	Height (in.)
2	35
3	38.5
4	41.5
5	44
6	46
7	48
8	50.5
9	53

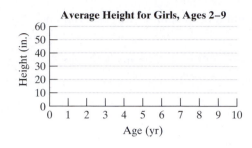

36. The data shown here give the average height for boys based on age. (*Source:* National Parenting Council) Make a line graph to illustrate these data. For each age value, plot a point for the corresponding height.

Age	Height (in.)
2	36
3	39
4	42
5	44
6	46.75
7	49
8	51
9	53.5

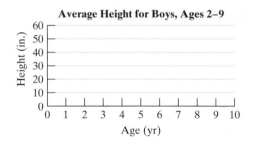

Expanding Your Skills

All packaged food items have to display nutritional facts so that the consumer can make informed choices. For Exercises 37–40, refer to the nutritional chart for Breyers French Vanilla ice cream.

37. How many servings are there per container? How much total fat is in one container of this ice cream?

38. How much total sodium is in one container of this ice cream?

39. If 8 g of fat is 13% of the daily value, what is the daily value of fat? Round to 1 decimal place.

40. If 50 mg of cholesterol is 17% of the daily value, what is the daily value of cholesterol? Round to the nearest whole unit.

Nutrition Facts		
Serving Size $\frac{1}{2}$ cup (68 g)		
Servings per Container 14		
Amount per Serving		
Calories 150		**Calories from Fat** 80
		% Daily Value
Total Fat	8 g	13%
Saturated fat	5 g	25%
Cholesterol	50 mg	17%
Sodium	45 mg	2%
Total Carbohydrate		
Dietary fiber	0 g	
Sugars	15 g	
Protein	3 g	

Frequency Distributions and Histograms

1. Frequency Distributions

The ages at the time of inauguration for 42 Presidents of the United States are given.

57	57	49	52	51	51	51	56	46
61	61	64	56	47	56	60	61	54
57	54	50	46	55	55	62	52	
57	68	48	54	54	51	43	69	
58	51	65	49	49	54	55	64	

The youngest President to take office to date was John F. Kennedy at 43 yr old. The oldest was Ronald Reagan at 69 yr old. Suppose we wanted to organize this information further by age groups. One way is to create a frequency distribution. A **frequency distribution** is a table displaying the number of values that fall within categories called **class intervals**. This is demonstrated in Example 1.

Example 1 Creating a Frequency Distribution

Complete the table to form a frequency distribution for the ages of U.S. Presidents.

Class Intervals, Age (yr)	Tally	Frequency (Number of Presidents)
40–44		
45–49		
50–54		
55–59		
60–64		
65–69		

Solution:

The classes represent different age groups. Go through the list of ages, and use tally marks to track the number of Presidents that fall within each class. Tally marks are shown in red for the first column of data: 57, 61, 57, 57, and 58.

Table 9-5

Class Intervals, Age (yr)	Tally	Frequency (Number of Presidents)
40–44	I	1
45–49	IIII II	7
50–54	IIII IIII III	13
55–59	IIII IIII I	11
60–64	IIII II	7
65–69	III	3

The frequency is a count of the tally marks within each class. See Table 9-5.

Objectives

1. Frequency Distributions
2. Histograms

Skill Practice

1. The ages (in years) of individuals arrested on a certain day in Galveston, Texas, are listed.

18	20	35	46
19	26	24	32
28	25	30	34
22	29	39	19
18	19	26	40

Complete the table to form a frequency distribution.

Class (Age)	Tally	Frequency
18–23		
24–29		
30–35		
36–41		
42–47		

Answer

1.

Class (Age)	Tally	Frequency
18–23	IIII II	7
24–29	IIII I	6
30–35	IIII	4
36–41	II	2
42–47	I	1

Skill Practice

For Exercises 2–4, consider the frequency distribution from margin Exercise 1.

2. Which class (age group) has the most values?
3. How many values are represented in the table?
4. What percent of the people arrested are in the 42–47 age group?

Example 2 **Interpreting a Frequency Distribution**

Consider the frequency distribution in Table 9-5.

a. Which class had the most values?

b. How many values are represented in the table?

c. What percent of the Presidents were 60 yr old or older at the time of inauguration?

Solution:

a. The 50–54 class had 13 data values, which is the greatest frequency.

b. The number of data values is given by the sum of the frequencies.

$$\text{Total number of values} = 1 + 7 + 13 + 11 + 7 + 3$$
$$= 42$$

c. The number of Presidents 60–64 yr old is 7. The number between 65 and 69 is 3. This means that there are 10 presidents who were 60 yr or older. The percent is

$$\frac{10}{42} \approx 0.238 \quad \text{or} \quad 23.8\%$$

Approximately 23.8% of Presidents were 60 yr old or older at the time of inauguration.

When creating a frequency distribution, keep these important guidelines in mind.

- The classes should be equally spaced. For instance, in Example 1, we would not want one class to represent a 5-yr interval and another to represent a 20-yr interval.
- The classes should not overlap. That is, a value should belong to one and only one class.
- In general, we usually create a frequency distribution with between 5 and 15 classes.

2. Histograms

A **histogram** is a special bar graph that illustrates data given in a frequency distribution. The class intervals are given on the horizontal scale. The height of each bar in a histogram represents the frequency for each class.

Example 3 **Constructing a Histogram**

Construct a histogram for the frequency distribution given in Example 1.

Solution:

Class Intervals, Age (yr)	Frequency (Number of Presidents)
40–44	1
45–49	7
50–54	13
55–59	11
60–64	7
65–69	3

Answers

2. 18–23 yr **3.** 20 **4.** 5%

To create a histogram of these data, we list the classes (ages of Presidents) on the horizontal scale. On the vertical scale we measure the frequency (Figure 9-8).

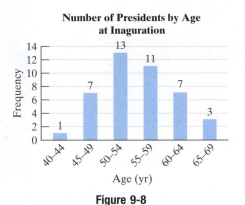

Number of Presidents by Age at Inaguration

Figure 9-8

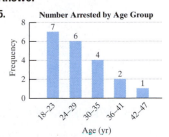
Section 9.2 Practice Exercises

Boost your GRADE at ALEKS.com!

ALEKS
version 3.0

- Practice Problems
- Self-Tests
- NetTutor
- e-Professors
- Videos

Study Skills Exercises

1. It is always helpful to read the material in a section and make notes before it is presented in class. Writing notes ahead of time will free you to listen more in class and to pay special attention to the concepts that need clarification. Refer to your class syllabus and list the next two sections that will be covered in class and a time that you can read them beforehand.

2. Define the key terms.

   a. Frequency distribution   **b. Class intervals**   **c. Histogram**

Objective 1: Frequency Distributions

3. From the frequency distribution, determine the total number of data.

Class Intervals	Frequency
1–4	14
5–8	18
9–12	24
13–16	10
17–20	6

4. From the frequency distribution, determine the total number of data.

Class Intervals	Frequency
1–50	29
51–100	12
101–150	6
151–200	22
201–250	56
251–300	60

5. For the table in Exercise 3, which category contains the most data?

6. For the table in Exercise 4, which category contains the most data?

7. The retirement age (in years) for 20 college professors is given. Complete the frequency distribution. (See Examples 1 and 2.)

| 67 | 56 | 68 | 70 | 60 | 65 | 73 | 72 | 56 | 65 |
| 71 | 66 | 72 | 69 | 65 | 65 | 63 | 65 | 68 | 70 |

Class Intervals (Age in Years)	Tally	Frequency (Number of Professors)
56–58		
59–61		
62–64		
65–67		
68–70		
71–73		

a. Which class has the most values?

b. How many data values are represented in the table?

c. What percent of the professors retire when they are 68 to 70 yr old?

8. The number of miles run in one day by 16 selected runners is given. Complete the frequency distribution.

| 2 | 4 | 7 | 3 | 8 | 4 | 5 | 7 |
| 4 | 6 | 4 | 3 | 4 | 2 | 4 | 10 |

Class Intervals (Number of Miles)	Tally	Frequency (Number of Runners)
1–2		
3–4		
5–6		
7–8		
9–10		

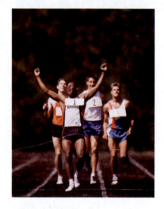

a. Which class has the most values?

b. How many data values are represented in the table?

c. What percent of the runners run 3 to 4 mi/day?

9. The number of gallons of gas purchased by 16 customers at a certain gas station is given. Complete the frequency distribution.

| 12.7 | 13.1 | 9.8 | 12.0 | 10.4 | 9.8 | 14.2 | 8.6 |
| 19.2 | 8.1 | 14.0 | 15.4 | 12.8 | 18.2 | 15.1 | 13.0 |

Class Intervals (Amount in Gal)	Tally	Frequency (Number of Customers)
8.0–9.9		
10.0–11.9		
12.0–13.9		
14.0–15.9		
16.0–17.9		
18.0–19.9		

a. Which class has the most values?

b. How many data values are represented in the table?

c. What percent of the customers purchased 18 to 19.9 gal of gas?

10. The hourly salaries (in dollars) for 15 student employees at Miami-Dade College are given. Complete the frequency distribution.

7.95	8.00	9.20	8.15	7.85
7.95	8.50	9.00	8.25	8.95
9.25	9.50	10.05	10.00	8.30

Class Intervals (Hourly Wage, $)	Tally	Frequency (Number of Employees)
7.50–7.99		
8.00–8.49		
8.50–8.99		
9.00–9.49		
9.50–9.99		
10.00–10.49		

a. Which class has the most values?

b. How many data values are represented in the table?

c. What percent of the employees earn $9.00 or more?

11. Explain what is wrong with the following class intervals.

Class	Tally	Frequency
0–4		
5–10		
11–17		
18–25		
26–34		

12. Explain what is wrong with the following class intervals.

Class	Tally	Frequency
1–6		
7–11		
12–17		
18–23		
24–28		

13. Explain what is wrong with the following class intervals.

Class	Tally	Frequency
1–20		
21–40		

14. Explain what is wrong with the following class intervals.

Class	Tally	Frequency
1–33		
34–66		
67–99		

15. Explain what is wrong with the following class intervals.

Class	Tally	Frequency
10–12		
12–14		
14–16		
16–18		
18–20		

16. Explain what is wrong with the following class intervals.

Class	Tally	Frequency
1–5		
5–10		
10–15		
15–20		
20–25		
25–30		

17. The heights of 20 students at Valencia Community College are given. Complete the frequency distribution.

70	71	73	62	65	70	69	70
64	66	73	63	68	67	69	72
64	66	67	69				

Class Interval (Height, in.)	Frequency (Number of Students)
62–63	
64–65	
66–67	
68–69	
70–71	
72–73	

18. The amount withdrawn in dollars from a certain ATM is given for 20 customers. Construct a frequency distribution.

40	50	200	200	100	120	200
50	100	60	100	100	30	40
100	100	50	200	150	200	

Class Interval (Amount, $)	Frequency (Number of Customers)
0–49	
50–99	
100–149	
150–199	
200–249	

Objective 2: Histograms

19. Construct a histogram for the frequency table in Exercise 17. **(See Example 3.)**

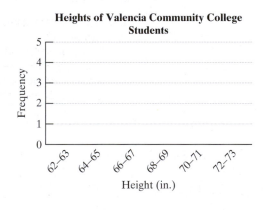

20. Construct a histogram for the frequency table in Exercise 18.

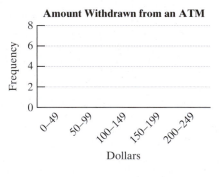

21. Construct a histogram, using the given data. Each number represents the number of Calories in a 100-g serving for selected fruits.

59	65	48	49	161	47	92	43
52	59	56	49	35	55	72	30
67	44	32	32	61	29	30	

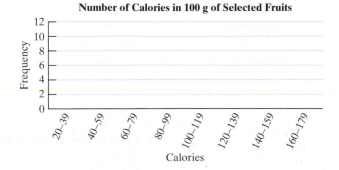

22. The list of data gives the number of children of the Presidents of the United States (in no particular order). Construct a histogram.

0	4	3	3	2	2	5	10	0	5	2
4	5	7	4	3	6	4	0	7	5	6
1	4	3	0	4	3	2	6	4	6	8
3	2	1	0	2	6	0	2	2		

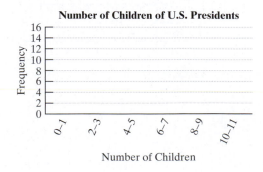

Circle Graphs

1. Interpreting Circle Graphs

Thus far we have used bar graphs, line graphs, and histograms to visualize data. A **circle graph** (or pie graph) is another type of graph used to show how a whole amount is divided into parts. Each part of the circle, called a **sector**, is like a slice of pie. The size of each piece relates to the fraction of the whole it represents.

Objectives

1. Interpreting Circle Graphs
2. Circle Graphs and Percents
3. Constructing Circle Graphs

Example 1 Interpreting a Circle Graph

The grade distribution for a math test is shown in the circle graph (Figure 9-9).

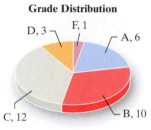

Grade Distribution

Figure 9-9

a. How many total grades are represented?

b. How many grades are B's?

c. How many times more C's are there than D's?

d. What percent of the grades were A's?

Skill Practice

A used car dealership sells cars and trucks. Use the circle graph to answer Exercises 1–4.

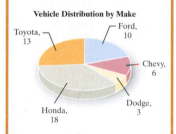

Vehicle Distribution by Make

1. How many total vehicles are represented?
2. How many are Toyotas?
3. How many times more Hondas are there than Dodges?
4. What percent are Fords?

Solution:

a. The total number of grades is equal to the sum of the number of grades from each category.

$$\text{Total number of grades} = 6 + 10 + 12 + 3 + 1$$
$$= 32$$

b. The number of B's is represented by the red portion of the graph. There are 10 B's.

c. There are 12 C's and 3 D's. The ratio of C's to D's is $\frac{12}{3} = 4$. Therefore, there are 4 times as many C's as D's.

d. There are 6 A's. The percent of A's is given by

$$\frac{6}{32} = 0.1875$$

Therefore, the percent of A's is 18.75%.

2. Circle Graphs and Percents

Sometimes circle graphs show data in percent form. This is illustrated in Example 2.

Example 2   Calculating Amounts by Using a Circle Graph

A certain video rental store carries 2000 different videos. It groups its video collection by the categories shown in the graph (Figure 9-10).

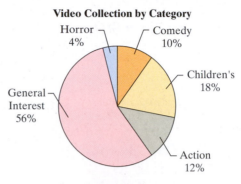

Video Collection by Category

Figure 9-10

  a. How many videos are comedy?

  b. How many videos are action or horror?

Solution:

  a. First note that the store carries 2000 different videos. From the graph we know that 10% are comedies. Therefore, this question can be interpreted as

    What is 10% of 2000?

$$x = (0.10) \cdot (2000)$$

$$= 200 \qquad \text{There are 200 comedies.}$$

  b. From the graph we know that 12% of the videos are action and 4% are horror. This accounts for 16% of the total video collection. Therefore, this question asks

    What is 16% of 2000?

$$x = (0.16) \cdot (2000)$$

$$= 320 \qquad \text{There are 320 videos that are action or horror.}$$

3. Constructing Circle Graphs

Recall that a full circle is a 360° arc. To draw a circle graph, we must compute the number of degrees of arc for each sector. In Example 2, 10% of the videos are comedies. To draw the sector for this category, we must determine 10% of 360°.

$$10\% \text{ of } 360° = 0.10(360°) = 36°$$

The sector representing comedies should be drawn with a 36° angle. To do this, we can use a protractor (Figure 9-11).

To draw a sector with a 36° arc, first draw a circle. Place the hole in the protractor over the center of the circle. Using the inner scale on the protractor, place a tick mark at 0° and at 36°. Use a straightedge to draw two line segments from the center of the circle to each tick mark. See Figure 9-11.

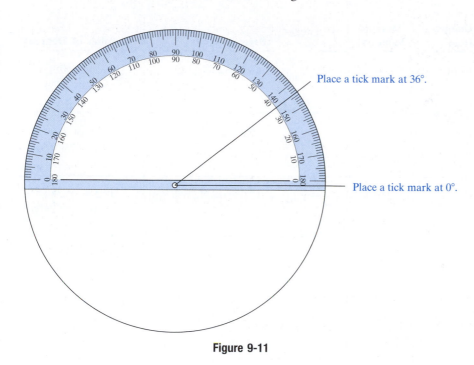

Place a tick mark at 36°.

Place a tick mark at 0°.

Figure 9-11

In Example 3, we use this technique to construct a circle graph.

Example 3 Constructing a Circle Graph

A teacher earns a monthly salary of $2400 after taxes. Her monthly budget is broken down in Table 9-6.

Table 9-6

Budget Item	Monthly Value ($)
Rent	840
Utilities	210
Car expenses	510
Groceries	360
Savings	300
Other	180

Construct a circle graph illustrating the information in this table. Label each sector of the graph with the percent that it represents.

Skill Practice

9. Voters in Oregon were asked to identify the political party to which they belonged. Construct a circle graph. Label each sector of the graph with the percent that it represents.

Political Affiliation	Number
Democrat	900
Republican	720
Libertarian	36
Green Party	144

Solution:

This problem calls for two types of calculations: (1) For each budget item, we must compute the percent of the whole that it represents. (2) We must determine the number of degrees for each category. We can use a table to help organize our calculations.

Budget Item	Monthly Value ($)	Percent	Number of Degrees
Rent	840	$= \dfrac{840}{2400} = 0.35$ or 35%	35% of 360° $= 0.35(360°)$ $= 126°$
Utilities	210	$= \dfrac{210}{2400} = 0.0875$ or 8.75%	8.75% of 360° $= 0.0875(360°)$ $= 31.5°$
Car	510	$= \dfrac{510}{2400} = 0.2125$ or 21.25%	21.25% of 360° $= 0.2125(360°)$ $= 76.5°$
Groceries	360	$= \dfrac{360}{2400} = 0.15$ or 15%	15% of 360° $= 0.15(360°)$ $= 54°$
Savings	300	$= \dfrac{300}{2400} = 0.125$ or 12.5%	12.5% of 360° $= 0.125(360°)$ $= 45°$
Other	180	$= \dfrac{180}{2400} = 0.075$ or 7.5%	7.5% of 360° $= 0.075(360°)$ $= 27°$

Now construct the circle graph. Use the degree measures found in the table for each sector. Label the graph with the percent for each sector (Figure 9-12).

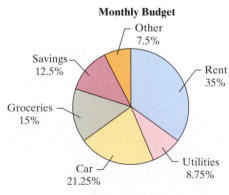

Monthly Budget

Other 7.5%
Savings 12.5%
Groceries 15%
Car 21.25%
Utilities 8.75%
Rent 35%

Figure 9-12

Answer

9.

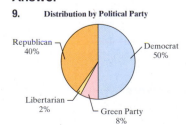

Distribution by Political Party

Republican 40%
Democrat 50%
Libertarian 2%
Green Party 8%

Section 9.3 | Practice Exercises

Study Skills Exercises

1. Some instructors are available to answer questions during evening hours via e-mail. Find out if you can contact your instructor by e-mail during evening hours or weekends, and write down the e-mail address.

2. Define the key terms.

   a. Circle graph   **b. Sector**

Objective 1: Interpreting Circle Graphs

For Exercises 3–10, refer to the graph. The graph represents the number of highway fatalities by age group in the U.S. (*Source:* U.S. Bureau of the Census) **(See Example 1.)**

Number of U.S. Traffic Fatalities

65 yr and older
(9600)

55–64 yr
(6400)

45–54 yr
(9600)

35–44 yr
(10,880)

25–34 yr
(11,520)

15–24 yr
(16,000)

3. What is the total number of traffic fatalities?

4. Which of the age groups has the most fatalities?

5. How many more people died in the 25–34 age group than in the 35–44 age group?

6. How many more people died in the 45–54 age group than in the 55–64 group?

7. What percent of the deaths were from the 15–24 group?

8. What percent of the deaths were from the 65 and older age group?

9. How many times more deaths were from the 15–24 age group than the 55–64 age group?

10. How many times more deaths were from the 25–34 age group than from the 65 and older group?

For Exercises 11–16, refer to the figure. The figure represents the average daily number of viewers for five daytime dramas. (*Source:* Nielsen Media Research)

11. How many viewers are represented?

12. How many viewers does the most popular daytime drama have?

13. How many times more viewers does *The Young and the Restless* have than *Guiding Light*?

14. How many times more viewers does *The Young and the Restless* have than *General Hospital*?

15. What percent of the viewers watch *General Hospital*?

16. What percent of the viewers watch *The Bold and the Beautiful*?

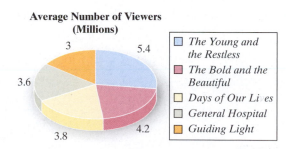

**Average Number of Viewers
(Millions)**

3   5.4

3.6

3.8   4.2

☐ *The Young and the Restless*
☐ *The Bold and the Beautiful*
☐ *Days of Our Lives*
☐ *General Hospital*
☐ *Guiding Light*

Objective 2: Circle Graphs and Percents

For Exercises 17–20, use the graph representing the type of music CDs found in a store containing approximately 8000 CDs. **(See Example 2.)**

17. How many CDs are musica Latina?

18. How many CDs are rap?

19. How many CDs are jazz or classical?

20. How many CDs are *not* Pop/R&B?

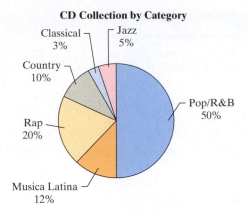

CD Collection by Category

For Exercises 21–24, use the graph representing the states that hosted Super Bowl I through Super Bowl XXXVI (a total of 36 Super Bowls).

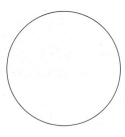 21. How many Super Bowls were played in Louisiana?

22. How many Super Bowls were played in Florida? Round to the nearest whole number.

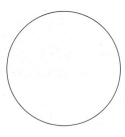 23. How many Super Bowls were played in Georgia? Round to the nearest whole number.

24. How many Super Bowls were played in Michigan? Round to the nearest whole number.

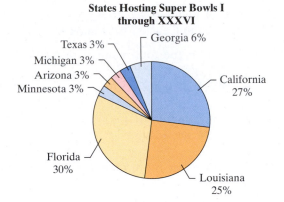

States Hosting Super Bowls I through XXXVI

Objective 3: Constructing Circle Graphs

For Exercises 25–32, use a protractor to construct an angle of the given measure.

25. 20°

26. 70°

27. 125°

28. 270°

29. 195°

30. 5°

31. 300°

32. 90°

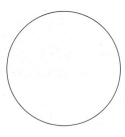 33. Draw a circle and divide it into sectors of 30°, 60°, 100°, and 170°.

34. Draw a circle and divide it into sectors of 125°, 180°, and 55°.

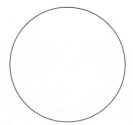

35. The table provided gives the expenses for one semester at college. **(See Example 3.)**

    a. Complete the table.

	Expenses	Percent	Number of Degrees
Tuition	$9000		
Books	600		
Housing	2400		

    b. Construct a circle graph to display the college expenses. Label the graph with percents.

College Expenses for a Semester

36. The table provided gives the number of establishments of the three largest pizza chains.

    a. Complete the table.

	Number of Stores	Percent	Number of Degrees
Pizza Hut	8100		
Domino's	7200		
Papa Johns	2700		

    b. Construct a circle graph. Label the graph with percents.

Percent of Pizza Establishments

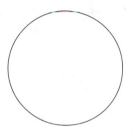

37. The Sunshine Nursery sells flowering plants, shrubs, ground cover, trees, and assorted flower pots. Construct a pie graph to show the distribution of the types of purchases.

Sunshine Nursery Distribution of Sales

Types of Purchases	Percent of Distribution
Flowering plants	45%
Shrubs	13%
Ground cover	18%
Trees	20%
Flower pots	4%

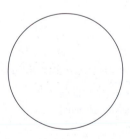

38. The party affiliation of registered Latino voters for a recent year is as follows:

45% Democrat   20% Republican

13% Other   22% Independent

Construct a circle graph from this information.

Party Affiliation of Latino Voters

| Section 9.4 | Mean, Median, and Mode |

Objectives

1. **Mean**
2. **Median**
3. **Mode**
4. **Weighted Mean**

1. Mean

When given a list of numerical data, it is often desirable to obtain a single number that represents the central value of the data. In this section, we introduce three such values called the mean, median, and mode. The first calculation we present is the mean (or average) of a list of data values.

> **DEFINITION   Mean**
>
> The **mean** (or average) of a set of numbers is the sum of the values divided by the number of values. We can write this as a formula.
>
> $$\text{Mean} = \frac{\text{sum of the values}}{\text{number of values}}$$

Skill Practice

Housing prices for five homes in one neighborhood are given.

$108,000 $149,000
$164,000 $118,000
$144,000

1. Find the mean of these five houses.
2. Suppose a new home is built in the neighborhood for $1.3 million ($1,300,000). Find the mean price of all six homes.

Avoiding Mistakes

When computing a mean remember that the data are added first before dividing.

Example 1   Finding the Mean of a Data Set

A small business employs five workers. Their yearly salaries are

$42,000 $36,000 $45,000 $35,000 $38,000

a. Find the mean yearly salary for the five employees.

b. Suppose the owner of the business makes $218,000 per year. Find the mean salary for all six individuals (that is, include the owner's salary).

Solution:

a. Mean salary of five employees

$$= \frac{42{,}000 + 36{,}000 + 45{,}000 + 35{,}000 + 38{,}000}{5}$$

$$= \frac{196{,}000}{5} \qquad \text{Add the data values.}$$

$$= 39{,}200 \qquad \text{Divide.}$$

The mean salary for employees is $39,200.

b. Mean of all six individuals

$$= \frac{42{,}000 + 36{,}000 + 45{,}000 + 35{,}000 + 38{,}000 + 218{,}000}{6}$$

$$= \frac{414{,}000}{6}$$

$$= 69{,}000$$

The mean salary with the owner's salary included is $69,000.

Answers

1. $136,600   **2.** $330,500

2. Median

In Example 1, you may have noticed that the mean salary was greatly affected by the unusually high value of $218,000. For this reason, you may want to use a different measure of "center" called the median. The **median** is the "middle" number in an ordered list of numbers.

PROCEDURE **Finding the Median**

To compute the median of a list of numbers, first arrange the numbers in order from least to greatest.

- If the number of data values in the list is *odd*, then the median is the middle number in the list.

- If the number of data values is *even*, there is no single middle number. Therefore, the median is the mean (average) of the two middle numbers in the list.

Example 2 **Finding the Median of a Data Set**

Consider the salaries of the five workers from Example 1.

$42,000 $36,000 $45,000 $35,000 $38,000

a. Find the median salary for the five workers.

b. Find the median salary including the owner's salary of $218,000.

Solution:

a. 35,000   36,000   **38,000**   42,000   45,000   Arrange the data in order.

Because there are five data values (an *odd* number), the median is the middle number.

The median is $38,000.

b. Now consider the scores of all six individuals (including the owner). Arrange the data in order.

35,000   36,000   **38,000   42,000**   45,000   218,000

There are six data values (an *even* number). The median is the average of the two middle numbers.

$$\frac{38,000 + 42,000}{2}$$

Add the two middle numbers.

$$= \frac{80,000}{2}$$

$$= 40,000$$

Divide.

The median of all six salaries is $40,000.

Skill Practice

3. Find the median of the five housing prices given in margin Exercise 1.

$108,000 $149,000
$164,000 $118,000
$144,000

4. Find the median of the six housing prices given in margin Exercise 2.

$108,000 $149,000
$164,000 $118,000
$144,000 $1,300,000

Answers

3. $144,000   **4.** $146,500

In Examples 1 and 2, the mean of all six salaries is $69,000, whereas the median is $40,000. These examples show that the median is a better representation for a central value when the data list has an unusually high (or low) value.

Skill Practice

5. The monthly rainfall for Houston, Texas, is given in the table. Find the median rainfall amount.

Month	Rainfall (in.)
Jan.	4.5
Feb.	3.0
March	3.2
April	3.5
May	5.1
June	6.8
July	4.3
Aug.	4.5
Sept.	5.6
Oct.	5.3
Nov.	4.5
Dec.	3.8

Example 3 Finding the Median of a Data Set

For a recent year, the student-to-teacher ratio for elementary schools is shown in Table 9-7. Find the median student-to-teacher ratio.

Table 9-7

State	Student-to-Teacher Ratio
California	20.6
Illinois	16.1
Indiana	16.1
Maine	12.5
Mississippi	16.1
New Hampshire	14.5
North Dakota	13.4
Rhode Island	14.8
Utah	21.9
Wisconsin	14.1

Source: National Center for Education Statistics

Solution:

First arrange the numbers in order from least to greatest:

ME	ND	WI	NH	RI	IL	IN	MS	CA	UT
12.5	13.4	14.1	14.5	14.8	16.1	16.1	16.1	20.6	21.9

$$\text{Median} = \frac{14.8 + 16.1}{2} = 15.45$$

There are 10 data values (an *even* number). Therefore, the median is the average of the middle two numbers. The median student-to-teacher ratio is 15.45. This indicates that there are approximately 15 or 16 students per teacher.

Note: The median may not be one of the original data values. This was true in Example 3.

3. Mode

A third representative value for a list of data is called the mode.

DEFINITION Mode

The **mode** of a set of data is the value or values that occur most often.

- If two values occur most often we say the data are **bimodal**.
- If more than two values occur most often, we say there is no mode.

Answer

5. 4.5 in.

Example 4 Finding the Mode of a Data Set

Find the mode of the student-to-teacher ratios from Example 3.

Solution:

12.5   13.4   14.1   14.5   14.8   16.1   16.1   16.1   20.6   21.9

The data value 16.1 appears the most often. Therefore, the mode is 16.1.

Example 5 Finding the Mode of a Data Set

Find the mode of the list of average monthly temperatures for Albany, New York. Values are in °F.

Jan.	Feb.	March	April	May	June	July	Aug.	Sept.	Oct.	Nov.	Dec.
22	25	35	47	58	66	71	69	61	49	39	26

Solution:

No data value occurs most often. There is no mode for this set of data.

Example 6 Finding the Mode of a Data Set

The grades for a quiz in college algebra are as follows. The scores are out of a possible 10 points. Find the mode.

9	4	6	9	9	8	2	1	4	9
5	10	10	5	7	7	9	8	7	3
9	7	10	7	10	1	7	4	5	6

Solution:

Sometimes arranging the data in order makes it easier to find the repeated values.

1	1	2	3	4	4	4	5	5	5
6	6	7	7	7	7	7	7	8	8
9	9	9	9	9	9	10	10	10	10

The score of 9 occurs 6 times. The score of 7 occurs 6 times. There are two modes, 9 and 7, because these scores both occur more than any other score. We say that these data are *bimodal*.

TIP: To remember the difference between median and mode, think of the *median* of a highway that goes down the *middle*. Think of the word *mode* as sounding similar to the word *most*.

4. Weighted Mean

Sometimes data values in a list appear multiple times. In such a case, we can compute a **weighted mean**. In Example 7, each data value is "weighted" by the number of times it appears in the list.

Skill Practice

9. People throw coins in wishing wells, and often the money is used for charity. The number of coins collected from one wishing well is shown in the table. Compute the mean dollar amount per coin. Round to three decimal places.

Number of Coins

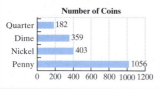

Example 7 Computing a Weighted Mean

Donations are made to a certain charitable organization in increments of $25, $50, $75, and $100, as shown in Figure 9-13. Find the mean amount donated.

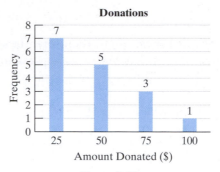

Figure 9-13

Solution:

Notice that there are 16 data values represented in the graph:

7 of these
$\begin{cases} \$25 \\ \$25 \\ \$25 \\ \$25 \\ \$25 \\ \$25 \\ \$25 \end{cases}$
5 of these
$\begin{cases} \$50 \\ \$50 \\ \$50 \\ \$50 \\ \$50 \end{cases}$
3 of these
$\begin{cases} \$75 \\ \$75 \\ \$75 \end{cases}$
1 of these $\{\$100$

The data value $25 occurs seven times. Rather than adding $25 seven times, we can find the sum by multiplying $25(7) = $175. Similarly, the value $50 occurs 5 times for a sum of $50(5) = $250, and so on. To find the sum of all the values, multiply each data value by the number of times it occurs (its frequency). Then add the results. This process can be organized easily in a table.

Amount Donated ($)	Frequency	Product ($)
25	7	(25)(7) = 175
50	5	(50)(5) = 250
75	3	(75)(3) = 225
100	1	(100)(1) = 100
Total:	**16**	**750** ← sum of all data values

↑ total number of donations made

The mean is the sum of all donations divided by the total number of donations made (total frequency).

$$\text{Mean} = \frac{750}{16} = 46.875$$

The mean amount donated is $46.88.

Answer

9. $0.056

Section 9.4   Practice Exercises

Study Skills Exercises

1. Most people cannot concentrate on studying for more than 1 hr without taking a break. To make the most of your time, write down a schedule for your next study session. Include breaks where you can eat a meal, walk the dog, or perform other simple tasks that need to be completed.

2. Define the key terms.

   a. Mean   b. Median   c. Mode   d. Bimodal   e. Weighted mean

Objective 1: Mean

For Exercises 3–8, find the mean of each set of numbers. (See Example 1.)

3. 4, 6, 5, 10, 4, 5, 8

4. 3, 8, 5, 7, 4, 2, 7, 4

5. 0, 5, 7, 4, 7, 2, 4, 3

6. 7, 6, 5, 10, 8, 4, 8, 6, 0

7. 10, 13, 18, 20, 15

8. 22, 14, 12, 16, 15

9. The wingspan of five butterflies is given in the table. Find the mean wingspan.

Butterfly	Wingspan (in.)
Queen Alexandra's birdwing	11.0
African giant swallowtail	9.1
Goliath birdwing	8.3
Buru opalescent birdwing	7.9
Chimaera birdwing	7.5

10. The number of wins in the American Baseball League, Central Division, for a recent year is given in the table. Find the mean number of wins.

Team	Number of Wins
Minnesota Twins	96
Chicago White Sox	90
Kansas City Royals	62
Cleveland Indians	78
Detroit Tigers	95

11. The flight times in hours for six flights between New York and Los Angeles are given. Find the mean flight time. Round to the nearest tenth of an hour.

   5.5, 6.0, 5.8, 5.8, 6.0, 5.6

12. A nurse takes the temperature of a patient every 10 min and records the temperatures as follows: 98°F, 98.4°F, 98.9°F, 100.1°F, and 99.2°F. Find the patient's mean temperature.

13. The number of Calories for six different chicken sandwiches and chicken salads is given in the table.

   a. What is the mean number of Calories for a chicken sandwich? Round to the nearest whole unit.

   b. What is the mean number of Calories for a salad with chicken? Round to the nearest whole unit.

   c. What is the difference in the means?

Chicken Sandwiches	Salads with Chicken
360	310
370	325
380	350
400	390
400	440
470	500

14. The heights of the players from two NBA teams are given in the table. All heights are in inches.

   a. Find the mean height for the players on the Philadelphia 76ers.

   b. Find the mean height for the players on the Milwaukee Bucks.

   c. What is the difference in the mean heights?

Philadelphia 76ers' Height (in.)	Milwaukee Bucks' Height (in.)
83	70
83	83
72	82
79	72
77	82
84	85
75	75
76	75
82	78
79	77

15. Zach received the following scores for his first four tests: 98%, 80%, 78%, 90%.

   a. Find Zach's mean test score.

   b. Zach got a 59% on his fifth test. Find the mean of all five tests.

   c. How did the low score of 59% affect the overall mean of five tests?

16. The prices of four steam irons are $50, $30, $25, and $45.

   a. Find the mean of these prices.

   b. An iron that costs $140 is added to the list. What is the mean of all five irons?

   c. How does the expensive iron affect the mean?

Objective 2: Median

For Exercises 17–22, find the median for each set of numbers. (See Examples 2–3.)

17. 16, 14, 22, 13, 20, 19, 17

18. 32, 35, 22, 36, 30, 31, 38

19. 109, 118, 111, 110, 123, 100

20. 134, 132, 120, 135, 140, 118

21. 58, 55, 50, 40, 40, 55

22. 82, 90, 99, 82, 88, 87

23. The infant mortality rates for five countries are given in the table. Find the median.

Country	Infant Mortality Rate (Deaths per 1000)
Sweden	3.93
Japan	4.10
Finland	3.82
Andorra	4.09
Singapore	3.87

24. The inflation rates for five countries are given in the table. Find the median.

Country	Inflation Rate (%)
Angola	1700
Sudan	133
Turkey	80
Venezuela	103
Bulgaria	311

25. The ages (in years) of the last 10 Presidents at the time of their inauguration are given. Find the median age.

46, 64, 69, 52, 61, 56, 55, 43, 62, 60

26. A list of the number of commuter rail stations from eight systems is given. Find the median number of stations.

124, 227, 108, 167, 121, 177, 49, 18

27. The number of passengers (in millions) on 9 leading airlines for a recent year is listed. Find the median number of passengers. (*Source:* International Airline Transport Association)

48.3, 42.4, 91.6, 86.8, 46.5, 71.2, 45.4, 56.4, 51.7

28. For a recent year the number of albums sold (in millions) is listed for the 10 best sellers. Find the median number of albums sold.

2.7, 3.0, 4.8, 7.4, 3.4, 2.6, 3.0, 3.0, 3.9, 3.2

Objective 3: Mode

For Exercises 29–34, find the mode(s) for each set of numbers. **(See Examples 4–6.)**

29. 4, 5, 3, 8, 4, 9, 4, 2, 1, 4

30. 12, 14, 13, 17, 19, 18, 19, 17, 17

31. 90%, 89%, 91%, 77%, 88%

32. 132, 253, 553, 255, 552, 234

33. 28, 21, 24, 23, 24, 30, 21

34. 45, 42, 40, 41, 49, 49, 42

35. The table gives the price of seven "smart" cell phones. Find the mode.

Brand and Model	Price ($)
Samsung	600
Kyocera	400
Sony Ericsson	800
PalmOne	450
Motorola	300
Siemens	600

36. The table gives the number of hazardous waste sites for selected states. Find the mode.

State	Number of Sites
Florida	51
New Jersey	112
Michigan	67
Wisconsin	39
California	96
Pennsylvania	94
Illinois	39
New York	90

37. The unemployment rates in percent for nine countries are given. Find the mode.

6.3%, 7.0%, 5.8%, 9.1%, 5.2%, 8.8%, 8.4%, 5.4%, 5.2%

38. The list gives the number of children who were absent from class for a 10-day period. Find the mode.

1, 6, 2, 2, 4, 4, 2, 2, 3, 2

39. The prices for five different brands of paper towel are given in the list. Find the mode.

$2.49, $2.39, $2.51, $2.49, $2.51

40. The length of time (in minutes) of eight TV commercials is given. Find the mode.

1.00, 0.50, 1.00, 1.25, 2.00, 0.50, 1.00, 0.50

Mixed Exercises

41. Six test scores for Jonathan's history class are listed. Find the mean and median. Round to the nearest tenth if necessary. Did the mean or median give a better overall score for Jonathan's performance?

92%, 98%, 43%, 98%, 97%, 85%

42. Nora's math test results are listed. Find the mean and median. Round to the nearest tenth if necessary. Did the mean or median give a better overall score for Nora's performance?

52%, 85%, 89%, 90%, 83%, 89%

43. Listed below are monthly costs for seven health insurance companies for a self-employed person, 55 yr of age, and in good health. Find the mean, median, and mode (if one exists). Round to the nearest dollar. (*Source:* eHealth Insurance Company, 2007)

$312, $225, $221, $256, $308, $280, $147

44. The salaries for seven Associate Professors at the University of Michigan are listed. These are salaries for 9-month contracts in 2006. Find the mean, median, and mode (if one exists). Round to the nearest dollar. (*Source:* University of Michigan, University Library Volume 2006, Issue 1)

$104,000, $107,000, $67,750, $82,500, $73,500, $88,300, $104,000

45. The 2007 prices of 10 single-family, 3-bedroom homes for sale in Santa Rosa, California, are listed. Find the mean, median, and mode (if one exists).

$850,000, $835,000, $839,000, $829,000,

$850,000, $850,000, $850,000, $847,000,

$1,850,000, $825,000

46. The 2007 prices of 10 single-family, 3-bedroom homes for sale in Boston, Massachusetts, are listed. Find the mean, median, and mode (if one exists).

$300,000, $2,495,000, $2,120,000, $220,000,

$194,000, $391,000, $315,000, $330,000,

$435,000, $250,000

Objective 4: Weighted Mean

47. There are 20 students enrolled in a 12th-grade math class. The graph displays the number of students by age. First complete the table, and then find the mean. **(See Example 7.)**

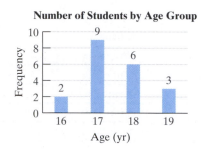

Number of Students by Age Group

Age (yr)	Number of Students	Product
16		
17		
18		
19		
Total:		

48. A survey was made in a neighborhood of 37 houses. The graph represents the number of residents who live in each house. Complete the table and determine the mean number of residents per house.

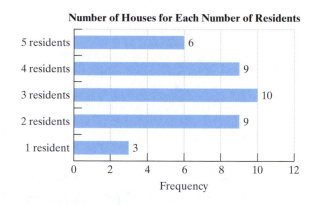

Number of Houses for Each Number of Residents

Number of Residents in Each House	Number of Houses	Product
1		
2		
3		
4		
5		
Total:		

49. Several instructors were asked the number of students who were initially enrolled in their classes. The results are represented in the graph. Find the weighted mean of the number of students initially enrolled in class. Round to the nearest whole unit.

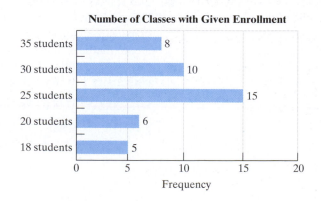

Number of Classes with Given Enrollment

At most colleges and universities, weighted means are used to compute students' grade point averages (GPAs). At one college, the grades A–F are assigned numerical values as follows:

$$A = 4.0 \quad C = 2.0$$
$$B+ = 3.5 \quad D+ = 1.5$$
$$B = 3.0 \quad D = 1.0$$
$$C+ = 2.5 \quad F = 0.0$$

Grade point average is a weighted mean where the "weights" for each grade are the number of credit-hours for that class. Use this information to answer Exercises 50–53.

50. Compute the GPA for the following grades. Round to the nearest hundredth.

Course	Grade	Number of Credit-Hours (Weights)
Intermediate Algebra	B	4
Theater	C	1
Music Appreciation	A	3
World History	D	5

51. Compute the GPA for the following grades. Round to the nearest hundredth.

Course	Grade	Number of Credit-Hours (Weights)
General Psychology	B+	3
Beginning Algebra	A	4
Student Success	A	1
Freshman English	B	3

52. Compute the GPA for the following grades. Round to the nearest hundredth.

Course	Grade	Number of Credit-Hours (Weights)
Business Calculus	B+	3
Biology	C	4
Library Research	F	1
American Literature	A	3

53. Compute the GPA for the following grades. Round to the nearest hundredth.

Course	Grade	Number of Credit-Hours (Weights)
University Physics	C+	5
Calculus I	A	4
Computer Programming	D	3
Swimming	A	1

Introduction to Probability

Section 9.5

1. Basic Definitions

The probability of an event measures the likelihood of the event to occur. It is of particular interest because of its application to everyday life.

- The probability of picking the winning six-number combination for the New York lotto grand prize is $\frac{1}{45,057,474}$.

- Genetic DNA analysis can be used to determine the risk that a child will be born with cystic fibrosis. If both parents test positive, the probability is 25% that a child will be born with cystic fibrosis.

Objectives

1. **Basic Definitions**
2. **Probability of an Event**
3. **Estimating Probabilities from Observed Data**
4. **Complementary Events**

To begin our discussion, we must first understand some basic definitions.

An activity with observable outcomes such as flipping a coin or rolling a die is called an **experiment**. The collection (or set) of all possible outcomes of an experiment is called the **sample space** of the experiment.

Skill Practice

1. Suppose one ball is selected from those shown and the color is recorded. Write the sample space for this experiment.

2. For an individual birth, the gender of the baby is recorded. Determine the sample space for this experiment.

Example 1 Determining the Sample Space of an Experiment

a. Suppose a single die is rolled. Determine the sample space of the experiment.

b. Suppose a coin is flipped. Determine the sample space of the experiment.

Solution:

a. A die is a single six-sided cube on which each side has between 1 and 6 dots painted on it. When the die is rolled, any of the six sides may come up.

The sample space is {1, 2, 3, 4, 5, 6}.   Notice that the symbols { } (called *set braces*) are used to enclose the elements.

b. The coin may land as a head H or as a tail T.
The sample space is {H, T}.

2. Probability of an Event

Any part of a sample space is called an **event**. For example, if we roll a die, the event of rolling number 5 or a greater number consists of the outcomes 5 and 6. In mathematics, we measure the likelihood of an event to occur by its probability.

Avoiding Mistakes

From the definition, a probability value can never be negative or greater than 1.

DEFINITION Probability of an Event

$$\text{Probability of an event} = \frac{\text{number of elements in event}}{\text{number of elements in sample space}}$$

Skill Practice

3. Find the probability of selecting a yellow ball from those shown.

4. Find the probability of a random birth resulting in a girl.

Example 2 Finding the Probability of an Event

a. Find the probability of rolling a 5 or greater on a die.

b. Find the probability of flipping a coin and having it land as heads.

Solution:

a. The event can occur in 2 ways: The die lands as a 5 or 6.
The sample space has 6 elements: 1, 2, 3, 4, 5, and 6.

The probability of rolling a 5 or greater: $\dfrac{2}{6}$ ← number of ways to roll a 5 or greater
← number of elements in the sample space

$= \dfrac{1}{3}$ Simplify to lowest terms.

Answers

1. {red, green, blue, yellow}
2. {male, female}
3. $\frac{1}{4}$ or 0.25   4. $\frac{1}{2}$ or 0.5

b. The event can occur in 1 way (the coin lands head side up).
The sample space has 2 outcomes: heads or tails.

The probability of flipping a head on a coin: $\dfrac{1}{2}$ ← number of ways to get heads

← number of elements in the sample space

The value of a probability can be written as a fraction, as a decimal, or as a percent. For example, the probability of a coin landing as heads is $\frac{1}{2}$ or 0.5 or 50%. In words, this means that if we flip a coin many times, theoretically we expect one-half (50%) of the outcomes to land as heads.

Example 3 **Finding Probabilities**

A class has 4 freshmen, 12 sophomores, and 6 juniors. If one individual is selected at random from the class, find the probability of selecting

a. A sophomore

b. A junior

c. A senior

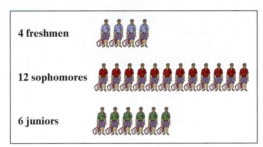

4 freshmen

12 sophomores

6 juniors

Skill Practice

A group of registered voters has 9 Republicans, 8 Democrats, and 3 Independents. Suppose one person from the group is selected at random.

5. What is the probability that the person is a Democrat?
6. What is the probability that the person is an Independent?
7. What is the probability that the person is registered with the Libertarian Party?

Solution:

In this case, there are 22 members of the class (4 freshmen + 12 sophomores + 6 juniors). This means that the sample space has 22 elements.

a. There are 12 sophomores in the class. The probability of selecting a sophomore is

$\dfrac{12}{22}$ There are 12 sophomores out of 22 people in the sample space.

$=\dfrac{6}{11}$ Simplify to lowest terms.

b. There are 6 juniors out of 22 people in the sample space. The probability of selecting a junior is

$\dfrac{6}{22}$ or $\dfrac{3}{11}$

c. There are no seniors in the class. The probability of selecting a senior is

$\dfrac{0}{22}$ or   0

A probability of 0 indicates that the event is impossible. It is impossible to select a senior from a class that has no seniors.

Answers

5. $\dfrac{8}{20}=\dfrac{2}{5}$ or 0.4   **6.** $\dfrac{3}{20}$ **7.** 0

3. Estimating Probabilities from Observed Data

We were able to compute the probabilities in Examples 2 and 3 because the sample space was known. Sometimes we need to collect information to help us estimate probabilities.

Skill Practice

Refer to Table 9-8 in Example 4.

8. What is the probability of selecting a patient who brushes twice a day?
9. What is the probability of selecting a patient who brushes more than twice a day?

Example 4   Estimating Probabilities from Observed Data

A dental hygienist records the number of times a day her patients say that they brush their teeth. Table 9-8 displays the results.

Table 9-8

Number of Times of Brushing Teeth per Day	Frequency
1	6
2	10
3	4
More than 3	1

If one of her patients is selected at random,

a. What is the probability of selecting a patient who brushes only one time a day?

b. What is the probability of selecting a patient who brushes more than once a day?

Solution:

a. The table shows that there are 6 patients who brush once a day. To get the total number of patients we add all of the frequencies ($6 + 10 + 4 + 1 = 21$). The probability of selecting a patient who brushes only once a day is

$$\frac{6}{21} \quad \text{or} \quad \frac{2}{7}$$

b. To find the number of patients who brush more than once a day, we add the frequencies for the patients who brush 2 times, 3 times, and more than 3 times ($10 + 4 + 1 = 15$). The probability of selecting a patient who brushes more that once a day is

$$\frac{15}{21} \quad \text{or} \quad \frac{5}{7}$$

4. Complementary Events

The events in Example 4(a) and 4(b) are called complementary events. The **complement of an event** is the set of all elements in the sample space that are not in the event. In this case, the number of patients who brush once a day and the number of patients who brush more than once a day make up the entire sample space, yet do not overlap. For this reason, the probability of an event plus the probability of its complement is 1. For Example 4, we have $\frac{2}{7} + \frac{5}{7} = \frac{7}{7} = 1$.

Example 5 **Finding the Probability of Complementary Events**

Find the indicated probability.

a. The probability of getting a winter cold is $\frac{3}{10}$. What is the probability of *not* getting a winter cold?

b. If the probability that a washing machine will break before the end of the warranty period is 0.0042, what is the probability that a washing machine will *not* break before the end of the warranty period?

Solution:

a. The probability of an event plus the probability of its complement must add up to 1. Therefore, we have an addition problem with a missing addend. This may also be expressed as subtraction.

$$\frac{3}{10} + ? = 1 \qquad \text{or equivalently} \qquad 1 - \frac{3}{10} = ?$$

$$\frac{10}{10} - \frac{3}{10} = \frac{7}{10} \qquad \text{Find a common denominator and subtract.}$$

There is a $\frac{7}{10}$ chance (70% chance) of *not* getting a winter cold.

b. The probability that a washing machine will break before the end of the warranty period is 0.0042. Then the probability that a machine will *not* break before the end of the warranty period is given by

$$1 - 0.0042 = 0.9958 \text{ or equivalently } 99.58\%$$

Skill Practice

10. For one particular medicine, the probability that a patient will experience side effects is $\frac{1}{20}$. What is the probability that a patient will *not* experience side effects?

11. The probability that a flight arrives on time is 0.18. What is the probability that a flight will *not* arrive on time?

Answers

10. $\frac{19}{20}$ **11.** 0.82

Section 9.5 Practice Exercises

Boost *your* GRADE at ALEKS.com!

- Practice Problems
- Self-Tests
- NetTutor
- e-Professors
- Videos

Study Skills Exercises

1. A good way to determine what will be on a test is to look at both your notes and the exercises assigned by your instructor. List five kinds of problems that you think will be on the test for this chapter.

2. Define the key terms.

    a. Experiment   **b. Sample space**   **c. Event**

    d. Probability of an event   **e. Complement of an event**

Review Exercises

For Exercises 3–8, find the mean, median, and mode (if one exists).

3. 13, 16, 22, 25, 10

4. 62, 64, 62, 67, 40

5. 8, 9, 10, 7, 8, 8, 11, 10

6. 96%, 88%, 89%, 90%, 88%, 50%

7. 20, 20, 18, 17, 19, 5

8. 100, 90, 95, 98, 90, 10

Objectives 1: Basic Definitions

9. A card is chosen from a deck consisting of 10 cards numbered 1–10. Determine the sample space of this experiment. **(See Example 1.)**

10. A marble is chosen from a jar containing a yellow marble, a red marble, a blue marble, a green marble, and a white marble. Determine the sample space of this experiment.

11. Two dice are thrown, and the sum of the top sides is observed. Determine the sample space of this experiment.

12. A coin is tossed twice. Determine the sample space of this experiment.

13. If a die is rolled, in how many ways can an odd number come up?

14. If a die is rolled, in how many ways can a number less than 6 come up?

Objective 2: Probability of an Event

15. Which of the values can represent the probability of an event?

    a. 1.62   **b.** $-\dfrac{7}{5}$ **c.** 0   **d.** 1

    e. 200%   **f.** 4.5   **g.** 4.5%   **h.** 0.87

16. Which of the values can represent the probability of an event?

    a. 1.5   **b.** 0   **c.** $\dfrac{2}{3}$ **d.** 1

    e. 150%   **f.** 3.7   **g.** 3.7%   **h.** 0.92

17. If a single die is rolled, what is the probability that it will come up as a number less than 3? **(See Example 2.)**

18. If a single die is rolled, what is the probability that it will come up as a number greater than 5?

19. If a single die is rolled, what is the probability that it will come up with an even number?

20. If a single die is rolled, what is the probability that it will come up as an odd number?

For Exercises 21–24, refer to the figure. A sock drawer contains 2 white socks, 5 black socks, and 1 blue sock. **(See Example 3.)**

21. What is the probability of choosing a black sock from the drawer?

22. What is the probability of choosing a white sock from the drawer?

23. What is the probability of choosing a blue sock from the drawer?

24. What is the probability of choosing a purple sock from the drawer?

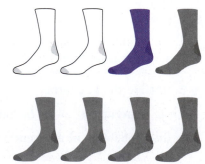

25. If a die is tossed, what is the probability that a number from 1 to 6 will come up?

26. If a die is tossed, what is the probability of getting a 7?

27. What is the probability of an impossible event?

28. What is the sum of the probabilities of an event and its complement?

29. In a deck of cards there are 12 face cards and 40 cards with numbers. What is the probability of selecting a face card from the deck?

30. In a deck of cards, 13 are diamonds, 13 are spades, 13 are clubs, and 13 are hearts. Find the probability of selecting a diamond from the deck.

31. A jar contains 7 yellow marbles, 5 red marbles, and 4 green marbles. What is the probability of selecting a red marble or a yellow marble?

32. A jar contains 10 black marbles, 12 white marbles, and 4 blue marbles. What is the probability of selecting a blue marble or a black marble?

Objective 3: Estimating Probabilities from Observed Data

33. The table displays the length of stay for vacationers at a small motel. **(See Example 4.)**

Length of Stay in Days	Frequency
2	14
3	13
4	18
5	28
6	11
7	30
8	6

a. What is the probability that a vacationer will stay for 4 days?

b. What is the probability that a vacationer will stay for less than 4 days?

c. Based on the information from the table, what percent of vacationers stay for more than 6 days?

34. A number of students at a large university were asked if they owned a car. The table shows the results.

	Number of Car Owners	Number Who Do Not Own a Car
Dorm resident	32	88
Lives off campus	59	26

a. What is the probability that a student selected at random lives in a dorm?

b. What is the probability that a student selected at random does not own a car?

35. A survey was made of 60 participants, asking if they drive an American-made car, a Japanese car, or a car manufactured in another foreign country. The table displays the results.

	Frequency
American	21
Japanese	30
Other	9

a. What is the probability that a randomly selected car is manufactured in America?

b. What percent of cars is manufactured in some country other than Japan?

36. The number of customer complaints for service representatives is given in the table. If one representative is picked at random, find the probability that the representative received

a. Exactly 3 complaints

b. Between 1 and 5 complaints, inclusive

c. At least 4 complaints

d. More than 5 complaints or fewer than 2 complaints

Number of Complaints	Number of Representatives
0	4
1	2
2	14
3	10
4	16
5	18
6	10
7	6

37. Mr. Gutierrez noted the times in which his students entered his classroom and constructed the following chart.

a. What is the probability that a student will be early to class?

b. What is the probability that a student will be late to class?

c. What percent of students arrive on time or early? Round to the nearest whole percent.

Time	Number of Students
About 10 min early	1
About 5 min early	6
On time	11
About 5 min late	7
About 10 min late	3
About 15 min late	1

38. Each person at an office party purchased a raffle ticket. The graph shows the results of the raffle.

a. How many people bought raffle tickets?

b. What is the probability of winning the TV set?

c. What percent of people won some type of prize?

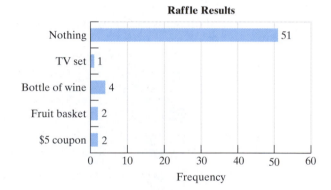

Objective 4: Complementary Events

39. If the probability of the horse Sugar 'N Spice to win is $\frac{2}{11}$, what is the probability that he will lose? **(See Example 5.)**

40. If the probability of being hit by lighting is $\frac{1}{1,000,000}$, what is the probability of not getting hit by lighting?

41. If the probability of having twins is 1.2%, what is the probability of not having twins?

42. The probability of a woman's surviving breast cancer is 88%. What is the probability that a woman would not survive breast cancer?

Group Activity

Creating a Statistical Report

Materials: A computer with Internet access or the local newspaper

Estimated time: 20–30 minutes

Group Size: 4

The group members will collect numerical data from the Internet or the newspaper. The data will be analyzed using the statistical techniques learned in this chapter. Here is one suggested project.

1. Record the age and gender of the individuals who were arrested in your town during the past week. This can often be found in the local section of the newspaper. For example, you can visit the website for the Daytona Beach *News-Journal* and select "local news" and then "news of record." Record 20 or 30 data values.

2. Compute the mean, median, and mode for the ages of men arrested. Compute the mean, median, and mode for the ages of women arrested. Do the statistics suggest a difference in the average age of arrest for men versus women?

3. Determine the percentage of men and the percentage of women in the sample. Does there appear to be a significant difference?

4. Organize the data by age group and construct a frequency distribution and histogram.

Note: The steps given in this project offer suggestions for organizing and analyzing the data you collect. These steps outline standard statistical techniques that apply to a variety of data sets. You might consider doing a different project that investigates a topic of interest to you. Here are some other ideas.

• Collect the weight and gender of babies born in the local hospital.

• Collect the age and gender of students who take classes at night versus those who take classes during the day.

• Collect stock prices for a 2- or 3-week period.

Can you think of other topics for a project?

Chapter 9   Summary

Section 9.1   Tables, Bar Graphs, Pictographs, and Line Graphs

Key Concepts

Statistics is the branch of mathematics that involves collecting, organizing, and analyzing **data** (information). Information can often be organized in tables and graphs. The individual entries within a table are called **cells**.

Example 1

The data in the table give the number of Calories for a 1-c serving of selected vegetables.

Vegetable (1 c)	Number of Calories
Corn	85
Green beans	35
Eggplant	25
Peas	125
Spinach	40

A **pictograph** uses an icon or small image to convey a unit of measurement.

Example 3

What is the value of each icon in the graph?

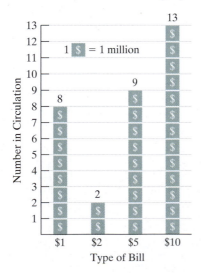

... icon is worth 1,000,000 bills in circulation.

Examples

Example 2

Construct a bar graph for the data in Example 1.

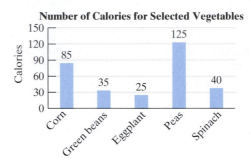

Line graphs are often used to track how one variable changes with respect to the change in a second variable.

Example 4

In what year were there 52.3 million married-couple households?

From the graph, the year 1990 corresponds to 52.3 million married-couple households.

Section 9.2   Frequency Distributions and Histograms

Key Concepts

A **frequency distribution** is a table displaying the number of data values that fall within specified intervals called **class intervals**.

When constructing a frequency distribution, keep these important guidelines in mind.

- The classes should be equally spaced.
- The classes should not overlap.
- In general, use between 5 and 15 classes.

A **histogram** is a special bar graph that illustrates data given in a frequency distribution. The class intervals are given on the horizontal scale. The height of each bar in a histogram measures the frequency for each class.

Examples

Example 1

Create a frequency distribution for the following data.

50	53
54	51
50	40
50	47
53	36
44	34
52	32
42	30

Class Intervals	Tally	Frequency
30–34	III	3
35–39	I	1
40–44	III	3
45–49	I	1
50–54	IIII III	8

Example 2

Create a histogram for the data in Example 1.

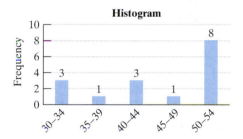

Section 9.3   Circle Graphs

Key Concepts

A **circle graph** (or pie graph) is a type of graph used to show how a whole amount is divided into parts. Each part of the circle, called a **sector**, is like a slice of pie.

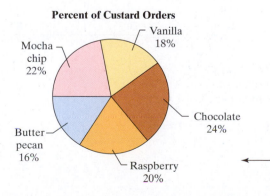

Percent of Custard Orders

Examples

Example 1

At Ritter's Frozen Custard, the flavors for the day are given with the number of orders for each flavor.

Flavor	Number of Orders
Vanilla	180
Chocolate	240
Raspberry	200
Butter pecan	160
Mocha chip	220

Construct a circle graph for the data given. Label the sectors with percents.

| Section 9.4 | Mean, Median, and Mode |

Key Concepts

The **mean** (or average) of a set of numbers is the sum of the values divided by the number of values.

$$\text{Mean} = \frac{\text{sum of the values}}{\text{number of values}}$$

The **median** is the "middle" number in an ordered list of numbers. For an ordered list of numbers:

- If the number of data values is *odd*, then the median is the middle number in the list.
- If the number of data values is *even*, the median is the mean of the two middle numbers in the list.

The **mode** of a set of data is the value or values that occur most often.

When data values in a list appear multiple times, we can compute a **weighted mean**.

Examples

Example 1

Find the mean test score: 92, 100, 86, 60, 90

$$\text{Mean} = \frac{92 + 100 + 86 + 60 + 90}{5}$$

$$= \frac{428}{5} = 85.6$$

Example 2

Find the median: 12 18 6 10 5

First order the list: 5 6 10 12 18
The median is the middle number, 10.

Example 3

Find the median: 15 20 20 32 40 45

The median is the average of 20 and 32:

$$\frac{20 + 32}{2} = \frac{52}{2} = 26 \qquad \text{The median is 26.}$$

Example 4

Find the mode: 7 2 5 7 7 4 6 10

The value 7 is the mode because it occurs most often.

Example 5

The ages of children in a day-care center are given in the table. Find the mean age.

Age (yr)	Frequency	Product
3	5	$3 \cdot 5 = 15$
4	10	$4 \cdot 10 = 40$
5	3	$5 \cdot 3 = 15$
6	2	$6 \cdot 2 = 12$
Total	**20**	82

$$\text{Mean} = \frac{82}{20} = 4.1$$

The mean age is 4.1 yr.

| Section 9.5 | **Introduction to Probability** |

Key Concepts

The collection (or set) of all possible outcomes of an experiment is called the **sample space**.

The **probability of event** is given by:

$$\frac{\text{number of elements in event}}{\text{number of elements in sample space}}$$

The probability of an event cannot be greater than 1.

The **complement of an event** is the set of all elements in the sample space that are not in the event.

Examples

Example 1

Define the sample space for selecting a colored ball.

Sample space = {red, blue, yellow, green}

Example 2

What is the probability of selecting a yellow ball from Example 1?

Let A represent the event of picking a yellow ball. Then A = {yellow}.

$$P(A) = \frac{1}{4} \quad \leftarrow \text{number of yellow balls}$$
$$\leftarrow \text{number of balls in box}$$

Example 3

49 CDs are in a shopping cart.

   10 Rap
   24 Rock
   12 Latina
   3 Classical

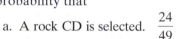

If one CD is selected at random, find the probability that

a. A rock CD is selected. $\dfrac{24}{49}$

b. A rock CD is *not* selected. This is the complementary event to part (a).

$$1 - \frac{24}{49} = \frac{25}{49}$$

Chapter 9   Review Exercises

Section 9.1

For Exercises 1–4, refer to the table. The table gives the number of Calories and the amount of fat, cholesterol, sodium, and total carbohydrate for one $\frac{1}{2}$-c serving of chocolate ice cream.

Ice Cream	Calories	Fat(g)	Cholesterol (mg)	Sodium (mg)	Carbohydrate (g)
Breyers	150	8	20	35	17
Häagen-Dazs	270	18	115	60	22
Edy's Grand	150	8	25	35	17
Blue Bell	160	8	35	70	18
Godiva	290	18	65	50	28

1. Which ice cream has the most calories?

2. Which ice cream has the least amount of cholesterol?

3. How many more times the sodium does Blue Bell have per serving than Edy's Grand?

4. What is the difference in the amount of carbohydrate for Godiva and Blue Bell?

Since 1940 the number of U.S. farms has decreased. However, the average size of the farms has increased. The graph shows the average size of U.S. farms for selected years. Refer to the graph for Exercises 5–8. (*Source:* U.S. Department of Agriculture)

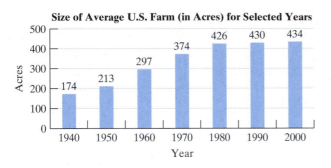

Size of Average U.S. Farm (in Acres) for Selected Years

5. What was the average size of the farms in 1970?

6. What is the difference between the average size farm in the year 2000 compared to 1940?

7. What is the difference between the average size farm in the year 1990 compared to 1980?

8. In which 10-year interval was the increase the greatest?

For Exercises 9–12, refer to the pictograph. The graph represents the number of tornadoes during four months with active weather.

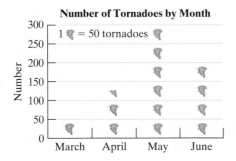

Number of Tornadoes by Month

9. What does each icon represent?

10. From the graph, estimate the number of tornadoes in May.

11. Which month had approximately 200 tornadoes?

12. Estimate the difference in the number of tornadoes in April and the number in March.

For Exercises 13–16, refer to the graph. The graph represents the number of liver transplants in the United States for selected years. (*Source:* U.S. Department of Health and Human Services)

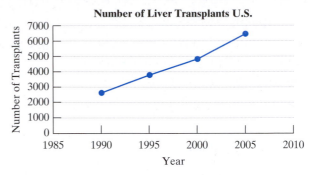

Number of Liver Transplants U.S.

13. In which year did the greatest number of liver transplants occur?

14. Approximate the number of liver transplants for the year 2000.

15. Does the trend appear to be increasing or decreasing?

16. Extend the graph to predict the number of liver transplants for the year 2007.

17. The table shows several movies that grossed over 100 million dollars in the United States. Construct a bar graph using horizontal bars. The length of each bar should represent the amount of money in millions that each movie grossed. (*Source: Washington Post*)

Movie Title	Gross (in millions)	Year
Titanic	601	1997
Star Wars (1)	461	1977
Shrek 2	436	2004
Lord of the Rings: The Fellowship of the Ring	314	2001
Harry Potter and the Goblet of Fire	290	2005
Wedding Crashers	209	2005

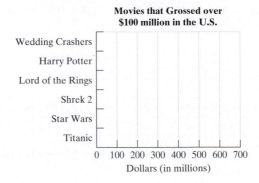

Movies that Grossed over $100 million in the U.S.

Section 9.2

The ages of students in a Spanish class are given.

18 22 19 26 31 20 40 24 43 22
29 28 35 42 29 30 24 31 23 21

Use these data for Exercises 18–19.

18. Complete the frequency table.

Class Intervals (Age)	Frequency
18–21	
22–25	
26–29	
30–33	
34–37	
38–41	
42–45	

19. Construct a histogram of the data in Exercise 18.

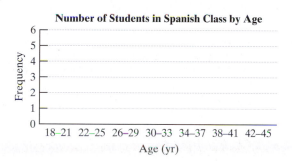

Number of Students in Spanish Class by Age

Section 9.3

The pie graph describes the types of subs offered at Larry's Sub Shop. Use the information in the graph for Exercises 20–22.

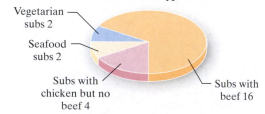

Number of Certain Types of Subs

Vegetarian subs 2
Seafood subs 2
Subs with chicken but no beef 4
Subs with beef 16

20. How many types of subs are offered at Larry's?

21. What fraction of the subs at Larry's is made with beef?

22. What fraction of the subs at Larry's is not made with beef?

23. A survey was conducted with 200 people, and they were asked their highest level of education. The results of the survey are given in the table.

a. Complete the table.

Education Level	Number of People	Percent	Number of Degrees
Grade school	10		
High school	50		
Some college	60		
Four-year degree	40		
Post graduate	40		

b. Construct a circle graph using percents from the information in the table.

Percent by Education Level

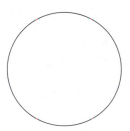

Section 9.4

24. For the list of quiz scores, find the mean, median, and mode(s).

20, 20, 18, 16, 18, 17, 16, 10, 20, 20, 15, 20

25. Juanita kept track of how many milligrams of calcium she took each day through vitamins and dairy products. Determine the mean number of milligrams of calcium per day. Round to the nearest 10 (mg).

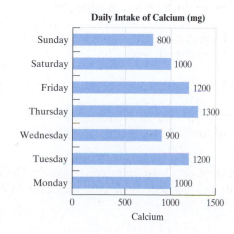

Daily Intake of Calcium (mg)

Day	Calcium
Sunday	800
Saturday	1000
Friday	1200
Thursday	1300
Wednesday	900
Tuesday	1200
Monday	1000

26. The seating capacity for five arenas used by the NBA is given in the table. Find the median number of seats.

Arena	Number of Seats
Phelps Arena, Atlanta	20,000
Fleet Center, Boston	18,624
Chevrolette Coliseum, Charlotte	23,799
United Center, Chicago	21,500
Gund Arena, Cleveland	20,562

27. The manager of a restaurant had his customers fill out evaluations on the service that they received. A scale of 1 to 5 was used, where 1 represents very poor service and 5 represents excellent service. Given the list of responses, determine the mode(s).

4 5 3 4 4 3 2 5 5 1 4 3 4 4 5
2 5 4 4 3 2 5 5 1 4

28. There are 20 children participating in an afternoon fitness program. The graph displays the number of children by age. Complete the table and then find the mean age of the children.

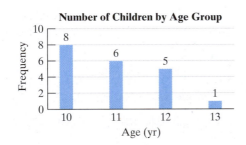

Number of Children by Age Group

Age (yr)	Number of Children	Product
10		
11		
12		
13		

Section 9.5

29. Roberto has six pairs of socks, each a different color: blue, green, brown, black, gray, and white. If Roberto randomly chooses a pair of socks, write the sample space for this event.

30. Refer to Exercise 29. What is the probability that Roberto will select a pair of gray socks?

31. Which of the following numbers could represent a probability?

a. $\frac{1}{2}$ **b.** $\frac{5}{4}$ **c.** 0 **d.** 1

e. 25% **f.** 2.5 **g.** 6% **h.** 6

32. A bicycle shop sells a child's tricycle in three colors: red, blue, and pink. In the warehouse there are 8 red tricycles, 6 blue tricycles, and 2 pink tricycles.

a. If Kevin selects a tricycle at random, what is the probability that he will pick a red tricycle?

b. What is the probability that he will not pick a red tricycle?

c. What is the probability of Kevin's selecting a green tricycle?

Chapter 9   Test

1. The table represents the world's major producers of primary energy for a recent year. All measurements are in quadrillions of Btu.

    Note: 1 quadrillion = 1,000,000,000,000,000. (*Source:* Energy Information Administration, U.S. Dept. of Energy)

Country	Amount of Energy Produced (quadrillions of BTUs)
United States	72
Russia	43
China	35
Saudi Arabia	43
Canada	18

    Construct a bar graph using horizontal bars. The length of each bar corresponds to the amount of energy produced for each country.

    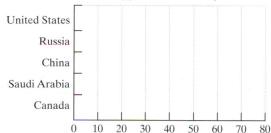

    World's Major Producers of Primary Energy (Quadrillions of Btu)

2. Of the approximately 2.9 million workers in 1820 in the United States, 71.8% were employed in farm occupations. Since then, the percent of U.S. workers in farm occupations has declined. The table shows the percent of total U.S. workers who worked in farm-related occupations for selected years. (*Source:* U.S. Department of Agriculture)

Year	Percent of U.S. Workers in Farm Occupations
1820	72%
1860	59%
1900	38%
1940	17%
1980	3%

a. Which year had the greatest percent of U.S. workers employed in farm occupations? What is the value of the greatest percent?

b. Make a line graph with the year on the horizontal scale and the percent on the vertical scale.

Percent of U.S. Workers in Farm Occupations

c. Based on the graph, estimate the percent of U.S. workers employed in farm occupations for the year 1960.

For Exercises 3–5, refer to the pictograph. The pictograph shows the flower sales for the first 5 months of the year for a flower shop.

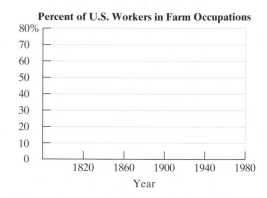

Flower Shop Sales ($)

= $1000 in sales

3. What is the value of each flower icon?

4. From the graph, estimate the sales for the month of April.

5. Which month brought in sales of $5000?

For Exercises 6–8, refer to the table. The rainfall amounts for Salt Lake City, Utah, and Seattle, Washington, are given in the table for selected months. All values are in inches. (*Source:* National Oceanic and Atmospheric Administration)

	April	May	June	July
Salt Lake City	2.02	2.09	0.77	0.72
Seattle	2.75	2.03	2.5	0.92

6. Which city is generally wetter?

7. What is the difference in the amount of rainfall in Seattle and Salt Lake City during June?

8. Find the mean amount of rainfall for these months for each city.

9. A cellular phone company questioned 20 people at a mall, to determine approximately how many minutes each individual spent on the cell phone each month. Using the list of results, complete the frequency distribution and construct a histogram.

100 120 250 180 300 200 250 175
110 280 330 280 300 325 60   75
100 350 60   90

Number of Minutes Used Monthly	Tally	Frequency
51–100		
101–150		
151–200		
201–250		
251–300		
301–350		

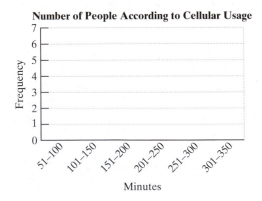

Number of People According to Cellular Usage

For Exercises 10–12, refer to the circle graph. The circle graph shows the percent of homes having different types of flooring in the living room area.

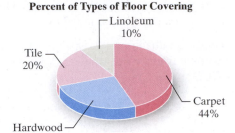

Percent of Types of Floor Covering

10. If 150 people were questioned, how many would be expected to have carpet on their living room floor?

11. If 200 people were questioned, how many would be expected to have tile on their living room floor?

12. If 300 people were questioned, how many would not be expected to have linoleum on their living room floor?

For Exercises 13–15, refer to the table. The table represents the heights of the Seven Summits (the highest peaks from each continent).

Mountain	Continent	Height (ft)
Mt. Kilimanjaro	Africa	19,340
Elbrus	Europe	18,510
Aconcagua	South America	22,834
Denali	North America	20,320
Vinson Massif	Antarctica	16,864
Mt. Kosciusko	Australia	7,310
Mt. Everest	Asia	29,035

13. What is the mean height of the Seven Summits? Round to the nearest whole unit.

14. What is the median height?

15. Is there a mode?

16. Mike and Darcy listed the amount of money paid for going to the movies for the past 3 months. This list contains the amount for 2 tickets. Find the mean, median, and mode.

$11 $14 $11 $16 $15 $16 $12 $16 $15 $20

17. A board game has a die with eight sides with the numbers 1–8 printed on each side.

  a. What is the sample space for rolling the die one time?

  b. What is the probability of rolling a 6?

  c. What is the probability of rolling an even number?

  d. What is the probability of rolling a number less than 3?

18. At a party there is a cooler filled with ice and soft drinks: 6 cans of diet cola, 4 cans of ginger ale, 2 cans of root beer, and 2 cans of cream soda. A person takes a can of soda at random.

  a. What is the probability that the person selects a can of ginger ale?

  b. What is the probability of the person not selecting a can of ginger ale?

19. Compute the GPA for the following grades. Round to the nearest hundredth. Use this scale:

A	= 4.0	C	= 2.0
B+	= 3.5	D+	= 1.5
B	= 3.0	D	= 1.0
C+	= 2.5	F	= 0.0

Course	Grade	Number of Credit-Hours (Weights)
Art Appreciation	B	4
College Algebra	A	3
English II	C	3
Physical Fitness	A	1

20. Which of the following is not a reasonable value for a probability?

  a. 0.36   **b.** $\dfrac{3}{4}$ **c.** 1.5

Chapters 1–9   Cumulative Review Exercises

1. Identify the place value of the underlined digit.

  a. 23,990,192   **b.** 5,981,902   **c.** 3,019,226

2. Add. $2087 + 53 + 10{,}499 + 6$

3. Estimate the product by first rounding each number to the nearest hundred.

$$687 \times 1243$$

4. Divide 651 by 23. Identify the divisor, dividend, whole part of the quotient, and remainder.

5. What fraction of this circle is shaded?

For Exercises 6–8, multiply or divide as indicated. Reduce the answers to lowest terms.

  6. $\dfrac{12}{7} \cdot \dfrac{14}{36}$ **7.** $\dfrac{105}{96} \div \dfrac{7}{16}$ **8.** $\dfrac{5}{8} \div \dfrac{6}{15} \cdot \dfrac{24}{25}$

For Exercises 9–11, add or subtract as indicated. Reduce to lowest terms.

  9. $\dfrac{97}{102} - \dfrac{63}{102}$ **10.** $\dfrac{3}{10} + \dfrac{7}{100}$ **11.** $\dfrac{1}{2} + \dfrac{5}{3} - \dfrac{1}{6}$

12. Simplify, using the order of operations.

$$2\dfrac{1}{8} \div 17 + \dfrac{7}{12} \cdot \dfrac{9}{14} - \dfrac{1}{3}$$

13. The table gives the prices of certain stocks and their increase or decrease from one day to the next. Complete the table.

Stock	Yesterday's Closing Price ($)	Increase/ Decrease	Today's Closing Price ($)
RylGold	13.28	0.27	
NetSolve	9.51	−0.17	
Metals USA	14.35	0.10	
PAM Transpt	18.09	0.09	
Steel Tch	21.63	−0.37	

For Exercises 14–16, multiply or divide by powers of 10.

14. 68.412×100 **15.** 68.412×0.1

16. $68.412 \div 0.001$

17. The estimated forest cover in the Brazilian Amazon in 1970 was approximately 3.7 million square kilometers. By 2005, the amount dropped to 3.0 million km^2. (*Source:* National Geographic Society)

 a. By how many square kilometers had the Brazilian Amazon forest cover decreased?

 b. Compute the percent decrease in forest cover. Round to the nearest tenth of a percent.

18. Quick Cut Lawn Company can service 5 customers in $2\frac{3}{4}$ hr. Speedy Lawn Company can service 6 customers in 3 hr. Find the unit rate in time per customer for both lawn companies and decide which company is faster.

19. If Rosa can type a 4-page English paper in 50 min, how long will it take her to type a 10-page term paper?

20. Find the values of x and y, assuming that the two triangles are similar.

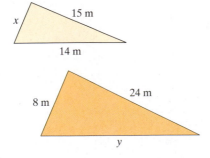

21. Out of a group of people, 95 said that they brushed their teeth twice a day. If this number represents 78% of the people surveyed, how many were surveyed? Round to the nearest whole unit.

22. The number of children accompanying their parents on business trips has jumped 230% in the last 10 years. If the number of children 10 years ago was 7.4 million, determine the present number.

23. Out of 120 people, 78 wear glasses. What percent does this represent?

24. A savings account pays 3.4% simple interest. If $1200 is invested for 5 years, what will be the balance?

25. Convert 2 ft 5 in. to inches.

26. Convert $4\frac{1}{2}$ gal to quarts.

27. Add. 3 yd 2 ft + 5 yd 2 ft

28. Subtract. 12 km − 2360 m

29. Divide 16 lb 12 oz by 4.

For Exercises 30–32, identify the type of angle. Choose from acute, obtuse, right, or straight.

30.    **31.**

32.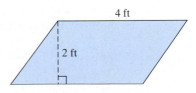

33. Find the area.

4 ft

2 ft

34. Find the volume. Use $\pi \approx \frac{22}{7}$.

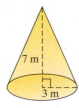

7 m

3 m

35. The data shown in the table give the average weight for boys based on age. (*Source:* National Parenting Council) Make a line graph to illustrate these data.

Age	Weight (lb)
5	44.5
6	48.5
7	54.5
8	61.25
9	69
10	74.5
11	85
12	89

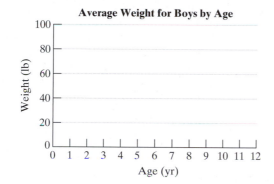

Average Weight for Boys by Age

36. The monthly number of deaths resulting from tornados for a recent year are given. Find the mean and median. Round to the nearest whole unit.

33, 10, 62, 132, 123, 316, 138,

123, 133, 18, 150, 26

37. Simplify the expression.

$$30 - 3(5 - 2)^2$$

A game has a spinner with four sections of equal size. Refer to the spinner for Exercises 38–40.

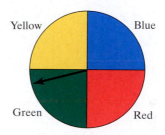

38. If a person spins the pointer once, determine the sample space.

39. What is the probability of the pointer landing on green?

40. What is the probability of the pointer *not* landing on green?

Real Numbers

10

CHAPTER OUTLINE

Chapter 10

In this chapter we begin our study of algebra by learning how to add, subtract, multiply, and divide positive and negative numbers.

As you work through this chapter, try working through this puzzle. Notice that negative signs also go in the blanks. They may appear to the left of a number written horizontally. They may also appear above a number written vertically.

Across

2. $|-24|$
4. -4^2
5. $30 - (-40)$
7. $(-84)(-11)$
8. $-315 + 604$
9. $(3 - 14)^2$

Down

1. $(-4)^2$
3. $2000 \div 5 \cdot 10 + \sqrt{81}$
4. $-14{,}668 - 5823$
6. $\dfrac{5010 - (-3500)}{-2}$

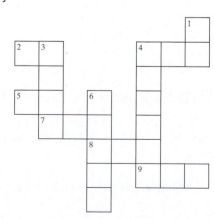

585

Section 10.1   Real Numbers and the Real Number Line

1. Integers

Thus far in the text we have worked with the number zero and numbers greater than zero. Numbers greater than zero are called **positive numbers**. Positive numbers lie to the right of zero on a number line (Figure 10-1).

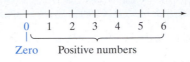

Zero   Positive numbers

Figure 10-1

In some applications of mathematics, we need to use *negative* numbers. For example:

- On a winter day in Detroit, the low temperature was 5 degrees below zero: $-5°$
- Tiger Woods' golf score in the U.S. Open was 3 below par: -3
- Maria is $128 overdrawn on her checking account. Her balance is: $-\$128$

The values $-5°$, -3, and $-\$128$ are negative numbers. **Negative numbers** lie to the left of zero on a number line (Figure 10-2).

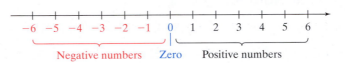

Negative numbers   Zero   Positive numbers

Figure 10-2

The numbers $\dots -3, -2, -1, 0, 1, 2, 3, \dots$ and so on are called **integers**.

Example 1   Writing Integers

Write an integer that denotes each numerical value.

a. Liquid nitrogen freezes at 346°F below zero.

b. The shoreline of the Dead Sea on the border of Israel and Jordan is the lowest land area on earth. The "altitude" is 1300 ft below sea level.

Solution:

a. $-346°F$ b. -1300 ft

---

2. Rational and Irrational Numbers

A number that can be written as a ratio of two integers is called a **rational number** (division by zero is excluded). For example, the following numbers are rational numbers.

$\dfrac{2}{3}$ because it is a ratio of 2 and 3.

$\dfrac{-5}{7}$ because it is a ratio of -5 and 7.

8	because it is a ratio of 8 and 1. That is, $8 = \frac{8}{1}$. This shows that an integer is also a rational number.
0.25	because it is a ratio of 25 and 100. That is, $0.25 = \frac{25}{100}$. This shows that a terminating decimal is a rational number.
$0.\overline{3}$	because it is a ratio of 1 and 3. That is, $0.\overline{3} = \frac{1}{3}$. This shows that a repeating decimal is a rational number.

Rational numbers consist of all numbers that can be expressed as a terminating decimal or as a repeating decimal.

- All numbers that can be expressed as *repeating decimals* are rational numbers.

  For example, $0.\overline{3} = 0.3333\ldots$ is a rational number.

- All numbers that can be expressed as *terminating decimals* are rational numbers.

  For example, 0.25 is a rational number.

- A number that *cannot* be expressed as a repeating or terminating decimal is not a rational number. These are called **irrational numbers**. An example of an irrational number is $\sqrt{2}$.

  For example, $\sqrt{2} \approx 1.\underline{41421356237}\ldots$

  The digits never repeat
and never stop.

Every rational number can be matched with a point on the number line.

Example 2   **Locating Numbers on a Number Line**

Locate each number on a number line.

a. -4.2 **b.** $\dfrac{15}{4}$ **c.** $-\dfrac{1}{3}$

Solution:

a. Because -4.2 is negative, it is located to the left of zero on the number line. Draw a dot two-tenths of a unit beyond -4.

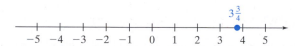

b. Write $\frac{15}{4}$ as a mixed number or decimal.

$$\frac{15}{4} = 3\frac{3}{4} \text{ or } 3.75$$

Draw a dot three-fourths of a unit beyond 3 on the number line.

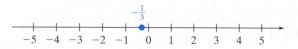

c. $-\dfrac{1}{3}$

Draw a dot one-third of a unit to the left of zero.

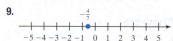

 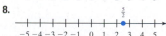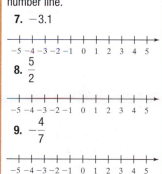

3. Real Numbers and the Real Number Line

So far we have talked about rational numbers (including integers) and irrational numbers. All of these numbers combined are called the **real numbers**. Furthermore, every real number corresponds to a point on a number line. For this reason, we sometimes call the number line the **real number line**.

- A number a is less than b (denoted $a < b$) if a lies to the left of b on the number line. See Figure 10-3.

- A number a is greater than b (denoted $a > b$) if a lies to the right of b on the number line. See Figure 10-4.

a is less than b a is greater than b

$a < b$ $a > b$

Figure 10-3   Figure 10-4

Example 3 Determining Order

Fill in the blank with $<$ or $>$.

a. $-3 \;\square\; -5$ **b.** $-2.7 \;\square\; 4.1$ **c.** $-\dfrac{4}{7} \;\square\; -\dfrac{3}{5}$

Solution:

a. $-3 \;\boxed{>}\; -5$ because -3 lies to the right of -5

b. $-2.7 \;\boxed{<}\; 4.1$ because -2.7 lies to the left of 4.1

c. To determine order between two fractions, write the fractions with a common denominator.

$$-\frac{4 \cdot 5}{7 \cdot 5} \;\square\; -\frac{3 \cdot 7}{5 \cdot 7}$$ The common denominator is 35.

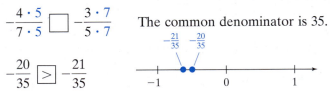

$$-\frac{20}{35} \;\boxed{>}\; -\frac{21}{35}$$

Notice that on the negative side of zero, -20 lies to the right of -21. Therefore, $-\frac{20}{35}$ is greater than $-\frac{21}{35}$.

4. Absolute Value

Notice on the number line that pairs of numbers such as 4 and -4 are the same distance from 0 (Figure 10-5). The distance between a number and zero on the number line is called its **absolute value**.

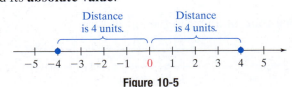

Figure 10-5

> **DEFINITION** Absolute Value
>
> The absolute value of a number a is denoted $|a|$. The value of $|a|$ is the distance between a and 0 on the number line.

From the number line we see that $|-4| = 4$ and $|4| = 4$.

Example 4 Finding Absolute Value

Determine the absolute value.

a. $|-5|$ b. $|2.1|$ c. $\left|-\dfrac{3}{2}\right|$ d. $|0|$

Solution:

a. $|-5| = 5$ The number -5 is 5 units from 0 on the number line.

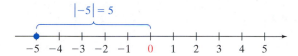

b. $|2.1| = 2.1$ The number 2.1 is 2.1 units from 0 on the number line.

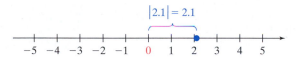

c. $\left|-\dfrac{3}{2}\right| = \dfrac{3}{2}$ The number $-\dfrac{3}{2}$ is $\dfrac{3}{2}$ units (or 1.5 units) from 0.

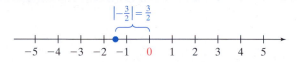

d. $|0| = 0$ The number 0 is 0 units from 0 on the number line.

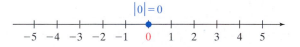

> **TIP:** The absolute value of a nonzero number is always positive. The absolute value of 0 is 0.

5. Opposite

Two numbers that are the same distance from zero on the number line, but on opposite sides of zero are called **opposites**. For example, the numbers -2 and 2 are opposites (see Figure 10-6).

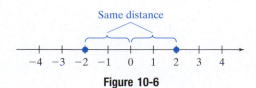

Figure 10-6

590 Chapter 10 Real Numbers

The opposite of a number a is denoted $-(a)$.

Original Number a	Opposite $-(a)$	Simplified Form	
5	$-(5)$	-5	The opposite of a positive number is a negative number.
2.4	$-(2.4)$	-2.4	
-7	$-(-7)$	7	The opposite of a negative number is a positive number.
$-\frac{1}{2}$	$-(-\frac{1}{2})$	$\frac{1}{2}$	

The opposite of a negative number is a positive number. Thus, if a is positive,

$$-(-a) = a \qquad \text{This is sometimes called the } \textit{double-negative property.}$$

Find the opposite of each number.

17. 8   **18.** -4.8

19. -2 **20.** $\frac{5}{3}$

Example 5 Finding the Opposite of a Real Number

Find the opposite of each number.

a. 4   **b.** 6.3   **c.** -11 **d.** $-\frac{5}{2}$

Solution:

a. The opposite of 4 is -4.

b. The opposite of 6.3 is -6.3.

c. The opposite of -11 is 11.

d. The opposite of $-\frac{5}{2}$ is $\frac{5}{2}$.

TIP: To find the opposite of a number, we can simply change the sign.

Simplify.

21. $-|-23|$ **22.** $-(-23)$

Example 6 Simplifying Expressions

Simplify.

a. $-|-12|$ **b.** $-(-12)$

Solution:

a. $-|-12|$ This expression represents the opposite of the absolute value of -12.

$= -12$

b. $-(-12)$ This expression represents the opposite of -12.

$= 12$

Answers

17. -8 **18.** 4.8   **19.** 2

20. $-\frac{5}{3}$ **21.** -23 **22.** 23

Section 10.1 Practice Exercises

Study Skills Exercises

1. When working with signed numbers, keep a simple example in your mind, such as temperature. We understand that 10 degrees below zero is colder than 2 degrees below zero so the inequality $-10 < -2$ makes sense. Write down another example involving signed numbers that you can easily remember.

2. Define the key terms.

   a. Absolute value   **b.** Integers   **c.** Irrational numbers

   d. Negative numbers   **e.** Opposite   **f.** Positive numbers

   g. Rational numbers   **h.** Real numbers   **i.** Real number line

Objective 1: Integers

For Exercises 3–12, write an integer that represents each numerical value. **(See Example 1.)**

3. Death Valley, California, has elevation of 86 m below sea level.

4. In a card game, Jack lost $45.

5. Playing *Wheel of Fortune,* Sally won $3800.

6. Jim's golf score is 5 over par.

7. Rena lost $500 in the stock market in one month.

8. LaTonya earned $23 in interest on her saving account.

9. Patrick lost 14 lb on a diet.

10. Emilie's score on a video game was 5000 points higher than that of the past record holder.

11. The number of Internet users rose by about 140,000.

12. A small business experienced a loss of $20,000 last year.

Objective 2: Rational and Irrational Numbers

For Exercises 13–24, locate the numbers on the number line. **(See Example 2.)**

13. $-2, 4$

14. $3, -1$

15. $2, -3$

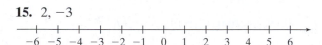

16. $-1, 5$

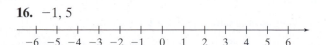

17. $-\frac{7}{2}, \frac{17}{4}$

18. $\frac{4}{3}, -\frac{2}{3}$

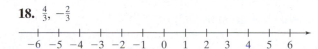

19. $\frac{11}{4}, -\frac{7}{8}$

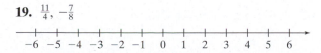

20. $-\frac{1}{4}, \frac{27}{5}$

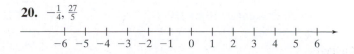

21. 4.1, −0.8

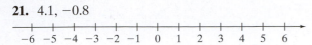

22. 0, −3.1

23. −2.5, 1.6

24. −1.9, 4.2

For Exercises 25–36, identify the number as rational or irrational.

25. $-\dfrac{2}{5}$

26. $-\dfrac{1}{9}$

27. 5

28. 3

29. 3.5

30. 1.1

31. $\sqrt{7}$

32. $\sqrt{11}$

33. π

34. 2π

35. $\sqrt{4}$

36. $\sqrt{9}$

Objective 3: Real Numbers and the Real Number Line

For Exercises 37–52, place the correct symbol, $>$ or $<$, between the two numbers. (See Example 3.)

37. 0 ☐ −3

38. −1 ☐ 0

39. −8 ☐ −9

40. −5 ☐ −2

41. −9.1 ☐ 2.2

42. −1.5 ☐ 1.5

43. −3.35 ☐ −3.3

44. 0.9 ☐ −0.5

45. $-\dfrac{2}{3}$ ☐ $-\dfrac{5}{6}$

46. $-\dfrac{1}{5}$ ☐ $-\dfrac{1}{4}$

47. $\dfrac{7}{8}$ ☐ $-\dfrac{1}{9}$

48. $\dfrac{1}{3}$ ☐ $-\dfrac{3}{2}$

49. 0 ☐ $\dfrac{1}{10}$

50. $-\dfrac{8}{7}$ ☐ 1

51. $-\dfrac{6}{5}$ ☐ −1

52. −1 ☐ $-\dfrac{10}{11}$

Objectives 4: Absolute Value

For Exercises 53–64, determine the absolute value. (See Example 4.)

53. $|-2|$

54. $|-6|$

55. $|4.5|$

56. $|2.9|$

57. $\left|-\dfrac{5}{2}\right|$

58. $\left|-\dfrac{4}{9}\right|$

59. $|0|$

60. $|6|$

61. $|-3.2|$

62. $|-0.4|$

63. $|21|$

64. $|8|$

65. a. Which is greater, −12 or −8?

   b. Which is greater, $|-12|$ or $|-8|$?

66. a. Which is greater, −14 or −20?

   b. Which is greater, $|-14|$ or $|-20|$?

67. a. Which is greater, 5.2 or 7.8?

   b. Which is greater, $|5.2|$ or $|7.8|$?

68. a. Which is greater, 3.89 or 4.29?

   b. Which is greater, $|3.89|$ or $|4.29|$?

69. Which is greater, −5 or $|-5|$?

70. Which is greater, −9 or $|-9|$?

71. Which is greater, 10 or $|10|$?

72. Which is greater, 256, or $|256|$?

Objective 5: Opposite

For Exercises 73–84, find the opposite. **(See Example 5.)**

73. 5   **74.** 31   **75.** −12   **76.** −25

77. $-\dfrac{1}{6}$ **78.** $-\dfrac{4}{7}$ **79.** $\dfrac{2}{11}$ **80.** $\dfrac{14}{15}$

81. 8.1   **82.** 9.5   **83.** −1.14   **84.** −2.25

For Exercises 85–96, write in symbols, do not simplify.

85. The opposite of 6   **86.** The opposite of 23

87. The opposite of negative 2   **88.** The opposite of negative 9

89. The absolute value of 7   **90.** The absolute value of 11

91. The absolute value of negative 3   **92.** The absolute value of negative 10

93. The opposite of the absolute value of 14   **94.** The opposite of the absolute value of 42

95. The opposite of the absolute value of negative 30   **96.** The opposite of the absolute value of negative 5

For Exercises 97–108, simplify the expression. **(See Example 6.)**

97. $-|2|$ **98.** $-|9|$ **99.** $-|-5.3|$ **100.** $-|-6.9|$

101. $-(-15)$ **102.** $-(-4)$ **103.** $|-4.7|$ **104.** $|-9.5|$

105. $-\left|\dfrac{12}{17}\right|$ **106.** $-\left|-\dfrac{1}{7}\right|$ **107.** $-\left(-\dfrac{3}{8}\right)$ **108.** $-\left(-\dfrac{4}{9}\right)$

Addition of Real Numbers

Section 10.2

1. Addition of Integers by Using a Number Line

Addition of real numbers can be visualized on a number line. To do so, we locate the first addend on the number line. Then to add a positive number, we move to the right on the number line. To add a negative number, we move to the left on the number line. This is demonstrated in Example 1.

Objectives

1. Addition of Integers by Using a Number Line
2. Addition of Real Numbers
3. Translating English Phrases to Mathematical Expressions

Example 1   Using a Number Line to Add Integers

Use a number line to add.

a. $5 + 3$ **b.** $-5 + 3$

Solution:

a. $5 + 3 = 8$

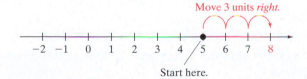

Move 3 units *right.*

Start here.

Begin at 5. Then because we are adding *positive* 3, move to the *right* 3 units. The sum is 8.

Skill Practice

Use a number line to add.

1. 3 + 2
2. −3 + 2

b. −5 + 3 = −2

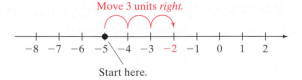

Begin at −5. Then because we are adding *positive* 3, move to the *right* 3 units. The sum is −2.

Skill Practice

Use a number line to add.

3. 3 + (−2)
4. −3 + (−2)

Example 2   **Using a Number Line to Add Integers**

Use a number line to add.

a. 5 + (−3)   **b.** −5 + (−3)

Solution:

a. 5 + (−3) = 2

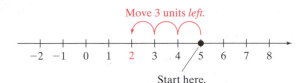

Begin at 5. Then because we are adding *negative* 3, move to the *left* 3 units. The sum is 2.

b. −5 + (−3) = −8

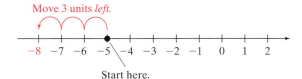

Begin at −5. Then since we are adding *negative* 3, move to the *left* 3 units. The sum is −8.

TIP: In Example 2(a) and 2(b), parentheses are inserted for clarity. The parentheses separate the number −3 from the + symbol for addition.

5 + (−3)   and   −5 + (−3)

2. Addition of Real Numbers

It is inconvenient to draw a number line each time we want to add signed numbers. Therefore, we offer two rules for adding real numbers. The first rule is used when the addends have the *same* sign (that is, if the addends are both positive or both negative).

PROCEDURE **Adding Numbers with the Same Sign**

To add two numbers with the *same* sign, add their absolute values and apply the common sign.

Answers

1. 5   **2.** −1   **3.** 1   **4.** −5

Example 3 Adding Real Numbers with the Same Sign

Add.

a. $-2 + (-4)$ **b.** $-12 + (-37)$ **c.** $10 + 66$

Solution:

a. $-2 + (-4)$ First find the absolute value of the addends.
$|-2| = 2$ and $|-4| = 4$.

$= -(2 + 4)$ Add their absolute values and apply the common sign
(in this case, the common sign is negative).
Common sign is negative.

$= -6$ The sum is -6.

b. $-12 + (-37)$ First find the absolute value of the addends.
$|-12| = 12$ and $|-37| = 37$.

$= -(12 + 37)$ Add their absolute values and apply the common sign
(in this case, the common sign is negative).
Common sign is negative.

$= -49$ The sum is -49.

c. $10 + 66$ First find the absolute value of the addends.
$|10| = 10$ and $|66| = 66$.

$= +(10 + 66)$ Add their absolute values and apply the common sign
(in this case, the common sign is positive).
Common sign is positive.

$= 76$ The sum is 76.

Skill Practice

Add.
5. $-6 + (-8)$
6. $-84 + (-27)$
7. $14 + 31$

The next rule helps us add two numbers with different signs.

PROCEDURE Adding Numbers with Different Signs

To add two numbers with *different* signs, subtract the smaller absolute value
from the larger absolute value. Then apply the sign of the number having the
larger absolute value.

Concept Connections

State the sign of the sum.
8. $-9 + 11$
9. $-9 + 7$

Example 4 Adding Real Numbers with Different Signs

Add.

a. $2 + (-7)$ **b.** $-6 + 24$ **c.** $-8 + 8$

Solution:

a. $2 + (-7)$ First find the absolute value of the addends.
$|2| = 2$ and $|-7| = 7$.

Note: The absolute value of -7 is greater than the absolute
value of 2. Therefore, the sum is negative.

$= -(7 - 2)$ Next, subtract the smaller absolute value from the larger
absolute value.
Apply the sign of the number with the larger absolute value.

$= -5$

Skill Practice

Add.
10. $5 + (-8)$
11. $-12 + 37$
12. $-4 + 4$

Answers

5. -14 **6.** -111 **7.** 45
8. Positive **9.** Negative **10.** -3
11. 25 **12.** 0

b. $-6 + 24$ First find the absolute value of the addends. $|-6| = 6$ and $|24| = 24$.

Note: The absolute value of 24 is greater than the absolute value of -6. Therefore, the sum is positive.

$= +(24 - 6)$ Next, subtract the smaller absolute value from the larger absolute value.

└ Apply the sign of the number with the larger absolute value.

$= 18$

c. $-8 + 8$ First find the absolute value of the addends. $|-8| = 8$ and $|8| = 8$.

$= (8 - 8)$ The absolute values are equal. Therefore, their difference is 0. The number zero is neither positive nor negative.

$= 0$

Example 4(c) illustrates that the sum of a number and its opposite is zero. For example,

$$-8 + 8 = 0 \qquad -12 + 12 = 0 \qquad 6 + (-6) = 0$$

Example 5 **Applying the Order of Operations**

Simplify.

a. $-10 + 4 + (-16)$ **b.** $-30 + (-12) + 4 + (-10) + 6$

Solution:

a. $-10 + 4 + (-16)$

$= -6 + (-16)$ Apply the order of operations by

$= -22$ adding from left to right.

b. $-30 + (-12) + 4 + (-10) + 6$ Add from left to right.

$= -42 + 4 + (-10) + 6$

$= -38 + (-10) + 6$

$= -48 + 6$

$= -42$

TIP: When several numbers are added, we can reorder and regroup the addends by using the commutative property and associative property of addition. In particular, we can group all the positive addends, and we can group all the negative addends. This makes the arithmetic easier. For example:

$$-30 + (-12) + 4 + (-10) + 6 = \overbrace{4 + 6}^{\substack{\text{positive} \\ \text{addends}}} + \overbrace{(-30) + (-12) + (-10)}^{\substack{\text{negative} \\ \text{addends}}}$$

$$= 10 + (-52)$$

$$= -42$$

Example 6   **Adding Real Numbers**

Add. $-2.73 + 4.81$

Solution:

$-2.73 + 4.81$ Find the absolute value of the addends.
$|-2.73| = 2.73$ and $|4.81| = 4.81$.

The sum is positive because 4.81 has a greater absolute value than -2.73.

$= +(4.81 - 2.73)$ Subtract the smaller absolute value from the larger absolute value.

    Apply the sign of the number with the larger absolute value.

$= 2.08$

Skill Practice

Add.

15. $-16.8 + 14.3$

Example 7   **Adding Real Numbers**

Add.   **a.** $-\dfrac{6}{11} + \left(-\dfrac{3}{11}\right)$ **b.** $\dfrac{2}{15} + \left(-\dfrac{4}{5}\right)$

Solution:

a. $-\dfrac{6}{11} + \left(-\dfrac{3}{11}\right)$ Find the absolute value of the addends.

$\left|-\dfrac{6}{11}\right| = \dfrac{6}{11}$ and $\left|-\dfrac{3}{11}\right| = \dfrac{3}{11}$.

$= -\left(\dfrac{6}{11} + \dfrac{3}{11}\right)$ Add their absolute values and apply the common sign (in this case, the common sign is negative).

Common sign is negative.

$= -\dfrac{9}{11}$

b. $\dfrac{2}{15} + \left(-\dfrac{4}{5}\right)$ The least common denominator is 15.

$= \dfrac{2}{15} + \left(-\dfrac{4 \cdot 3}{5 \cdot 3}\right)$ Write each fraction with the LCD.

$= \dfrac{2}{15} + \left(-\dfrac{12}{15}\right)$ Find the absolute value of the addends.

$\left|\dfrac{2}{15}\right| = \dfrac{2}{15}$ and $\left|-\dfrac{12}{15}\right| = \dfrac{12}{15}$.

The absolute value of $-\dfrac{12}{15}$ is greater than the absolute value of $\dfrac{2}{15}$. Therefore, the sum is negative.

$= -\left(\dfrac{12}{15} - \dfrac{2}{15}\right)$ Next, subtract the smaller absolute value from the larger absolute value.

    Apply the sign of the number with the larger absolute value.

$= -\dfrac{10}{15}$ Subtract.

$= -\dfrac{2}{3}$ Simplify to lowest terms. $-\dfrac{\overset{2}{\cancel{10}}}{\underset{3}{\cancel{15}}} = -\dfrac{2}{3}$

Skill Practice

Add.

16. $-\dfrac{6}{8} + \dfrac{5}{8}$

17. $\dfrac{3}{10} + \left(-\dfrac{3}{5}\right)$

Avoiding Mistakes
Do not forget to get a common denominator before adding or subtracting fractions.

Answers

15. -2.5 **16.** $-\dfrac{1}{8}$ **17.** $-\dfrac{3}{10}$

3. Translating English Phrases to Mathematical Expressions

Recall from Section 1.2 that several key words imply addition: *sum; added to; increased by; more than; plus;* and *total of.*

Example 8 Translating to a Mathematical Expression

Translate to a mathematical expression. Then simplify the expression.

$$\text{The sum of } -14.1, 8.7, \text{ and } 12.9$$

Solution:

$$-14.1 + 8.7 + 12.9 \qquad \text{Translate the English phrase.}$$

$$= -5.4 + 12.9 \qquad \text{Simplify. Add from left to right.}$$

$$= \qquad 7.5$$

Example 9 Translating to a Mathematical Expression

Translate to a mathematical expression. Then simplify the expression.

$$\text{The number } -\frac{1}{6} \text{ added to } \frac{2}{3}$$

Solution:

$$\frac{2}{3} + \left(-\frac{1}{6}\right) \qquad \text{Translate the English phrase.}$$

$$= \frac{2 \cdot 2}{3 \cdot 2} + \left(-\frac{1}{6}\right) \qquad \text{Write the first fraction with a least common denominator of 6.}$$

$$= \frac{4}{6} + \left(-\frac{1}{6}\right) \qquad \text{Find the absolute value of the addends.}$$

$$\left|\frac{4}{6}\right| = \frac{4}{6} \text{ and } \left|-\frac{1}{6}\right| = \frac{1}{6}.$$

The absolute value of $\frac{4}{6}$ is greater than the absolute value of $-\frac{1}{6}$. Therefore, the sum is positive.

$$= +\left(\frac{4}{6} - \frac{1}{6}\right) \qquad \text{Subtract the smaller absolute value from the larger absolute value.}$$

Apply the sign of the number with the larger absolute value.

$$= \frac{3}{6} \qquad \text{Subtract.}$$

$$= \frac{1}{2} \qquad \text{Simplify to lowest terms.}$$

Section 10.2   Practice Exercises

Boost *your* GRADE at ALEKS.com!   ALEKS® version 3.0
- Practice Problems
- Self-Tests
- NetTutor
- e-Professors
- Videos

Study Skills Exercise

1. When you are trying to memorize rules for signed numbers, 3×5 cards come in handy. Write the rule you want to memorize on one side and an example of the rule on the other side. Keep these cards handy so that when you have a few minutes (such as waiting for the doctor), you can pull them out and quiz yourself. Write the rules from this section that would be good to put on cards.

Review Exercises

For Exercises 2–8, place the correct symbol ($>$, $<$, or $=$) between the two numbers.

2. $-6 \ \square \ -5$

3. $-\dfrac{2}{3} \ \square \ -\dfrac{11}{12}$

4. $|-2.4| \ \square \ -|2.4|$

5. $|6| \ \square \ |-6|$

6. $0 \ \square \ -0.6$

7. $-|-10| \ \square \ 10$

8. $-(-2) \ \square \ 2$

Objective 1: Addition of Integers by Using a Number Line

For Exercises 9–20, refer to the number line to add the integers. **(See Examples 1–2.)**

9. $2 + (-4)$

10. $5 + (-1)$

11. $-3 + 5$

12. $-6 + 3$

13. $-4 + (-4)$

14. $-2 + (-5)$

15. $-3 + 9$

16. $-1 + 5$

17. $0 + (-7)$

18. $(-5) + 0$

19. $-1 + (-3)$

20. $-4 + 3$

Objective 2: Addition of Real Numbers

21. Explain the process to add two numbers with the same sign.

For Exercises 22–29, add the numbers with the same sign. **(See Example 3.)**

22. $23 + 12$

23. $12 + 3$

24. $-70 + (-15)$

25. $-40 + (-33)$

26. $-6 + (-10)$

27. $-100 + (-24)$

28. $23 + 50$

29. $44 + 45$

30. Explain the process to add two numbers with different signs.

For Exercises 31–42, add the numbers with different signs. **(See Example 4.)**

31. $75 + (-23)$

32. $26 + (-14)$

33. $-34 + 12$

34. $-88 + 35$

35. $-90 + 66$

36. $-23 + 49$

37. $78 + (-33)$

38. $10 + (-23)$

39. $2 + (-2)$

40. $-6 + 6$

41. $-1.3 + 1.3$

42. $4.5 + (-4.5)$

Mixed Exercises

For Exercises 43–66, simplify. **(See Example 5.)**

43. $12 + (-3)$

44. $-33 + (-1)$

45. $-23 + (-3)$

46. $-5 + 15$

47. $4 + (-45)$

48. $-13 + (-12)$

49. $(-103) + (-47)$

50. $119 + (-59)$

51. $0 + (-17)$

52. $-29 + 0$

53. $-19 + (-22)$

54. $-300 + (-24)$

55. $6 + (-12) + 8$

56. $20 + (-12) + (-5)$

57. $-33 + (-15) + 18$

58. $3 + 5 + (-1)$

59. $7 + (-3) + 6$

60. $12 + (-6) + (-9)$

61. $-10 + (-3) + 5$

62. $-23 + (-4) + (-12) + (-5)$

63. $-18 + (-5) + 23$

64. $14 + (-15) + 20 + (-42)$

65. $4 + (-12) + (-30) + 16 + 10$

66. $24 + (-5) + (-19)$

For Exercises 67–84, add the real numbers. **(See Examples 6–7.)**

67. $23.9 + 2.1$

68. $10.9 + 6.3$

69. $-34.2 + (-4.1)$

70. $-8.6 + (-12)$

71. $-\dfrac{3}{4} + \left(-\dfrac{5}{4}\right)$

72. $-\dfrac{2}{5} + \left(-\dfrac{1}{10}\right)$

73. $-\dfrac{7}{8} + \left(-\dfrac{1}{4}\right)$

74. $-\dfrac{1}{2} + \left(-\dfrac{7}{12}\right)$

75. $34.8 + (-45)$

76. $90 + (-12.3)$

77. $-23.1 + 24.5$

78. $-12.2 + 10.9$

79. $\dfrac{3}{8} + \left(-\dfrac{3}{16}\right)$

80. $\dfrac{1}{3} + \left(-\dfrac{7}{9}\right)$

81. $\left(-\dfrac{5}{6}\right) + \dfrac{1}{4}$

82. $\left(-\dfrac{4}{5}\right) + \dfrac{7}{20}$

83. $-\dfrac{1}{7} + \left(-\dfrac{2}{5}\right)$

84. $\left(-\dfrac{3}{8}\right) + \left(-\dfrac{2}{9}\right)$

Objective 3: Translating English Phrases to Mathematical Expressions

85. Give at least two words or phrases that would indicate addition.

For Exercises 86–95, translate to a mathematical expression. Then simplify the expression. **(See Examples 8–9.)**

86. The sum of -23 and 49

87. The sum of 89 and -11

88. The total of 3, -10, and 5

89. The total of -2, -4, 14, and 20

90. The number -4.2 is added to -2.2

91. The number -4.5 is added to -12

92. 8 more than $-\dfrac{1}{4}$

93. 2 more than $-\dfrac{1}{3}$

94. $-\dfrac{3}{4}$ increased by 6

95. $-\dfrac{1}{5}$ increased by 1

96. At 6:00 A.M. the temperature was −4°F. By noon, the temperature had risen by 12°F. What was the temperature at noon?

97. At noon the temperature was 14°F. By midnight, the temperature had fallen 20°F. What was the temperature at midnight?

98. Jorge's checking account is overdrawn. His beginning balance was −$56.52. If he deposits his paycheck for $389.81, what is his new balance?

99. Ellen's checking account balance is $23.89. If she writes a check for $40.00, what is her balance?

100. Ron bought a new pair of glasses. A bill for $320.50 was sent to his insurance company. If his insurance paid only $150, what is Ron's balance?

101. Savannah was in a minor accident with her car. To repair the damaged car, she paid $570.32. Her insurance company paid for the repair except for a $250 deductible. How much did the insurance company pay?

Expanding Your Skills

102. Find two integers whose sum is −10. Answers may vary.

103. Find two integers whose sum is −14. Answers may vary.

104. Find two integers whose sum is −2. Answers may vary.

105. Find two integers whose sum is 0. Answers may vary.

Calculator Connections

Topic: Adding Real Numbers on a Calculator

To enter negative numbers on a calculator, use the $(-)$ key or the $+\circ-$ key. To use the $(-)$ key, enter the number the same way that it is written. That is, enter the negative sign first and then the number, such as $(-)$ 5. If your calculator has the $+\circ-$ key, type the number first, followed by the $+\circ-$ key. Thus, −5 is entered as 5 $+\circ-$.

Try entering the expressions below to determine which method your calculator uses.

Expression	Keystrokes	Result
$-10 + (-3)$	$(-)$ 10 $+$ $(-)$ 3 **ENTER** or 10 $+\circ-$ $+$ 3 $+\circ-$ $=$	−13
$-4.2 + 6.7$	$(-)$ 4.2 $+$ 6.7 **ENTER** or 4.2 $+\circ-$ $+$ 6.7 $=$	2.5

Calculator Exercises

For Exercises 106–111, add by using a calculator.

106. $302 + (-422)$

107. $-900 + 334$

108. $-23.991 + (-44.23)$

109. $-103.4 + (-229.1)$

110. $23 + (-125) + 912 + (-99)$

111. $891 + 12 + (-223) + (-341)$

| Section 10.3 | **Subtraction of Real Numbers** |

Objectives

1. **Definition of Subtraction**
2. **Subtraction of Rational Numbers**
3. **Applying the Order of Operations**
4. **Applications of Subtraction**

1. Definition of Subtraction

In Section 10.2, we learned the rules for adding real numbers. Subtraction of real numbers is defined in terms of the addition process. For example, consider the following subtraction problem. The corresponding addition problem produces the same result.

$$6 - 4 = 2 \quad \Leftrightarrow \quad 6 + (-4) = 2$$

In each case, we start at 6 on the number line and move to the *left* 4 units. Adding the *opposite* of 4 produces the same result as subtracting 4. This is true in general. To subtract two real numbers, add the opposite of the second number to the first number.

Concept Connections

Fill in the blank to change subtraction to addition of the opposite.

1. $9 - 3 = 9 + \square$
2. $-9 - 3 = -9 + \square$
3. $9 - (-3) = 9 + \square$
4. $-9 - (-3) = -9 + \square$

PROCEDURE Subtracting Real Numbers

If a and b are real numbers, then $a - b = a + (-b)$.
Therefore, to perform subtraction, follow these steps:

Step 1 Leave the first number (the minuend) unchanged.
Step 2 Change the subtraction sign to an addition sign.
Step 3 Add the opposite of the second number (the subtrahend).

For example,

$$10 - 4 = 10 + (-4) = 6$$
$$-10 - 4 = -10 + (-4) = -14$$

Subtracting 4 is the same as adding -4.

$$10 - (-4) = 10 + (4) = 14$$
$$-10 - (-4) = -10 + (4) = -6$$

Subtracting -4 is the same as adding 4.

Skill Practice

Subtract.
5. $12 - 19$
6. $-8 - 14$
7. $30 - (-3)$

Example 1 Subtracting Real Numbers

Subtract.

a. $15 - 20$ **b.** $-7 - 12$ **c.** $40 - (-8)$

Solution:

Add the opposite of 20.

a. $15 - 20 = 15 + (-20) = -5$ Rewrite the subtraction in terms of addition of the opposite.

Change subtraction to addition.

b. $-7 - 12 = -7 + (-12)$ Rewrite as addition of the opposite.
$$= -19$$

c. $40 - (-8) = 40 + (8)$ Rewrite as addition of the opposite.
$$= 48$$

Answers
1. -3 2. -3 3. 3 4. 3
5. -7 6. -22 7. 33

Recall from Section 1.3 that several key words imply subtraction.

Word or Phrase	Example	In Symbols
a minus b	−15 minus 10	$-15 - 10$
the difference of a and b	the difference of 10 and −2	$10 - (-2)$
a decreased by b	9 decreased by 1	$9 - 1$
a less than b	−12 less than 5	$5 - (-12)$
subtract a from b	subtract −3 from 8	$8 - (-3)$

Example 2 **Translating to a Mathematical Expression**

Translate each English phrase to a mathematical expression. Then simplify.

a. The difference of −52 and 10

b. −35 decreased by −6

Solution:

a. the difference of

$-52 - 10$ Translate: The difference of −52 and 10

$= -52 + (-10)$ Rewrite as addition of the opposite.

$= -62$

b. decreased by

$-35 - (-6)$ Translate: −35 decreased by −6.

$= -35 + (6)$ Rewrite subtraction in terms of addition.

$= -29$

Example 3 **Translating to a Mathematical Expression**

Translate each English phrase to a mathematical expression. Then simplify.

a. 12 less than −8

b. Subtract 27 from 5.

Solution:

a. To translate "12 less than −8," we must *start* with −8 and then subtract 12.

$-8 - 12$ Translate: 12 less than −8.

$= -8 + (-12)$ Rewrite as addition of the opposite.

$= -20$

b. To translate "Subtract 27 from 5," we must *start* with 5, then subtract 27.

$5 - 27$ Translate: Subtract 27 from 5.

$= 5 + (-27)$ Rewrite as addition of the opposite.

$= -22$

2. Subtraction of Rational Numbers

Example 4   Subtracting Rational Numbers

Subtract.

a. $-6.7 - 4.2$ **b.** $\dfrac{9}{4} - \dfrac{7}{4}$ **c.** $\dfrac{3}{10} - \left(-\dfrac{7}{30}\right)$

Solution:

a. $-6.7 - 4.2$

$= -6.7 + (-4.2)$ Rewrite as addition of the opposite.

$= -10.9$

b. $\dfrac{9}{4} - \dfrac{7}{4}$

$= \dfrac{9}{4} + \left(-\dfrac{7}{4}\right)$ Rewrite as addition of the opposite.

$= \dfrac{2}{4}$ Add.

$= \dfrac{1}{2}$ Simplify.

c. $\dfrac{3}{10} - \left(-\dfrac{7}{30}\right)$

$= \dfrac{3}{10} + \left(\dfrac{7}{30}\right)$ Rewrite as addition of the opposite.

$= \dfrac{3 \cdot 3}{10 \cdot 3} + \left(\dfrac{7}{30}\right)$ The least common denominator is 30.

$= \dfrac{9}{30} + \dfrac{7}{30}$

$= \dfrac{16}{30}$ Add.

$= \dfrac{8}{15}$ Simplify to lowest terms. $\dfrac{\overset{8}{\cancel{16}}}{\underset{15}{\cancel{30}}} = \dfrac{8}{15}$

3. Applying the Order of Operations

In Example 5, we revisit the order of operations.

Example 5 Applying the Order of Operations

Simplify.   **a.** $-4 - 6 + (-3) - 5 + 8$ **b.** $[3 - (-4)]^2$

Solution:

a. $-4 - 6 + (-3) - 5 + 8$

$= -4 + (-6) + (-3) + (-5) + 8$ Rewrite all subtractions in terms of addition.

$= -10 + (-3) + (-5) + 8$ Add from left to right.

$= -13 + (-5) + 8$

$= -18 + 8$

$= -10$

b. $[3 - (-4)]^2$ Perform operations inside grouping symbols.

$= [3 + (4)]^2$ Rewrite as addition of the opposite.

$= [7]^2$ Add.

$= 49$ Simplify.

Skill Practice

Simplify.

15. $-8 - 10 + (-6) - (-1)$

16. $(-4 - 2)^2$

4. Applications of Subtraction

Example 6 Applying Subtraction of Real Numbers

A helicopter is hovering at a height of 200 ft above the ocean. A submarine is directly below the helicopter 125 ft below sea level. Find the difference in elevation between the helicopter and the submarine.

200 ft

-125 ft

Skill Practice

17. The highest point in California is Mt. Whitney at 14,494 ft above sea level. The lowest point in California is Death Valley at an "altitude" of -282 ft (282 ft below sea level). Find the difference in the elevation between the highest point and lowest point in California.

Solution:

$$\begin{pmatrix} \text{Difference between} \\ \text{elevation of helicopter} \\ \text{and submarine} \end{pmatrix} = \begin{pmatrix} \text{Elevation of} \\ \text{helicopter} \end{pmatrix} - \begin{pmatrix} \text{"Elevation" of} \\ \text{submarine} \end{pmatrix}$$

$= 200 \text{ ft} - (-125 \text{ ft})$

$= 200 \text{ ft} + (125 \text{ ft})$ Rewrite as addition.

$= 325 \text{ ft}$

The helicopter and submarine are 325 ft apart.

Answers

15. -23 **16.** 36   **17.** 14,776 ft

Section 10.3   Practice Exercises

Boost your GRADE at ALEKS.com!      • Practice Problems   • e-Professors
• Self-Tests   • Videos
• NetTutor

Study Skills Exercise

1. In this section we find that the symbol − has more than one meaning. It can mean minus, opposite, or negative. Write yourself an explanation and example for each of these meanings of −.

Minus: Opposite: Negative:

Review Exercises

For Exercises 2–9, add the numbers.

2. $34 + (-13)$

3. $-34 + (-13)$

4. $-34 + 13$

5. $-\dfrac{5}{9} + \dfrac{7}{12}$

6. $\dfrac{5}{9} + \left(-\dfrac{7}{12}\right)$

7. $-\dfrac{5}{9} + \left(-\dfrac{7}{12}\right)$

8. $3 + (-4.2) + \left(-\dfrac{2}{5}\right) + 5$

9. $\left(-\dfrac{1}{2}\right) + 6.5 + (-8) + 2 + (-4)$

Objective 1: Definition of Subtraction

10. Explain the process to subtract signed numbers.

For Exercises 11–18, rewrite the subtraction problem as an equivalent addition problem. Then simplify.
(See Example 1.)

11. $2 - 9 = $ ____ + ____ = ____

12. $5 - 11 = $ ____ + ____ = ____

13. $4 - (-3) = $ ____ + ____ = ____

14. $12 - (-8) = $ ____ + ____ = ____

15. $-3 - 15 = $ ____ + ____ = ____

16. $-7 - 21 = $ ____ + ____ = ____

17. $-11 - (-13) = $ ____ + ____ = ____

18. $-23 - (-9) = $ ____ + ____ = ____

For Exercises 19–42, subtract the integers.

19. $35 - (-17)$

20. $23 - (-12)$

21. $-24 - 9$

22. $-5 - 15$

23. $50 - 62$

24. $38 - 46$

25. $-17 - (-25)$

26. $-2 - (-66)$

27. $-8 - (-8)$

28. $-14 - (-14)$

29. $120 - (-41)$

30. $91 - (-62)$

31. $-15 - 19$

32. $-82 - 44$

33. $3 - 25$

34. $6 - 33$

35. $-13 - 13$

36. $-43 - 43$

37. $24 - 25$

38. $43 - 98$

39. $-6 - (-38)$

40. $-75 - (-21)$

41. $-48 - (-33)$

42. $-29 - (-32)$

43. State at least two words or phrases that would indicate subtraction.

44. Is subtraction commutative? For example, does $3 - 7 = 7 - 3$?

For Exercises 45–56, translate each English phrase to a mathematical expression. Then simplify. **(See Examples 2–3.)**

45. 14 minus 23

46. 27 minus 40

 **47.** Subtract 12 from 5.

48. Subtract 10 from 16.

49. The difference of 105 and 110

50. The difference of 70 and 98

51. 320 decreased by −20

52. 150 decreased by 75

53. Subtract 24 from −35.

54. Subtract 189 from 175.

55. 21 less than −34

56. 22 less than −90

Objective 2: Subtraction of Rational Numbers

For Exercises 57–72, subtract the rational numbers. **(See Example 4.)**

57. $5.2 - 13.5$

58. $4.4 - 10.2$

59. $-2.3 - 1.9$

60. $-4.1 - 2.1$

61. $-3.6 - (-9.1)$

62. $-8.9 - (-10.5)$

63. $5.5 - (-2.8)$

64. $11.9 - (-4.3)$

65. $\dfrac{2}{3} - \left(-\dfrac{1}{6}\right)$

66. $\dfrac{5}{9} - \left(-\dfrac{2}{9}\right)$

67. $-\dfrac{3}{10} - \left(-\dfrac{7}{10}\right)$

68. $-\dfrac{5}{8} - \left(-\dfrac{3}{4}\right)$

69. $\dfrac{3}{14} - \dfrac{12}{7}$

70. $\dfrac{2}{5} - \dfrac{9}{10}$

71. $-\dfrac{1}{2} - \dfrac{5}{4}$

72. $-\dfrac{11}{15} - \dfrac{3}{5}$

Objective 3: Applying the Order of Operations

For Exercises 73–84, simplify by using the order of operations. **(See Example 5.)**

73. $2 + 5 - (-3) - 10$

74. $4 - 8 + 12 - (-1)$

 **75.** $-5 + 6 + (-7) - 4 - (-9)$

76. $-2 - 1 + (-11) + 6 - (-8)$

77. $[-2 - (-6)]^2$

78. $[-1 - (-4)]^2$

79. $[-5 - (-6)]^3$

80. $[-3 - (-5)]^3$

81. $25 - 13 - (-40)$

82. $-35 + 15 - (-28)$

83. $5.5 - \left(\dfrac{1}{2} - \dfrac{1}{5}\right)$

84. $-6.8 - \left(\dfrac{2}{5} + \dfrac{3}{10}\right)$

Objective 4: Applications of Subtraction

85. The liquid hydrogen in the Space Shuttle main engine is −423°F. The temperature in the engine's combustion chamber reaches 6000°F. Find the difference between the temperature in the combustion chamber and the temperature of the liquid hydrogen. **(See Example 6.)**

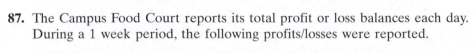 **86.** The moon does not have an atmosphere, so temperatures range from −184°C during its night to 214°C during its day. Find the difference between the highest temperature on the moon and the lowest temperature.

87. The Campus Food Court reports its total profit or loss balances each day. During a 1 week period, the following profits/losses were reported.

If the Campus Food Court's profit/loss balance was $17,476.55 at the beginning of the week, what is the balance at the end of the reported week?

Monday	+$1,786.84
Tuesday	−$2,342.47
Wednesday	−$754.32
Thursday	+$321.63
Friday	+$1597.28

88. For nine holes of golf, Tiger Woods made the following scores: −1, 0, 0, −1, 0, 1, 0, −2, 0. A negative number represents a score *below* par and a positive number represents a score *above* par. What is his total score after nine holes?

For Exercises 89–90, refer to the graph indicating the change in value of a particular stock for one week.

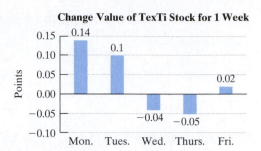

Change Value of TexTi Stock for 1 Week

89. What is the difference between the change in value of TexTi stock on Monday compared to its change in value on Wednesday?

90. What is the difference between the change in value of TexTi stock on Tuesday and its change in value on Thursday?

91. Ivan owes $320 on his credit card; that is, his balance is −$320. If he charges $55 for a night out, what is his new balance?

92. If Justin's balance on his credit card was −$210 and he made the minimum payment of $25, what is his new balance?

93. The height of Mount Everest is 29,029 ft. The lowest point on the surface of the Earth is −35,798 ft (that is, 35,798 ft below sea level), occurring at the Mariana Trench on the Pacific Ocean floor. What is the difference in altitude between the height of Mt. Everest and the Mariana Trench?

94. Mt. Rainier is 4392 m at its highest point. Death Valley, California, is 86 m below sea level (−86) at the basin, Badwater. What is the difference between the altitude of Mt. Rainier and the altitude at Badwater?

In a statistics class, a student learns that the **range** of a set of data is the difference between the highest and lowest values. That is, range = highest − lowest.

For Exercises 95–96, find the range.

95. Low temperatures (°C) for one week in Anchorage, Alaska: −4°, −8°, 0°, 3°, −8°, −1°, 2°

96. Low temperatures (°C) for one week in Fargo, North Dakota: −6°, −2°, −10°, −4°, −12°, −1°, −3°

97. Find two integers whose difference is −6. Answers may vary.

98. Find two integers whose difference is −20. Answers may vary.

Expanding Your Skills

For Exercises 99–102, write the next three numbers in the sequence.

99. 5, 1, −3, −7, ___, ___, ___

100. −13, −18, −23, −28, ___, ___, ___

101. $\frac{1}{3}$, 0, $-\frac{1}{3}$, $-\frac{2}{3}$, ___, ___, ___

102. $\frac{1}{4}$, 0, $-\frac{1}{4}$, $-\frac{1}{2}$, ___, ___, ___

For Exercises 103–110, assume $a > 0$ (this means that a is positive) and $b < 0$ (this means that b is negative). Find the sign of each expression.

103. $a - b$

104. $b - a$

105. $|a| + |b|$

106. $|a + b|$

107. $-|a|$

108. $-|b|$

109. $-(a)$

110. $-(b)$

Problem Recognition Exercises

Addition and Subtraction of Real Numbers

For Exercises 1–40, perform the indicated operations.

1. $-7 - 5$ **2.** $-7 - (-5)$ **3.** $-7 + (-5)$ **4.** $-7 + 5$

5. $10 - (-45)$ **6.** $10 - 45$ **7.** $10 + (-45)$ **8.** $10 + 45$

9. $-31.2 - (-52.6)$ **10.** $-31.2 + (-52.6)$ **11.** $-31.2 - 52.6$ **12.** $-31.2 + (52.6)$

13. $-19.5 + (21.5)$ **14.** $-19.5 - 21.5$ **15.** $-19.5 + (-21.5)$ **16.** $-19.5 - (-21.5)$

17. $|-12 + 8|$ **18.** $|12 - 8|$ **19.** $|-12 - 8|$ **20.** $|12 - (-8)|$

21. $\frac{1}{8} - \frac{5}{4}$ **22.** $-\frac{1}{8} - \frac{5}{4}$ **23.** $\frac{1}{8} + \left(-\frac{5}{4}\right)$ **24.** $-\frac{1}{8} - \left(-\frac{5}{4}\right)$

25. $-\frac{7}{9} - \frac{1}{6}$ **26.** $-\frac{7}{9} + \frac{1}{6}$ **27.** $-\frac{7}{9} - \left(-\frac{1}{6}\right)$ **28.** $\frac{7}{9} - \frac{1}{6}$

29. $2\frac{1}{4} - 5\frac{1}{2}$ **30.** $4\frac{1}{3} + \left(-\frac{5}{6}\right)$ **31.** $-1\frac{2}{5} - 3\frac{1}{10}$ **32.** $2\frac{5}{6} - \left(-1\frac{1}{6}\right)$

33. $-\frac{3}{4} + 3$ **34.** $-\frac{4}{5} - 1$ **35.** $-2 + 0.001$ **36.** $4 - 5.987$

37. $-56 + 56$ **38.** $14 + (-14)$ **39.** $-56 - 56$ **40.** $14 - (-14)$

Section 10.4 | **Multiplication and Division of Real Numbers**

Objectives

1. **Multiplication of Real Numbers**
2. **Multiplying Many Factors**
3. **Exponential Expressions**
4. **Division of Real Numbers**

Concept Connections

1. Write $4(-5)$ as repeated addition.

1. Multiplication of Real Numbers

We know from our knowledge of arithmetic that the product of two positive numbers is a positive number. This can be shown by using repeated addition.

For example: $3(4) = 4 + 4 + 4 = 12$

Now consider a product of numbers with different signs.

For example: $3(-4) = -4 + (-4) + (-4) = -12$ (3 times -4)

These examples suggest that the product of a positive number and a negative number is *negative*.

Now what if we have a product of two negative numbers? To determine the sign, consider the following pattern of products.

$$3 \times -4 = -12$$
$$2 \times -4 = -8$$
$$1 \times -4 = -4$$
$$0 \times -4 = 0$$
$$-1 \times -4 = 4$$
$$-2 \times -4 = 8$$
$$-3 \times -4 = 12$$

The pattern increases by 4 with each row.

The product of two negative numbers is *positive*.

From the first four rows, we see that the product increases by 4 for each row. For the pattern to continue, it follows that the product of two negative numbers must be *positive*.

Avoiding Mistakes

Do not confuse the rule for multiplying two negative numbers with the rule for adding two negative numbers.

- The product of two negative numbers is positive.
- The sum of two negative numbers is negative.

PROCEDURE Multiplying Real Numbers

- The product of two real numbers with the *same* sign is positive.

  Examples: $(5)(6) = 30$

  $(-5)(-6) = 30$

- The product of two real numbers with *different* signs is negative.

  Examples: $4(-10) = -40$

  $-4(10) = -40$

- The product of any real number and zero is zero.

  Examples: $3(0) = 0$

  $0(5) = 0$

Answer

1. $-5 + (-5) + (-5) + (-5)$

Example 1 **Multiplying Real Numbers**

Multiply.   **a.** $-8(-7)$ **b.** $-5 \cdot 10$ **c.** $(18)(-2)$

Solution:

a. $-8(-7) = 56$ Same signs. Product is positive.

b. $-5 \cdot 10 = -50$ Different signs. Product is negative.

c. $(18)(-2) = -36$ Different signs. Product is negative.

Example 2 **Multiplying Real Numbers**

Multiply.   **a.** $(-3.1)(-4.6)$ **b.** $-\frac{4}{7} \cdot \left(5\frac{5}{6}\right)$ **c.** $-12 \cdot \left(-\frac{7}{3}\right)$ **d.** $-\frac{3}{4} \cdot 0$

Solution:

a. $(-3.1)(-4.6) = 14.26$ Same signs. Product is positive.

b. $-\frac{4}{7} \cdot \left(5\frac{5}{6}\right) = -\frac{4}{7} \cdot \frac{35}{6}$ Convert the mixed number to an improper fraction.

$$= -\frac{\overset{2}{\cancel{4}}}{\underset{1}{\cancel{7}}} \cdot \frac{\overset{5}{\cancel{35}}}{\underset{3}{\cancel{6}}}$$ Multiply fractions.

$$= -\frac{10}{3} \text{ or } -3\frac{1}{3}$$ Different signs. Product is negative.

c. $-12 \cdot \left(-\frac{7}{3}\right) = -\frac{12}{1} \cdot \left(-\frac{7}{3}\right)$ Write the whole number as a fraction.

$$= -\frac{\overset{4}{\cancel{12}}}{1} \cdot \left(-\frac{7}{\underset{1}{\cancel{3}}}\right)$$

$$= \frac{28}{1}$$ Same signs. Product is positive.

$$= 28$$

d. $-\frac{3}{4} \cdot 0 = 0$ The product of any number and zero is zero.

Recall that the terms *product*, *multiply*, and *times* imply multiplication.

Example 3 **Translating to an Algebraic Expression**

Translate each English phrase to an algebraic expression. Then simplify.

   a. The product of 7 and -5 **b.** -3 times -11

Solution:

a. $7(-5)$ Translate: The product of 7 and -5.

$= -35$ Different signs. Product is negative.

b. $(-3)(-11)$ Translate: -3 times -11.

$= 33$ Same signs. Product is positive.

2. Multiplying Many Factors

In each of the following products, we can apply the order of operations and multiply from left to right.

two negative factors

$(-2)(-2)$

$= 4$

Product is positive.

three negative factors

$(-2)(-2)(-2)$

$= 4(-2)$

$= -8$

Product is negative.

four negative factors

$(-2)(-2)(-2)(-2)$

$= 4(-2)(-2)$

$= (-8)(-2)$

$= 16$

Product is positive.

five negative factors

$(-2)(-2)(-2)(-2)(-2)$

$= 4(-2)(-2)(-2)$

$= (-8)(-2)(-2)$

$= 16(-2)$

$= -32$

Product is negative.

These products indicate the following rules.

- The product of an *even* number of negative factors is *positive*.

- The product of an *odd* number of negative factors is *negative*.

Example 4 Multiplying Several Factors

Multiply.

a. $(-2)(-5)(-7)$ **b.** $(-4)(2)(-1)(5)$

Solution:

a. $(-2)(-5)(-7)$ This product has an odd number of negative factors.

$= -70$ The product is negative.

b. $(-4)(2)(-1)(5)$ This product has an even number of negative factors.

$= 40$ The product is positive.

3. Exponential Expressions

Be particularly careful when evaluating exponential expressions involving negative numbers. An exponential expression with a negative base is written with parentheses around the base, such as $(-3)^4$.

To evaluate $(-3)^4$, the base -3 is multiplied 4 times:

$$(-3)^4 = (-3)(-3)(-3)(-3) = 81$$

If parentheses are *not* used, the expression -3^4 has a different meaning:

- The expression -3^4 has a base of 3 (not -3) and can be interpreted as $-1 \cdot 3^4$. Hence,

$$-3^4 = -1 \cdot (3)(3)(3)(3) = -81$$

- The expression -3^4 can also be interpreted as "the opposite of 3^4." Hence,

$$-3^4 = -(3 \cdot 3 \cdot 3 \cdot 3) = -81$$

Example 5 Simplifying Exponential Expressions

Simplify.

   a. $(-4)^2$ **b.** -4^2 **c.** $-(-4)^2$ **d.** $\left(-\dfrac{1}{2}\right)^4$

Solution:

  a. $(-4)^2 = (-4)(-4)$ The base is -4.

         $= 16$ Multiply.

  b. $-4^2 = -(4)(4)$ The base is 4. This is equal to

         $= -16$ $-1 \cdot 4^2 = -1 \cdot (4)(4)$.

  c. $-(-4)^2 = -(-4)(-4)$ The base is -4. This is equal to

          $= -16$ $-1 \cdot (-4)^2 = -1 \cdot (-4)(-4)$.

  d. $\left(-\dfrac{1}{2}\right)^4 = \left(-\dfrac{1}{2}\right)\left(-\dfrac{1}{2}\right)\left(-\dfrac{1}{2}\right)\left(-\dfrac{1}{2}\right)$ The base is $\left(-\dfrac{1}{2}\right)$.

            $= \dfrac{1}{16}$ Multiply.

Example 6 Simplifying Exponential Expressions

Simplify.

   a. $(-5)^3$ **b.** -5^3 **c.** $-(-5)^3$

Solution:

  a. $(-5)^3 = (-5)(-5)(-5)$ The base is -5.

         $= -125$ Multiply.

  b. $-5^3 = -(5)(5)(5)$ The base is 5. This is equal to

                        $-1 \cdot 5^3 = -1 \cdot (5)(5)(5)$.

         $= -125$ Multiply.

  c. $-(-5)^3 = -(-5)(-5)(-5)$ The base is -5. This is equal to

         $= -(-125)$ $-1 \cdot (-5)^3 = -1 \cdot (-5)(-5)(-5)$

         $= 125$ Multiply.

4. Division of Real Numbers

Recall from Section 2.5 that two numbers are **reciprocals** if their product is 1. In particular, the reciprocal of a negative number must also be a negative number to form a product of 1.

Number	Reciprocal	Product
$\dfrac{2}{3}$	$\dfrac{3}{2}$	$\dfrac{2}{3} \cdot \dfrac{3}{2} = 1$
$-\dfrac{4}{7}$	$-\dfrac{7}{4}$	$-\dfrac{4}{7} \cdot \left(-\dfrac{7}{4}\right) = 1$
-5	$-\dfrac{1}{5}$	$-5 \cdot \left(-\dfrac{1}{5}\right) = 1$

Also recall from Section 2.5 that division can be expressed in terms of multiplication. To divide two real numbers, we multiply the first number (the dividend) by the reciprocal of the second number (the divisor).

$$12 \div (-2) = 12 \cdot \left(-\frac{1}{2}\right) = \frac{\overset{6}{\cancel{12}}}{1} \cdot \left(-\frac{1}{\underset{1}{\cancel{2}}}\right) = -6$$

multiply by the reciprocal of the divisor

Because division of real numbers can be expressed in terms of multiplication, the sign rules that apply to multiplication also apply to division.

PROCEDURE   **Dividing Real Numbers**

- The quotient of two real numbers with the *same* sign is positive.

  Examples: $\dfrac{20}{5} = 4$ and $\dfrac{-20}{-5} = 4$

- The quotient of two real numbers with different signs is negative.

  Examples: $\dfrac{16}{-8} = -2$ and $\dfrac{-16}{8} = -2$

- Zero divided by any nonzero number is zero.

  Examples: $\dfrac{0}{-3} = 0$ and $\dfrac{0}{12} = 0$

- Any nonzero number divided by zero is undefined.

  Example: $\dfrac{5}{0}$ is undefined.

Skill Practice

Divide.

29. $-40 \div 10$ **30.** $\dfrac{-36}{-12}$

31. $\dfrac{18}{-2}$

Answers

24. They are reciprocals.   **25.** 7
26. $-\dfrac{5}{3}$ **27.** $-\dfrac{1}{6}$ **28.** Not possible
29. -4 **30.** 3   **31.** -9

Example 7   **Dividing Real Numbers**

Divide.

a. $50 \div (-5)$ **b.** $\dfrac{-42}{-7}$ **c.** $\dfrac{-39}{3}$

Solution:

a. $50 \div (-5) = -10$ Different signs. The quotient is negative.

b. $\dfrac{-42}{-7} = 6$ Same signs. The quotient is positive.

c. $\dfrac{-39}{3} = -13$ Different signs. The quotient is negative.

Example 8 Dividing Real Numbers

Divide.

a. $-9.45 \div (-2.7)$ **b.** $\dfrac{4}{15} \div \left(-\dfrac{8}{25}\right)$

Solution:

a. $-9.45 \div (-2.7) = 3.5$ Same signs. The quotient is positive.

b. $\dfrac{4}{15} \div \left(-\dfrac{8}{25}\right) = \dfrac{4}{15} \cdot \left(-\dfrac{25}{8}\right)$ Multiply by the reciprocal of the divisor.

$= \dfrac{\overset{1}{4}}{\underset{3}{15}} \cdot \left(-\dfrac{\overset{5}{25}}{\underset{2}{8}}\right)$ Multiply fractions.

$= -\dfrac{5}{6}$ Different signs. The quotient is negative.

Avoiding Mistakes

Remember, a number and its reciprocal have the same sign. That is, the reciprocal of $-\frac{8}{25}$ is $-\frac{25}{8}$.

When we use fraction notation to divide a positive and negative number, there are several forms in which we can write the quotient. For example, the following are all equal.

$$\dfrac{-2}{3} = \dfrac{2}{-3} = -\dfrac{2}{3}$$ By convention, we usually write the quotient with the negative sign in front of the fraction, $-\frac{2}{3}$.

If we use fraction notation to divide two negative numbers, we know that the quotient must be positive. For example:

$$\dfrac{-5}{-9} = \dfrac{5}{9}$$ The quotient is positive.

Example 9 Dividing Real Numbers

Divide, if possible.

a. $\dfrac{-7}{-4}$ **b.** $15 \div (-25)$ **c.** $\dfrac{0}{-7}$ **d.** $-6.1 \div 0$

Solution:

a. $\dfrac{-7}{-4} = \dfrac{7}{4}$ Same signs. The quotient is positive.

In this example, 7 and 4 share no common factors. Therefore, the fraction cannot be simplified further. However, the quotient may be written in several forms: $\frac{7}{4}$ or $1\frac{3}{4}$ or 1.75.

b. $15 \div (-25)$ The number 25 does not divide evenly into 15. However, the expression can be written as a fraction and then simplified to lowest terms.

$$= \frac{15}{-25}$$

$$= \frac{\overset{3}{\cancel{15}}}{\underset{5}{\cancel{-25}}}$$ Simplify to lowest terms.

$$= -\frac{3}{5}$$ Different signs. The quotient is negative.

c. $\frac{0}{-7} = 0$ Zero divided by any nonzero number is 0.

d. $-6.1 \div 0$ Any number divided by zero is undefined.

Section 10.4 Practice Exercises

Study Skills Exercises

1. Often students learn a rule about signs that states, "Two negatives are a positive." This rule is incomplete and therefore not always true. Note the following combinations of two negatives:

$-2 + (-4)$ the sum of two negatives

$-(-5)$ the opposite of a negative

$-|-10|$ the opposite of an absolute value

$(-3)(-6)$ the product of two negatives

Determine which of the following are negative and which are positive. Then write the rule for multiplying two numbers with the same sign.

$-2 + (-4)$ _____

$-(-5)$ _____

$-|-10|$ _____

$(-3)(-6)$ _____

When multiplying two numbers with the same sign, the product is _____.

2. Define the key term **reciprocal**.

Review Exercises

For Exercises 3–8, add or subtract as indicated.

3. $14 - (-5)$ **4.** $-24 - 50$ **5.** $-33 + (-11)$

6. $-7 - (-23)$ **7.** $23 - 12 + (-4) - (-10)$ **8.** $9 + (-12) - 17 - 4 - (-15)$

Objective 1: Multiplication of Real Numbers

For Exercises 9–36, multiply the real numbers. **(See Examples 1–2.)**

9. $-3(5)$ **10.** $-2(13)$ **11.** $-12 \cdot 4$ **12.** $-6 \cdot 11$

13. $-15(-3)$ **14.** $-3(-25)$ **15.** $9(-8)$ **16.** $8(-3)$

17. $(-1.2)(-3.2)$ **18.** $(-3.3)(-2.5)$ **19.** $-6(0.4)$ **20.** $-8(1.3)$

21. $7(-1.1)$ **22.** $5(-3.4)$ **23.** $-14 \cdot 0$ **24.** $-8 \cdot 0$

25. $\left(-\dfrac{2}{3}\right)\left(-\dfrac{6}{7}\right)$ **26.** $\left(-\dfrac{8}{9}\right)\left(-\dfrac{3}{4}\right)$ **27.** $\dfrac{3}{5}\left(-\dfrac{5}{21}\right)$ **28.** $\dfrac{5}{12}\left(-\dfrac{4}{7}\right)$

29. $6 \cdot \left(-\dfrac{5}{12}\right)$ **30.** $4 \cdot \left(-\dfrac{1}{16}\right)$ **31.** $\left(-2\dfrac{3}{5}\right)\left(-1\dfrac{2}{3}\right)$ **32.** $\left(-3\dfrac{1}{3}\right)\left(-2\dfrac{1}{5}\right)$

33. $\left(-\dfrac{8}{9}\right) \cdot 0$ **34.** $\left(-\dfrac{2}{11}\right) \cdot 0$ **35.** $(-3.5)(-1.4)$ **36.** $(-1.6)(-6.5)$

For Exercises 37–42, translate the English phrase to an algebraic expression. Then simplify. **(See Example 3.)**

37. Multiply -3 and -1.   **38.** Multiply -12 and -4.   **39.** The product of -5 and 3

40. The product of 9 and -2 **41.** 1.3 times -3 **42.** -2.3 times 6

Objective 2: Multiplying Many Factors

For Exercises 43–52, multiply. **(See Example 4.)**

43. $(5)(-2)(4)(-10)$ **44.** $(-3)(-5)(-2)(4)$ **45.** $(-11)(-4)(-2)$

46. $(20)(-3)(-1)$ **47.** $(24)(-2)(0)(-3)$ **48.** $(3)(0)(-13)(22)$

49. $(-1)(-1)(-1)(-1)(-1)(-1)$ **50.** $(-1)(-1)(-1)(-1)(-1)(-1)(-1)$

51. $(-1)(-1)(-1)(-1)(-1)$ **52.** $(-1)(-1)(-1)(-1)$

Objective 3: Exponential Expressions

For Exercises 53–68, simplify. **(See Examples 5–6.)**

53. -10^2 **54.** -8^2 **55.** $(-10)^2$ **56.** $(-8)^2$

57. -3^3 **58.** -4^3 **59.** $(-3)^3$ **60.** $(-4)^3$

61. -0.2^3 **62.** -0.4^3 **63.** $\left(-\dfrac{2}{3}\right)^3$ **64.** $\left(-\dfrac{3}{5}\right)^3$

65. $(-6)^2$ **66.** $(-6)^3$ **67.** $-(-6)^2$ **68.** $-(-6)^3$

Objective 4: Division of Real Numbers

For Exercises 69–92, divide the real numbers, if possible. **(See Examples 7–9.)**

69. $\dfrac{-15}{5}$

70. $\dfrac{30}{-6}$

71. $\dfrac{56}{-8}$

72. $\dfrac{-48}{3}$

73. $\dfrac{-25}{-15}$

74. $\dfrac{-6}{-18}$

75. $\dfrac{-2}{-3}$

76. $\dfrac{-9}{-8}$

77. $\dfrac{13}{0}$

78. $\dfrac{-41}{0}$

79. $\dfrac{0}{-2}$

80. $\dfrac{0}{5}$

81. $(-20) \div (-5)$

82. $(-10) \div (-2)$

83. $-0.91 \div -0.7$

84. $-1.3 \div -0.5$

85. $\left(\dfrac{8}{7}\right) \div \left(-\dfrac{4}{5}\right)$

86. $\left(-\dfrac{2}{5}\right) \div \left(\dfrac{8}{15}\right)$

87. $\left(-\dfrac{1}{6}\right) \div 0$

88. $\left(\dfrac{2}{11}\right) \div 0$

89. $\dfrac{-5}{-8}$

90. $\dfrac{-1}{-3}$

91. $-18 \div 24$

92. $12 \div (-30)$

For Exercises 93–98, translate the English phrase into a mathematical expression. Then simplify.

93. The quotient of -100 and 20

94. The quotient of 46 and -23

95. -32 divided by -64

96. 108 divided by -24

97. 13 divided into -52

98. -15 divided into -45

Mixed Exercises

For Exercises 99–114, perform the indicated operation.

99. $8 + (-6)$

100. $8 - (-6)$

101. $8(-6)$

102. $8 \div (-6)$

103. $-9 - (-12)$

104. $(-9)(-12)$

105. $-36 \div (-12)$

106. $-36 + (-12)$

107. $(-5)(-4)$

108. $-90 \div (-6)$

109. $0 + (-15)$

110. $0 - (-15)$

111. $\dfrac{1}{3} \div \left(-\dfrac{5}{6}\right)$

112. $\dfrac{1}{3} + \left(-\dfrac{5}{6}\right)$

113. $\dfrac{1}{3} - \left(-\dfrac{5}{6}\right)$

114. $\dfrac{1}{3}\left(-\dfrac{5}{6}\right)$

Expanding Your Skills

115. Which is greater, $(-2)^{50}$ or $(-2)^{51}$?

116. Which is greater, $(-3)^{20}$ or $(-3)^{21}$?

117. Which is greater, $(5)^{40}$ or $(5)^{41}$?

118. Which is greater, $(6)^{10}$ or $(6)^{11}$?

For Exercises 119–122, assume $a > 0$ (this means that a is positive) and $b < 0$ (this means that b is negative). Find the sign of each expression.

119. $a \cdot b$

120. $b \div a$

121. $-a \div (b)$

122. $a(-b)$

Calculator Connections

Topic: Multiplying and Dividing Real Numbers on a Calculator

Knowing the sign of the product or quotient can make using the calculator easier. For example, look at $\frac{-78}{-26}$. Note the keystrokes if we enter this into the calculator as written:

78 `+○-` `÷` 26 `+○-` `=` | 3 |

But since we know that the quotient of two negative numbers is positive, we can simply enter:

78 `÷` 26 `=` | 3 |

Calculator Exercises:

For Exercises 123–128, use a calculator to perform the indicated operations.

123. $(-413)(871)$

124. $-6125 \cdot (-97)$

125. $(-52.12)(-101.5)$

126. $\dfrac{-576,828}{-10,682}$

127. $5,945,308 \div (-9452)$

128. $\dfrac{-301,224}{9128}$

Problem Recognition Exercises

Operations on Real Numbers

For Exercises 1–40, perform the indicated operations.

1. $15 - (-5)$

2. $15(-5)$

3. $15 + (-5)$

4. $15 \div (-5)$

5. $-36(-2)$

6. $-36 - (-2)$

7. $\dfrac{-36}{-2}$

8. $-36 + (-2)$

9. $20(-4)$

10. $-20(-4)$

11. $-20(4)$

12. $20(4)$

13. $-5 - 9 - 2$

14. $-4(-9)(-2)$

15. $10 + (-3) + (-12)$

16. $10 - (-3) - (-12)$

17. $(-1)(-2)(-3)(-4)$

18. $(-1)(-2)(3)(4)$

19. $(-1)(-2)(-3)(4)$

20. $(-1)(2)(3)(4)$

21. $\dfrac{3}{5} \div \left(-\dfrac{10}{9}\right)$

22. $\dfrac{3}{5}\left(-\dfrac{10}{9}\right)$

23. $\dfrac{3}{5} + \left(-\dfrac{10}{9}\right)$

24. $\dfrac{3}{5} - \left(-\dfrac{10}{9}\right)$

25. $-\dfrac{2}{3} + \left(-\dfrac{7}{9}\right)$

26. $-\dfrac{2}{3} \div \left(-\dfrac{7}{9}\right)$

27. $\left(-2\dfrac{1}{4}\right)\left(1\dfrac{4}{9}\right)$

28. $-2\dfrac{1}{4} + 1\dfrac{4}{9}$

29. $41.5 - (-13.6)$

30. $-13.56 + 4.12$

31. $-60.41 - 33.50$

32. $-0.06 - (-0.04)$

33. $\dfrac{-12}{-11}$

34. $\dfrac{5}{-30}$

35. $\dfrac{0}{-8}$

36. $-4.5 \div 0$

37. $42 \div (-0.002)$

38. $-360 \div (-0.009)$

39. $-44 - (-44)$

40. $-60 \cdot \left(-\dfrac{1}{60}\right)$

Section 10.5 | Order of Operations

Objective

1. Order of Operations

1. Order of Operations

The order of operations was first introduced in Section 1.7 and then used throughout the text. The order of operations still applies when simplifying expressions with real numbers.

> **PROCEDURE** Order of Operations
>
> **Step 1** Perform all operations inside parentheses and other grouping symbols first.
> **Step 2** Simplify any expressions containing exponents or square roots.
> **Step 3** Perform multiplication or division in the order that they appear from left to right.
> **Step 4** Perform addition or subtraction in the order that they appear from left to right.

Skill Practice

Simplify.
1. $8 - 2(3 - 10)$
2. $8^2 - 2^3 \div (-7 + 5)$

Example 1 Applying the Order of Operations

Simplify.

a. $-12 - 6(7 - 5)$ **b.** $3^2 - 10^2 \div (-1 - 4)$

Solution:

a. $-12 - 6(7 - 5)$

$= -12 - 6(2)$ Simplify within parentheses first.

$= -12 - 12$ Multiply before subtracting.

$= -24$ Subtract. *Note:* $-12 - 12 = -12 + (-12) = -24$.

b. $3^2 - 10^2 \div (-1 - 4)$ Simplify within parentheses.
  Note: $-1 - 4 = -1 + (-4) = -5$.

$= 3^2 - 10^2 \div (-5)$ Simplify exponents.
  Note: $3^2 = 3 \cdot 3 = 9$ and $10^2 = 10 \cdot 10 = 100$.

$= 9 - 100 \div (-5)$ Divide before subtracting.
  Note: $100 \div (-5) = -20$.

$= 9 - (-20)$ Subtract. *Note:* $9 - (-20) = 9 + (20) = 29$.

$= 29$

Answers

1. 22 **2.** 68

Example 2 Applying the Order of Operations

Simplify. $\dfrac{1}{30} - \left(-\dfrac{1}{3}\right)^2 \cdot \dfrac{3}{5}$

Skill Practice
Simplify.
3. $-\dfrac{2}{9} + \left(-\dfrac{5}{6}\right)^2 \cdot \dfrac{2}{5}$

Solution:

$\dfrac{1}{30} - \left(-\dfrac{1}{3}\right)^2 \cdot \dfrac{3}{5}$

$= \dfrac{1}{30} - \dfrac{1}{9} \cdot \dfrac{3}{5}$ Simplify exponents. *Note:* $\left(-\dfrac{1}{3}\right)\cdot\left(-\dfrac{1}{3}\right)=\dfrac{1}{9}$.

$= \dfrac{1}{30} - \dfrac{1}{\overset{}{\underset{3}{9}}} \cdot \dfrac{\overset{1}{3}}{5}$ Multiply fractions.

$= \dfrac{1}{30} - \dfrac{1}{15}$ The least common denominator is 30.

$= \dfrac{1}{30} - \dfrac{1 \cdot 2}{15 \cdot 2}$ Write each fraction with the LCD.

$= \dfrac{1}{30} - \dfrac{2}{30}$

$= \dfrac{1}{30} + \left(-\dfrac{2}{30}\right)$ Write the subtraction in terms of addition.

$= -\dfrac{1}{30}$

Example 3 Applying the Order of Operations

Simplify. $3 - [-6 - (5 - 7)]$

Skill Practice
Simplify.
4. $-10 - [-2 + (3 - 5)]$

Solution:

$3 - [-6 - (5 - 7)]$

$= 3 - [-6 - (5 + (-7))]$ Write the subtraction in terms of addition.

$= 3 - [-6 - (-2)]$ Simplify within the innermost parentheses.

$= 3 - [-6 + (2)]$ Write the subtraction in terms of addition.

$= 3 - [-4]$ Simplify within the parentheses by adding.

$= 3 + (4)$ Write the subtraction in terms of addition.

$= 7$ Add.

Answers
3. $\dfrac{1}{18}$ **4.** -6

┌─── **Skill Practice** ───┐

Simplify.

5. $\dfrac{23 - |2 - 5|}{-6 - 4}$

└──────────────────────────┘

TIP: Absolute value bars act as grouping symbols. Perform the operations inside first.

Answer

5. -2

Example 4 Applying the Order of Operations

Simplify. $\dfrac{|-7 + 3| - 9}{3 - 2 \cdot 4}$

Solution:

$\dfrac{|-7 + 3| - 9}{3 - 2 \cdot 4}$ Simplify the numerator and denominator separately.

$= \dfrac{|-4| - 9}{3 - 8}$ Numerator: Add within the absolute values.
Denominator: Multiply before subtracting.

$= \dfrac{4 + (-9)}{3 + (-8)}$ Numerator: Evaluate the absolute value and write the subtraction as addition.
Denominator: Write the subtraction as addition.

$= \dfrac{-5}{-5}$ Add.

$= 1$ Divide.

Section 10.5   Practice Exercises

Boost your GRADE at ALEKS.com!

- Practice Problems
- Self-Tests
- NetTutor
- e-Professors
- Videos

Study Skills Exercise

1. Make a list of resources that are available to you at times when you are not on campus and you need help with your algebra studies (for example, websites, online tutoring, and classmates). Write down Web addresses and phone numbers and keep them handy.

Review Exercises

For Exercises 2–7, multiply or divide as indicated.

2. $-100 \div (-4)$

3. $\left(-\dfrac{2}{9}\right) \div \left(\dfrac{8}{27}\right)$

4. $10 \cdot \left(-\dfrac{3}{5}\right)$

5. $-2.8(-1.1)$

6. $5.5 \div (-0.5)$

7. $(-1)(-5)(-8)(3)$

8. a. What number must be multiplied by -5 to obtain -35?

    b. What number must be multiplied by -5 to obtain 35?

Objective 1: Order of Operations

For Exercises 9–62, simplify by using the order of operations. **(See Examples 1–4.)**

9. $5 + 2(3 - 5)$

10. $6 - 4(8 - 10)$

11. $-8 - 6^2$

12. $-10 - 5^2$

13. $4 + (3 - 8)^2$

14. $5 + (2 - 9)^2$

15. $120 \div (-4)(5)$

16. $36 \div (-2)(3)$

17. $-2.1 - 6 \div 5$

18. $-8.3 - 10 \div 8$

19. $[5.3 - (-2.7)]^2$

20. $(-7.1 - 1.9)^2$

21. $-2(3 - 6) + 10$

22. $-4 \div (1 - 3) - 8$

23. $-16 \div (-4)(-5)$

24. $-12(-1) \div 6$

25. $8 - (-3)(-2)^3$

26. $1 - (-5)(-3)^2$

27. $12 + (14 - 16)^2 \div (-4)$

28. $-7 + (1 - 5)^2 \div 4$

29. $-48 \div 12 \div (-2)$

30. $-100 \div (-5) \div (-5)$

31. $90 \div (-3)(-1) \div (-6)$

32. $64 \div (-4)2 \div (-16)$

33. $[9^2 - (-7)^2] \div (-4)$

34. $|(-8)^2 - 5^2| \div (-3)$

35. $2 + 2^3 - |10 - 12|$

36. $14 - 4^2 + 2 - 10$

37. $-6(48 \div 12)^2$

38. $-5(35 \div 5)^2$

39. $\left(-\dfrac{1}{2}\right) \cdot \dfrac{1}{3} \div \dfrac{1}{12}$

40. $\dfrac{2}{9} \div \left(-\dfrac{1}{3}\right) \cdot \dfrac{6}{5}$

41. $\dfrac{1}{6} + \left(-\dfrac{5}{4}\right) \cdot \dfrac{4}{3}$

42. $-\dfrac{3}{8} + \dfrac{5}{24} \div \left(-\dfrac{5}{6}\right)$

43. $\left(-\dfrac{2}{3}\right)^2 - \left(\dfrac{5}{21}\right) \div \dfrac{15}{7}$

44. $\left(-\dfrac{1}{3}\right)^3 + \left(\dfrac{2}{9}\right) \cdot \dfrac{5}{6}$

45. $\dfrac{5}{2} \cdot \left(\dfrac{3}{2}\right)^2 + \left(-\dfrac{1}{2}\right)^3$

46. $\left(-\dfrac{7}{3}\right) \div \left(\dfrac{4}{3}\right)^2 - \left(\dfrac{1}{2}\right)^4$

47. $21 - [4 - (5 - 8)]$

48. $15 - [10 - (20 - 25)]$

49. $-17 - 2[18 \div (-3)]$

50. $-8 - 5(-45 \div 15)$

51. $4 + 2[9 + (-4 + 12)]$

52. $-13 + 3[11 + (-15 + 10)]$

53. $2^2 - |-3 + 9|$

54. $5^2 - |10 + (-8)|$

55. $\dfrac{|3 + (-5)|}{4 - (3)(-2)}$

56. $\dfrac{|-11 + 7|}{8 - 4(-1)}$

57. $\dfrac{13 - (2)(4)}{-1 - 2^2}$

58. $\dfrac{10 - (-3)(5)}{-9 - 4^2}$

59. $\dfrac{1 - 4(3 - 5)}{5^2 - 2^2}$

60. $\dfrac{-3 - (2 + 4)}{6^2 - 3^2}$

61. $\dfrac{6 - 3^2}{(5 - 2)^2}$

62. $\dfrac{-2 - 5^2}{(6 - 3)^2}$

63. Find the average temperature: $-8°, -11°, -4°, 1°, 9°, 4°, -5°$

64. Find the average temperature: $15°, 12°, 10°, 3°, 0°, -2°, -3°$

65. Find the average golf score: $-8, -8, -6, -5, -2, 3, 3, 0, -4$

66. Find the average golf score: $-6, -2, 5, 1, 0, -3, 7, 2, -4$

67. According to Ask A Scientist©, the coldest temperature ever recorded was in Vostok, Antarctica, on July 31, 1973. The temperature was recorded at $-89.6°C$. Use the following formula to convert this temperature to Fahrenheit.

$$F = \frac{9}{5}C + 32$$

68. The BBC reports that the coldest temperature in the United Kingdom was $-27.2°C$ in Altnaharra, Highland, in 1995. Using the formula from Exercises 67, convert this temperature to Fahrenheit.

Expanding Your Skills

For Exercises 69–72, simplify the expressions containing both fractions and decimals.

69. $\left(\dfrac{1}{2}\right)^2 \div 0.05 + \left(-\dfrac{3}{4} \cdot \dfrac{8}{3}\right)$

70. $-0.8 - \dfrac{19}{20} + \left(\dfrac{4}{5} \div \dfrac{1}{2}\right) - (-0.15)$

71. $2\left(\dfrac{7}{8} - \dfrac{1}{4}\right) - (-1.5)^2$

72. $-2.1 + 4\left(\dfrac{7}{16} - 0.0375\right)$

Group Activity

Checking Weather Predictions

Materials: A computer with online access

Estimated time: 2–3 minutes each day for 10 days

Group Size: 3

1. Go to a website such as http://www.weather.com/ to find the predicted high and low temperatures for a 10-day period for a city and state of your choice.

2. Record the predicted high and low temperatures for each of the 10 days. Record these values in the second column of each table.

Day	Predicted **High**	Actual **High**	Difference (error)
1			
2			
3			
4			
5			
6			
7			
8			
9			
10			

Day	Predicted **Low**	Actual **Low**	Difference (error)
1			
2			
3			
4			
5			
6			
7			
8			
9			
10			

3. For the next 10 days, record the actual high and low temperatures for your chosen city for that day. Record these values in the third column of each table.

4. For each day, compute the difference between the predicted and actual temperature and record the results in the fourth column of each table. We will call this difference the *error*.

$$\text{Error} = (\text{Predicted temperature}) - (\text{Actual temperature})$$

5. If the error is *negative*, does this mean that the weather service overestimated or underestimated the temperature?

6. If the error is *positive*, does this mean that the weather service overestimated or underestimated the temperature?

7. Find the mean (average) error for the high temperature predictions.

8. Find the mean (average) error for the low temperature predictions.

9. Which set of predictions was most accurate, the high temperature predictions or the low temperature predictions?

Chapter 10   Summary

Section 10.1   Real Numbers and the Real Number Line

Key Concepts

The numbers . . . $-3, -2, -1, 0, 1, 2, 3, . . .$ and so on are called **integers**. The negative integers lie to the left of zero on the number line.

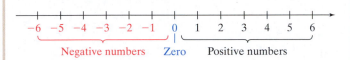

A number that can be written as a ratio of two integers is called a **rational number** (division by zero is excluded).

Irrational numbers are numbers that cannot be written as a ratio of integers.

The set of all rational numbers and irrational numbers together forms the **real numbers**.

The **absolute value** of a number a is denoted $|a|$. The value of $|a|$ is the distance between a and 0 on the number line.

Two numbers that are the same distance from zero on the number line, but on opposite sides of zero are called **opposites**.

The double-negative property states that the opposite of a negative number is a positive number. That is, $-(-a) = a$, for $a > 0$.

Examples

Example 1

The temperature 5° below zero can be represented by a negative number: $-5°$.

Example 2

The following numbers are rational because they can be written as a ratio of two integers.

a. $\dfrac{1}{2}$ is a ratio of 1 and 2

b. -0.75 is a ratio of -3 and 4

c. -4 is a ratio of -4 and 1

Example 3

a. $|5| = 5$ b. $|-13| = 13$ c. $|0| = 0$

Example 4

The opposite of 12 is $-(12) = -12$.

Example 5

The opposite of -23 is $-(-23) = 23$.

Section 10.2 Addition of Real Numbers

Key Concepts

To add integers by using a number line, locate the first addend on the number line. Then to add a positive number, move to the right on the number line. To add a negative number, move to the left on the number line.

Examples

Example 1

Add $-2 + (-4)$ by using the number line.

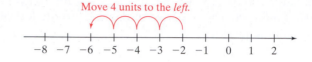

Move 4 units to the *left*.

$-2 + (-4) = -6$

Integers can be added by using the following rules:

Adding Numbers with the Same Sign

To add two numbers with the same sign, add their absolute values and apply the common sign.

Example 2

a. $5 + 2 = 7$

b. $-5 + (-2) = -7$

Adding Numbers with Different Signs

To add two numbers with different signs, subtract the smaller absolute value from the larger absolute value. Then apply the sign of the number having the larger absolute value.

Example 3

a. $6 + (-5) = 1$

b. $(-6) + 5 = -1$

Section 10.3 Subtraction of Real Numbers

Key Concepts

Subtraction of Real Numbers

If a and b are real numbers, then

$$a - b = a + (-b)$$

To perform subtraction, follow these steps:

1. Leave the first number (the minuend) unchanged.
2. Change the subtraction sign to an addition sign.
3. Add the opposite of the second number (the subtrahend).

Examples

Example 1

a. $3 - 9 = 3 + (-9) = -6$

b. $-3 - 9 = -3 + (-9) = -12$

c. $3 - (-9) = 3 + (9) = 12$

d. $-3 - (-9) = -3 + (9) = 6$

Section 10.4 Multiplication and Division of Real Numbers

Key Concepts

Multiplication of Real Numbers

1. The product of two real numbers with the same sign is positive.
2. The product of two real numbers with different signs is negative.
3. The product of any real number and zero is zero.

The product of an *even* number of negative factors is *positive*.

The product of an *odd* number of negative factors is *negative*.

When evaluating an exponential expression, attention must be paid when parentheses are used. That is,

$(-2)^4 = (-2)(-2)(-2)(-2) = 16$, while

$-2^4 = -1 \cdot (2)(2)(2)(2) = -16$.

Two numbers are **reciprocals** if their product is 1.

Division of Real Numbers

1. The quotient of two real numbers with the same sign is positive.
2. The quotient of two real numbers with different signs is negative.

Division by zero is undefined.
Zero divided by a nonzero number is 0.

Examples

Example 1

a. $-8(-3) = 24$

b. $8(-3) = -24$

c. $-8(0) = 0$

Example 2

a. $(-5)(-4)(-1)(-3) = 60$

b. $(-2)(-1)(-6)(-3)(-2) = -72$

Example 3

a. $\left(-\dfrac{2}{3}\right)^2 = \left(-\dfrac{2}{3}\right)\left(-\dfrac{2}{3}\right) = \dfrac{4}{9}$

b. $-\left(\dfrac{2}{3}\right)^2 = -1 \cdot \left(\dfrac{2}{3}\right)\left(\dfrac{2}{3}\right) = -\dfrac{4}{9}$

Example 4

The reciprocal of -4 is $-\dfrac{1}{4}$ because

$-4 \cdot -\dfrac{1}{4} = 1$

Example 5

a. $-36 \div (-9) = 4$

b. $\dfrac{42}{-6} = -7$

Example 6

a. $\dfrac{-15}{0}$ is undefined.

b. $\dfrac{0}{-3} = 0$

Section 10.5   Order of Operations

Key Concepts

Order of Operations

1. Perform all operations inside parentheses and other grouping symbols first.
2. Simplify any expressions containing exponents or square roots.
3. Perform multiplication or division in the order that they appear from left to right.
4. Perform addition or subtraction in the order that they appear from left to right.

Examples

Example 1

$$-15 - 2(8 - 11)^2 = -15 - 2(-3)^2$$
$$= -15 - 2 \cdot 9$$
$$= -15 - 18$$
$$= -33$$

Example 2

$$\frac{1}{5} \div \left(-\frac{7}{20}\right) \cdot \left(-\frac{9}{2}\right) = \frac{1}{5} \cdot \left(-\frac{20}{7}\right) \cdot \left(-\frac{9}{2}\right)$$
$$= \frac{1}{\overset{}{\underset{1}{5}}} \cdot \left(-\frac{\overset{4}{20}}{7}\right) \cdot \left(-\frac{9}{2}\right)$$
$$= -\frac{\overset{2}{4}}{7} \cdot \left(-\frac{9}{\underset{1}{2}}\right)$$
$$= \frac{18}{7}$$

Chapter 10   Review Exercises

Section 10.1

For Exercises 1–4, write an integer that represents each numerical value.

1. The population of Detroit, Michigan, decreased by 64,599 from 2000 to 2005. (Source: U.S. Census Bureau)

2. The country's deficit is $5 billion.

3. The temperature rose 15° in one day.

4. Cecelia's bank account earned $10 in interest.

For Exercises 5–8, locate the numbers on the number line.

5. -2 6. $-5\frac{1}{3}$ 7. 0 8. 3.8

$$\begin{array}{ccccccccccccc} & & & & & & & & & & & & \\ \hline -6 & -5 & -4 & -3 & -2 & -1 & 0 & 1 & 2 & 3 & 4 & 5 & 6 \end{array}$$

For Exercises 9–12, determine the opposite and the absolute value for each number.

9. -4 10. $-\frac{1}{2}$ 11. 3.5 12. 6

13. Evaluate.
   a. $-(-9)$ b. $-|-9|$

14. Evaluate.
   a. $-(1.5)$ b. $-|1.5|$

For Exercises 15–20, place the correct symbol, $>$ or $<$, between the two numbers.

15. $-\frac{5}{6} \,\square\, -1$ 16. $-0.5 \,\square\, 0.5$

17. $|3| \,\square\, -|3|$ 18. $8 \,\square\, |-9|$

19. $-2.8 \,\square\, -2$ 20. $\left|-\frac{3}{2}\right| \,\square\, -\frac{1}{2}$

Section 10.2

For Exercises 21–24, add the integers by using the number line.

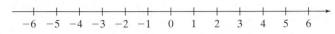

21. $6 + (-2)$ **22.** $-3 + 6$

23. $-3 + (-2)$ **24.** $-3 + 0$

25. State the rule for adding two numbers with the same sign.

26. State the rule for adding two numbers with different signs.

For Exercises 27–34, add the real numbers.

27. $35 + (-22)$ **28.** $-105 + 90$

29. $-29 + (-41)$ **30.** $3.22 + (-4.1)$

31. $-6.5 + (-4.16)$ **32.** $\left(-\dfrac{7}{4}\right) + \left(\dfrac{5}{2}\right)$

33. $\left(-\dfrac{1}{5}\right) + \left(-\dfrac{7}{10}\right)$ **34.** $2 + \left(-\dfrac{7}{3}\right)$

For Exercises 35–40, translate each phrase to a mathematical expression. Then simplify the expression.

35. The sum of 23 and -35

36. 57 plus -10

37. The total of -5, -13, and 20

38. -42 increased by 12

39. 3 more than -12

40. -89 plus -22

For Exercises 41–42, add.

41. $-3 + (-10) + 12 + 14 + (-10)$

42. $9 + (-15) + 2 + (-7) + (-4)$

Section 10.3

43. State the steps for subtracting two numbers.

For Exercises 44–51, subtract the real numbers.

44. $4 - (-23)$ **45.** $19 - 44$

46. $-2 - (-24)$ **47.** $-289 - 130$

48. $-2.9 - 4.5$ **49.** $3.8 - 4.5$

50. $\left(-\dfrac{5}{3}\right) - \left(-\dfrac{5}{12}\right)$ **51.** $0 - \left(-\dfrac{20}{21}\right)$

For Exercises 52–55, translate the mathematical statement to an English phrase. Answers will vary.

52. $4 - 6$ **53.** $23 - (-6)$

54. $-2 - 14$ **55.** $-25 - (-7)$

56. The temperature in Fargo, North Dakota, rose from $-6°$F to $-1°$F. By how many degrees did the temperature rise?

57. Sam's balance in his checking account was $-\$40$, so he deposited \$132. What is his present balance?

58. Find the average of the golf scores: $-3, 4, 0, 9, -2,$ $-1, 0, 5, -3$ (These scores are the number of holes above or below par.)

Section 10.4

For Exercises 59–70, multiply or divide as indicated.

59. $6(-3)$ **60.** $\dfrac{-12}{4}$

61. $\dfrac{-900}{-60}$ **62.** $(-7)(-8)$

63. $-2.8 \div 0.04$ **64.** $(-62.6)(2.5)$

65. $\left(-\dfrac{2}{3}\right)\left(-\dfrac{21}{8}\right)$ **66.** $\left(-2\dfrac{1}{8}\right) \div \left(1\dfrac{1}{4}\right)$

67. $\left(-\dfrac{1}{5}\right) \div 0$ **68.** $\dfrac{0}{-5}$

69. $(-1)(-8)(2)(1)(-2)$ **70.** $\dfrac{-9}{-5}$

For Exercises 71–76, simplify.

71. $(-6)^2$ **72.** -6^2

73. $\left(-\dfrac{3}{4}\right)^3$ **74.** $-\left(\dfrac{3}{4}\right)^3$

75. $(-1)^{10}$ **76.** $(-1)^{21}$

77. What is the sign of the product of three negative factors?

78. What is the sign of the product of four negative factors?

For Exercises 79–82, translate the English phrase into a mathematical expression. Then simplify.

79. The quotient of -45 and -15

80. The product of -4 and 19

81. 30 times -5

82. -136 divided by -8

Section 10.5

For Exercises 83–91, simplify by using the order of operations.

83. $28 \div (-7) \cdot 3 - (-1)$

84. $(-4)^3 \div 8 - (-6)$

85. $|10 - (-3)^2| \cdot (-11) + 4$

86. $[-9 - (-7)]^3 \cdot 3 \div (-6)$

87. $18 - (-5)^2 + 14 \div 2$

88. $\left(\dfrac{1}{15}\right) \div \left(-\dfrac{7}{10}\right) \cdot \left(\dfrac{3}{2}\right) + \left(-\dfrac{6}{7}\right)$

89. $\left(-\dfrac{3}{8}\right)^2 - \left(-\dfrac{1}{2}\right)^3$

90. $6 - [5 - (2 - 8)]$

91. $\dfrac{3 - |2 + (-7)|}{3^2 - 5^2}$

92. Find the average temperature for one week; $2°, 4°, -6°, -1°, 0°, -4°, -2°$

Chapter 10   Test

1. Write an integer that represents the numerical value.

  a. Dwayne lost $220 during in his last trip to Las Vegas.

  b. Garth Brooks has 26 more platinum albums than Elvis Presley.

For Exercises 2–4, refer to these numbers: $-3, -\frac{3}{5}, 0,$ $\sqrt{7}, 4, -1, \frac{4}{7}, -\pi$

2. List all the numbers that are integers.

3. List all the rational numbers.

4. List all the irrational numbers.

For Exercises 5–10, place the correct symbol, $>$ or $<$, between the two numbers.

5. $-5 \; \square \; -2$

6. $|-5| \; \square \; |-2|$

7. $0 \; \square \; -2.4$

8. $\dfrac{4}{5} \; \square \; -\dfrac{2}{3}$

9. $-|-9| \; \square \; 9$

10. $-|33.1| \; \square \; |-33.1|$

For Exercises 11–18, add or subtract as indicated.

11. $9 + (-14)$

12. $-23 + (-5)$

13. $-4 - (-13)$

14. $-30 - 11$

15. $-1.5 + 2.1$

16. $0.5 - 2.8$

17. $-\dfrac{2}{3} - \dfrac{4}{7}$

18. $\dfrac{5}{4} + \left(-\dfrac{7}{8}\right)$

For Exercises 19–26, multiply or divide as indicated.

19. $6(-12)$

20. $(-11)(-8)$

21. $\dfrac{-24}{-12}$

22. $\dfrac{54}{-3}$

23. $\dfrac{-44}{0}$

24. $(-91)(0)$

25. $\dfrac{3}{10} \div \left(-\dfrac{4}{5}\right)$ **26.** $\dfrac{-13}{-6}$

27. a. What is the sign of the product of an even number of negative factors?

    b. What is the sign of the product of an odd number of negative factors?

28. Simplify the exponential expressions.

    a. $(-8)^2$ **b.** -8^2 **c.** $(-4)^3$ **d.** -4^3

For Exercises 29–34, translate to a mathematical expression. Then simplify the expression.

29. The product of -3 and -7

30. 8 more than -13

31. Subtract -4 from 18.

32. The quotient of 6 and $-\dfrac{2}{3}$

33. -8.1 increased by 5

34. The total of -3, 15, -6, and -1

For Exercises 35–42, simplify.

35. $-14 + 22 - (-5) + (-10)$

36. $(-3)(-1)(-4)(-1)(-5)$

37. $-20 \div (-2)^2 + (-14)$

38. $12 \cdot (-6) + [20 - (-12)] - 15$

39. $-\dfrac{2}{15} + \left(-\dfrac{20}{21} \cdot \dfrac{7}{5}\right)$ **40.** $\left(-\dfrac{1}{3}\right)^2 \div \left(\dfrac{5}{6} - \dfrac{1}{9}\right)$

41. $16 - 2[5 - (1 - 4)]$ **42.** $\dfrac{15 - 2|3 - 9|}{8 - 2^2}$

43. Find the average temperature:
    $4°, -3°, -1°, 5°, -2°, 0°, 4°$

Chapters 1–10 Cumulative Review Exercises

For Exercises 1–4, add, subtract, or multiply as indicated.

1.   3490
    $+123$

2.   2901
    -332

3.   23
    34
    98
    $+22$

4.   790
    $\times 24$

5. Write the prime factorization of 720.

6. Write a fraction that represents the shaded region of the figure.

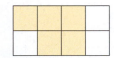

7. On a quiz, Harold missed 3 out of 14 questions. Write a fraction representing the fraction of the quiz questions that he answered *correctly*.

8. Amy has a box that contains 20 oz of snack crackers. How many individual packages will she get if she puts $2\frac{1}{2}$ oz in each package?

9. Find the LCM of 16, 40, and 10.

10. Simplify. $\dfrac{3}{16} + \dfrac{33}{40} - \dfrac{7}{10}$

11. Add. $3\dfrac{3}{5} + 2\dfrac{13}{15}$ **12.** Subtract. $16\dfrac{1}{2} - 12\dfrac{13}{14}$

13. Round the numbers to the indicated place.

    a. 34.2298 Thousandths   **b.** 9.0314 Tenths

14. Convert cents to dollars. 209¢

15. Multiply. 204.55(2.4)

16. Divide. $402.5 \div 3.5$

17. Write a ratio of the shortest side to the longest side of the rectangle.

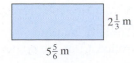

$5\frac{5}{6}$ m

18. A DC-10 aircraft used 9964 gal of fuel in 4 hr. Find the unit rate in gallons per hour.

19. On a map 1 in. represents 6 mi. What is the distance between two cities that measure $3\frac{1}{2}$ in. on the map?

20. Given that the two triangles are similar, find sides x and y.

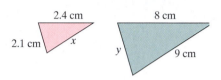

For Exercises 21–23, solve the percent equations.

21. What is 32% of 600?

22. What percent of 300 is 336?

23. 15 is 6% of what?

24. A pair of shoes was discounted 20%. If the original price was $86, what is the sale price?

For Exercises 25–28, convert the units of measure.

25. 2 ft 4 in. = _____ in.

26. 20 qt = _____ gal

27. 60 mL = _____ L

28. 30 oz = _____ lb

29. A car travels 6 mi due north and then turns and travels 8 mi due east. What is the distance of the car from the point of origin?

For Exercises 30–31, find the area.

30. Parallelogram

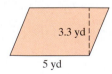

3.3 yd

5 yd

31. Square

$2\frac{1}{4}$ m

32. Find the area, A, and the circumference, C, of the circle. Use 3.14 for π.

1.5 km

For Exercises 33–35, use the following data.

The number of miles walked in one day by 10 selected people is given.

4   4   4   3   6   4   6   5   3   4

33. Complete the frequency distribution for the data.

Number of Miles	Tally	Frequency (Number of Walkers)
3		
4		
5		
6		

34. Construct a horizontal bar graph from the frequency distribution in Exercise 33. Label the vertical axis with the number of miles and the horizontal axis with the frequency.

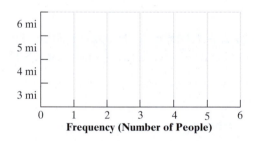

35. What is the mean number of miles walked per day?

36. Refer to the circle graph. If the monthly budget for a small business is $1200, how much will be spent on postage?

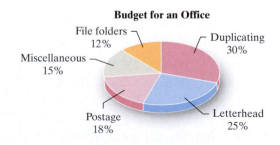

Budget for an Office

File folders 12%

Duplicating 30%

Miscellaneous 15%

Postage 18%

Letterhead 25%

For Exercises 37–40, simplify.

37. $43 - (-12)$

38. $-12 + (-5) - 3 - (-8)$

39. $(-4)^2 - 6^2$

40. $\frac{8}{9} \cdot \left(\frac{1}{3} - \frac{5}{6}\right)^3 \div \left(-\frac{2}{3}\right)$

Solving Equations

CHAPTER OUTLINE

Chapter 11

In this chapter, we learn how to simplify algebraic expressions by clearing parentheses and combining like terms. Then we move on to solving linear equations and using equations to solve applications problems.

Each line below forms an equation whose solution is the number shown. Each statement contains two blanks. One must be filled with an $=$ sign to form the equation. Fill in the other blank with either $+$ or $-$ to form an equation with the solution shown to the right. The first equation is done for you.

Equation

$3x$	$=$	5	$+$	1
4		x		8
$-3a$		-1		10
-2		y		10
$5t$		1		-11
2		$4x$		6

Solution

2

4

-3

8

-2

-1

Table 11-3 **Distributive Property of Multiplication over Addition**

Property	In Symbols/Examples	Comments/Notes
Distributive property of multiplication over addition*	$a \cdot (b + c) = a \cdot b + a \cdot c$ Ex.: $-3(2 + x) = -3(2) + (-3)(x)$ $= -6 + (-3)x$ $= -6 - 3x$	The factor outside the parentheses is multiplied by each term in the sum.

*Note that the distributive property of multiplication over addition is sometimes referred to as just the *distributive property*.

Example 3 demonstrates the use of the commutative properties.

Skill Practice

Apply the commutative property of addition or multiplication to rewrite the expression.

6. $x + 3$

7. $z(2)$

8. $-12 + p$

9. ab

Example 3 **Applying the Commutative Properties of Real Numbers**

Apply the commutative property of addition or multiplication to rewrite the expression.

a. $6 + p$ **b.** $y(7)$ **c.** $-5 + n$ **d.** xy

Solution:

a. $6 + p = p + 6$ Commutative property of addition

b. $y(7) = 7y$ Commutative property of multiplication

c. $-5 + n = n + (-5)$ Commutative property of addition

         $= n - 5$

d. $xy = yx$ Commutative property of multiplication

Recall from Section 1.3 that subtraction is not a commutative operation. However, if we rewrite the difference of two numbers $a - b$ as $a + (-b)$, then we can apply the commutative property of addition. For example,

$$x - 9 = x + (-9) \qquad \text{Rewrite as addition of the opposite.}$$

$$= -9 + x \qquad \text{Apply the commutative property of addition.}$$

Example 4 demonstrates the associative properties of addition and multiplication.

Skill Practice

Use the associative property of addition or multiplication to rewrite the expression. Then simplify the expression.

10. $2(-8w)$

11. $6.2 + (14.7 + x)$

12. $-\dfrac{1}{3}(3y)$

Example 4 **Applying the Associative Properties of Real Numbers**

Use the associative property of addition or multiplication to rewrite each expression. Then simplify the expression.

a. $5(7w)$ **b.** $1.2 + (4.5 + y)$ **c.** $-\dfrac{2}{5}\left(-\dfrac{5}{2}z\right)$

Solution:

a. $5(7w) = (5 \cdot 7)w$ Apply the associative property of multiplication.

      $= 35w$ Simplify.

b. $1.2 + (4.5 + y) = (1.2 + 4.5) + y$ Apply the associative property of addition.

      $= 5.7 + y$ Simplify.

Answers

6. $3 + x$ **7.** $2z$

8. $p + (-12)$ or $p - 12$

9. ba **10.** $[2 \cdot (-8)]w; -16w$

11. $(6.2 + 14.7) + x; 20.9 + x$

12. $\left(-\dfrac{1}{3} \cdot 3\right)y; -y$

c. $-\dfrac{2}{5}\left(-\dfrac{5}{2}z\right) = \left(-\dfrac{2}{5} \cdot -\dfrac{5}{2}\right)z$ Apply the associative property of multiplication.

$\qquad\qquad = 1 \cdot z$ Simplify. Note that the factors within the parentheses are reciprocals. Therefore, their product is 1.

$\qquad\qquad = z$

Note that in most cases, a detailed application of the associative properties will not be given. Instead the process will be written in one step, such as

$$5(7w) = 35w \qquad 1.2 + (4.5 + y) = 5.7 + y \qquad -\dfrac{2}{5}\left(-\dfrac{5}{2}z\right) = z$$

Example 5 demonstrates the use of the distributive property.

Example 5   **Applying the Distributive Property**

Apply the distributive property.

a. $3(x + 4)$ **b.** $2(3y - 5z + 1)$ **c.** $\dfrac{2}{3}\left(6p + \dfrac{1}{4}\right)$

Solution:

a. $3(x + 4) = 3(x) + 3(4)$ Apply the distributive property.

$\qquad\qquad = 3x + 12$ Simplify.

b. $2(3y - 5z + 1) = 2(3y + (-5z) + 1)$ First write the subtraction as addition of the opposite.

$\qquad\qquad = 2(3y + (-5z) + 1)$ Apply the distributive property.

$\qquad\qquad = 2(3y) + 2(-5z) + 2(1)$

$\qquad\qquad = 6y + (-10z) + 2$ Simplify.

$\qquad\qquad = 6y - 10z + 2$

TIP: In Example 5(b), we rewrote the expression by writing the subtraction as addition of the opposite. Often this step is not shown, and fewer steps are shown overall. For example:

$$2(3y - 5z + 1) = 2(3y) + 2(-5z) + 2(1)$$
$$= 6y - 10z + 2$$

c. $\dfrac{2}{3}\left(6p + \dfrac{1}{4}\right) = \dfrac{2}{3}(6p) + \dfrac{2}{3}\left(\dfrac{1}{4}\right)$ Apply the distributive property.

$\qquad\qquad = \dfrac{2}{3}\left(\dfrac{6p}{1}\right) + \dfrac{2}{3}\left(\dfrac{1}{4}\right)$ Write the whole number as an improper fraction.

$\qquad\qquad = \dfrac{2}{\underset{1}{3}}\left(\dfrac{\overset{2}{6p}}{1}\right) + \dfrac{2}{3}\left(\dfrac{1}{\underset{2}{4}}\right)$ Multiply fractions.

$\qquad\qquad = 4p + \dfrac{1}{6}$ Simplify.

Skill Practice

Apply the distributive property.

13. $4(2 + m)$

14. $6(5p - 3q + 1)$

15. $\dfrac{5}{2}\left(4t + \dfrac{1}{3}\right)$

Answers

13. $8 + 4m$

14. $30p - 18q + 6$

15. $10t + \dfrac{5}{6}$

Skill Practice

Apply the distributive property.

16. $-4(6 - 10x)$

17. $-(2x - 3y + 4z)$

Example 6   Applying the Distributive Property

Apply the distributive property.

a. $-8(2 - 5y)$ **b.** $-(-4a + b + 3c)$

Solution:

a. $-8(2 - 5y)$

$= -8[2 + (-5y)]$ Write the subtraction as addition of the opposite.

$= -8[2 + (-5y)]$ Apply the distributive property.

$= -8(2) + (-8)(-5y)$

$= -16 + 40y$ Simplify.

b. $-(-4a + b + 3c)$ The negative sign preceding the parentheses indicates that we take the opposite of the expression within parentheses. This is equivalent to multiplying the expression within parentheses by -1.

$= -1 \cdot (-4a + b + 3c)$

$= -1(-4a) + (-1)(b) + (-1)(3c)$ Apply the distributive property.

$= 4a - b - 3c$ Simplify.

TIP: Notice that a negative factor outside the parentheses changes the signs of all terms to which it is multiplied.

$$-1 \cdot (-4a + b + 3c)$$
$$= +4a - b - 3c$$

Answers

16. $-24 + 40x$

17. $-2x + 3y - 4z$

Section 11.1   Practice Exercises

Study Skills Exercises

1. When beginning a study of algebra, some students do not understand the concept of a variable. A variable is a letter that represents an unknown value. When trying to find what number added to 5 equals -12, we write $5 + x = -12$. Rewrite the following, using variables:

What number times 4 equals 6? _____

What number divided by 5 equals 3? _____

2. Define the key terms.

    a. Constant   **b. Expression**   **c. Variable**

Objective 1: Algebraic Expressions

3. Maria needs to buy 8 wine glasses. Write an expression for the cost of 8 glasses at p dollars each.
(See Example 1.)

4. Carolyn sells homemade candles. Write an expression for her total revenue if she sells 5 candles for r dollars each.

5. Jonathan is 4 in. taller than his brother. Write an expression for Jonathan's height if his brother is t in. tall.

6. It takes Perry $\frac{1}{2}$ hr longer than David to mow the lawn. If it takes David l hr to mow the lawn, write an expression for the amount of time it takes Perry to mow the lawn.

7. A sedan travels 6 mph slower than a sports car. Write an expression for the speed of the sedan if the sports car travels v mph.

8. Bill's daughter is 30 yr younger than he is. Write an expression for his daughter's age if Bill is A yr old.

9. A piece of ribbon is cut into n pieces of equal length. If the length of the ribbon is 4 yd, write an expression for the length of each piece.

10. A party is planned for 20 people. If there is p oz of punch available, write an expression for the amount of punch for each person, assuming that each person drinks an equal amount.

11. The price of gas has doubled over the last 3 yr. If gas cost g dollars per gallon 3 yr ago, write an expression for the current price per gallon.

12. Suppose that the amount of rain that fell on Monday was twice the amount that fell on Sunday. Write an expression for the amount of rain on Monday, if Sunday's amount was t in.

Objective 2: Evaluating Expressions

For Exercises 13–20, evaluate the expression for the given values. **(See Example 2.)**

13. $-6x$ for

    a. $x = 2$

    b. $x = -5$

14. $-2y^2$ for

    a. $y = 3$

    b. $y = -3$

15. $3p + 5q$ for

    a. $p = 2, q = -\dfrac{1}{5}$

    b. $p = -5, q = 0$

16. $9c - 2d$ for

    a. $c = -1, d = \dfrac{1}{2}$

    b. $c = 3, d = -2$

17. $-a^2$ for

    a. $a = -7$

    b. $a = 7$

18. $-b^3$ for

    a. $b = -3$

    b. $b = 3$

19. $-4(r - s)^2$ for

    a. $r = 8, s = 6$

    b. $r = 3, s = -1$

20. $-5(u + v)^2$ for

    a. $u = 10, v = -7$

    b. $u = 0, v = -2$

For Exercises 21–24, evaluate the expression when $x = -2$, $y = \frac{2}{3}$, $z = 4$, and $w = -\frac{1}{2}$.

21. $y(x - 4)$ **22.** $w(-x - 4)$ **23.** $z^2 - x + 6$ **24.** $x^3 - w - \frac{3}{2}$

For Exercises 25–28, evaluate the expression when $a = 12$, $b = -3$, and $c = -2$.

25. $bc \div a$ **26.** $5b - c$ **27.** $b^2 - c^2$ **28.** $b^2 + c^2$

For Exercises 29–32, find the area A or perimeter P.

29. $P = 2l + 2w$ for $l = 6$ in. and $w = 2.3$ in. (perimeter of a rectangle)

30. $A = lw$ for $l = \frac{3}{2}$ ft and $w = 4$ ft (area of a rectangle)

31. $A = \pi r^2$ for $\pi = \frac{22}{7}$ m and $r = \frac{7}{2}$ m (area of a circle)

32. $A = \frac{1}{2}bh$ for $b = \frac{4}{5}$ yd and $h = \frac{10}{11}$ yd (area of a triangle)

Objective 3: Properties of Real Numbers

For Exercises 33–44, apply the commutative property of addition or multiplication to rewrite each expression.
(See Example 3.)

33. $5 + w$ **34.** $t + 2$ **35.** $-\frac{1}{3} + b$ **36.** $-\frac{1}{2} + c$

37. $r(2)$ **38.** $a(-4)$ **39.** $t(-s)$ **40.** $d(-c)$

41. xy **42.** ab **43.** $7 - p$ **44.** $8 - q$

For Exercises 45–56, apply the associative property of addition or multiplication to rewrite each expression. Then simplify the expression. **(See Example 4.)**

45. $-2(6b)$ **46.** $-3(2c)$ **47.** $3 + (8 + t)$ **48.** $7 + (5 + p)$

49. $-4.2 + (2.5 + r)$ **50.** $1.1 + (-0.8 + w)$ **51.** $3(6x)$ **52.** $9(5k)$

53. $-\frac{4}{7}\left(-\frac{7}{4}d\right)$ **54.** $\frac{5}{6}\left(\frac{6}{5}m\right)$ **55.** $-9 + (-12 + h)$ **56.** $-11 + (-4 + s)$

For Exercises 57–76, apply the distributive property. **(See Examples 5–6.)**

57. $4(x + 8)$ **58.** $5(3 + w)$ **59.** $-2(p + 4)$ **60.** $-6(k + 2)$

61. $-10(t - 3)$ **62.** $-7(p - 4)$ **63.** $-5(-2 + x)$ **64.** $-8(-3 + y)$

65. $4(a + 4b - c)$ **66.** $2(3q - r + s)$ **67.** $4\left(\frac{2}{3} + g\right)$ **68.** $8\left(\frac{5}{6} + m\right)$

69. $-(3 - n)$ **70.** $-(13 - t)$ **71.** $-(-a - 8)$ **72.** $-(-d - 10)$

73. $-(3x + 9 - 5y)$ **74.** $-(a - 8b + 4c)$ **75.** $-(-5q - 2s - 3t)$ **76.** $-(-10p - 12q + 3)$

Mixed Exercises

For Exercises 77–96, apply the appropriate property to simplify the expression.

77. $6(2x)$ **78.** $-3(12k)$ **79.** $6(2 + x)$ **80.** $-3(12 + k)$

81. $-6(-1 - k)$ **82.** $-4(-8 + h)$ **83.** $-6 + (-1 - k)$ **84.** $-4 + (-8 + h)$

85. $-8 + (4 - p)$ **86.** $3 + (25 - m)$ **87.** $-8(4 - p)$ **88.** $3(25 - m)$

89. $8\left(\dfrac{1}{2}a\right)$ **90.** $-20\left(\dfrac{1}{5}b\right)$ **91.** $8\left(\dfrac{1}{2} + a\right)$ **92.** $-20\left(\dfrac{1}{5} + b\right)$

93. $\dfrac{5}{9}(9 + y)$ **94.** $-\dfrac{3}{4}(8 - b)$ **95.** $\dfrac{5}{9}(9y)$ **96.** $-\dfrac{3}{4}(8b)$

Simplifying Expressions

1. Definition of *Like* Terms

Objectives

1. **Definition of *Like* Terms**
2. **Combining *Like* Terms**
3. **Clearing Parentheses and Combining *Like* Terms**

An algebraic expression is the sum of one or more terms. A **term** is a number or a product or quotient of numbers and variables. For example, the expression

$$-8x^3 + xy - 40 \quad \text{can be written as} \quad -8x^3 + xy + (-40)$$

This expression consists of the terms $-8x^3$, xy, and -40. The terms $-8x^3$ and xy are called **variable terms**, and the term -40 is called a **constant term**.

It is important to distinguish between a term and the factors within a term. For example, the quantity xy is one term, and the values x and y are factors within the term. The constant factor in a term is called the **coefficient** of the term.

Term	Coefficient of the term
$-8x^3$	-8
xy or $1xy$	1
-40	-40

Avoiding Mistakes

Variables without a coefficient explicitly written have a coefficient of 1. Thus, the term xy is equal to $1xy$. The 1 is understood.

Terms are said to be *like terms* if they each have the same variables and the corresponding variables are raised to the same powers. For example,

Like Terms	Unlike Terms	
$-4x$ and $6x$	$-4x$ and $6y$	(different variables)
$18ab$ and $4ba$	$18ab$ and $4a$	(different variables)
$7m^2n^5$ and $3m^2n^5$	$7m^2n^5$ and $3mn^5$	(different powers of m)
$5p$ and $-3p$	$5p$ and 3	(different variables)
8 and 10	8 and $10x$	(different variables)

Skill Practice

Given:

$-4.8y + 8.1y^2 - \dfrac{1}{2}y - 8$

1. List the terms of the expression.
2. List the coefficients of the expression.
3. Which two terms are *like* terms?

Example 1 Identifying Terms, Coefficients, and *Like* Terms

a. List the terms of the expression: $1.4x^3 - 6x^2 + x + 5$

b. Identify the coefficient of each term: $1.4x^3 - 6x^2 + x + 5$

c. Which two terms are *like* terms? $-6x, 5, -3y$, and $4x$

Solution:

a. The expression $1.4x^3 - 6x^2 + x + 5$ can be written as
$1.4x^3 + (-6x^2) + x + 5$

Therefore, the terms are $1.4x^3, -6x^2, x$, and 5.

b. The coefficients are $1.4, -6, 1$, and 5.

c. The terms $-6x$ and $4x$ are *like* terms.

2. Combining *Like* Terms

Two terms may be combined if they are *like* terms. To add or subtract *like* terms, we use the distributive property, as shown in Example 2.

Skill Practice

Add or subtract as indicated.
4. $-4w + 11w$
5. $z - 4z + 22z$

Example 2 Using the Distributive Property to Add and Subtract *Like* Terms

Add or subtract as indicated.

a. $8y + 6y$ **b.** $-15w + 4w - w$

Solution:

a. $8y + 6y = (8 + 6)y$ Apply the distributive property.

$= 14y$ Simplify.

b. $-15w + 4w - w = -15w + 4w - 1w$ First note that $w = 1w$.

$= (-15 + 4 - 1)w$ Apply the distributive property.

$= (-12)w$ Simplify within parentheses.

$= -12w$

Although the distributive property is used to add and subtract *like* terms, it is tedious to write each step. Observe that adding or subtracting *like* terms is a matter of adding or subtracting the coefficients and leaving the variable factors unchanged. This can be shown in one step.

$$8y + 6y = 14y \qquad \text{and} \qquad -15w + 4w - 1w = -12w$$

This shortcut will be used throughout the text.

Answers

1. $-4.8y, 8.1y^2, -\frac{1}{2}y, -8$
2. $-4.8, 8.1, -\frac{1}{2}, -8$
3. $-4.8y$ and $-\frac{1}{2}y$
4. $7w$ **5.** $19z$

| **Example 3** | **Adding and Subtracting *Like* Terms** |

Simplify by combining *like* terms.

$$-3x + 8y + 4x - 19 - 10y$$

Solution:

$-3x + 8y + 4x - 19 - 10y$

$= -3x + 4x + 8y - 10y - 19$ Arrange *like* terms together.

$= 1x - 2y - 19$ Combine *like* terms.

$= x - 2y - 19$ Note that $1x = x$. Also note that the remaining terms cannot be combined because they are not *like* terms. The variable factors are different.

Skill Practice

Simplify.

6. $4a - 10b - a + 16b + 9$

| **Example 4** | **Adding and Subtracting *Like* Terms** |

Simplify by combining *like* terms.

a. $\frac{2}{5}m + \frac{1}{8}n - \frac{1}{5}m + \frac{3}{8}n$ **b.** $0.2a - 1.4 + 1.4a - 6b - 2.1$

Solution:

a. $\frac{2}{5}m + \frac{1}{8}n - \frac{1}{5}m + \frac{3}{8}n$

$= \frac{2}{5}m - \frac{1}{5}m + \frac{1}{8}n + \frac{3}{8}n$ Arrange *like* terms together.

$= \frac{1}{5}m + \frac{4}{8}n$ Combine *like* terms.

$= \frac{1}{5}m + \frac{1}{2}n$ Simplify fractions.

b. $0.2a - 1.4 + 1.4a - 6b - 2.1$

$= 0.2a + 1.4a - 6b - 1.4 - 2.1$ Arrange *like* terms together.

$= 1.6a - 6b - 3.5$ Combine *like* terms.

Skill Practice

Simplify.

7. $\frac{5}{9} - \frac{2}{3}w + \frac{5}{3}w - \frac{4}{9}$

8. $6.3x - 4.1y - 2.4 + 2.1y + 1.1$

3. Clearing Parentheses and Combining *Like* Terms

Notice that when the distributive property is applied, the original parentheses are dropped. This is often called *clearing parentheses*.

Answers

6. $3a + 6b + 9$ **7.** $w + \frac{1}{9}$

8. $6.3x - 2y - 1.3$

Skill Practice

Simplify.

9. $8 - 6(w + 4)$

Example 5   Clearing Parentheses and Combining *Like* Terms

Simplify by clearing parentheses and combining *like* terms. $6 - 3(2y + 9)$

Solution:

$6 - 3(2y + 9)$ The order of operations indicates that we must perform multiplication before subtraction.

It is also important to understand that a factor of -3 (not 3) will be multiplied by all terms within the parentheses. To see why, rewrite the subtraction in terms of addition of the opposite.

$$6 - 3(2y + 9) = 6 + (-3)(2y + 9)$$ Rewrite subtraction as addition of the opposite.

$$= 6 + (-3)(2y) + (-3)(9)$$ Apply the distributive property.

$$= 6 + (-6y) + (-27)$$ Simplify.

$$= -6y + 6 + (-27)$$ Arrange *like* terms together.

$$= -6y - 21$$ Combine *like* terms.

Skill Practice

Simplify.

10. $-5(10 - m) - 2(m + 1)$

Example 6   Clearing Parentheses and Combining *Like* Terms

Simplify by clearing parentheses and combining *like* terms.
$-8(x - 4) - 5(x + 7)$

Solution:

$-8(x - 4) - 5(x + 7)$

$$= -8[x + (-4)] + (-5)(x + 7)$$ Rewrite subtraction as addition of the opposite.

$$= -8[x + (-4)] + (-5)(x + 7)$$ Apply the distributive property.

$$= -8(x) + (-8)(-4) + (-5)(x) + (-5)(7)$$

$$= -8x + 32 - 5x - 35$$ Simplify.

$$= -8x - 5x + 32 - 35$$ Arrange *like* terms together.

$$= -13x - 3$$ Combine *like* terms.

Answers

9. $-6w - 16$ **10.** $3m - 52$

Section 11.2   Practice Exercises

Boost your GRADE at ALEKS.com!

 version 3.0

- Practice Problems
- Self-Tests
- NetTutor
- e-Professors
- Videos

Study Skills Exercises

1. Two important concepts in this section are *terms* and *factors*. Consider the expression $2x + 5y$. The quantities $2x$ and $5y$ are terms of the expression. Now consider the single term expression $2xy$. In this expression, 2, x, and y are factors. Write in your own words the difference between a term and a factor.

2. Define the key terms.

    a. Coefficient   **b. Constant term**   **c. *Like* terms**   **d. Term**   **e. Variable term**

Review Exercises

For Exercises 3–8, simplify the expression, using the associative or distributive properties to clear parentheses.

3. $6(p + 3)$

4. $(-7p + 2) + 10$

5. $4(-6q)$

6. $-3(t - 2)$

7. $13 + (-4 - h)$

8. $-(x - 20y - 14z)$

Objective 1: Definition of *Like* Terms

For Exercises 9–16, for each expression, list the terms and identify each term as a variable term or a constant term. **(See Example 1.)**

9. $2a + 5b^2 + 6$

10. $-5x - 4 + 7y$

11. $8 + 9a$

12. $12 - 8k$

13. $4pq - 9p$

14. $9t - 8st$

15. $10h^2 - 15 - 4h$

16. $w^2 - 3 - 5w$

For Exercises 17–24, identify the coefficients for each term. **(See Example 1.)**

17. $6p - 4q$

18. $-5a^3 - 2a$

19. $-14h + 12$

20. $8x + 9$

21. $x - y$

22. $p - q$

23. $5t - 8s - 3$

24. $6g - 16h - 2$

For Exercises 25–36, determine if the two terms are *like* terms or unlike terms. **(See Example 1.)**

25. $3a, -2a$

26. $8b, 12b$

27. $4x, 4y$

28. $-9k, -9h$

29. $7xy, -3yx$

30. $-5ab, ba$

31. $6a, 13a^2$

32. $20k^3, 3k$

33. $14, 14y$

34. $25x, 25$

35. $17, -32$

36. $8, -22$

Objective 2: Combining *Like* Terms

For Exercises 37–62, combine the *like* terms. **(See Examples 2–4.)**

37. $6rs + 8rs$

38. $4x + 21x$

39. $-4h + 12h$

40. $9p - 13p$

41. $4x^2 + 9 - x^2$

42. $13t^2 - t^2 + 4$

43. $10x - 12y - 4x - 3y$

44. $14a - 5b + 3a - b$

45. $-6k - 9k + 12k$

46. $-11p + 23p - p$

47. $-8uv + 6u + 12uv$

48. $9pq - 9p + 13pq$

49. $6 - 14m - 15 - 2m$

50. $1 - 8n + 5 - 3n$

51. $18 - 3a + 5b - 6a + 2$

52. $13 + w - 5z - 4 + 7w$

53. $-5p^2 + 6p - p^2 + 7 - 8p$

54. $-3q^2 - 10q + q^2 - 15 + 5q$

55. $\frac{1}{2}y + \frac{3}{2}y - \frac{5}{6}$

56. $-\frac{4}{5}p + \frac{2}{5}p + \frac{4}{7}$

57. $\frac{3}{4}a + 3 - \frac{1}{8}a + 6$

58. $\frac{1}{3}b - 4 + \frac{2}{9}b - 4$

59. $2.3x^2 + 4.1x - 5.3x^2 - 6x$

60. $1.2y - 0.4y^2 - 0.3y - 1.5y^2$

61. $4.4 - 0.9a + 3.2$

62. $9.7 - 8.8b - 3.2$

Objective 3: Clearing Parentheses and Combining *Like* Terms

For Exercises 63–86, clear parentheses and combine *like* terms. **(See Examples 5–6.)**

63. $5(t - 6) + 2$

64. $7(a - 4) + 8$

65. $-3(2x + 1) - 13$

66. $-2(4b + 3) - 10$

67. $4 + 6(y - 3)$

68. $11 + 2(p - 8)$

69. $21 - 7(3 - q)$

70. $10 - 5(2 - 5m)$

71. $-3 - (2n + 1)$

72. $-13 - (6s + 5)$

73. $-2(a + 3b) - (4a - 5b)$

74. $-(2m - 7n) - 3(6m - n)$

75. $10(x + 5) - 3(2x + 9)$

76. $6(y - 9) - 5(2y - 5)$

77. $-(12z + 1) + 2(7z - 5)$

78. $-(8w + 5) + 3(w - 15)$

79. $3(w + 3) - (4w + y) - 3y$

80. $2(s + 6) - (8s - t) + 6t$

81. $20a - 4(b + 3a) - 5b$

82. $16p - 3(2p - q) + 7q$

83. $6 - (3m - n) - 2(m + 8) + 5n$

84. $12 - (5u + v) - 4(u - 6) + 2v$

85. $15 + 2(w - 4) - (2w - 5z) + 7z$

86. $7 + 3(2a - 5) - (6a - 8b) - 2b$

Expanding Your Skills

For Exercises 87–94, clear parentheses and combine *like* terms in expressions involving fractions and decimals.

87. $6\left(\dfrac{1}{2}x - \dfrac{2}{3}\right) - 4\left(\dfrac{5}{2}x + \dfrac{3}{4}\right)$

88. $-12\left(\dfrac{5}{6}p + \dfrac{1}{4}\right) + 9\left(\dfrac{2}{9}p - \dfrac{1}{3}\right)$

89. $\dfrac{2}{3}(9y + 6) - \dfrac{3}{2}(18y - 16)$

90. $-\dfrac{1}{4}(4w - 8) + \dfrac{1}{2}(4w + 10)$

91. $10(0.2q - 3) - 100(0.04q - 0.5)$

92. $100(0.14b + 0.2) - 10(1.3b - 4)$

93. $100(1.04a - 2.1b) - 10(21.1a + 0.3b)$

94. $10(-7.2x - y) + 1000(0.023x + 0.004y)$

Section 11.3 Addition and Subtraction Properties of Equality

Objectives

1. Definition of an Equation
2. Addition and Subtraction Properties of Equality

1. Definition of an Equation

An **equation** is a statement that indicates that two quantities are equal. The following are equations.

$$x = 7 \qquad z + 3 = 8 \qquad -6p = 18$$

All equations have an equal sign. Furthermore, notice that the equal sign separates the equation into two parts, the left-hand side and the right-hand side. A **solution** to an equation is a value of the variable that makes the equation a true statement. Substituting a solution to an equation for the variable makes the right-hand side equal to the left-hand side.

Equation	Solution	Check
$x = 7$	7	$x = 7$

$$7 = 7 \checkmark$$

Substitute 7 for x.
Right-hand side equals left-hand side.

$z + 3 = 8$ 5 $z + 3 = 8$

$$5 + 3 = 8$$

Substitute 5 for z.
Right-hand side equals left-hand side.

$$8 = 8 \checkmark$$

$-6p = 18$ -3 $-6p = 18$

$$-6(-3) = 18$$

Substitute -3 for p.
Right-hand side equals left-hand side.

$$18 = 18 \checkmark$$

Concept Connections

Which of the following are equations?

1. $-2w = 6$
2. $4 = t - 3$
3. $x + 8$

Example 1 Determining Whether a Number Is a Solution to an Equation

Determine whether the given number is a solution to the equation.

a. $2x - 9 = 3$; 6   **b.** $8 = 8p - 4$; $-\frac{1}{2}$

Solution:

a. $2x - 9 = 3$

$2(6) - 9 \stackrel{?}{=} 3$ Substitute 6 for x.

$12 - 9 \stackrel{?}{=} 3$ Simplify.

$3 = 3 \checkmark$ The right-hand side equals the left-hand side.
Thus, 6 is a solution to the equation $2x - 9 = 3$.

b. $8 = 8p - 4$

$8 \stackrel{?}{=} 8\left(-\frac{1}{2}\right) - 4$ Substitute $-\frac{1}{2}$ for p.

$8 \stackrel{?}{=} -4 - 4$ Simplify.

$8 \neq -8$ The right-hand side does not equal left-hand side.
Thus, $-\frac{1}{2}$ is *not* a solution to the equation $8 = 8p - 4$.

Skill Practice

Determine whether the given number is a solution to the equation.

4. $2 + 3x = 23$; 7
5. $-4x + 1 = 9$; 2

In the study of algebra, you will encounter a variety of equations. In this chapter, we focus on a specific type of equation called a linear equation in one variable.

DEFINITION Linear Equation in One Variable

Let a and b be real numbers such that $a \neq 0$. A **linear equation in one variable** is an equation that can be written in the form

$$ax + b = 0$$

Answers

1. Equation   2. Equation
3. Not an equation
4. Yes, 7 is a solution.
5. No, 2 is not a solution.

2. Addition and Subtraction Properties of Equality

Given the equation $x = 3$, we can easily determine that the solution is 3. The solution to the equation $2x + 14 = 20$ is also 3. These two equations are called **equivalent equations** because they have the same solution. However, while the solution to $x = 3$ is obvious, the solution to $2x + 14 = 20$ is not. Our goal in this chapter is to learn how to *solve* equations.

> ┌ **Concept Connections** ┐
>
> **6.** Are the two equations equivalent equations?
>
> $x = 5$ and $2x = 10$

To solve an equation, we use algebraic principles to write an equation such as $2x + 14 = 20$ in an equivalent but simpler form, such as $x = 3$. The addition and subtraction properties of equality are the first tools we will learn to solve an equation.

> **PROPERTY**   **The Addition and Subtraction Properties of Equality**
>
> Let a, b, and c represent algebraic expressions.
>
> **1.** The **addition property of equality:**   If $a = b$
>   then $a + c = b + c$
>
> **2.** The **subtraction property of equality:**   If $a = b$
>   then $a - c = b - c$

The addition and subtraction properties of equality indicate that adding or subtracting the same quantity to each side of an equation results in an equivalent equation. This is true because if two quantities are increased (or decreased) by the same amount, then the resulting quantities will also be equal (Figure 11-2).

Figure 11-2

> ┌ **Skill Practice** ┐
>
> Solve the equation and check the solution.
>
> **7.** $y - 4 = 12$
>
> **8.** $7.9 = -3.2 + p$

> **Example 2**   Applying the Addition Property of Equality

Solve the equations and check the solution.

   a. $x - 6 = 18$ **b.** $3.8 = -4.1 + x$

Solution:

To solve an equation, the goal is to isolate the variable on one side of the equation. That is, we want to create an equivalent equation of the form $x = $ number. To accomplish this, we can use the fact that the sum of a number and its opposite is zero.

   a. $x - 6 = 18$

       $x - 6 + 6 = 18 + 6$ To isolate x, add 6 to both sides, because $-6 + 6 = 0$.

           $x + 0 = 24$ Simplify.

              $x = 24$ The variable is isolated (by itself) on the left-hand side of the equation. The solution is 24.

Answers

6. Yes, because the solution to each equation is the same. The solution is 5 for both.

7. 16   **8.** 11.1

Check: $x - 6 = 18$ Original equation

$(24) - 6 \stackrel{?}{=} 18$ Substitute 24 for x.

$18 = 18$ ✓   The right-hand side = the left-hand side.

b. $3.8 = -4.1 + x$

$3.8 + 4.1 = -4.1 + 4.1 + x$ To isolate x, add 4.1 to both sides, because $-4.1 + 4.1 = 0$.

$7.9 = 0 + x$ Simplify.

$7.9 = x$ The solution is 7.9.

Check: $3.8 = -4.1 + x$ Original equation

$3.8 \stackrel{?}{=} -4.1 + (7.9)$ Substitute 7.9 for x.

$3.8 = 3.8$ ✓

TIP: Notice that the variable may be isolated on *either* side of the equal sign. In Example 2(a), the variable appears on the left. In Example 2(b), the variable appears on the right.

In Example 3, we apply the subtraction property of equality. This indicates that we can subtract the same quantity from both sides of the equation to obtain an equivalent equation.

Example 3   Applying the Subtraction Property of Equality

Solve the equations and check.

a. $z + 11 = 14$ **b.** $-8 = 2 + q$

Solution:

a. $z + 11 = 14$

$z + 11 - 11 = 14 - 11$ Subtract 11 from both sides, because $11 - 11 = 0$.

$z + 0 = 3$ Simplify.

$z = 3$ The solution is 3.

Check: $z + 11 = 14$ Original equation

$(3) + 11 \stackrel{?}{=} 14$ Substitute 3 for z.

$14 = 14$ ✓

b. $-8 = 2 + q$

$-8 - 2 = 2 - 2 + q$ Subtract 2 from both sides, because $2 - 2 = 0$.

$-10 = 0 + q$ Simplify.

$-10 = q$ The solution is -10.

Check: $-8 = 2 + q$ Original equation

$-8 \stackrel{?}{=} 2 + (-10)$ Substitute -10 for q.

$-8 = -8$ ✓

Skill Practice

Solve the equation and check the solution.

9. $m + 8 = 21$

10. $-16 = 1 + z$

Answers

9. 13   **10.** -17

Example 4 Applying the Addition and Subtraction Properties of Equality

Solve the equations.

a. $\dfrac{3}{10} + m = -\dfrac{2}{3}$ **b.** $8.54 = p + 1.96$

Solution:

a.
$$\dfrac{3}{10} + m = -\dfrac{2}{3}$$

To isolate m, we subtract $\frac{3}{10}$ because $\frac{3}{10} - \frac{3}{10} = 0$.

$$\dfrac{3}{10} - \dfrac{3}{10} + m = -\dfrac{2}{3} - \dfrac{3}{10}$$

Subtract $\frac{3}{10}$ from both sides.

$$0 + m = -\dfrac{2 \cdot 10}{3 \cdot 10} - \dfrac{3 \cdot 3}{10 \cdot 3}$$

To subtract the fractions, first obtain a common denominator. The LCD is 30.

$$m = -\dfrac{20}{30} - \dfrac{9}{30}$$

$$m = -\dfrac{29}{30}$$

The solution is $-\frac{29}{30}$ and checks in the original equation.

b.
$$8.54 = p + 1.96$$

To isolate p, we subtract 1.96 because $1.96 - 1.96 = 0$.

$$8.54 - 1.96 = p + 1.96 - 1.96$$

Subtract 1.96 from each side.

$$6.58 = p + 0$$

$$6.58 = p$$

The solution is 6.58 and checks in the original equation.

Section 11.3 Practice Exercises

Study Skills Exercises

1. Up to this point we have been simplifying expressions. We will now begin solving equations. Consider the two lists:

Expressions	Equations
$3x + 2y$	$5x + 2 = 6$
$6(8 + x) + 2$	$2(x - 5) = 14$
$7y$	$7 = y$

Explain the difference between an expression and an equation.

2. Define the key terms.

    a. Addition property of equality   **b.** Equation   **c.** Equivalent equations

    d. Linear equation in one variable   **e.** Solution   **f.** Subtraction property of equality

Review Exercises

For Exercises 3–8, simplify the expression.

3. $-10a + 3b - 3a + 13b$ **4.** $4 - 23y + 11 - 16y$ **5.** $-(-8h + 2k - 13)$

6. $3(-4m + 3) - 12$ **7.** $5z - 8(z - 3) - 20$ **8.** $-(7p - 12) - 10(1 - p) + 6$

Objective 1: Definition of an Equation

For Exercises 9–20, determine whether the given number is a solution to the equation. **(See Example 1.)**

9. $5x + 3 = -2$; -1 **10.** $3y - 2 = 4$; 2 **11.** $10 = p - 16$; 26

12. $-14 = q - 1$; -13 **13.** $-z + 8 = 20$; 12 **14.** $-7 - w = -10$; -3

15. $6m - 3 = -6$; $-\dfrac{1}{2}$ **16.** $-12n + 2 = -1$; $\dfrac{1}{4}$ **17.** $13 = 13 + 6t$; 0

18. $-\dfrac{1}{5} = r - \dfrac{1}{5}$; 0 **19.** $25 = -5q - 5$; 4 **20.** $39 = -7p - 4$; 5

Objective 2: Addition and Subtraction Properties of Equality

For Exercises 21–26, fill in the blank with the appropriate number.

21. $13 + (-13) = \underline{\hspace{1cm}}$ **22.** $6 + \underline{\hspace{1cm}} = 0$ **23.** $\underline{\hspace{1cm}} + (-7) = 0$

24. $1 + (-1) = \underline{\hspace{1cm}}$ **25.** $3.2 + \underline{\hspace{1cm}} = 0$ **26.** $\underline{\hspace{1cm}} + (-0.3) = 0$

For Exercises 27–38, solve the equation using the addition property of equality. **(See Example 2.)**

27. $g - 23 = 14$ **28.** $h - 12 = 30$ **29.** $-4 + k = 12$

30. $-16 + m = 4$ **31.** $-18 = n - 3$ **32.** $-9 = t - 6$

33. $-\dfrac{5}{6} + p = \dfrac{1}{3}$ **34.** $-\dfrac{3}{4} + q = \dfrac{3}{2}$ **35.** $k - 4.3 = -1.2$

36. $a - 0.04 = -2.04$ **37.** $13 = -21 + w$ **38.** $2 = -17 + w$

For Exercises 39–44, fill in the blank with the appropriate number.

39. $52 - \underline{\hspace{1cm}} = 0$ **40.** $2 - 2 = \underline{\hspace{1cm}}$ **41.** $18 - 18 = \underline{\hspace{1cm}}$

42. $\underline{\hspace{1cm}} - 15 = 0$ **43.** $\underline{\hspace{1cm}} - 100 = 0$ **44.** $21 - \underline{\hspace{1cm}} = 0$

For Exercises 45–56, solve the equation using the subtraction property of equality. **(See Example 3.)**

45. $x + 34 = 6$ **46.** $y + 12 = 4$ **47.** $17 + b = 20$

48. $5 + c = 14$ **49.** $-32 = t + 14$ **50.** $-23 = k + 11$

51. $8.2 = 21.8 + m$ **52.** $16.01 = 20.88 + n$ **53.** $a + \dfrac{3}{5} = -\dfrac{7}{10}$

54. $b + \dfrac{1}{4} = -\dfrac{3}{8}$ **55.** $21 = 14 + w$ **56.** $9 = 8 + u$

Mixed Exercises

For Exercises 57–77, solve the equation by using the appropriate property. **(See Example 4.)**

57. $1 + p = 0$

58. $r - 12 = 13$

59. $-34 + t = -40$

60. $7 + q = 4$

61. $\dfrac{2}{3} = y - \dfrac{5}{12}$

62. $\dfrac{7}{11} = z + \dfrac{3}{11}$

63. $-2.5 = -1.1 + m$

64. $-4.1 = -3.5 + n$

65. $w - 23 = -11$

66. $p - 10 = -9$

67. $x + 21 = 16$

68. $y + 18 = -4$

69. $-2 = a - 15$

70. $-1 = b - 49$

71. $4.01 + p = 3.22$

72. $2.8 + q = 6.1$

73. $t + \dfrac{3}{8} = 2$

74. $r - \dfrac{4}{7} = -1$

75. $27 = z - 22$

76. $109 = x + 49$

77. $-70 = -55 + w$

Expanding Your Skills

For Exercises 78–83, first simplify each side of the equation. Then solve the equation.

78. $5h - 4h + 4 = 3$

79. $10x - 9x - 11 = 15$

80. $9 + (-2) = 4 + t$

81. $-13 + 15 = p + 5$

82. $3(r - 2) - 2r = 6 + (-2)$

83. $4(k + 2) - 3k = -6 + 9$

Section 11.4 — Multiplication and Division Properties of Equality

Objectives

1. Multiplication and Division Properties of Equality
2. Using the Properties of Equality

1. Multiplication and Division Properties of Equality

Adding or subtracting the same quantity to both sides of an equation results in an equivalent equation. In a similar way, multiplying or dividing both sides of an equation by the same nonzero quantity also results in an equivalent equation. This is stated formally as the multiplication and division properties of equality.

> **PROPERTY   The Multiplication and Division Properties of Equality**
>
> Let a, b, and c represent algebraic expressions.
>
> 1. The **multiplication property of equality:**   If $a = b$
>   then $a \cdot c = b \cdot c$
>
> 2. The **division property of equality:**   If $a = b$
>   then $\dfrac{a}{c} = \dfrac{b}{c}$ (provided $c \neq 0$)

To understand the multiplication property of equality, suppose we start with a true equation such as $10 = 10$. If both sides of the equation are multiplied by a constant such as 3, the result is also a true statement (Figure 11-3).

$$10 = 10$$
$$3 \cdot 10 = 3 \cdot 10$$
$$30 = 30$$

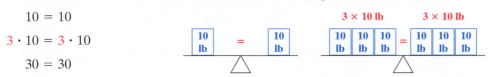

Figure 11-3

To solve an equation in the variable x, the goal is to write the equation in the form $x =$ number. In particular, notice that we desire the coefficient of x to be 1. That is, we want to write the equation as $1 \cdot x =$ number. To solve an equation such as $3x = 12$, we can multiply both sides of the equation by the reciprocal of the x-term coefficient. In this case, multiply both sides by the reciprocal of 3, which is $\frac{1}{3}$.

$$3x = 12$$

$\frac{1}{3} \cdot (3x) = \frac{1}{3} \cdot (12)$ Multiply by the reciprocal of 3, which is $\frac{1}{3}$.

$1 \cdot x = 4$ The coefficient of the x-term is now 1.

$x = 4$ Simplify. The solution is 4.

> **TIP:** Recall that the product of a number and its reciprocal is 1. For example:
> $$\frac{1}{5} \cdot (5) = 1$$
> $$\frac{3}{2} \cdot \frac{2}{3} = 1$$
> $$-\frac{7}{2} \cdot \left(-\frac{2}{7}\right) = 1$$

The division property of equality can also be used to solve the equation $3x = 12$ by dividing both sides by the coefficient of the x-term. In this case, divide both sides by 3 to make the coefficient of x equal to 1.

$$3x = 12$$

$\dfrac{3x}{3} = \dfrac{12}{3}$ Divide by the coefficient of x which is 3.

$1 \cdot x = 4$ The coefficient on the x-term is now 1.

$x = 4$ Simplify. The solution is 4.

> **TIP:** Recall that the quotient of a nonzero real number and itself is 1. For example:
> $$\frac{5}{5} = 1 \text{ and } \frac{-3.5}{-3.5} = 1$$

Example 1   **Applying the Multiplication and Division Properties of Equality**

Solve the equations by using the multiplication or division property of equality.

   a. $10x = 50$ **b.** $28 = -4p$

Solution:

   a. $10x = 50$

$\dfrac{10x}{10} = \dfrac{50}{10}$ To obtain a coefficient of 1 for the x-term, divide both sides by 10.

$1x = 5$ Simplify.

$x = 5$ The solution is 5.

         Check: $10x = 50$ Original equation

                $10(5) \stackrel{?}{=} 50$ Substitute 5 for x.

                $50 = 50$ ✔

Skill Practice

Solve.
   1. $4x = 32$
   2. $18 = -2w$

Answers
1. 8   **2.** -9

b. $28 = -4p$

$$\frac{28}{-4} = \frac{-4p}{-4}$$ To obtain a coefficient of 1 for the x-term, divide both sides by -4. This is also equivalent to multiplying by $-\frac{1}{4}$.

$-7 = 1p$ Simplify.

$-7 = p$ The solution is -7 and checks in the original equation.

Example 2   **Applying the Multiplication and Division Properties of Equality**

Solve the equation by using the multiplication or division property of equality.

a. $-y = 3.4$ **b.** $23.18 = 6.1w$

Solution:

a. $-y = 3.4$ Note that $-y$ is the same as $-1 \cdot y$.

$-1y = 3.4$

$$\frac{-1 \cdot y}{-1} = \frac{3.4}{-1}$$ To obtain a coefficient of 1 for the y-term, divide both sides by -1.

$1y = -3.4$ Simplify.

$y = -3.4$ The solution is -3.4 and checks in the original equation.

b. $23.18 = 6.1w$

$$\frac{23.18}{6.1} = \frac{6.1w}{6.1}$$ To obtain a coefficient of 1 on the w-term, divide both sides by 6.1.

$3.8 = 1w$ Simplify.

$3.8 = w$ The solution is 3.8 and checks in the original equation.

Example 3 Applying the Multiplication and Division Properties of Equality

Solve the equation by using the multiplication or division property of equality.

a. $-\dfrac{2}{3}p = -\dfrac{4}{7}$ **b.** $5 = \dfrac{d}{8}$

Skill Practice

Solve.

5. $-\dfrac{5}{9}p = \dfrac{1}{3}$

6. $\dfrac{c}{12} = 3$

Solution:

a. $-\dfrac{2}{3}p = -\dfrac{4}{7}$

$-\dfrac{3}{2}\left(-\dfrac{2}{3}p\right) = -\dfrac{3}{2}\left(-\dfrac{4}{7}\right)$ To obtain a coefficient of 1 for the p-term, multiply by the reciprocal of $-\frac{2}{3}$ which is $-\frac{3}{2}$.

$1p = \dfrac{12}{14}$ Simplify.

$p = \dfrac{6}{7}$ The solution is $\frac{6}{7}$.

TIP: For Example 3(a) you may have first thought of dividing by $-\frac{2}{3}$. Recall from Section 2.5 that dividing by a fraction is the same as multiplying by its reciprocal.

b. $5 = \dfrac{d}{8}$

$5 = \dfrac{1}{8}d$ The expression $\frac{d}{8}$ is equivalent to $\frac{1}{8}d$.

$8(5) = 8\left(\dfrac{1}{8}d\right)$ To obtain a coefficient of 1 on the d-term, multiply both sides by the reciprocal of $\frac{1}{8}$, which is 8.

$40 = 1d$ Simplify.

$40 = d$ The solution is 40.

TIP: When applying the multiplication or division properties, we will generally use the following conventions:
- If the coefficient of the variable term is expressed as a fraction, multiply both sides by its reciprocal.
- Otherwise, divide both sides by the coefficient itself.

2. Using the Properties of Equality

It is important to distinguish between cases where the addition or subtraction property of equality should be used to isolate a variable versus where the multiplication or division property of equality should be used. Compare the equations:

$$4 + x = 12 \quad \text{and} \quad 4x = 12$$

In the first equation, the operation between 4 and x is addition. Therefore, we want to reverse the process by *subtracting* 4 from both sides. In the second equation, the operation between 4 and x is multiplication. To isolate x, we reverse the process by *dividing* by 4 or, equivalently, by multiplying by the reciprocal $\frac{1}{4}$.

$$4 + x = 12 \quad \text{and} \quad 4x = 12$$

$$4 - 4 + x = 12 - 4 \qquad \dfrac{4x}{4} = \dfrac{12}{4}$$

$$x = 8 \qquad\qquad x = 3$$

Answers

5. $-\dfrac{3}{5}$ **6.** 36

In Example 4, we practice distinguishing which property of equality to use.

Skill Practice

Solve.

7. $\dfrac{t}{5} = -8$

8. $-4.6 + x = 12.9$

9. $5 = -2p$

10. $z + \dfrac{1}{3} = \dfrac{5}{6}$

Example 4   Solving Linear Equations

Solve the equations.

a. $\dfrac{m}{12} = -3$ **b.** $3.2 = x + 19.5$ **c.** $6 = -4t$ **d.** $y - \dfrac{5}{9} = \dfrac{2}{3}$

Solution:

a. $\dfrac{m}{12} = -3$ The operation between m and 12 is division. To obtain a coefficient of 1 for the m-term, multiply both sides by 12.

$12\left(\dfrac{m}{12}\right) = 12(-3)$ Multiply both sides by 12.

$m = -36$ Simplify both sides. The solution -36 checks in the original equation.

b. $3.2 = x + 19.5$ The operation between x and 19.5 is addition. To isolate the x-term, we can subtract 19.5 from both sides because $19.5 - 19.5 = 0$.

$3.2 - 19.5 = x + 19.5 - 19.5$ Subtract 19.5 from both sides.

$-16.3 = x$ Simplify. The solution -16.3 checks in the original equation.

c. $6 = -4t$ The operation between t and -4 is multiplication. To obtain a coefficient of 1 on the t-term, we can divide both sides by -4.

$\dfrac{6}{-4} = \dfrac{-4t}{-4}$ Divide both sides by -4.

$-\dfrac{6}{4} = t$

$-\dfrac{3}{2} = t$ Simplify. The solution $-\frac{3}{2}$ checks in the original equation.

d. $y - \dfrac{5}{9} = \dfrac{2}{3}$ The operation between y and $\frac{5}{9}$ is subtraction. Therefore, we can add $\frac{5}{9}$ to both sides to isolate y.

$y - \dfrac{5}{9} + \dfrac{5}{9} = \dfrac{2}{3} + \dfrac{5}{9}$ Add $\frac{5}{9}$ to both sides.

$y = \dfrac{2 \cdot 3}{3 \cdot 3} + \dfrac{5}{9}$ Obtain a common denominator. The LCD is 9.

$y = \dfrac{6}{9} + \dfrac{5}{9}$ Add the fractions.

$y = \dfrac{11}{9}$ The solution $\frac{11}{9}$ checks in the original equation.

Answers

7. -40 **8.** 17.5

9. $-\dfrac{5}{2}$ **10.** $\dfrac{1}{2}$

Section 11.4   Practice Exercises

Study Skills Exercises

1. One way to know that you really understand a concept is to try to explain it to someone else. In your own words, explain when you would apply the multiplication property of equality or the division property of equality.

2. Define the key terms.

   a. Division property of equality   b. Multiplication property of equality

Review Exercises

For Exercises 3–10, solve the equation.

3. $p - 12 = 33$

4. $-8 = 10 + k$

5. $16 = h - 5$

6. $-4 + w = 22$

7. $p - 6 = -19$

8. $\dfrac{1}{6} = -\dfrac{11}{6} + m$

9. $n + \dfrac{1}{2} = -\dfrac{2}{3}$

10. $2.4 + z = -12$

Objective 1: Multiplication and Division Properties of Equality

For Exercises 11–18, fill in the blank with the appropriate number.

11. $3 \cdot \underline{\hspace{1cm}} = 1$

12. $-6 \cdot \underline{\hspace{1cm}} = 1$

13. $-\dfrac{4}{7} \cdot \underline{\hspace{1cm}} = 1$

14. $\dfrac{3}{10} \cdot \underline{\hspace{1cm}} = 1$

15. $-7 \div \underline{\hspace{1cm}} = 1$

16. $2 \div \underline{\hspace{1cm}} = 1$

17. $5.1 \div \underline{\hspace{1cm}} = 1$

18. $-6.8 \div \underline{\hspace{1cm}} = 1$

For Exercises 19–50, solve the equation by using the multiplication or division property of equality. (See Examples 1–3.)

19. $14b = -42$

20. $-6p = 12$

21. $-8k = 56$

22. $5y = -25$

23. $-t = -13$

24. $-h = -17$

25. $\dfrac{2}{3}m = 14$

26. $\dfrac{5}{9}n = 40$

27. $\dfrac{b}{7} = -3$

28. $\dfrac{a}{4} = -12$

29. $-2.8 = -0.7t$

30. $-3.3 = -3r$

31. $-\dfrac{u}{2} = -15$

32. $-\dfrac{v}{10} = -4$

33. $6 = -18w$

34. $4 = -32g$

35. $1.3x = 5.33$

36. $8.1y = 17.82$

37. $\dfrac{5}{4}k = -\dfrac{1}{2}$

38. $-\dfrac{11}{12}h = -\dfrac{1}{6}$

39. $0 = \dfrac{3}{8}m$

40. $0 = \dfrac{1}{10}n$

41. $-\dfrac{9}{4}x = -\dfrac{3}{5}$

42. $-\dfrac{15}{14}y = \dfrac{1}{2}$

43. $100 = 5k$

44. $95 = 19h$

45. $31 = -p$

46. $11 = -q$

47. $3p = \dfrac{5}{2}$

48. $2q = \dfrac{7}{5}$

49. $-4a = 0$

50. $-7b = 0$

Objective 2: Using the Properties of Equality

51. In your own words explain how to determine when to use the addition property of equality.

52. In your own words explain how to determine when to use the subtraction property of equality.

53. Explain how to determine when to use the division property of equality.

54. Explain how to determine when to use the multiplication property of equality.

Mixed Exercises

For Exercises 55–86, solve the equation. **(See Example 4.)**

55. $4 + x = -12$

56. $6 + z = -18$

57. $4y = -12$

58. $6p = -18$

59. $q - 4 = -12$

60. $p - 6 = -18$

61. $\dfrac{h}{4} = -12$

62. $\dfrac{w}{6} = -18$

63. $\dfrac{2}{3} + t = 1$

64. $\dfrac{3}{4} + q = 1$

65. $-9a = -12$

66. $-8b = -44$

67. $7 = r - 23$

68. $11 = s - 4$

69. $-\dfrac{y}{3} = 5$

70. $-\dfrac{h}{5} = 1$

71. $2p = \dfrac{5}{6}$

72. $4q = \dfrac{3}{5}$

73. $-\dfrac{3}{7}x = \dfrac{9}{10}$

74. $-\dfrac{2}{11}y = \dfrac{4}{15}$

75. $t - 12.9 = 15$

76. $c - 4.11 = 1.2$

77. $5 + u = 3.2$

78. $3 + v = 1.7$

79. $50 = a + 72$

80. $23 = w + 41$

81. $-1 = b - 16$

82. $-5 = y - 8$

83. $-12 = 30x$

84. $-10 = 12h$

85. $-6 = -\dfrac{1}{2}q$

86. $4 = -\dfrac{1}{6}k$

Expanding Your Skills

For Exercises 87–92, first simplify each side of the equation. Then solve the equation.

87. $5x - 2x = -15$

88. $13y - 10y = -18$

89. $3p + 4p = 25 - 4$

90. $2q + 3q = 54 - 9$

91. $-2(a + 3) - 6a + 6 = 8$

92. $-(b - 11) - 3b - 11 = -16$

Section 11.5 | Solving Equations with Multiple Steps

Objectives

1. Solving Equations with Multiple Steps
2. Solving Linear Equations Involving Parentheses

1. Solving Equations with Multiple Steps

In Sections 11.3 and 11.4 we studied a one-step process to solve linear equations. We used the addition, subtraction, multiplication, and division properties of equality. In this section we combine these properties to solve equations that require multiple steps. This is shown in Example 1.

Example 1 Solving a Linear Equation

Solve. $2x - 3 = 15$

Solution:

Remember that our goal is to isolate x. Therefore, in this equation, we first isolate the *term* containing x. This can be done by adding 3 to both sides.

$2x - 3 + 3 = 15 + 3$ Add 3 to both sides.

$2x = 18$ The term containing x is now isolated (by itself). The resulting equation now requires only one step to solve.

$\dfrac{2x}{2} = \dfrac{18}{2}$ Divide both sides by 2 to make the coefficient on x equal to 1.

$x = 9$ Simplify. The solution is 9.

       <u>Check</u>: $2x - 3 = 15$ Original equation

                $2(9) - 3 \stackrel{?}{=} 15$ Substitute 9 for x.

                  $18 - 3 = 15$ ✓

> **Skill Practice**
>
> **1.** Solve.
>
> $3x + 7 = 25$

As Example 1 shows, we will generally apply the addition (or subtraction) property of equality to isolate the variable term first. Then we will apply the multiplication (or division) property of equality to obtain a coefficient of 1 on the variable term.

Example 2 Solving a Linear Equation

Solve. $14 = \dfrac{y}{4} + 8$

Solution:

$14 = \dfrac{y}{4} + 8$

$14 - 8 = \dfrac{y}{4} + 8 - 8$ Subtract 8 from both sides. This will isolate the term containing the variable y.

$6 = \dfrac{y}{4}$ Simplify.

$6(4) = \dfrac{y}{4}(4)$ Multiply both sides by 4.

$24 = y$ Simplify. The solution is 24.

       <u>Check</u>: $14 = \dfrac{y}{4} + 8$ Original equation

                 $14 \stackrel{?}{=} \dfrac{(24)}{4} + 8$ Substitute 24 for y.

                 $14 = 6 + 8$ ✓

> **Skill Practice**
>
> Solve.
>
> **2.** $-13 = -9 + \dfrac{y}{2}$

Answers

1. 6   **2.** -8

Example 3 Solving a Linear Equation

Solve. $2z - 9.2 = 2.6$

Solution:

$$2z - 9.2 = 2.6$$

$$2z - 9.2 + 9.2 = 2.6 + 9.2 \qquad \text{Add 9.2 to both sides. This will isolate the term containing the variable } z.$$

$$2z = 11.8 \qquad \text{Simplify.}$$

$$\frac{2z}{2} = \frac{11.8}{2} \qquad \text{Divide both sides by 2.}$$

$$z = 5.9 \qquad \text{The solution is 5.9 and checks in the original equation.}$$

In Example 4, the variable x appears on both sides of the equation. In this case, apply the addition or subtraction properties of equality to collect the variable terms on one side of the equation and the constant terms on the other side.

Example 4 Solving a Linear Equation with Variables on Both Sides

Solve. $4x + 5 = -2x - 13$

Solution:

To isolate x, we must first "move" all x-terms to one side of the equation. For example, suppose we add $2x$ to both sides. This would "remove" the x-term from the right-hand side because $-2x + 2x = 0$. The term $2x$ is then combined with $4x$ on the left-hand side.

$$4x + 5 = -2x - 13$$

$$4x + 2x + 5 = -2x + 2x - 13 \qquad \text{Add } 2x \text{ to both sides.}$$

$$6x + 5 = -13 \qquad \text{Simplify. Next, we want to isolate the term containing } x.$$

$$6x + 5 - 5 = -13 - 5 \qquad \text{Subtract 5 from both sides to isolate the } x\text{-term.}$$

$$6x = -18 \qquad \text{Simplify.}$$

$$\frac{6x}{6} = \frac{-18}{6} \qquad \text{Divide both sides by 6 to obtain a coefficient of 1 for the } x\text{-term.}$$

$$x = -3 \qquad \text{The solution is } -3 \text{ and checks in the original equation.}$$

TIP: Note that the variable may be isolated on either side of the equation. In Example 4, for instance, we could have isolated the x-terms on the right-hand side of the equation.

$$4x + 5 = -2x - 13$$

$4x - 4x + 5 = -2x - 4x - 13$ Subtract $4x$ from both sides. This "removes" the
$$5 = -6x - 13$$ x-term from the left-hand side.

$5 + 13 = -6x - 13 + 13$ Add 13 to both sides to isolate the x-term.

$18 = -6x$ Simplify.

$$\frac{18}{-6} = \frac{-6x}{-6}$$ Divide both sides by -6.

$-3 = x$ This is the same solution as in Example 4.

Often we can simplify both sides of an equation before applying the properties of equality. This is demonstrated in Example 5.

Example 5   Solving a Linear Equation by Simplifying First

Solve. $6 - 8y + 3 = y + 3y + 6$

Solution:

Notice that *like* terms can be combined on both sides of the equation first, to make the equation simpler.

$6 - 8y + 3 = y + 3y + 6$

$9 - 8y = 4y + 6$ Combine *like* terms. On the left-hand side $6 + 3 = 9$. On the right-hand side, $y + 3y = 4y$.

$9 - 8y - 4y = 4y - 4y + 6$ We can collect all variable terms on the left-hand side by subtracting $4y$ from both sides.

$9 - 12y = 6$ Simplify.

$9 - 9 - 12y = 6 - 9$ To isolate the y-term on the left, subtract 9 from both sides.

$-12y = -3$ Simplify.

$$\frac{-12y}{-12} = \frac{-3}{-12}$$ Divide both sides by -12 to make the coefficient on the y-term 1.

$$y = \frac{1}{4}$$ The solution is $\frac{1}{4}$ and checks in the original equation.

Skill Practice

Solve.

5. $14 - 3w + 2$
$= 4w + 21 - 2w$

---

2. Solving Linear Equations Involving Parentheses

In Examples 1–5, we used multiple steps to solve equations. We also learned how to collect the variable terms on one side of the equation so that the variable could be isolated. The following guidelines summarize the steps to solve a linear equation.

Answer

5. -1

> **PROCEDURE   Solving a Linear Equation in One Variable**
>
> **Step 1** Simplify both sides of the equation.
>
> - Clear parentheses if necessary.
> - Combine *like* terms if necessary.
>
> **Step 2** Use the addition or subtraction property of equality to collect the variable terms on one side of the equation.
>
> **Step 3** Use the addition or subtraction property of equality to collect the constant terms on the *other* side of the equation.
>
> **Step 4** Use the multiplication or division property of equality to make the coefficient of the variable term equal to 1.
>
> **Step 5** Check the answer in the original equation.

Skill Practice

Solve.

6. $6(z + 4) - 9 = -12 - 3z$

Example 6   Solving a Linear Equation

Solve. $2(y - 6) + 32 = 8 - 4y$

Solution:

$$2(y - 6) + 32 = 8 - 4y$$

$$2y - 12 + 32 = 8 - 4y$$ **Step 1:** Simplify both sides of the equation. Clear parentheses.

$$2y + 20 = 8 - 4y$$ Combine *like* terms on the left-hand side. Note that $-12 + 32 = 20$.

$$2y + 4y + 20 = 8 - 4y + 4y$$ **Step 2:** Add $4y$ to both sides to collect the variable terms on the left.

$$6y + 20 = 8$$ Simplify.

$$6y + 20 - 20 = 8 - 20$$ **Step 3:** Subtract 20 from both sides to collect the constants on the right.

$$6y = -12$$ Simplify.

$$\frac{6y}{6} = \frac{-12}{6}$$ **Step 4:** Divide both sides by 6 to obtain a coefficient of 1 on the y-term.

$$y = -2$$ Simplify. The solution is -2.

Check: $2(y - 6) + 32 = 8 - 4y$ **Step 5:** Check the solution in the original equation.

$$2(-2 - 6) + 32 \overset{?}{=} 8 - 4(-2)$$ Substitute -2 for y.

$$2(-8) + 32 \overset{?}{=} 8 - (-8)$$

$$-16 + 32 \overset{?}{=} 16$$

$$16 = 16 \checkmark$$ The solution checks.

Answer

6. -3

Example 7   Solving a Linear Equation

Solve. $2x + 3x + 2 = -4(3 - x)$

Solution:

$2x + 3x + 2 = -4(3 - x)$

$\quad 5x + 2 = -12 + 4x$ **Step 1:** Simplify both sides of the equation. On the left, combine *like* terms. On the right, clear parentheses.

$5x - 4x + 2 = -12 + 4x - 4x$ **Step 2:** Subtract $4x$ from both sides to collect the variable terms on the left.

$\quad\quad x + 2 = -12$ Simplify.

$x + 2 - 2 = -12 - 2$ **Step 3:** Subtract 2 from both sides to collect the constants on the right.

$\quad\quad\quad x = -14$ **Step 4:** The coefficient on the x-term is already 1. The solution is -14.

Check: $2x + 3x + 2 = -4(3 - x)$ **Step 5:** Check in the original equation.

$2(-14) + 3(-14) + 2 \overset{?}{=} -4[3 - (-14)]$ Substitute -14 for x.

$-28 - 42 + 2 \overset{?}{=} -4(17)$

$-70 + 2 \overset{?}{=} -68$

$-68 = -68 \checkmark$ The solution checks.

── **Skill Practice** ──

Solve.

7. $2 - y - 4$
$\quad = 6 - 2(y - 8)$

Answer
7. 24

Section 11.5   Practice Exercises

Study Skills Exercise

1. When you are solving multistep equations, it is recommended that you write an explanation for each step along the way. In Example 6, an equation is solved with each step shown. Your job is to write an explanation for each step.

$-7x + 2 = 4(x - 5)$ **Explanation**

$-7x + 2 = 4x - 20$

$-7x - 4x + 2 = 4x - 4x - 20$

$-11x + 2 = -20$

$-11x + 2 - 2 = -20 - 2$

$-11x = -22$

 $\dfrac{-11x}{-11} = \dfrac{-22}{-11}$

$x = 2$ The solution is 2.

Review Exercises

For Exercises 2–8, solve the equation.

2. $4c = -\dfrac{1}{3}$

3. $\dfrac{1}{3}b = -4$

4. $-\dfrac{1}{5} + t = \dfrac{6}{5}$

5. $-\dfrac{3}{8} = w + \dfrac{1}{4}$

6. $-p = -\dfrac{7}{10}$

7. $-8h = 0$

8. $5 + q = 0$

Objective 1: Solving Equations with Multiple Steps

For Exercises 9–28, solve the equation. **(See Examples 1–3.)**

9. $3m + 2 = 14$

10. $-2n + 5 = -15$

11. $-8c - 12 = 36$

12. $5t - 1 = -11$

13. $1 = -4z + 21$

14. $-4 = -3p + 14$

15. $9 = 12x - 7$

16. $-7 = 5y - 8$

17. $3.4 - 2d = 8.2$

18. $2.9 - 4g = 23.3$

19. $-0.57 = 15h + 16.23$

20. $1.9 = 8k + 4.06$

21. $\dfrac{b}{3} - 12 = -9$

22. $\dfrac{c}{5} + 2 = 4$

23. $-9 = \dfrac{w}{2} - 3$

24. $-16 = \dfrac{t}{4} - 14$

25. $3x + \dfrac{1}{2} = \dfrac{5}{4}$

26. $9z - \dfrac{3}{8} = \dfrac{9}{16}$

27. $10 - y = 37$

28. $25 - c = -3$

For Exercises 29–46, solve the equation. **(See Examples 4–5.)**

29. $8 + 4b = 2 + 2b$

30. $2w + 10 = 5w - 5$

31. $7 - 5t = 3t - 2$

32. $4 - 2p = 8 + 5p$

33. $4 - 3d = 5d - 4$

34. $-3k + 14 = -4 + 3k$

35. $12p = 3p + 21$

36. $2x + 10 = 4x$

37. $-z - 2 = -2z$

38. $9y = -y + 25$

39. $1 + \dfrac{1}{4}p = 2 + \dfrac{3}{4}p$

40. $\dfrac{4}{3} + \dfrac{2}{3}q = -\dfrac{5}{3} - \dfrac{1}{3}q - 4$

41. $4 + 2a - 7 = 3a + a + 3$

42. $4b + 2b - 7 = 2 + 4b + 5$

43. $-8w + 8 + 3w = 2 - 6w + 2$

44. $-12 + 5m + 10 = -2m - 10 - m$

45. $6y + 2y - 2 = 14 + 3y - 12$

46. $-7t - 20 + 7 = -7 - 3t$

Objective 2: Solving Linear Equations Involving Parentheses

For Exercises 47–60, solve the equation. **(See Examples 6–7.)**

47. $3n - 4(n - 1) = 16$

48. $4p - 3(p + 2) = 18$

49. $9q - 5(q - 3) = 5q$

50. $6h - 2(h + 6) = 10h$

51. $2(1 - m) = 5 - 3m$

52. $3(2 - g) = 12 - g$

53. $-4(k - 2) + 14 = 3k - 20$

54. $-3(x + 4) - 9 = -2x + 12$

55. $3z - 9 = 3(5z - 1)$

56. $4y - 9 = 8(y - 2)$

57. $6w + 2(w - 1) = 14 - (3w + 1)$

58. $-3t - 3(t - 4) = 2 - (2t - 1)$

59. $6(u - 1) + 5u + 1 = 5(u + 6) - u$

60. $2(2v + 3) + 8v = 6(v - 1) + 3v$

2. Translat

We begin solvi
sentence and a
key words that
Table 11-4.

Table

Ad
the
a p
b a
b n
a i
the

Mu
the
a t
a n

Example 1

A number dec

Solution:

Let x represent

A number dec

x

The number is

Example 2

Two times the

Solution:

Problem Recognition Exercises

Solving Equations

For Exercises 1–6, identify the problem as an expression or as an equation.

1. $-5 + 4x - 6x = 7$

2. $-8(4 - 7x) + 4$

3. $4 - 6(2x - 3) + 1$

4. $10 - x = 2x + 19$

5. $6 - 3(x + 4) = 6$

6. $9 - 6(x + 1)$

For Exercises 7–30, solve the equation.

7. $5t = 20$

8. $6x = 36$

9. $5 + t = 20$

10. $6 + x = 36$

11. $5(t - 3) = 20$

12. $6(x - 2) = 36$

13. $5x - 3 = 20$

14. $6x - 2 = 36$

15. $5 + 3p - 2 = 0$

16. $16 - 2k + 2 = 0$

17. $0 = 2x + 5x + 1$

18. $0 = 7y - 3y + 8$

19. $-\dfrac{2}{3}p - \dfrac{1}{6} = -\dfrac{2}{3}$

20. $\dfrac{1}{5}b + \dfrac{7}{10} = \dfrac{3}{5}$

21. $-14 = \dfrac{r}{6} - 12$

22. $11 = \dfrac{s}{4} + 3$

23. $2.3u + 0.2 = -1.2u + 7.2$

24. $7.5w - 2.7 = 1.4w + 15.6$

25. $6a + 3a - 21 = 4 - 5a - 1$

26. $7 - 5k + 10 = 3k - 2k + 9$

27. $-2(x - 3) + 14 = 10 - (x + 4)$

28. $26 - (3x + 12) = 11 - 4(x + 1)$

29. $2 - 3(y + 1) = -4y + 7$

30. $5(t + 5) - 3t = t - 9$

Sectio

Section 9.3   Th... ...ic Equations by ...   667

Objectives

1. Problem-Solving
2. Translating Ver...
   Statements into
3. Applications of
   Equations

Avoiding Mist

It is always a good
your answer in the
problem to determi
answer is reasonab

Student Answer Appendix

Chapter 1

Chapter Opener Puzzle

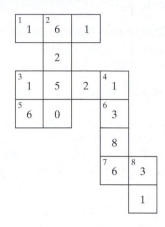

Section 1.1 Practice Exercises, pp. 6–8

3. 7: ones; 5: tens; 4: hundreds; 3: thousands;
1: ten-thousands; 2: hundred-thousands; 8: millions
5. Tens **7.** Ones **9.** Hundreds **11.** Thousands
13. Hundred-thousands **15.** Billions
17. Ten-thousands **19.** Millions **21.** Ten-millions
23. Billions **25.** 5 tens + 8 ones
27. 5 hundreds + 3 tens + 9 ones
29. 5 hundreds + 3 ones
31. 1 ten-thousand + 2 hundreds + 4 tens + 1 one
33. 524 **35.** 150 **37.** 1,906 **39.** 85,007
41. Ones, thousands, millions, billions
43. Two hundred forty-one
45. Six hundred three
47. Thirty-one thousand, five hundred thirty
49. One hundred thousand, two hundred thirty-four
51. Nine thousand, five hundred thirty-five
53. Twenty thousand, three hundred twenty
55. One thousand, three hundred seventy-seven
57. 6,005 **59.** 672,000 **61.** 1,484,250
63.

```
        d              a      c              b
  +--+--+--+--+--+--+--+--+--+--+--+--+--+-->
  0  1  2  3  4  5  6  7  8  9  10 11 12 13
```

65. 10 **67.** 4 **69.** 8 is greater than 2, or 2 is less than 8
71. 3 is less than 7, or 7 is greater than 3 **73.** <
75. > **77.** < **79.** > **81.** < **83.** <
85. False **87.** 99 **89.** There is no greatest whole
number. **91.** 7 **93.** 964

Section 1.2 Practice Exercises, pp. 16–20

3. 3 hundreds + 5 tens + 1 one
5. 1 hundred + 7 ones **7.** 4012

9.

+	0	1	2	3	4	5	6	7	8	9
0	0	1	2	3	4	5	6	7	8	9
1	1	2	3	4	5	6	7	8	9	10
2	2	3	4	5	6	7	8	9	10	11
3	3	4	5	6	7	8	9	10	11	12
4	4	5	6	7	8	9	10	11	12	13
5	5	6	7	8	9	10	11	12	13	14
6	6	7	8	9	10	11	12	13	14	15
7	7	8	9	10	11	12	13	14	15	16
8	8	9	10	11	12	13	14	15	16	17
9	9	10	11	12	13	14	15	16	17	18

11. Addends: 2, 8; sum: 10
13. Addends: 11, 10; sum: 21
15. Addends: 5, 8, 2; sum: 15
17. 74 **19.** 58 **21.** 48 **23.** 19 **25.** 588
27. 798 **29.** 237 **31.** 198 **33.** 84 **35.** 115
37. 937 **39.** 850 **41.** 41 **43.** 29 **45.** 1003
47. 836 **49.** 24,004 **51.** 132,658 **53.** 21 + 30
55. 13 + 8 **57.** 23 + (9 + 10) **59.** (41 + 3) + 22
61. The sum of any number and 0 is that number.
a. 423 **b.** 25 **c.** 67 **63.** 100 + 42; 142
65. 23 + 81; 104 **67.** 76 + 2; 78 **69.** 1320 + 448; 1768
71. For example: The sum of 54 and 24
73. For example: 88 added to 12
75. For example: The total of 4, 23, and 77
77. For example: 10 increased by 8 **79.** 276 people
81. 74,283,000 viewers **83.** $45,500 **85.** 423 desks
87. 13,538 participants **89.** 821,024 nonteachers
91. 104 cm **93.** 110 m **95.** 42 yd **97.** 288 ft

Section 1.2 Calculator Connections, p. 20

99. 9,536,940 **100.** 908,788 **101.** 8,163,940
102. 192,780 **103.** 21,491,394 **104.** 5,257,179
105. 67,342,000 viewers **106.** 121,480,019 votes

Section 1.3 Practice Exercises, pp. 27–30

3. 1151 **5.** 899 **7.** 0 < 10
9. Minuend: 12; subtrahend: 8; difference: 4
11. Minuend: 21; subtrahend: 12; difference: 9
13. Minuend: 9; subtrahend: 6; difference: 3
15. 18 + 9 = 27 **17.** 27 + 75 = 102 **19.** 5
21. 3 **23.** 6 **25.** 45 **27.** 61 **29.** 1126
31. 321 **33.** 10,004 **35.** 1103 **37.** 17 **39.** 49
41. 104 **43.** 521 **45.** 23 **47.** 4764 **49.** 1,403
51. 2217 **53.** 378 **55.** 713 **57.** 30,941
59. 5,662,119 **61.** 78 − 23; 55 **63.** 78 − 6; 72
65. 422 − 100; 322 **67.** 1090 − 72; 1018
69. 50 − 13; 37 **71.** 103 − 35; 68
73. For example: 93 minus 27

75. For example: Subtract 85 from 165.
77. The expression $7 - 4$ means 7 minus 4, yielding a difference of 3. The expression $4 - 7$ means 4 minus 7 which results in a difference of -3. (This is a mathematical skill we have not yet learned.)   **79.** $33   **81.** 55 more hits
83. 8 plants   **85.** 1979 times   **87.** 13 m   **89.** 10 yd
91. 7748   **93.** 195,489

Section 1.3 Calculator Connections, p. 30

95. 4,447,302   **96.** 897,058,513   **97.** 2,906,455
98. 49,408 mi^2 **99.** 17,139 mi^2 **100.** The difference in land area between Colorado and Rhode Island is 102,673 mi^2.
101. 13,093 mi^2

Section 1.4 Practice Exercises, pp. 35–37

3. 26   **5.** 5007   **7.** Ten-thousands
9. If the digit in the tens place is 0, 1, 2, 3, or 4, then change the tens and ones digits to 0. If the digit in the tens place is 5, 6, 7, 8, or 9, increase the digit in the hundreds place by 1 and change the tens and ones digits to 0.
11. 340   **13.** 730   **15.** 9400   **17.** 8500   **19.** 35,000
21. 3000   **23.** 10,000   **25.** 109,000   **27.** 490,000
29. $148,000,000   **31.** 239,000 mi   **33.** 160   **35.** 180
37. 500   **39.** 2100   **41.** $151,000,000
43. $11,000,000 more   **45.** $10,000,000
47. a. 2003; $3,500,000   **b.** 2005; $2,000,000
49. Massachusetts; 79,000 students   **51.** 71,000 students
53. Answers may vary.   **55.** 10,000 mm   **57.** 440 in.

Section 1.5 Practice Exercises, pp. 46–49

3. 1,010,000   **5.** 5400   **7.** 6×5; 30
9. 3×9; 27   **11.** Factors: 13, 42; product: 546
13. Factors: 3, 5, 2; product: 30
15. For example: 5×12; $5 \cdot 12$; $5(12)$
17. d   **19.** e   **21.** c   **23.** 8×14
25. $(6 \times 2) \times 10$ **27.** $(5 \times 7) + (5 \times 4)$
29. 144   **31.** 52   **33.** 655   **35.** 1376
37. 11,280   **39.** 23,184   **41.** 378,126   **43.** 448
45. 1632   **47.** 864   **49.** 2431   **51.** 6631
53. 19,177   **55.** 186,702   **57.** 21,241,448
59. 4,047,804   **61.** 24,000   **63.** 2,100,000
65. 72,000,000   **67.** 36,000,000   **69.** 60,000,000
71. 2,400,000,000   **73.** $1000   **75.** $1,370,000
77. 4000 minutes   **79.** $1665   **81.** 287,500 sheets
83. 372 mi   **85.** 276 ft^2 **87.** 5329 cm^2
89. 105,300 mi^2 **91. a.** 2400 $in.^2$ **b.** 42 windows
c. 100,800 $in.^2$ **93.** 128 ft^2

Section 1.6 Practice Exercises, pp. 57–60

3. 4944   **5.** 1253   **7.** 664,210   **9.** 902
11. 9; the dividend is 72; the divisor is 8; the quotient is 9
13. 8; the dividend is 64; the divisor is 8; the quotient is 8
15. 5; the dividend is 45; the divisor is 9; the quotient is 5
17. You cannot divide a number by zero (the quotient is undefined). If you divide zero by a number (other than zero), the quotient is always zero.
19. 15   **21.** 0   **23.** Undefined   **25.** 1
27. Undefined   **29.** 0   **31.** $2 \times 3 = 6$, $2 \times 6 \neq 3$
33. Multiply the quotient and the divisor to get the dividend.
35. 13   **37.** 41   **39.** 486   **41.** 409   **43.** 203
45. 822   **47.** Correct   **49.** Incorrect; 253 R2

51. Correct   **53.** Incorrect; 25 R3   **55.** 7 R5
57. 10 R2   **59.** 27 R1   **61.** 197 R2   **63.** 42 R4
65. 1557 R1   **67.** 751 R6   **69.** 835 R2   **71.** 479 R9
73. 43 R19   **75.** 308   **77.** 1259 R26   **79.** 229 R96
81. 302   **83.** $497 \div 71$; 7   **85.** $877 \div 14$; 62 R9
87. $42 \div 6$; 7   **89.** 14 classrooms
91. 5 cases; 8 cans left over   **93.** 52 mph   **95.** 22 lb
97. $1200 \div 20 = 60$; approximately 60 words per minute
99. Yes, they can all attend if they sit in the second balcony.
101. a. 12 loads   **b.** 2 oz

Section 1.6 Calculator Connections, p. 60

103. 7,665,000,000 bbl   **104.** 13,000 min
105. $211 billion   **106.** Each crate weighs 255 lb.

Chapter 1 Problem Recognition Exercises, p. 61

1. 133   **2.** 89   **3.** 67   **4.** 116   **5.** 3094
6. 3650   **7.** 64   **8.** 87   **9.** 334   **10.** 5783
11. 4963   **12.** 5500   **13.** 328   **14.** 547   **15.** 246,000
16. 2,820,000   **17.** 172   **18.** 9154   **19.** 82
20. 6561   **21.** 230 R 4   **22.** 425 R12   **23.** 20,000
24. 540,000   **25.** 22,718   **26.** 76   **27.** 11,643
28. 6867   **29.** 548   **30.** 843   **31.** 34,524
32. 22,761   **33.** 4180   **34.** 41,800   **35.** 418,000
36. 4,180,000   **37.** 35,000   **38.** 3500   **39.** 350
40. 35   **41.** 506   **42.** 8742   **43.** 612   **44.** 1034

Section 1.7 Practice Exercises, pp. 66–69

3. True   **5.** False   **7.** True   **9.** 9^4 **11.** 2^7
13. 3^6 **15.** $4^4 \cdot 2^3$ **17.** $8 \cdot 8 \cdot 8 \cdot 8$
19. $4 \cdot 4 \cdot 4 \cdot 4 \cdot 4 \cdot 4 \cdot 4 \cdot 4$ **21.** 8   **23.** 9   **25.** 27
27. 125   **29.** 32   **31.** 81   **33.** 1   **35.** 1
37. The number 1 raised to any power equals 1.   **39.** 1000
41. 100,000   **43.** 2   **45.** 6   **47.** 10   **49.** 0
51. No, addition and subtraction should be performed in the order in which they appear from left to right.   **53.** 26
55. 1   **57.** 49   **59.** 3   **61.** 2   **63.** 53   **65.** 8
67. 45   **69.** 24   **71.** 4   **73.** 40   **75.** 90   **77.** 25
79. 4   **81.** 81   **83.** 18   **85.** 0   **87.** 5   **89.** 6
91. 3   **93.** 201   **95.** 109   **97.** 18
99. 38¢ per pound   **101** 121 mm per month
103. 24   **105.** 70

Section 1.7 Calculator Connections, p. 69

107. 24,336   **108.** 174,724   **109.** 248,832
110. 1,500,625   **111.** 79,507   **112.** 357,911
113. 8028   **114.** 293,834   **115.** 66,049   **116.** 1728
117. 35   **118.** 43

Section 1.8 Practice Exercises, pp. 75–79

3. $71 + 14$; 85   **5.** $2 \cdot 14$; 28   **7.** $102 - 32$; 70
9. $10 \cdot 13$; 130   **11.** $24 \div 6$; 4   **13.** $5 + 13 + 25$; 43
15. For example: sum, added to, increased by, more than, plus, total of   **17.** For example: difference, minus, decreased by, less, subtract   **19.** Denali is 6074 ft higher than White Mountain Peak.   **21.** 921,500,000 metric tons
23. The whole screen has 12,096 pixels.
25. There will be 120 classes of Beginning Algebra.
27. There will be 9 gal used.
29. Jeannette will pay $29,560 for 1 year.
31. The Prius can go 1100 mi.

33. The maximum capacity is 3150 seats.
35. Jackson's monthly payment was $390.
37. Each trip will take 2 hr.   **39.** Perimeter
41. The cost will be $86.   **43.** The cost is $1020.
45. There will be $36 left in Gina's account.
47. The total bill was $154,032.
49. a. Latayne will receive $48.   **b.** She can buy 6 CDs.
51. Michael Jordan scored 33,454 points with the Bulls.
53. a. One bottle will last for 30 days.   **b.** The owner should order a refill no later than September 28.
55. a. The distance is 360 mi.   **b.** 14 in. represents 840 mi.
57. 104 boxes will be filled completely with 2 books left over.
59. a. Marc needs five $20 bills.   **b.** He will receive $16 in change.   **61.** He earned $520.

Chapter 1 Review Exercises, pp. 88–92

1. Ten-thousands   **3.** 92,046
5. 3 millions + 4 hundred-thousands + 8 hundreds + 2 tens
7. Two hundred forty-five   **9.** 3602
11.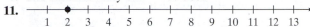
13. True   **15.** Addends: 105, 119; sum: 224
17. 71   **19.** 17,410
21. a. Commutative property   **b.** Associative property
c. Commutative property
23. 44 + 92; 136   **25.** 23 + 6; 29
27. 45,797 thousand seniors
29. Minuend: 14; subtrahend: 8; difference: 6
31. 26   **33.** 121   **35.** 31,019   **37.** 38 − 31; 7
39. 251 − 42; 209   **41.** 71,892,438 tons
43. 2336 thousand visitors   **45.** 9,330,000
47. 1500   **49.** 163,000 m^3
51. Factors: 33, 40; product: 1320   **53.** c   **55.** d
57. b   **59.** 52,224   **61.** $429
63. 7; divisor: 6, dividend: 42, quotient: 7
65. 3   **67.** Undefined
69. Multiply the quotient and the divisor to get the dividend.
71. 58   **73.** 52 R3   **75.** $9\overline{)108}$; 12
77. a. 4 T-shirts   **b.** 5 hats   **79.** $2^4 \cdot 5^3$ **81.** 256
83. 1,000,000   **85.** 12   **87.** 75   **89.** 15   **91.** 55
93. $89   **95. a.** The Cincinnati Zoo has 13,000 more animals than the San Diego Zoo.   **b.** The San Diego Zoo has 50 more species than the Cincinnati Zoo.
97. He will receive $19,600,000 per year.

Chapter 1 Test, pp. 88–92

1. a. Hundreds   **b.** Thousands   **c.** Millions
d. Ten-thousands   **2. a.** 4,065,000   **b.** Twenty-one million, three hundred twenty-five thousand   **c.** Twelve million, two hundred eighty-seven thousand   **d.** 729,000   **e.** Eleven million, four hundred ten thousand
3. a. 14 > 6   **b.** 72 < 81   **4.** 129   **5.** 328
6. 113   **7.** 227   **8.** 2842   **9.** 447   **10.** 21 R9
11. 546   **12.** 8103   **13.** 20   **14.** 1,500,000,000
15. 336   **16.** 0   **17.** Undefined
18. a. The associative property of multiplication; the expression shows a change in grouping.
b. The commutative property of multiplication; the expression shows a change in the order of the factors.

19. a. 4900   **b.** 12,000   **c.** 8,000,000
20. There were approximately 1,430,000 people.
21. 4   **22.** 24   **23.** 48   **24.** 33
25. Jennifer has a higher average of 29. Brittany has an average of 28.
26. a. 23,418 thousand subscribers   **b.** The largest increase was between the years 2004 and 2005.
27. The North Side Fire Department is the busiest with an average of five calls per week.
28. 156 mm   **29.** Perimeter: 350 ft; area: 6016 ft^2,
30. 4,560,000 m^2

Chapter 2

Chapter Opener Puzzle

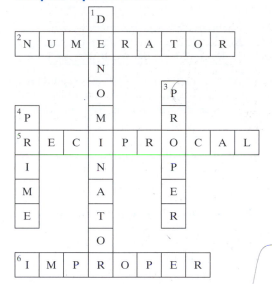

Section 2.1 Practice Exercises, pp. 102–106

3. Numerator: 2; denominator: 3
5. Numerator: 12; denominator: 11
7. 6 ÷ 1; 6   **9.** 2 ÷ 2; 1   **11.** 0 ÷ 3; 0
13. 2 ÷ 0; undefined   **15.** $\frac{3}{4}$ **17.** $\frac{5}{9}$ **19.** $\frac{1}{6}$
21. $\frac{3}{8}$ **23.** $\frac{3}{4}$ **25.** $\frac{1}{8}$ **27.** $\frac{41}{103}$ **29.** $\frac{10}{21}$
31. Proper   **33.** Improper   **35.** Improper
39. $\frac{5}{2}$ **41.** $\frac{12}{4}$ **43.** $\frac{7}{4}$; $1\frac{3}{4}$ **45.** $\frac{13}{8}$; $1\frac{5}{8}$ **47.** $\frac{7}{4}$
49. $\frac{38}{9}$ **51.** $\frac{24}{7}$ **53.** $\frac{29}{4}$ **55.** $\frac{137}{12}$ **57.** $\frac{171}{8}$
59. 19   **61.** 7   **63.** $4\frac{5}{8}$ **65.** $7\frac{4}{5}$ **67.** $2\frac{7}{10}$
69. $5\frac{7}{9}$ **71.** $12\frac{1}{11}$ **73.** $3\frac{5}{6}$ **75.** $44\frac{1}{7}$ **77.** $1056\frac{1}{5}$
79. $810\frac{3}{11}$ **81.** $12\frac{7}{15}$ **83.**
85.   **87.**

89.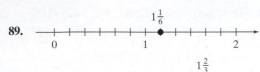

91.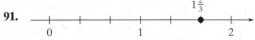

93. False   **95.** True

Section 2.2 Practice Exercises, pp. 111–113

3. $\frac{8}{12}; \frac{4}{12}$ **5.** $\frac{5}{4}; \frac{3}{4}$ **7.** $\frac{7}{12}$; proper   **9.** $4\frac{3}{5}$

11. For example: $2 \cdot 4$ and $1 \cdot 8$

13. For example: $4 \cdot 6$ and $2 \cdot 2 \cdot 2 \cdot 3$

15.

Product	36	42	30	15	81
Factor	12	7	30	15	27
Factor	3	6	1	1	3
Sum	15	13	31	16	30

17. A whole number is divisible by 2 if it is an even number.
19. A whole number is divisible by 3 if the sum of its digits is divisible by 3.   **21. a.** No **b.** Yes **c.** Yes **d.** No
23. a. No **b.** No **c.** No **d.** No
25. a. Yes **b.** Yes **c.** No **d.** No
27. a. Yes **b.** No **c.** Yes **d.** Yes
29. Yes   **31.** Prime   **33.** Composite   **35.** Composite
37. Prime   **39.** Neither   **41.** Composite
43. Prime   **45.** Composite
47. There are two whole numbers that are neither prime nor composite, 0 and 1.   **49.** False
51. 2, 3, 5, 7, 11, 13, 17, 19, 23, 29, 31, 37, 41, 43, 47
53. No, 9 is not a prime number.   **55.** Yes.
57. $2 \cdot 5 \cdot 7$.   **59.** $2 \cdot 2 \cdot 5 \cdot 13$ or $2^2 \cdot 5 \cdot 13$
61. $3 \cdot 7 \cdot 7$ or $3 \cdot 7^2$ **63.** $2 \cdot 3 \cdot 23$
65. $2 \cdot 2 \cdot 2 \cdot 7 \cdot 11$ or $2^3 \cdot 7 \cdot 11$ **67.** Prime.
69. 1, 2, 3, 4, 6, 12   **71.** 1, 2, 4, 8, 16, 32
73. 1, 3, 9, 27, 81   **75.** 1, 2, 3, 4, 6, 8, 12, 16, 24, 48
77. No   **79.** Yes   **81.** Yes   **83.** No   **85.** Yes
87. No   **89.** Yes   **91.** No

Section 2.3 Practice Exercises, pp. 118–122

3. $5 \cdot 29$ **5.** $2 \cdot 2 \cdot 23$ or $2^2 \cdot 23$ **7.** $5 \cdot 17$ **9.** $3 \cdot 5 \cdot 13$

11. 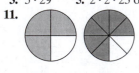   **13.**

15. False   **17.** ≠   **19.** =   **21.** =
23. ≠   **25.** $\frac{1}{2}$ **27.** $\frac{1}{3}$ **29.** $\frac{9}{5}$ **31.** $\frac{5}{4}$ **33.** $\frac{4}{5}$
35. 1   **37.** 2   **39.** 1   **41.** $\frac{3}{4}$ **43.** 3   **45.** $\frac{7}{10}$
47. $\frac{13}{15}$ **49.** $\frac{77}{39}$ **51.** $\frac{2}{5}$ **53.** $\frac{2}{7}$ **55.** 0
57. Undefined   **59.** $\frac{3}{5}$ **61.** $\frac{3}{4}$ **63.** $\frac{5}{3}$ **65.** $\frac{21}{11}$

67. $\frac{17}{100}$ **69.** Heads: $\frac{5}{12}$; tails: $\frac{7}{12}$ **71. a.** $\frac{3}{13}$ **b.** $\frac{10}{13}$
73. a. Jonathan: $\frac{5}{7}$; Jared: $\frac{6}{7}$ **b.** Jared sold the greater fractional part.   **75. a.** Raymond: $\frac{10}{11}$; Travis: $\frac{9}{11}$
b. Raymond read the greater fractional part.
77. a. 300,000,000 **b.** 36,000,000 **c.** $\frac{3}{25}$
79. For example: $\frac{6}{8}, \frac{9}{12}, \frac{12}{16}$ **81.** For example: $\frac{6}{9}, \frac{4}{6}, \frac{2}{3}$

Section 2.3 Calculator Connections, p. 122

83. $\frac{8}{9}$ **84.** $\frac{13}{14}$ **85.** $\frac{41}{51}$ **86.** $\frac{21}{10}$ **87.** $\frac{29}{30}$
88. $\frac{13}{7}$ **89.** $\frac{3}{2}$ **90.** $\frac{31}{19}$

Section 2.4 Practice Exercises, pp. 129–133

3. Numerator: 10; denominator: 14; $\frac{5}{7}$
5. Numerator: 25; denominator: 15; $\frac{5}{3}$ **7.**
9. 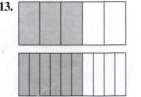   **11.** $\frac{1}{8}$ **13.** 6   **15.** $\frac{3}{16}$ **17.** $\frac{14}{81}$
19. $\frac{24}{35}$ **21.** $\frac{8}{11}$ **23.** $\frac{24}{5}$ **25.** $\frac{65}{36}$ **27.** $\frac{2}{15}$ **29.** $\frac{5}{8}$
31. $\frac{35}{4}$ **33.** $\frac{8}{3}$ **35.** $\frac{4}{5}$ **37.** 8   **39.** 12   **41.** $\frac{30}{7}$
43. 10   **45.** $\frac{5}{3}$ **47.** $\frac{3}{8}$ **49.** 24   **51.** $\frac{1}{1000}$
53. $\frac{1}{1,000,000}$ **55.** $\frac{1}{81}$ **57.** $\frac{27}{8}$ **59.** 27   **61.** $\frac{1}{225}$
63. 2   **65.** $\frac{2}{9}$ **67.**   **69.**
71. 44 cm^2 **73.** 32 m^2 **75.** 4 yd^2 **77.** $\frac{1}{4}$ cm^2
79. $\frac{195}{256}$ in.^{2} **81.** 48 yd^2 **83.** 9 cm^2
85. The amount left is 10 gal.
87. Jim ate $\frac{1}{8}$ of the pizza for breakfast.   **89.** 6,550,000
91. First place: $800; second place: $300; third place: $100
93. a. $\frac{1}{36}$ **b.** $\frac{1}{6}$ **95.** $\frac{1}{5}$ **97.** $\frac{8}{9}$ **99.** $\frac{1}{32}$
101. They are the same.

Section 2.5 Practice Exercises, pp. 139–142

3. $\frac{18}{5}$ **5.** 2   **7.** $\frac{5}{3}$ **9.** 1   **11.** 1   **13.** $\frac{8}{7}$
15. $\frac{9}{10}$ **17.** $\frac{1}{4}$ **19.** No reciprocal exists.   **21.** $\frac{1}{3}$

23. Multiplying   25. $\dfrac{8}{25}$ 27. $\dfrac{35}{26}$ 29. $\dfrac{35}{9}$ 31. 5

33. 1   35. $\dfrac{21}{2}$ 37. 20   39. 16   41. $\dfrac{3}{5}$

43. $\dfrac{1}{4}$ 45. $\dfrac{90}{13}$ 47. 20   49. $\dfrac{7}{2}$ 51. $\dfrac{5}{36}$

53. 8   55. $\dfrac{2}{5}$ 57. $\dfrac{40}{3}$ 59. 2   61. $\dfrac{55}{56}$ 63. $\dfrac{3}{2}$

65. $\dfrac{2}{3}\cdot 6$ multiplies $\dfrac{2}{3}$ by $\dfrac{6}{1}$, and $\dfrac{2}{3}\div 6$ multiplies $\dfrac{2}{3}$ by $\dfrac{1}{6}$. So $\dfrac{2}{3}\cdot 6=4$ and $\dfrac{2}{3}\div 6=\dfrac{1}{9}$.   67. $\dfrac{3}{7}$ 69. $\dfrac{7}{6}$ 71. $\dfrac{7}{32}$

73. $\dfrac{9}{400}$ 75. 49   77. $\dfrac{7}{16}$ 79. 18

81. Li wrapped 54 packages.   83. 24 cups of juice

85. The stack will be 12 in. high.

87. a. 27 commercials in 1 hr   b. 648 commercials in 1 day

89. a. Ricardo's mother will pay $16,000.
b. Ricardo will have to pay $8000.   c. He will have to finance $216,000.   91. a. She plans to sell $\frac{3}{4}$ acre.
b. She will keep $\frac{3}{2}$ or $1\frac{1}{2}$ acres

93. She can prepare 14 samples.

95. 12 ft, because $30 \div \dfrac{5}{2}=12$.

Chapter 2 Problem Recognition Exercises, p. 143

1. a. $\dfrac{16}{5}$ b. $\dfrac{16}{5}$ c. $\dfrac{20}{9}$ d. $\dfrac{9}{20}$

2. a. $\dfrac{40}{7}$ b. $\dfrac{40}{7}$ c. $\dfrac{35}{18}$ d. $\dfrac{18}{35}$

3. a. $\dfrac{27}{2}$ b. $\dfrac{27}{2}$ c. $\dfrac{32}{3}$ d. $\dfrac{3}{32}$

4. a. 9   b. 9   c. 25   d. $\dfrac{1}{25}$

5. a. $\dfrac{25}{36}$ b. 1   c. 1   d. $\dfrac{25}{36}$

6. a. 0   b. 0   c. Undefined   d. 0

7. a. $\dfrac{8}{189}$ b. $\dfrac{7}{96}$ c. $\dfrac{2}{21}$ d. $\dfrac{21}{128}$

8. a. $\dfrac{7}{27}$ b. $\dfrac{7}{12}$ c. $\dfrac{3}{7}$ d. $\dfrac{27}{28}$

9. a. $\dfrac{27}{20}$ b. $\dfrac{108}{5}$ c. $\dfrac{3}{80}$ d. $\dfrac{3}{5}$

10. a. $\dfrac{2}{5}$ b. $\dfrac{1}{250}$ c. 160   d. $\dfrac{8}{5}$

11. a. $\dfrac{2}{3}$ b. $\dfrac{2}{3}$ c. $\dfrac{2}{3}$ d. $\dfrac{3}{2}$

12. a. $\dfrac{3}{5}$ b. $\dfrac{5}{3}$ c. 60   d. 60

13. a. 32   b. 2   c. 2   d. 32

14. a. $\dfrac{1}{14}$ b. $\dfrac{2}{7}$ c. $\dfrac{1}{14}$ d. $\dfrac{2}{7}$

15. a. $\dfrac{8}{3}$ b. 96   c. $\dfrac{1}{9}$ d. 144

16. a. $\dfrac{1}{6}$ b. $\dfrac{3}{8}$ c. $\dfrac{2}{9}$ d. $\dfrac{9}{8}$

Section 2.6 Practice Exercises, pp. 148–150

3. $\dfrac{26}{9}$ 5. $\dfrac{12}{11}$ 7. $\dfrac{2}{9}$ 9. $\dfrac{17}{5}$ 11. $\dfrac{11}{7}$ 13. $12\dfrac{5}{6}$

15. $9\dfrac{3}{4}$ 17. $7\dfrac{2}{5}$ 19. $1\dfrac{2}{3}$ 21. 38   23. $27\dfrac{2}{3}$

25. $72\dfrac{1}{2}$ 27. 0   29. $7\dfrac{1}{2}$ 31. $2\dfrac{4}{25}$ 33. $\dfrac{34}{55}$

35. $4\dfrac{5}{12}$ 37. $2\dfrac{6}{17}$ 39. 2   41. 0   43. 17

45. $4\dfrac{2}{3}$ 47. $1\dfrac{3}{4}$ 49. Tabitha earned $38.

51. $642\frac{1}{2}$ lb   53. a. 7 weeks old   b. $8\frac{1}{2}$ weeks old.

55. a. Lucy earned $72 more than Ricky.   b. Together they earned $922.

57. 2   59. $5\dfrac{1}{3}$ 61. $1\dfrac{4}{5}$ 63. 0

65. $1\dfrac{1}{6}$ 67. 0   69. $1\dfrac{1}{2}$ 71. Undefined   73. $2\dfrac{5}{8}$

75. $2\dfrac{3}{8}$ 77. The total cost is $168.

Section 2.6 Calculator Connections, p. 150

79. $318\dfrac{1}{4}$ 80. $3\dfrac{1}{15}$ 81. $17\dfrac{18}{19}$ 82. $466\dfrac{1}{5}$

83. $2\dfrac{99}{146}$ 84. $2\dfrac{404}{753}$ 85. $480\dfrac{1}{8}$ 86. $280\dfrac{5}{27}$

Chapter 2 Review Exercises, pp. 157–160

1. $\dfrac{1}{2}$ 3. a. $\dfrac{5}{3}$ b. Improper   5. $\dfrac{7}{15}$ 7. $\dfrac{7}{6}$ or $1\dfrac{1}{6}$

9. $\dfrac{57}{5}$ 11. $5\dfrac{2}{9}$

13., 15.

17. $60\dfrac{11}{13}$ 19. 55, 140, 260, 1200   21. Prime

23. Neither   25. $2\cdot 2\cdot 2\cdot 2\cdot 2\cdot 2$ or 2^6

27. $2\cdot 2\cdot 3\cdot 3\cdot 5\cdot 5$ or $2^2\cdot 3^2\cdot 5^2$

29. 1, 2, 4, 5, 8, 10, 16, 20, 40, 80

31. $=$ 33. $\dfrac{2}{7}$ 35. $\dfrac{7}{3}$ 37. 2   39. $\dfrac{7}{10}$

41. a. $\dfrac{3}{5}$ b. $\dfrac{2}{5}$ 43. $\dfrac{32}{9}$ 45. 15   47. $\dfrac{12}{7}$

49. $\dfrac{1}{625}$ 51. $\dfrac{1}{17}$ 53. $A=lw$ 55. $\dfrac{10}{3}$ or $3\dfrac{1}{3}$ m^2

57. Maximus requires $\dfrac{7}{2}$ or $3\dfrac{1}{2}$ yd of lumber.

59. There are 300 Asian American students.

61. There are 750 Caucasian male students.

63. 1   65. $\dfrac{1}{7}$ 67. 6   69. multiplying   71. $\dfrac{7}{5}$

73. $\dfrac{1}{6}$ 75. 14   77. $\dfrac{4}{5}$ 79. $\dfrac{1}{52}$ 81. $18\div\dfrac{2}{3}$; 27

83. Amelia earned $576.

85. Yes. $9 \div \dfrac{3}{8} = 24$ so he will have 24 pieces, which is more than enough for his class.

87. $23\dfrac{2}{3}$ **89.** $22\dfrac{1}{2}$ **91.** $1\dfrac{1}{2}$ **93.** $4\dfrac{1}{2}$ **95.** $\dfrac{3}{5}$

97. It will take $3\dfrac{1}{8}$ gal.

Chapter 2 Test, pp. 160–161

1. a. $\dfrac{5}{8}$ **b.** Proper   **2. a.** $\dfrac{7}{3}$ **b.** Improper

3. $\dfrac{11}{2}$; $5\dfrac{1}{2}$

4. $\dfrac{7}{7}$ is an improper fraction because the numerator is greater than or equal to the denominator.

5. a. $3\dfrac{2}{3}$ **b.** $\dfrac{34}{9}$ **6.**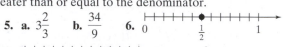

7.

8. ├┼┼┼┼┼┼●┼┼┼┼→
　　0　　　　$\frac{7}{12}$　　1

9.
$\frac{13}{5}$
├┼┼┼┼┼┼┼┼┼┼┼┼●┼┼┼┼┼→
0　　　1　　　2　　　3　　　4

10. a. Composite   **b.** Neither   **c.** Prime
d. Neither   **e.** Prime   **f.** Composite
11. a. $1, 3, 5, 9, 15, 45$ **b.** $3 \cdot 3 \cdot 5$ or $3^2 \cdot 5$
12. a. Add the digits of the number. If the sum is divisible by 3, then the original number is divisible by 3.
b. Yes.   **13. a.** No   **b.** Yes   **c.** Yes   **d.** No   **14.** $=$

15. $\neq$ **16.** $\dfrac{10}{7}$ or $1\dfrac{3}{7}$ **17.** $\dfrac{6}{7}$

18. a. Christine: $\dfrac{3}{5}$; Brad: $\dfrac{4}{5}$

b. Brad has the greater fractional part completed.

19. $\dfrac{19}{69}$ **20.** $\dfrac{25}{2}$ or $12\dfrac{1}{2}$ **21.** $\dfrac{4}{9}$ **22.** $\dfrac{1}{2}$ **23.** $\dfrac{4}{15}$

24. $\dfrac{3}{4}$ **25.** $\dfrac{4}{35}$ **26.** $9\dfrac{3}{5}$ **27.** $\dfrac{13}{12}$

28. $\dfrac{44}{3}$ or $14\dfrac{2}{3}$ cm^2 **29.** $20 \div \dfrac{1}{4}$

30. 48 quarter-pounders

31. 5 dogs are female pure breeds.

32. They can build on a maximum of $\dfrac{2}{5}$ acre.

Chapters 1–2 Cumulative Review Exercises, pp. 161–162

1.

Mountain	Height (ft)		
	Standard Form	Words
Mt. Foraker (Alaska)	17,400	Seventeen thousand, four hundred
Mt. Kilimanjaro (Tanzania)	19,340	Nineteen thousand, three hundred forty
El Libertador (Argentina)	22,047	Twenty-two thousand, forty-seven
Mont Blanc (France-Italy)	15,771	Fifteen thousand, seven hundred seventy-one

2. 1430   **3.** 139   **4.** 214,344   **5.** 24   **6.** 1863
7. 18 R2   **8.** 120,000,000,000   **9.** 184   **10.** 6
11. 22   **12.** 16   **13.** 4   **14.** d   **15.** c   **16.** b

17. e   **18.** a   **19. a.** $\dfrac{4}{7}$ **b.** $\dfrac{7}{3}$ or $2\dfrac{1}{3}$
20. a. Proper   **b.** Improper   **c.** Improper
21. a. $1, 2, 3, 5, 6, 10, 15, 30$ **b.** $2 \cdot 3 \cdot 5$.

22. a. $\dfrac{12}{7}$ **b.** $\dfrac{2}{5}$ **23.** $\dfrac{119}{171}$ **24.** $\dfrac{5}{6}$

25. Yes. $\dfrac{8}{13} \cdot \dfrac{5}{16} = \dfrac{5}{26}$ and $\dfrac{5}{16} \cdot \dfrac{8}{13} = \dfrac{5}{26}$

26. Yes. $\left(\dfrac{1}{2} \cdot \dfrac{2}{9}\right) \cdot \dfrac{5}{3} = \dfrac{1}{9} \cdot \dfrac{5}{3} = \dfrac{5}{27}$ and

$\dfrac{1}{2} \cdot \left(\dfrac{2}{9} \cdot \dfrac{5}{3}\right) = \dfrac{1}{2} \cdot \left(\dfrac{10}{27}\right) = \dfrac{5}{27}$

27. $\dfrac{6}{25}$ **28.** $\dfrac{11}{9}$ or $1\dfrac{2}{9}$ m^2 **29.** 50 ft^2

30. $\dfrac{3}{40}$ of the students are males from out of state.

Chapter 3
Chapter Opener Puzzles

An improper fraction is $\dfrac{\text{t}}{1}\ \dfrac{\text{o}}{2}\ \dfrac{\text{p}}{3}\ \ \dfrac{\text{h}}{4}\ \dfrac{\text{e}}{5}\ \dfrac{\text{a}}{6}\ \dfrac{\text{v}}{7}\ \dfrac{\text{y}}{8}$.

Section 3.1 Practice Exercises, pp. 167–170

3. 8 ft   **5.** 20 m   **7.** 7 fourths
9.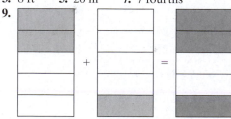

11. $\dfrac{13}{11}$ **13.** $\dfrac{9}{5}$ **15.** 1

17. $\dfrac{2}{3}$ **19.** $\dfrac{13}{10}$ **21.** $\dfrac{5}{2}$

23. Bethany has $\dfrac{5}{2}$ or $2\dfrac{1}{2}$ cups of bleach and water mixture.

25. 11 baskets   **27.** 6 fifths

29. 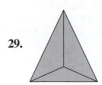 $-$ $=$

31. $\dfrac{3}{8}$ **33.** $\dfrac{3}{2}$ **35.** 2   **37.** $\dfrac{2}{3}$ **39.** $\dfrac{2}{5}$ **41.** $\dfrac{1}{4}$

43. $\dfrac{1}{4}$ g is left.   **45.** $\dfrac{3}{2}$ **47.** $\dfrac{12}{5}$ **49.** 1   **51.** $\dfrac{4}{5}$

53. $\dfrac{5}{2}$ **55.** $\dfrac{7}{4}$ **57.** $\dfrac{81}{100}$ **59.** $\dfrac{5}{3}$ **61.** $\dfrac{9}{5}$ **63.** $\dfrac{16}{7}$

65. $\dfrac{13}{3}$ **67.** $\dfrac{1}{3}$ **69.** $\dfrac{12}{7}$ or $1\dfrac{5}{7}$ m   **71.** $\dfrac{7}{2}$ or $3\dfrac{1}{2}$ in.

73. There was $\dfrac{1}{2}$ gal left over.   **75.** He used $\dfrac{3}{8}$ L.

77. a. Thilan walked $5\dfrac{1}{2}$ mi total.   **b.** He walked an average of $\dfrac{11}{12}$ mi per day.

79. Perimeter: 2 ft; area: $\dfrac{15}{64}$ ft^2

81. Perimeter: $\dfrac{70}{3}$ or $23\dfrac{1}{3}$ yd; area: $\dfrac{286}{9}$ or $31\dfrac{7}{9}$ yd^2

83. $\dfrac{3}{5} + \dfrac{2}{5}$; 1   **85.** $\dfrac{11}{15} - \dfrac{8}{15}$; $\dfrac{1}{5}$

Section 3.2 Practice Exercises, pp. 176–179

3. $\dfrac{1}{2}$ **5.** $\dfrac{5}{3}$ **7.** 6

9. a. 48, 72, 240   **b.** 4, 8, 12

11. a. 72, 360, 108   **b.** 6, 12, 9

13. 50   **15.** 48   **17.** 120   **19.** 72   **21.** 60

23. 75   **25.** 120   **27.** 210   **29.** 540   **31.** 60

33. 240   **35.** 180   **37.** 180

39. The shortest length of floor space is 60 in. (5 ft).

41. It will take 120 hr (5 days) for the satellites to be lined up again.

43. $\dfrac{14}{21}$ **45.** $\dfrac{10}{16}$ **47.** $\dfrac{12}{16}$ **49.** $\dfrac{12}{15}$ **51.** $\dfrac{49}{42}$

53. $\dfrac{121}{99}$ **55.** $\dfrac{15}{39}$ **57.** $\dfrac{11,000}{4000}$ **59.** $\dfrac{15}{70}$ **61.** $>$

63. $<$ **65.** $=$ **67.** $<$ **69.** $\dfrac{7}{8}$ **71.** $\dfrac{2}{3}, \dfrac{3}{4}, \dfrac{7}{8}$

73. $\dfrac{1}{4}, \dfrac{5}{16}, \dfrac{3}{8}$ **75.** $\dfrac{13}{12}, \dfrac{17}{15}, \dfrac{4}{3}$

77. The longest cut is above the left eye. The shortest cut is on the right hand.

79. The greatest amount is $\frac{2}{3}$ lb of turkey. The least amount is $\frac{3}{5}$ lb of ham.

81. a and b

Section 3.3 Practice Exercises, pp. 185–187

3. $\dfrac{12}{14}$ **5.** $\dfrac{14}{21}$ **7.** $\dfrac{25}{5}$ **9.** $\dfrac{8}{4}$ **11.** $\dfrac{80}{100}$ **13.** $\dfrac{5}{40}$

15. $\dfrac{19}{16}$ **17.** $\dfrac{1}{6}$ **19.** $\dfrac{1}{4}$ **21.** $\dfrac{83}{42}$ **23.** $\dfrac{3}{8}$ **25.** $\dfrac{1}{3}$

27. $\dfrac{25}{36}$ **29.** $\dfrac{5}{8}$ **31.** $\dfrac{25}{8}$ **33.** $\dfrac{8}{3}$ **35.** $\dfrac{17}{3}$ **37.** $\dfrac{2}{7}$

39. $\dfrac{89}{100}$ **41.** $\dfrac{1}{100}$ **43.** $\dfrac{391}{1000}$ **45.** $\dfrac{9}{8}$ **47.** $\dfrac{23}{60}$

49. $\dfrac{9}{16}$ **51.** $\dfrac{1}{36}$ **53.** $\dfrac{7}{12}$ **55.** $\dfrac{7}{3}$ **57.** $\dfrac{4}{3}$ **59.** $\dfrac{38}{35}$

61. $\dfrac{13}{125}$ **63.** $\dfrac{23}{24}$ **65.** Inez added $\dfrac{9}{8}$ or $1\dfrac{1}{8}$ cups.

67. The storm delivered $\dfrac{5}{32}$ in. of rain.

69. The trough now holds the original amount of 5 gal.

71. a. $\dfrac{13}{36}$ **b.** $\dfrac{23}{36}$ **73.** $\dfrac{13}{5}$ or $2\dfrac{3}{5}$ m

75. Perimeter: 3 ft   **77.** b

Section 3.4 Practice Exercises, pp. 193–197

3. $\dfrac{13}{6}$ **5.** $\dfrac{24}{5}$ **7.** $\dfrac{1}{2}$ **9.** $7\dfrac{4}{11}$ **11.** $15\dfrac{3}{7}$

13. $15\dfrac{9}{16}$ **15.** $10\dfrac{13}{15}$ **17.** 5   **19.** 2   **21.** $3\dfrac{1}{5}$

23. $8\dfrac{2}{3}$ **25.** $14\dfrac{1}{2}$ **27.** $23\dfrac{1}{8}$ **29.** $19\dfrac{17}{48}$ **31.** $9\dfrac{7}{8}$

33. $42\dfrac{2}{7}$ **35.** $11\dfrac{3}{5}$ **37.** $2\dfrac{2}{15}$ **39.** $12\dfrac{1}{6}$ **41.** $2\dfrac{5}{14}$

43. $\dfrac{3}{3}$ **45.** $\dfrac{12}{12}$ **47.** $11\dfrac{1}{2}$ **49.** $1\dfrac{3}{4}$ **51.** $7\dfrac{13}{14}$

53. $3\dfrac{1}{6}$ **55.** $2\dfrac{7}{9}$ **57.** $2\dfrac{3}{17}$ **59.** $6\dfrac{5}{14}$ **61.** $7\dfrac{7}{24}$

63. $6\dfrac{2}{15}$ **65.** $\dfrac{11}{16}$ **67.** $9\dfrac{7}{36}$ **69.** $\dfrac{29}{32}$ **71.** $10\dfrac{20}{21}$

73. $\dfrac{32}{35}$ **75.** $7\dfrac{13}{72}$ **77.** $7\dfrac{3}{4}$ in.   **79.** The index finger is longer.   **81.** The total is $16\dfrac{11}{12}$ hr.   **83.** $3\dfrac{5}{12}$ ft

85. $\dfrac{1}{32}$ in.   **87.** The printing area width is 6 in.

89. There is $2\dfrac{5}{6}$ hr remaining.   **91.** The blinds will hang $\dfrac{1}{3}$ ft below the window.   **93. a.** $3\dfrac{3}{8}$ L   **b.** $\dfrac{5}{8}$ L

95. 4   **97.** $4\dfrac{1}{4}$

Section 3.4 Calculator Connections, p. 198

98. $\dfrac{211}{168}$ or $1\dfrac{43}{168}$ **99.** $\dfrac{11}{30}$ **100.** $\dfrac{37}{132}$ **101.** $\dfrac{137}{391}$

102. $\dfrac{2509}{54}$ or $46\dfrac{25}{54}$ **103.** $\dfrac{2171}{84}$ or $25\dfrac{71}{84}$ **104.** $\dfrac{402}{77}$ or $5\dfrac{17}{77}$

105. $\dfrac{213}{68}$ or $3\dfrac{9}{68}$

Chapter 3 Problem Recognition Exercises, pp. 198–199

1. $\frac{9}{5}$ or $1\frac{4}{5}$ **2.** $\frac{14}{25}$ **3.** $\frac{7}{2}$ or $3\frac{1}{2}$ **4.** 1

5. $\frac{10}{9}$ or $1\frac{1}{9}$ **6.** $\frac{8}{5}$ or $1\frac{3}{5}$ **7.** $\frac{13}{6}$ or $2\frac{1}{6}$ **8.** $\frac{1}{2}$

9. $\frac{17}{4}$ or $4\frac{1}{4}$ **10.** $\frac{5}{4}$ or $1\frac{1}{4}$ **11.** $\frac{11}{6}$ or $1\frac{5}{6}$ **12.** $\frac{33}{8}$ or $4\frac{1}{8}$

13. $\frac{221}{18}$ or $12\frac{5}{18}$ **14.** $\frac{26}{17}$ or $1\frac{9}{17}$ **15.** $\frac{3}{2}$ or $1\frac{1}{2}$

16. $\frac{43}{6}$ or $7\frac{1}{6}$ **17.** $\frac{29}{8}$ or $3\frac{5}{8}$ **18.** $\frac{3}{2}$ or $1\frac{1}{2}$ **19.** $\frac{32}{3}$ or $10\frac{2}{3}$

20. $\frac{35}{8}$ or $4\frac{3}{8}$ **21.** $\frac{11}{6}$ or $1\frac{5}{6}$ **22.** $\frac{5}{3}$ or $1\frac{2}{3}$ **23.** $\frac{17}{3}$ or $5\frac{2}{3}$

24. $\frac{22}{3}$ or $7\frac{1}{3}$ **25.** $\frac{53}{15}$ or $3\frac{8}{15}$ **26.** $\frac{73}{13}$ or $4\frac{13}{15}$

27. $\frac{14}{5}$ or $2\frac{4}{5}$ **28.** $\frac{63}{10}$ or $6\frac{3}{10}$ **29.** $\frac{25}{18}$ or $1\frac{7}{18}$

30. $\frac{50}{9}$ or $5\frac{5}{9}$ **31.** $\frac{7}{9}$ **32.** $\frac{43}{9}$ or $4\frac{7}{9}$ **33.** 1

34. $\frac{106}{45}$ or $2\frac{16}{45}$ **35.** $\frac{81}{25}$ or $3\frac{6}{25}$ **36.** $\frac{56}{45}$ or $1\frac{11}{45}$

37. 1   **38.** 1   **39.** 1   **40.** 1

Section 3.5 Practice Exercises, pp. 204–207

3. $12\frac{2}{9}$ **5.** $2\frac{2}{3}$ **7.** $3\frac{13}{36}$ **9.** $\frac{67}{13}$ **11.** $\frac{39}{10}$ **13.** $5\frac{4}{5}$

15. $1\frac{11}{19}$ **17.** $2\frac{1}{4}$ **19.** $3\frac{2}{3}$ **21.** $7\frac{1}{2}$ **23.** $4\frac{2}{7}$

25. 13   **27.** $\frac{13}{25}$ **29.** $1\frac{10}{27}$ **31.** $3\frac{1}{4}$ **33.** $1\frac{3}{7}$

35. a. The difference is $\frac{3}{10}$ sec.   **b.** The average is $3\frac{3}{5}$ sec.

37. a. The total weight loss is 51 lb.   **b.** The average is $8\frac{1}{2}$ lb.

c. The difference is $6\frac{1}{2}$ lb.   **39.** The stock dropped $3\frac{7}{8}$.

41. George will receive \$26,750.

43. Each piece is $3\frac{13}{16}$ ft.   **45.** $2\frac{1}{4}$ lb of cheese was eaten.

47. 20 loaves can be made.   **49.** The new rate is $7\frac{1}{4}$ points.

51. Stephanie will need $11\frac{1}{4}$ yd for the dresses.

53. Wilma has $1\frac{1}{12}$ lb left.   **55.** Joan saves $152\frac{1}{2}$ gal.

57. She needs $15\frac{1}{3}$ ft more.   **59.** The perimeter is 100 in.

61. Matt needs $76\frac{1}{3}$ ft of gutter.

63. The area of the whole roof is $1022\frac{7}{16}$ ft².

65. a. The area is $247\frac{1}{2}$ m².   **b.** They will need 65 m.

67. $152\frac{3}{4}$ m²

Chapter 3 Review Exercises, pp. 212–214

1. 8 books   **3.** 12 mi   **5.** Fractions with the same denominators are considered like fractions.

7. $\frac{3}{2}$ **9.** $\frac{1}{2}$ **11.** $\frac{9}{7}$ **13.** $\frac{3}{4}$ **15.** $\frac{11}{13}$

17. 12 in. or 1 ft   **19. a.** 7, 14, 21, 28   **b.** 13, 26, 39, 52
c. 22, 44, 66, 88   **21. a.** 1, 2, 4, 5, 10, 20, 25, 50, 100
b. 1, 5, 13, 65   **c.** 1, 2, 5, 7, 10, 14, 35, 70   **23.** 150
25. 420   **27.** They will meet on the 12th day.

29. $\frac{63}{35}$ **31.** $\frac{170}{150}$ **33.** $>$ **35.** $\frac{8}{15}, \frac{72}{105}, \frac{7}{10}, \frac{27}{35}$

37. $\frac{29}{100}$ **39.** $\frac{1}{2}$ **41.** $\frac{43}{20}$ **43.** $\frac{1}{34}$ **45.** $\frac{17}{40}$

47. $\frac{1}{15}$ **49. a.** $\frac{35}{4}$ or $8\frac{3}{4}$ m   **b.** $\frac{315}{128}$ or $2\frac{59}{128}$ m²

51. $11\frac{11}{63}$ **53.** $2\frac{5}{8}$ **55.** $3\frac{1}{24}$ **57.** $12\frac{5}{14}$

59. $3\frac{2}{5}$ **61.** $63\frac{15}{16}$ **63.** 8, $8\frac{5}{18}$ **65.** 50, $50\frac{9}{40}$

67. Corry drove a total of $8\frac{1}{6}$ hr.   **69.** $12\frac{2}{5}$

71. $\frac{4}{27}$ **73.** 12   **75.** The appraised value is \$144,000.

Chapter 3 Test, p. 215

1. $\frac{7}{5}$ **2.** $\frac{1}{2}$ **3.** When subtracting like fractions, keep the same denominator and subtract the numerators. When multiplying fractions, multiply the denominators as well as the numerators.

4. a. 24, 48, 72, 96   **b.** 1, 2, 3, 4, 6, 8, 12, 24
c. $2 \cdot 2 \cdot 2 \cdot 3$ or $2^3 \cdot 3$

5. 240   **6.** $\frac{35}{63}$ **7.** $\frac{33}{63}$ **8.** $\frac{36}{63}$ **9.** $\frac{11}{21}, \frac{5}{9}, \frac{4}{7}$

10. $\frac{9}{16}$ **11.** $\frac{1}{3}$ **12.** $\frac{1}{3}$ **13.** $\frac{2}{3}$ **14.** $17\frac{3}{8}$

15. $2\frac{1}{11}$ **16.** $60\frac{5}{12}$ **17.** $1\frac{1}{2}$ **18.** $\frac{25}{6}$ or $4\frac{1}{6}$

19. 7   **20.** $\frac{12}{295}$ **21.** $\frac{10}{3}$ or $3\frac{1}{3}$ **22.** 1 lb is needed.

23. The Ford Expedition can tow 8950 lb.

24. Area: $25\frac{2}{25}$ m²; perimeter: $20\frac{1}{5}$ m

25. Justin has \$10,500 for cabinets.

26. The difference is $4\frac{2}{3}$ ft.

Chapters 1–3 Cumulative Review Exercises, p. 216

1. Twenty-three million, four hundred thousand, eight hundred six

2. 96   **3.** 48   **4.** 1728   **5.** 3   **6.** 1,500,000,000
7. $4^2 \cdot 5^4 \cdot 8^2$ **8.** 36   **9.** 17, 19, 23, 29, 31
10. $2 \cdot 5 \cdot 7$ **11.** Numerator: 21; denominator: 17

12. $\frac{4}{16}$ or $\frac{1}{4}$ **13.** $\frac{17}{22}$ had pepperoni and $\frac{5}{22}$ did not have pepperoni.   **14. a.** Improper   **b.** Proper   **c.** Improper

15. b   **16. a.** Composite   **b.** Composite   **c.** Prime

17. $2 \cdot 2 \cdot 2 \cdot 3 \cdot 3 \cdot 5$ or $2^3 \cdot 3^2 \cdot 5$ **18.** $\dfrac{1}{5}$ **19.** $\dfrac{3}{8}$

20. $\dfrac{4}{7}$ **21.** $\dfrac{3}{4}$ **22.** $\dfrac{33}{16}$ **23.** $\dfrac{2}{5}$ **24.** $\dfrac{305}{22}$ or $13\dfrac{19}{22}$

25. $\dfrac{26}{17}$ or $1\dfrac{9}{17}$ **26.** $\dfrac{10}{3}$ or $3\dfrac{1}{3}$

27. The distance around is approximately 88 cm.

28. $4\dfrac{1}{3}$ yd   **29.** $\dfrac{63}{8}$ or $7\dfrac{7}{8}$ m²   **30. a.** $2\dfrac{1}{2}$ **b.** $6\dfrac{3}{8}$

Chapter 4

Chapter Opener Puzzle

4	1	2	5	3
2	3	4	1	5
1	ᵃ2	5	3	ᶜ4
3	5	1	4	2
ᵈ5	4	ᵇ3	2	1

Answer

a. 20.25   **b.** 202.5   **c.** 0.2025   **d.** 2.025

Section 4.1 Practice Exercises, pp. 224–227

3. 100   **5.** 10,000   **7.** $\dfrac{1}{100}$ **9.** $\dfrac{1}{10,000}$

11. Tenths   **13.** Hundredths   **15.** Tens

17. Ten-thousandths   **19.** Thousandths   **21.** Ones

23. Nine-tenths   **25.** Twenty-three hundredths

27. Thirty-three thousandths

29. Four hundred seven ten-thousandths

31. Three and twenty-four hundredths

33. Five and nine-tenths   **35.** Fifty-two and three-tenths

37. Six and two hundred nineteen thousandths

39. 8472.014   **41.** 700.07   **43.** 2,469,000.506

45. $3\dfrac{7}{10}$ **47.** $2\dfrac{4}{5}$ **49.** $\dfrac{1}{4}$ **51.** $\dfrac{11}{20}$ **53.** $20\dfrac{203}{250}$

55. $15\dfrac{1}{2000}$ **57.** $\dfrac{42}{5}$ **59.** $\dfrac{157}{50}$ **61.** $\dfrac{47}{2}$ **63.** $\dfrac{1191}{100}$

65. 34.2, 34.25, 34.29, 34.3   **67.** 0.042, 0.043, $\dfrac{4}{10}$, 0.42, 0.43

69. <   **71.** >   **73.** >   **75.** <   **77.** a, b

79. 0.3444, 0.3493, 0.3558, 0.3585, 0.3664

81. These numbers are equivalent, but they represent different levels of accuracy.

83. 7.1   **85.** 49.9   **87.** 33.42   **89.** 9.096

91. 21.0   **93.** 7.000   **95.** 0.0079   **97.** 0.0036 mph

	Number	Hundreds	Tens	Tenths	Hun-dredths	Thou-sandths
99.	971.0948	1000	970	971.1	971.09	971.095
101.	21.9754	0	20	22.0	21.98	21.975

103. 0.972

Section 4.2 Practice Exercises, pp. 233–237

3. b, c   **5.** 23.5   **7.** 8.603   **9.** 2.8300   **11.** 63.2

13. 8.951   **15.** 15.991   **17.** 79.8005   **19.** 31.0148

21. 62.6032   **23.** 100.414   **25.** 128.44   **27.** 82.063

29. 14.24   **31.** 3.68   **33.** 12.32   **35.** 5.677

37. 1.877   **39.** 57.368   **41.** 21.6646   **43.** 14.765

45. 159.558   **47.** 15.347   **49.** 6.581   **51.** 19.912

53. 10.3327   **55.** 5.9156   **57.** 9.001

59. a. 321.724 days   **b.** 156.73 days

61. a. The water is rising 1.7 in./hr.   **b.** At 1:00 P.M. the level will be 11 in.   **c.** At 3:00 P.M. the level will be 14.4 in.

63.

Check No.	Description	Debit	Credit	Balance
				$ 245.62
2409	Electric bill	$ 52.48		193.14
2410	Groceries	72.44		120.70
2411	Department store	108.34		12.36
	Paycheck		$1084.90	1097.26
2412	Restaurant	23.87		1073.39
	Transfer from savings		200	1273.39

65. 1.35 million cells per microliter

67. The pile containing the two nickels and two pennies is higher.

69. $x = 8.9$ in.; $y = 15.4$ in.; the perimeter is 98.8 in.

71. $x = 2.075$ ft; $y = 2.59$ ft; the perimeter is 22.17 ft.

73. 27.2 mi   **75.** 7 mm

Section 4.2 Calculator Connections, p. 237

77. IBM increased by $5.90 per share.

78. FedEx decreased by $3.66 per share.

79. Between February and March, FedEx increased the most, by $2.27 per share.

80. Between April and May, IBM increased the most, by $7.96 per share.

81. Between March and April, FedEx decreased the most, by $8.09 per share.

82. Between February and March, IBM decreased the most, by $6.73 per share.

Section 4.3 Practice Exercises, pp. 243–245

3. 1000   **5.** 0.01   **7.** 0.4   **9.** 3.6   **11.** 8

13. 0.18   **15.** 17.904   **17.** 0.028   **19.** 100   **21.** 30

23. 0.07   **25.** 0.2   **27.** 37.35   **29.** 4.176

31. 4.736   **33.** 2.891   **35.** 114.88   **37.** 2.248

39. 0.00144   **41.** The decimal point will move to the right two places.   **43. a.** 51   **b.** 510   **c.** 5100   **d.** 51,000

45. The decimal point will move to the left one place.

47. 3490   **49.** 96,590   **51.** 0.933   **53.** 0.05403

55. 20.01   **57.** 0.00005   **59.** 2,600,000   **61.** 400,000

63. $20,549,000,000   **65. a.** 201.6 lb   **b.** 640 lb of CO_2

67. The bill was $312.17.   **69.** $2.81 can be saved.

71. 0.00115 km²   **73.** The area is 333 yd².   **75.** 324¢

77. 37¢   **79.** $3.47   **81.** $20.41   **83. a.** $1   **b.** $1.50

85. $(0.2)^2 = 0.04$, which is not equal to 0.4.

87. 0.16   **89.** 1.69   **91.** 0.001   **93.** 0.0016

95. a. 0.09   **b.** 0.3   **97.** 0.1   **99.** 0.6

Section 4.4 Practice Exercises, pp. 253–255

3. 5280 **5.** 3.776 **7.** 2.02 **9.** 0.9 **11.** 0.18
13. 0.53 **15.** 21.1 **17.** 1.96 **19.** 0.035
21. 16.84 **23.** 0.12 **25.** 0.16 **27.** 5.$\overline{3}$ **29.** 3.1$\overline{6}$
31. 2.$\overline{15}$ **33.** 503 **35.** 9.92 **37.** 56 **39.** 2.975
41. 208.$\overline{3}$ **43.** 48.5 **45.** 1100 **47.** 42,060
49. The decimal point will move to the left two places.
51. 0.03923 **53.** 9.802 **55.** 0.00027
57. 0.00102 **59. a.** 2.4 **b.** 2.44 **c.** 2.444
61. a. 1.9 **b.** 1.89 **c.** 1.889
63. a. 3.6 **b.** 3.63 **c.** 3.626 **65.** 0.26 **67.** 14.8
69. 20.667 **71.** 35.67 **73.** 111.3
75. Unreasonable; $960 **77.** Unreasonable; $140,000
79. The monthly payment is $42.50.
81. a. 13 bulbs would be needed (rounded up to the nearest whole unit). **b.** $9.75 **c.** The energy efficient fluorescent bulb would be more cost effective.
83. Babe Ruth's batting average was 0.342.
85. 2.2 mph **87.** 47.265 **89.** b, d

Section 4.4 Calculator Connections, p. 256

91. 1149686.166 **92.** 3411.4045 **93.** 1914.0625
94. 69,568.83693 **95.** 95.6627907 **96.** 293.5070423
97. Answers will vary. **98.** Answers will vary.
99. a. 0.27 **b.** Yes the claim is accurate. The decimal, 0.27 is close to 0.25, which is equal to $\frac{1}{4}$.
100. 272 people per square mile
101. a. 1,600,000 mi per day **b.** 66,666.$\overline{6}$ mph
102. When we say that 1 year is 365 days, we are ignoring the 0.256 day each year. In 4 years, that amount is $4 \times 0.256 = 1.024$, which is another whole day. This is why we add one more day to the calendar every 4 years.

Chapter 4 Problem Recognition Exercises, p. 257

1. a. 223.04 **b.** 12,304 **c.** 23.04 **d.** 1.2304 **e.** 123.05 **f.** 1.2304 **g.** 12,304 **h.** 123.03
2. a. 6078.3 **b.** 5,078,300 **c.** 4078.3 **d.** 5.0783 **e.** 5078.301 **f.** 5.0783 **g.** 5,078,300 **h.** 5078.299
3. a. 7.191 **b.** 7.191 **4. a.** 730.4634 **b.** 730.4634
5. a. 52.64 **b.** 52.64 **6. a.** 59.384 **b.** 59.384
7. a. 86.4 **b.** 5.4 **8. a.** 185 **b.** 46.25
9. a. 80 **b.** 448 **10. a.** 54 **b.** 496.8
11. 1 **12.** 1 **13.** 4000 **14.** 6,400,000
15. 200,000 **16.** 2700 **17.** 1,350,000,000
18. 1,700,000 **19.** 4.4001 **20.** 76.7001

Section 4.5 Practice Exercises, pp. 263–266

3. 0.39 **5.** 0.0071 **7.** $\frac{1}{625}$ **9.** $\frac{1}{8}$ **11.** $\frac{4}{10}$; 0.4
13. $\frac{98}{100}$; 0.98 **15.** 0.28 **17.** 0.632 **19.** 0.875
21. 3.2 **23.** 5.25 **25.** 1.2 **27.** 0.75
29. 3.3125 **31.** 7.45 **33.** 0.88 **35.** 3.$\overline{8}$
37. 0.4$\overline{6}$ **39.** 0.52$\overline{7}$ **41.** 0.$\overline{54}$ **43.** 0.$\overline{126}$
45. 1.1$\overline{36}$ **47.** 0.143 **49.** 0.08 **51.** 0.9 **53.** 0.71
55. 1.2 **57. a.** 0.$\overline{1}$ **b.** 0.$\overline{2}$ **c.** 0.$\overline{4}$ **d.** 0.$\overline{5}$
If we memorize that $\frac{1}{9} = 0.\overline{1}$, then $\frac{2}{9} = 2 \cdot \frac{1}{9} = 2 \cdot 0.\overline{1} = 0.\overline{2}$, and so on.

59.

	Decimal Form	Fraction Form
a.	0.45	$\frac{9}{20}$
b.	1.625	$\frac{13}{8}$ or $1\frac{5}{8}$
c.	0.$\overline{7}$	$\frac{7}{9}$
d.	0.$\overline{45}$	$\frac{5}{11}$

61.

	Decimal Form	Fraction Form
a.	0.$\overline{3}$	$\frac{1}{3}$
b.	2.125	$\frac{17}{8}$ or $2\frac{1}{8}$
c.	0.8$\overline{63}$	$\frac{19}{22}$
d.	1.68	$\frac{42}{25}$

63.

Stock	Closing Price ($) (Decimal)	Closing Price ($) (Fraction)
McGraw-Hill	69.25	$69\frac{1}{4}$
Walgreens	44.95	$44\frac{19}{20}$
Home Depot	38.50	$38\frac{1}{2}$
General Electric	37.44	$37\frac{11}{25}$

65. = **67.** < **69.** > **71.** < **73.** = **75.** <
77. $\frac{1}{10}$, 0.$\overline{1}$, $\frac{1}{5}$

79. 1.75, 1.$\overline{7}$, 1.8

81. $\frac{9}{9} = 1$ **83.** 7

Section 4.6 Practice Exercises, pp. 272–275

3. 313.72 **5.** $\frac{107}{27}$ **7.** $\frac{5}{4}$ **9.** 6.96 **11.** 6.25
13. 10 **15.** 8.77 **17.** 25.75 **19.** 2 **21.** 12.98
23. 67.35 **25.** 25.05 **27.** 23.4 **29.** 1.28
31. 10.83 **33.** 2.84 **35.** 0.935 **37.** 4.4$\overline{3}$
39. a. 471 mi **b.** 62.8 mph

41. Jorge will be charged $98.75.
43. She has 24.3 g left for dinner.
45. Caren should get $4.77 in change.
47. Duncan's average is 78.75.
49. The average snowfall per month is 14.54 in.
51. Answers will vary.
53. a. 29.8   **b.** Overweight   **55.** 3.475   **57.** 0.52

Section 4.6 Calculator Connections, p. 276

59. a. 237 shares   **b.** $13.90 will be leftover.
60. a. Approximately 921,800 homes could be powered.
b. Approximately 342,678 additional homes could be powered.
61. a. Marty will have to finance $120,000.   **b.** There are 360 months in 30 yr.   **c.** He will pay $287,409.60   **d.** He will pay $167,409.60 in interest.
62. a. Gwen needs to finance $94,000.   **b.** There are 180 months in 15 yr.   **c.** Gwen will pay $152,820.00.   **d.** She will pay $58,820.00 in interest.
63. Each person will get approximately $13,410.10.
64. The average price is $58.00.

Chapter 4 Review Exercises, pp. 283–286

1. The 3 is in the tens place, 2 is in the ones place, 1 is in the tenths place, and 6 is in the hundredths place.
3. Five and seven-tenths
5. Fifty-one and eight thousandths
7. 33,015.047   **9.** $4\frac{4}{5}$ **11.** $\frac{13}{10}$ **13.** >
15. 4.3875, 4.3953, 4.4839, 4.5000, 4.5142
17. 34.890   **19.** a, b   **21.** 49.743   **23.** 5.45
25. 197.96   **27.** 7.809   **29.** $x = 4.5$ in., $y = 5.07$ in.; the perimeter is 201 in.   **31.** 3.74 in.   **33.** 74.113
35. 346.5   **37.** 100.34   **39.** 1.0422
41. 432,000   **43.** The call will cost $1.61.
45. a. 7280 people   **b.** 18,000 people   **47.** 42.8
49. $8.7\overline{6}$ **51.** 0.03   **53.** 9.0234   **55.** 260
57. 11.62   **59. a.** $0.50 per roll.   **b.** $0.57 per roll.
c. The 12-pack is better.
61. $\frac{35}{100}$; 0.35   **63.** 2.4   **65.** 0.192   **67.** $0.58\overline{3}$
69. $4.3\overline{18}$ **71.** 0.29   **73.** 3.67   **75.** $\frac{2}{9}$ **77.** $3\frac{1}{3}$

79.

Stock	Closing Price ($) (Decimal)	Closing Price ($) (Fraction)
Sun	5.20	$5\frac{1}{5}$
Sony	55.53	$55\frac{53}{100}$
Verizon	41.16	$41\frac{4}{25}$

81. =   **83.** 0.28   **85.** 5   **87.** 78.5
89. $89.90 will be saved by buying the combo package.

Chapter 4 Test, pp. 287–288

1. a. Tens place   **b.** Hundredths place
2. Five hundred nine and twenty-four thousandths
3. $1\frac{13}{50}$; $\frac{63}{50}$ **4.** 0.4419, 0.4484, 0.4489, 0.4495
5. b is correct.   **6.** 52.832   **7.** 21.29   **8.** 126.45

26. 9.57   **27.** 47.25   **28. a.** 38.8 mi   **b.** 5.5 mi/day

Chapters 1–4 Cumulative Review Exercises, pp. 288–289

1. 14   **2.** 4039   **3.** 4840   **4.** 3872   **5.** 2,415,000
6. Dividend: 4530; divisor: 225; whole-number part of the quotient: 20; remainder: 30   **7.** To check a division problem, multiply the whole-number part of the quotient and the divisor. Then add the remainder to get the dividend. That is, $20 \times 225 + 30 = 4530$.
8. The difference between sales for Wal-Mart and Sears is $181,956 million.
9. $\frac{6}{55}$ **10.** $\frac{4}{7}$ **11.** $\frac{49}{100}$ **12.** 2   **13.** $\frac{2}{3}$ **14.** 0
15. There is $9000 left.   **16.** $\frac{2}{5}$ **17.** $\frac{97}{100}$ **18.** $\frac{38}{11}$
19. $\frac{33}{7}$ **20.** $\frac{3}{2}$ **21.** Area: $\frac{15}{64}$ ft²; perimeter: 2 ft
22. The average is $1\frac{3}{16}$ km.   **23.** 174.13
24. 668.79   **25.** 75.275   **26.** 16   **27.** 339.12
28. 46.48   **29. a.** 3.75248   **b.** 3.75248   **c.** Commutative property of multiplication
30.

Bone	Length (in.) (Decimal)	Length (in.) (Mixed Number)
Femur	19.875	$19\frac{7}{8}$
Fibula	15.9375	$15\frac{15}{16}$
Humerus	14.375	$14\frac{3}{8}$
Innominate bone (hip)	7.5	$7\frac{1}{2}$

[Handwritten note on sticky: khanacademy.org/join   Class code: ZNHP5XVW   free tutoring services   To make extra credit   2/4/19]

Chapter 5

Chapter Opener Puzzles

$\underset{4}{r}\ \underset{9}{a}\ \underset{1}{t}\ \underset{}{i}\ o$ $\underset{}{s}\ \underset{}{i}\ \underset{}{m}\ \underset{}{i}\ \underset{}{l}\ \underset{}{a}\ r$ $\underset{7}{r}\ \underset{6}{a}\ \underset{}{t}\ e$

$\underset{8}{p}\ \underset{}{r}\ \underset{}{o}\ \underset{}{p}\ \underset{}{o}\ \underset{5}{r}\ \underset{}{t}\ \underset{}{i}\ \underset{}{o}\ n$ $\underset{}{e}\ \underset{}{q}\ \underset{2}{u}\ \underset{3}{a}\ \underset{}{t}\ \underset{}{i}\ \underset{}{o}\ \underset{10}{n}$

It is difficult to move a square because it has a

$\underset{1}{s}\ \underset{2}{q}\ \underset{3}{u}\ \underset{4}{a}\ \underset{5}{r}\ \underset{6}{e}$ $\underset{7}{r}\ \underset{8}{o}\ \underset{9}{o}\ \underset{10}{t}$.

Section 5.1 Practice Exercises, pp. 296–298

3. $5:6$ and $\frac{5}{6}$ **5.** 11 to 4 and $\frac{11}{4}$ **7.** $1:2$ and 1 to 2

9. a. $\frac{3}{2}$ **b.** $\frac{2}{3}$ **c.** $\frac{3}{5}$ **11. a.** $\frac{21}{52}$ **b.** $\frac{21}{31}$ **13.** $\frac{2}{3}$

15. $\frac{1}{5}$ **17.** $\frac{4}{1}$ **19.** $\frac{11}{5}$ **21.** $\frac{6}{5}$ **23.** $\frac{1}{2}$ **25.** $\frac{3}{2}$

27. $\frac{6}{7}$ **29.** $\frac{8}{9}$ **31.** $\frac{7}{1}$ **33.** $\frac{1}{8}$ **35.** $\frac{5}{4}$ **37.** $\frac{4}{11}$

39. a. $\frac{6}{16} = \frac{3}{8}$ **b.** $\frac{\frac{1}{2}}{1\frac{1}{3}} = \frac{3}{8}$ **41.** $\frac{1}{11}$ **43.** $\frac{10}{1}$ **45.** $\frac{15}{32}$

47. $\frac{20}{61}$ **49.** $\frac{2}{3}$ **51.** $\frac{1}{4}$ **53.** 13 units   **55. a.** 1.5

b. $1.\overline{6}$ **c.** 1.6 **d.** 1.625; yes   **57.** Answers will vary.

Section 5.2 Practice Exercises, pp. 302–305

3. $3:5$ and $\frac{3}{5}$ **5.** $\frac{4}{3}$ **7.** $\frac{9}{17}$ **9.** $\frac{\$32}{5\ \text{ft}^2}$ **11.** $\frac{117\ \text{mi}}{2\ \text{hr}}$

13. $\frac{\$29}{4\ \text{hr}}$ **15.** $\frac{1\ \text{page}}{2\ \text{sec}}$ **17.** $\frac{65\ \text{calories}}{4\ \text{crakers}}$ **19.** $\frac{\$15}{2\ \text{trays}}$

21. a, c, d   **23.** 113 mi/day   **25.** 96 km/hr
27. \$55 per payment   **29.** \$0.38/lb   **31.** \$256,000 per person   **33.** 14.3 m/sec   **35.** \$0.050 per oz
37. \$0.995 per liter   **39.** \$52.50 per tire   **41.** \$5.417 per bodysuit   **43. a.** \$0.075/oz **b.** \$0.075/oz
c. Both sizes cost the same amount per ounce.
45. The larger can is \$0.041 per ounce. The smaller can is \$0.051 per ounce. The larger can is the better buy.
47. Coca-Cola: 3.25 g/fl oz; Mello Yellow: 3.92 g/fl oz; Ginger Ale: 3 g/fl oz; Mello Yellow has the greatest amount per fluid oz.
49. Coca-Cola: 12 cal/fl oz; Mello Yellow: 14.2 cal/fl oz; Ginger Ale: 11.25 cal/fl oz; Ginger Ale has the least number of calories per fluid oz.
51. 295,000 vehicles/year   **53. a.** 2.2 million per year
b. 2.04 million per year **c.** Mexico
55. Cheetah: 29 m/sec; antelope: 24 m/sec. The cheetah is faster.

Section 5.2 Calculator Connections, pp. 305–306

56. a. 9.9 wins/yr **b.** 8.6 wins/yr **c.** Shula
57. a. 2.1 wins/loss **b.** 1.5 wins/loss **c.** Shula
58. a. \$0.22 per ounce **b.** \$0.21 per ounce **c.** \$0.19 per ounce; The best buy is Irish Spring.
59. The unit prices are \$0.113 per ounce, \$0.137 per ounce, and \$0.149 per ounce. The best buy is the 32-oz jar.
60. a. \$0.299 per ounce **b.** \$0.208 per ounce

c. \$0.332 per ounce   The best buy is the 4-pack of 6-oz cans for \$4.99.   **61. a.** \$0.017 per ounce **b.** \$0.052 per ounce
The case of 24 twelve-oz cans for \$4.99 is the better buy.

Section 5.3 Practice Exercises, pp. 311–313

3. $\frac{1}{15}$ **5.** $\frac{3\ \text{apples}}{1\ \text{pie}}$ **7.** $\frac{22\ \text{mi}}{3\ \text{gal}}$ **9.** $\frac{4}{16} = \frac{5}{20}$

11. $\frac{25}{15} = \frac{10}{6}$ **13.** $\frac{2}{3} = \frac{4}{6}$ **15.** $\frac{30}{25} = \frac{12}{10}$

17. $\frac{\$6.25}{1\ \text{hr}} = \frac{\$187.50}{30\ \text{hr}}$ **19.** $\frac{1\ \text{in.}}{7\ \text{mi}} = \frac{5\ \text{in.}}{35\ \text{mi}}$

21. No   **23.** Yes   **25.** Yes   **27.** Yes   **29.** Yes
31. Yes   **33.** No   **35.** Divide by 2   **37.** Divide by 5
39. Divide by 8   **41.** Divide by 0.6   **43.** Yes
45. No   **47.** $x = 4$ **49.** $x = 3$ **51.** $p = 75$
53. $n = 12$ **55.** $t = 12$ **57.** $y = 36$ **59.** $x = 3$

61. $m = \frac{15}{2}$ or $7\frac{1}{2}$ or 7.5 **63.** $k = 30$ **65.** $h = 2.5$

67. $x = 4$ **69.** $z = \frac{1}{80}$

Section 5.4 Practice Exercises, pp. 318–321

3. $=$ **5.** $\neq$ **7.** $n = \frac{20}{3}$ or $6\frac{2}{3}$ or $6.\overline{6}$ **9.** $k = 6$

11. $y = 4.9$ **13.** Pam can drive 610 mi on 10 gal of gas.
15. 78 kg of crushed rock will be required.
17. The actual distance is about 80 mi.
19. There are 3800 male students.
21. Heads would come up about 315 times.
23. There would be approximately 3 earned runs for a 9-inning game.   **25.** Pierre can buy 666€.
27. 45 visits would be a result of falls.   **29.** There are approximately 357 bass in the lake.   **31.** There are approximately 4000 bison in the park.
33. $x = 24$ cm, $y = 36$ cm   **35.** $x = 1$ yd, $y = 10.5$ yd
37. The flagpole is 12 ft high.   **39.** The platform is 2.4 m tall.   **41.** $x = 17.5$ in.   **43.** $x = 6$ ft, $y = 8$ ft
45. $x = 21$ ft; $y = 21$ ft; $z = 53.2$ ft

Section 5.4 Calculator Connections, p. 321

47. There were approximately 166,005 crimes committed.
48. The Washington Monument is approximately 555 ft tall.
49. Approximately 15,400 women would be expected to have breast cancer.
50. Approximately 295,000 men would be expected to have prostate disease.

Chapter 5 Review Exercises, pp. 326–329

1. 5 to 4 and $\frac{5}{4}$ **3.** $8:7$ and 8 to 7   **5. a.** $\frac{4}{5}$

b. $\frac{5}{4}$ **c.** $\frac{5}{9}$ **7.** $\frac{4}{1}$ **9.** $\frac{2}{5}$ **11.** $\frac{9}{2}$ **13.** $\frac{4}{3}$

15. a. This year's enrollment is 1520 students.   **b.** $\frac{4}{19}$

17. $\frac{1}{5}$ **19.** $\frac{4\ \text{hot dogs}}{9\ \text{min}}$ **21.** $\frac{650\ \text{tons}}{9\ \text{ft}}$

23. All unit rates have a denominator of 1, and reduced rates may not.   **25.** 4° per hour   **27.** 11min/lawn

29. $3.333 per towel   31. a. $0.078/oz   b. $0.075/oz
c. The 48-oz jar is the best buy.   33. The difference is about 12¢ per roll or $0.12 per roll.
35. a. There was an increase of 120,000 hybrid vehicles.
  b. There will be 10,000 additional hybrid vehicles per month.
37. $\frac{16}{14} = \frac{12}{10\frac{1}{2}}$ 39. $\frac{5}{3} = \frac{10}{6}$ 41. $\frac{\$11}{1 \text{ hr}} = \frac{\$88}{8 \text{ hr}}$
43. No   45. Yes   47. Yes   49. No   51. $x = 4$
53. $b = 3$ 55. $h = 13.6$ 57. The human equivalent is 84 years.   59. Alabama had approximately 4,600,000 people.
61. $x = 10$ in., $y = 62.1$ in.   63. $x = 1.6$ yd, $y = 1.8$ yd

Chapter 5 Test, pp. 329–330

1. 25 to 521, 25 : 521, $\frac{25}{521}$ 2. a. $\frac{17}{23}$ b. $\frac{17}{6}$ 3. $\frac{13}{4}$
4. $\frac{9}{8}$ 5. $\frac{5}{8}$ 6. a. $\frac{21}{125}$ b. $\frac{9}{125}$ c. The poverty ratio was greater in New Mexico.   7. a. $\frac{\frac{1}{2}}{1\frac{1}{2}} = \frac{1}{3}$ b. $\frac{30}{90} = \frac{1}{3}$
8. $\frac{85 \text{ mi}}{2 \text{ hr}}$ 9. $\frac{10 \text{ lb}}{3 \text{ weeks}}$ 10. $\frac{1 \text{ g}}{2 \text{ cookies}}$
11. 21.45 g/cm³   12. 2.29 oz/lb   13. $0.22 per ounce
14. $0.50 per ring   15. Generic: $0.05/tablet; Aleve: $0.08/capsule. The generic pain reliever is the better buy.
16. They form equal ratios or rates.
17. $\frac{42}{15} = \frac{28}{10}$ 18. $\frac{20 \text{ pages}}{12 \text{ min}} = \frac{30 \text{ pages}}{18 \text{ min}}$
19. $\frac{\$15}{1 \text{ hr}} = \frac{\$75}{5 \text{ hr}}$ 20. No   21. $p = 35$ 22. $x = 12.5$
23. $n = 5$ 24. $y = 6$ 25. It will take 7.5 min.
26. Cherise spends 30 hr each week on homework outside of class.   27. There are approximately 27 fish in her pond.
28. $x = 1\frac{1}{2}$ mi, $y = 8$ mi   29. 16 cm

Chapters 1–5 Cumulative Review Exercises, pp. 330–331

1. Five hundred three thousand, forty-two
2. Approximately 1400   3. 22,600,000   4. 22 R 3
5. 22.1875   6. 6   7.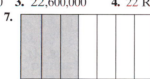
8. $\frac{7}{5}$ 9. $\frac{39}{14}$ 10. $\frac{9}{25}$
11. Bruce has $4\frac{1}{2}$ in. of sandwich left.   12. 2   13. $\frac{35}{9}$
14. $\frac{9}{13}$ 15. Emil needs $13\frac{1}{12}$ ft of wallpaper border.
16. It sold $61\frac{11}{16}$ acres, and $20\frac{9}{16}$ acres is left.
17. There are 59 ninths.   18. One thousand four and seven hundred one thousandths.   19. 28.057
20. $\frac{109}{25}$ 21. 4392.3   22. 2.379   23. 130.9 cm
24. $\frac{61}{44}$ or 61 : 44   25. $\frac{13}{1}$ 26. $\frac{7}{50}$; Approximately 7 out of 50 deaths are due to cancer.   27. 125 people/mi²
28. a. Yes b. No   29. $x = 4.5$
30. Jim can drive 100 mi on 4 gal.

Chapter 6

Chapter Opener Puzzle

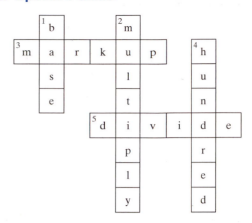

Section 6.1 Practice Exercises, pp. 339–342

3. 48%   5. 50%   7. 25%   9. 2%   11. 70%
13. Replace the symbol % by $\times \frac{1}{100}$ (or $\div$ 100). Then reduce the fraction to lowest terms.
15. $\frac{3}{100}$ 17. $\frac{21}{25}$ 19. $\frac{1}{4}$ 21. $\frac{17}{500}$ 23. $\frac{23}{20}$ or $1\frac{3}{20}$
25. $\frac{7}{4}$ or $1\frac{3}{4}$ 27. $\frac{1}{200}$ 29. $\frac{1}{400}$ 31. $\frac{2}{3}$ 33. $\frac{49}{200}$
35. Replace the % symbol by $\times 0.01$ (or $\div$ 100).
37. 0.72   39. 0.66   41. 0.129   43. 0.4105
45. 2.01   47. 0.0075   49. 0.1625   51. 0.622
53. 25%   55. 100%   57. 150%   59. d   61. b
63. a   65. d   67. b   69. c   71. 0.225; $\frac{9}{40}$
73. 0.043; $\frac{43}{1000}$ 75. 0.006; $\frac{3}{500}$ 77. 0.35; $\frac{7}{20}$
79. 40% = 0.4 or $\frac{2}{5}$; 42% = 0.42 or $\frac{21}{50}$; 59% = 0.59 or $\frac{59}{100}$; 73% = 0.73 or $\frac{73}{100}$

Section 6.2 Practice Exercises, pp. 347–349

3. $\frac{13}{10}$ or $1\frac{3}{10}$ 5. $\frac{1}{200}$ 7. $0.06\overline{3}$ 9. 0.003
11. 162%   13. 26%   15. 125%   17. 77%
19. 27%   21. 19%   23. 175%   25. 12.4%
27. 0.6%   29. 101.4%   31. 71%   33. 95%
35. 87.5% or $87\frac{1}{2}$%   37. 81.25% or $81\frac{1}{4}$%
39. $83.\overline{3}$% or $83\frac{1}{3}$%   41. $44.\overline{4}$% or $44\frac{4}{9}$%   43. 25%
45. 10%   47. $66.\overline{6}$% or $66\frac{2}{3}$%   49. 175%
51. 135%   53. $122.\overline{2}$% or $122\frac{2}{9}$%
55. $166.\overline{6}$% or $166\frac{2}{3}$%   57. 42.9%   59. 7.7%
61. 45.5%   63. 86.7%
65. The fraction $\frac{1}{2} = 0.5$ and $\frac{1}{2}$% = 0.5% = 0.005.
67. 25% = 0.25 and 0.25% = 0.0025
69. a, c   71. a, c